D0712977

LOGIC, METHODOLOGY AND PHILOSOPHY OF SCIENCE

STUDIES IN LOGIC

AND

THE FOUNDATIONS OF MATHEMATICS

Editors

L. E. J. BROUWER, *Laren (N.H.)*

A. HEYTING, *Amsterdam*

A. ROBINSON, *Los Angeles*

P. SUPPES, *Stanford*

NORTH-HOLLAND PUBLISHING COMPANY
AMSTERDAM

LOGIC, METHODOLOGY AND PHILOSOPHY OF SCIENCE

PROCEEDINGS OF THE 1964 INTERNATIONAL CONGRESS

Edited by

YEHOSHUA BAR-HILLEL

Professor of Logic and Philosophy of Science
The Hebrew University of Jerusalem, Israel

1972

NORTH-HOLLAND PUBLISHING COMPANY

AMSTERDAM

North-Holland ISBN 07204 2235 3

First edition 1965
Second printing 1972

PRINTED IN THE NETHERLANDS

PREFACE

This volume constitutes the Proceedings of the 1964 International Congress for Logic, Methodology and Philosophy of Science. The Congress was held at the Hebrew University of Jerusalem, Israel, from August 26 to September 2, 1964, under the joint auspices of the Division of Logic, Methodology and Philosophy of Science of the International Union of History and Philosophy of Science, and the Israel Academy of Sciences and Humanities. Its Organization was initiated by the Israel Association for Logic and Philosophy of Science.

The papers published here are the texts, at times considerably expanded and revised, of addresses presented by the invitees to the Congress. These papers fall in one of two categories: invited addresses and contributions to symposia.

Short communications presented by members of the Congress are not included in the *Proceedings*, but the titles of their papers are listed in the complete scientific program which appears at the end of this volume. Abstracts of these communications were printed in a little booklet issued to the members at the beginning of the Congress.

The Congress was organized by a committee consisting of S. Adler, A. A. Fraenkel, L. Guttman, H. Hanani, A. Katzir, M. Jammer, A. Lévy, E. I. J. Poznański, M. O. Rabin, S. Sambursky and Y. Bar-Hillel (Secretary).

The work of the 1964 Congress was planned to follow rather closely the precedent set by the highly successful 1960 Congress held in Stanford, California, U.S.A.

However, in lieu of the eleven sections of that Congress, the present one was divided into eight sections only. There was no section on Methodology and Philosophy of Linguistics, in view of the fact that a Colloquium on Algebraic Linguistics and Automata Theory was held in Jerusalem during the two days immediately preceding the Congress. In addition, the section on Methodology and Philosophy of Biological and Psychological Sciences and the one on Methodology and Philosophy of Social Sciences were combined this time into a section on Methodology and Philosophy of Life Sciences. Finally, the section on the Methodology and Philosophy of of Historical Sciences did not materialize because of a combination of various inadvertencies which the prospective chairman of the section was unable to overcome.

The chairmen of the various sections who, together with the Executive Committee, were responsible for the setting-up of the program of the Congress, were: A Mostowski, Section 1 (Mathematical Logic); A. Tarski, Section 2 (Foundations of Mathematical Theories); P. Lorenzen, Section 3 (Philosophy of Logic and Mathematics); I. Scheffler, Section 4 (General Problems of Methodology and Philosophy of Science); G. H. von Wright, Section 5 (Foundations of Probability and Induction); A. Grünbaum, Section 6 (Methodology and Philosophy of Physical Sciences); P. Suppes, Section 7 (Methodology and Philosophy of Life Sciences); W. Kneale, Section 8 (History of Logic and Methodology of Science). However, Mostowski and Grünbaum were unable to attend the Congress after having completed the organization of their sections, and were replaced by L. Kalmár and S. Sambursky, respectively, who were kind enough to take this task upon themselves at the last moment.

In addition to the subsidies received from the two sponsoring institutions, generous financial grants from the U.S. National Science Foundation, the U.S. Armed Forces, and the American Council of Learned Societies, and smaller grants from the Hebrew University of Jerusalem, which also put its lecture halls and students' dormitories at the disposal of the Congress, I.B.M.-Israel, the Israel Ministry for Foreign Affairs and the Jerusalem City Council, made possible the attendance of a substantial number of scholars from close to thirty countries throughout the world.

The editor has confined himself to arranging the volume and handling various technical matters relating to publication without attempting detailed editorial treatment, with the exception of some improvements in the style of papers written in English by authors from whom this was not their native language. Some effort was made to achieve uniformity in matters of typography, footnotes and references, but the choice of notation and symbolism was left to the individual authors.

On behalf of the Organizing Committee the editor wishes to thank the many persons, too numerous to name, who contributed their generous assistance in arranging for the Congress and preparing the present volume for publication.

The Editor

The Hebrew University of Jerusalem

April 1965

CONTENTS

I. MATHEMATICAL LOGIC

II. FOUNDATIONS OF MATHEMATICAL THEORIES

III. PHILOSOPHY OF LOGIC AND MATHEMATICS

IV. METHODOLOGY AND PHILOSOPHY OF SCIENCE

V. FOUNDATIONS OF PROBABILITY AND INDUCTION

VI. METHODOLOGY AND PHILOSOPHY OF PHYSICAL SCIENCES

VII. METHODOLOGY AND PHILOSOPHY OF LIFE SCIENCES

VIII. HISTORY OF LOGIC, METHODOLOGY, AND PHILOSOPHY OF SCIENCE

Section I

Mathematical Logic

TRANSFINITE AUTOMATA RECURSIONS AND WEAK SECOND ORDER THEORY OF ORDINALS*

J. RICHARD BÜCHI
Purdue University, Lafayette, Indiana, U.S.A.

1. Introduction. We identify the ordinal α with the set of all ordinals $x < \alpha$. The weak second order theory of $[\alpha, <]$ is the interpreted formalism WST $[\alpha, <]$ which makes use of: (a) the propositional connectives with usual interpretation; (b) a binary relation letter $<$ interpreted as ordering relation on α; (c) individual variables $t, x, y, z, \cdots$, ranging over α; (d) monadic predicate variables $i, j, s, r, \cdots$, ranging over finite subsets of α; (e) quantifiers $\forall$, $\exists$ for both types of variables. The purpose of this paper is to provide, for any ordinal α, a clear understanding of which relations R on finite subsets of α can be defined by formulas $\Sigma(i_1, \cdots, i_n)$ of WST$[\alpha, <]$. In addition we obtain a decision method for truth of sentences Σ in WST$[\alpha, <]$.

The binary expansion of natural numbers can be extended to ordinals. If $x < 2^\alpha$ the binary expansion ϕx is a finite subset i of α, namely $\phi x = \{u_1, \cdots, u_n\}$ if $x = 2^{u_1} + \cdots + 2^{u_n}$, $u_n < \cdots < u_1 < \alpha$. ϕ is a one-to-one map of the ordinal 2^α onto all finite subsets of α, and it yields a simple translation of the first order theory FT$[2^\alpha, +, E]$ onto WST$[\alpha, <]$. Here E stands for the binary relation "x is a power of 2, and occurs in the representation of y as decreasing sum of powers of 2," i.e., $(\exists u)[x = 2^u \wedge u \in \phi y]$. (Note that 2^α is not necessarily closed under $+$; in this case $+$ is to be interpreted as a ternary relation.) This translation in fact is simply a change in the choice of primitive notions. Our main results can be restated: 1. For any α we obtain a clear understanding of which relations R on 2^α are definable by formulas $\Sigma(x_1, \cdots, x_n)$ in the first order theory of $[2^\alpha, +, E]$. 2. For any α, there is a decision method for truth of first order sentences $[2^\alpha, +, E]$.

The key to the understanding of WST$[\alpha, <]$ is a simple sort of transfinite predicate recursions with finite predicates i on α as inputs. In case $\alpha = \omega$ these are just the finite automata recursions, and the methods used in the present paper specialize to those used in [1]. We refer the reader

* This work was supported in part by grant PG-2754 from the National Science Foundation, and by a travel grant from Purdue University.

to this paper and its review [5], if he desires more explicit explanations of notations, or is interested in analyzing by-products of the main results.

Results on definability of individuals in FT$[\omega^\alpha, +]$ have been obtained earlier by A. Ehrenfeucht [6]. His methods are quite different from ours; we refer the reader to [3], for a lucid presentation of this work. In [3] and [4] it is stated that Ehrenfeucht also knew a decision method for FT$[\omega^\alpha, +]$. However, ours is the first published proof of this result.

Let RST$[\alpha, <]$, the restricted second order theory of $[\alpha, <]$, be like WST$[\alpha, <]$, except that predicate variables now range over all subsets of α. That RST$[\omega, <]$ is decidable was shown in [2]. Whether this holds for RST$[\omega^2, <]$ is not known.*

2. Transfinite automata recursions. We show here how to extend the basic facts about the behavior of finite automata to a simple sort of transfinite recursion on monadic predicates, which we call "special recursions" for short. Lemma 4 says that the projection of the behavior of a special recursion is again the behavior of some special recursion. This is the crucial result of this paper, as it can be used to eliminate predicate quantifiers in formulas of WST$[\alpha, <]$.

We introduce the following abbreviations for limit ordinals of finite order p.

$$(1) \qquad \begin{aligned} &(\forall t)_x^y \Sigma &&: (\forall t)[x \leqq t < y \supset \Sigma] \\ &(\exists t)_x^y \Sigma &&: (\exists t)[x \leqq t < y \wedge \Sigma] \\ &\mathrm{Lm}_0 x &&: T \\ &\mathrm{Lm}_{p+1} x &&: (\forall y)_0^x (\exists z)_{y+1}^x \mathrm{Lm}_p z \\ &\mathrm{lm}_p x &&: \mathrm{Lm}_p x \wedge \sim \mathrm{Lm}_{p+1} x \end{aligned}$$

Thus Lm_0 is the set of all ordinals. Lm_1 consists of 0 and all limit ordinals. Lm_2 consists of 0 and limits of limit ordinals, etc. While lm_p consists of all ordinals of form $r + \omega^p$. Let x be a limit ordinal, $y < x$, and suppose that the predicate j is constant between y and x. Then $\lim_{t<x} jt$ stands for jy. I.e.

$$(2) \qquad \begin{aligned} &\lim_{t<x} jt \equiv T \quad \text{if } (\exists y)_0^x (\forall t)_y^x jt \\ &\lim_{t<x} jt \equiv F \quad \text{if } (\exists y)_0^x (\forall t)_y^x \sim jt \end{aligned}$$

Now consider the system of propositional formulas,

* Added in proof: RST$[\alpha, <]$ is decidable, for any countable α.

$$
(3)\quad
\begin{aligned}
r0 &\equiv E\\
r(x+1) &\equiv H[rx, ix]\\
rx &\equiv \lim_{t<x} U_1[rt], && \text{if } \mathrm{lm}_1 x\\
&\ \vdots\\
rx &\equiv \lim_{t<x} U_{p-1}[rt], && \text{if } \mathrm{lm}_{p-1} x\\
rx &\equiv \lim_{t<x} U_p[rt], && \text{if } \mathrm{Lm}_p x \text{ and } x \neq 0.
\end{aligned}
$$

We use here the vector notation explained in detail in [1]. Thus $\boldsymbol{i}$ stands for a vector $[i_1, \cdots, i_k]$ of predicate variables, $\boldsymbol{r}$ for a vector $[r_1, \cdots, r_n]$ of predicate variables, $\boldsymbol{E}$ for a vector $[E_1, \cdots, E_n]$ of truth values, $\boldsymbol{H}[\boldsymbol{Y}, \boldsymbol{X}]$ for a vector $[H_1, \cdots, H_n]$ of propositional expressions in $X_1, \cdots, X_k, Y_1, \cdots, Y_n$. Furthermore, each line of (3) abbreviates a conjunction of n equivalences. $\boldsymbol{i}$ will be called the *input predicates* of (3), k-vectors $\boldsymbol{X} = [X_1, \cdots, X_k]$ of truth values are called the *input states* of (3). $\boldsymbol{r}$ is called the *transition predicates*, its states $\boldsymbol{Y} = [Y_1, \cdots, Y_n]$ are called *transit states*. In the sequel we will usually not explicitly mention the number of components of a vector $\boldsymbol{A}$ of propositional expressions; it can be recovered from the context. $\boldsymbol{F}$ always stands for a vector $[F, \cdots, F]$, all of whose components are the truth value F.

DEFINITION 1: *A special predicate recursion of order* p, or shortly a *p-recursion*, is a system $\mathfrak{R} = \boldsymbol{E}, \boldsymbol{H}[\boldsymbol{Y}, \boldsymbol{X}], \boldsymbol{U}_1[\boldsymbol{Y}], \cdots, \boldsymbol{U}_p[\boldsymbol{Y}]$ of propositional expressions satisfying

$$
(4)\quad
\begin{aligned}
\boldsymbol{U}_a[\boldsymbol{U}_b[\boldsymbol{Y}]] &\equiv \boldsymbol{U}_a[\boldsymbol{Y}], \quad \text{for } 0 \leqq b < a \leqq p;\\
\boldsymbol{U}_p[\boldsymbol{U}_p[\boldsymbol{Y}]] &\equiv \boldsymbol{U}_p[\boldsymbol{Y}].
\end{aligned}
$$

Here $\boldsymbol{U}_0[\boldsymbol{Y}]$ stands for $\boldsymbol{H}[\boldsymbol{Y}, \boldsymbol{F}]$. An *output* to such a recursion is a propositional expression $O[\boldsymbol{Y}]$. A sequence consisting of a p-recursion and an output will also be called a *finite automaton of order* p, or a *p-automaton*.

Suppose now that $\boldsymbol{i}$ is a vector of specific predicates $i_1, \cdots, i_k$ on the ordinal α. In some cases (3) will determine, in recursive fashion, a vector $\boldsymbol{r} = \zeta \boldsymbol{i}$ on $\alpha + 1$. Namely, just in case $\lim_{t<x} \boldsymbol{U}_a[\boldsymbol{r}t]$ becomes meaningful at each limit $x \leqq \alpha$, i.e., if $1 \leqq a < p$ and $\mathrm{lm}_a x$ or $a = p$ and $\mathrm{Lm}_a x$, there must exist $y < x$ such that $\boldsymbol{U}_a[\boldsymbol{r}t]$ remains constant for $y \leqq t < x$. By transfinite induction one easily shows that this really happens, in case $\boldsymbol{i}$ consists of finite predicates, and (4) holds. Just note that "$\boldsymbol{i}$ is finite" means $(\forall x)_{\mathrm{Lm}_1}(\exists y)_0^x(\forall t)_y^x\ \boldsymbol{i}t \equiv \boldsymbol{F}$. Thus we can state

LEMMA 1. *If* $\mathfrak{R} = \boldsymbol{E}, \boldsymbol{H}, \boldsymbol{U}_1, \cdots, \boldsymbol{U}_p$ *is a special predicate recursion of order* p *and* α *is any ordinal, then* (3) *uniquely determines a recursive operator* $\boldsymbol{r} = \zeta_{\mathfrak{R}} \boldsymbol{i}$, *defined for all vectors* $\boldsymbol{i}$ *of finite predicates on* α, *and taking as values vectors of predicates on* $\alpha + 1$.

Note that the components of the values of ζ are not necessarily finite subsets of $\alpha + 1$. If $\boldsymbol{i}$ is finite and $\boldsymbol{r} = \zeta \boldsymbol{i}$ then $\boldsymbol{r}\alpha$ will be called the *terminal state* of $\mathfrak{R}$ under the *input signal* $\boldsymbol{i}$. If this terminal state belongs to the output O, we will say that $\boldsymbol{i}$ belongs to the behavior of $\mathfrak{R}, O$, i.e.

DEFINITION 2. Let $\mathfrak{A} = \boldsymbol{E}, \boldsymbol{H}, \boldsymbol{U}_1, \cdots, \boldsymbol{U}_p, O$ be an automaton of order p. Let $\zeta = \zeta_{\mathfrak{A}}$ be the recursive operator determined by $\mathfrak{A}$. The *α-behavior* of $\mathfrak{A}$ is the set $\mathrm{beh}^{\alpha}_{\mathfrak{A}}$ consisting of all vectors $\boldsymbol{i}$ of finite predicates on α such that $O[\boldsymbol{r}\alpha]$ holds, if $\boldsymbol{r} = \zeta \boldsymbol{i}$.

The purpose of this section is to show how the well known facts about behavior of finite automata extend to α-behaviors of p-automata. The following lemma is trivial.

LEMMA 2. *The complement (in the set of all finite subsets of* α*) of an* α*-behavior of a* p*-automaton is again the* α*-behavior of some* p*-automaton. In fact* $\mathrm{beh}^{\alpha}_{\mathfrak{R},\tilde{O}} = \sim \mathrm{beh}^{\alpha}_{\mathfrak{R},O}$, *for any* p*-recursion* $\mathfrak{R}$.

Equally trivial is the closure under $\cup$ and $\cap$ of the class of α-behaviors. Also the following lemma is quite obvious on intuitive grounds, even though its proof is somewhat cumbersome for typographic reasons.

DEFINITION 3. A p-recursion $\boldsymbol{E}, \boldsymbol{H}, \boldsymbol{U}_1, \cdots, \boldsymbol{U}_p$ is in *expanded form* if

$$\begin{aligned} \boldsymbol{H}[Y, X] \;&.\!\equiv\!.\; \boldsymbol{A}_1[X] Y_1 \vee \cdots \vee \boldsymbol{A}_n[X] Y_n; \\ \boldsymbol{U}_a[Y] \;&.\!\equiv\!.\; \boldsymbol{B}_a Y_1 \vee \cdots \vee \boldsymbol{B}_a Y_n, \quad \text{for } a = 1, \cdots, p. \end{aligned} \tag{5}$$

The output $O[Y]$ is in *expanded form* if

$$O[Y] \;.\!\equiv\!.\; C_1 Y_1 \vee \cdots \vee C_n Y_n. \tag{5'}$$

LEMMA 3. *To every* p*-automaton* $\mathfrak{A} = \boldsymbol{E}, \boldsymbol{H}, \boldsymbol{U}_1, \cdots, \boldsymbol{U}_p, O$ *one can construct a* p*-automaton* $\mathfrak{B} = \boldsymbol{G}, \boldsymbol{L}, \boldsymbol{W}_1, \cdots, \boldsymbol{W}_p, Q$ *which is in expanded form and has the same* α*-behavior as* $\mathfrak{A}$, *for any* α.

PROOF. Given the automaton $\mathfrak{A}$ we define for $a = 1, \cdots, n$; $b = 1, \cdots, p$; $\mu, \nu_1, \cdots, \nu_n = T, F$:

$$\begin{aligned} A_{\mu,a,\nu_1,\ldots,\nu_n}[X] \;&.\!\equiv\!.\; H^{\mu}_{a}[\nu_1, \cdots, \nu_n, X]; \\ B_{b,\mu,a,\nu_1,\ldots,\nu_n} \;&.\!\equiv\!.\; U^{\mu}_{b,a}[\nu_1, \cdots, \nu_n]; \\ C_{\nu_1,\ldots,\nu} \;&.\!\equiv\!.\; O[\nu_1, \cdots, \nu_n]. \end{aligned} \tag{6}$$

Here Z^{ν} stands for Z or $\tilde{Z}$ according to whether $\nu = T$ or $\nu = F$. Now we define $\mathfrak{B} = \boldsymbol{G}, \boldsymbol{L}, \boldsymbol{W}_1, \cdots, \boldsymbol{W}_p, \boldsymbol{Q}$ by

$$
(7)\quad
\begin{aligned}
G_{\nu_1,\ldots,\nu_n} &.\equiv. E_1^{\nu_1}\cdots E_n^{\nu_n};\\
L_{\sigma_1,\ldots,\sigma_n}[Z,X] &.\equiv. \bigvee_{\nu_1,\ldots,\nu_n} [\bigwedge_a A_{\sigma_a,a,\nu_1,\ldots,\nu_n}[X]]Z_{\nu_1,\ldots,\nu_n};\\
W_{b,\sigma_1,\ldots,\sigma_n}[Z] &.\equiv. \bigvee_{\nu_1,\ldots,\nu_n} [\bigwedge_a B_{b,\sigma_a,a,\nu_1,\ldots,\nu_n}]Z_{\nu_1,\ldots,\nu_n};\\
Q[Z] &.\equiv. \bigvee_{\nu_1,\ldots,\nu_n} C_{\nu_1,\ldots,\nu_n}Z_{\nu_1,\ldots,\nu_n}.
\end{aligned}
$$

Note that $\bigvee_{\nu_1,\ldots,\nu_n} D_{\nu_1,\ldots,\nu_n}Y_1^{\nu_1}\cdots Y_n^{\nu_n}.\equiv.D_{Y_1,\ldots,Y_n}$. Using this, it follows by (6) and (7) that

$$
(8)\quad
\begin{aligned}
L_{\sigma_1,\ldots,\sigma_n}[Z,X] &.\equiv. H_1^{\sigma_1}[Y,X]\cdots H_n^{\sigma_n}[Y,X];\\
W_{b,\sigma_1,\ldots,\sigma}[Z] &.\equiv. U_{b,1}^{\sigma_1}[Y]\cdots U_{b,n}^{\sigma_n}[Y], \text{ if } Z_{\nu_1,\ldots,\nu_n} \equiv Y_1^{\nu_1}\cdots Y_n^{\nu_n};\\
Q[Z] &.\equiv. O[Y].
\end{aligned}
$$

By assumption $\boldsymbol{E}, \boldsymbol{H}, \boldsymbol{U}_1, \cdots, \boldsymbol{U}_p$ is a special recursion. From Definition 3 and (8) one easily shows that,

$$
(9)\quad
\begin{aligned}
\boldsymbol{W}_b[\boldsymbol{L}[\boldsymbol{Z},\boldsymbol{F}]] &= \boldsymbol{W}_b[\boldsymbol{Z}], \quad \text{if } 1 \leqq b \leqq p;\\
\boldsymbol{W}_a[\boldsymbol{W}_b[\boldsymbol{Z}]] &= \boldsymbol{W}_a[\boldsymbol{Z}], \text{ if } 1 \leqq b < a \leqq p \text{ and } Z_{\nu_1\ldots,\nu_n} \equiv Y_1^{\nu_1}\cdots Y_n^{\nu_n};\\
\boldsymbol{W}_p[\boldsymbol{W}_p[\boldsymbol{Z}]] &= \boldsymbol{W}_p[\boldsymbol{Z}].
\end{aligned}
$$

For example, because $Z_{\nu_1,\ldots,\nu_n} \equiv Y_1^{\nu_1}\cdots Y_n^{\nu_n}$ it follows by (8) that also $L_{\nu_1\ldots,\nu_n}[\boldsymbol{Z},\boldsymbol{F}] \equiv H_1^{\nu_1}[Y,\mathrm{F}]\cdots H_n^{\nu_n}[Y,\boldsymbol{F}]$. Therefore, $W_{b,\sigma_1,\ldots,\sigma_n}[\boldsymbol{L}[\boldsymbol{Z},\mathrm{F}]] \overset{(8)}{\equiv} U_1^{\sigma_1}[\boldsymbol{H}[Y,\boldsymbol{F}]]\cdots U_n^{\sigma_n}[\boldsymbol{H}[Y,\boldsymbol{F}]] \overset{Def.3}{\equiv} U_1^{\sigma_1}[Y]\cdots U_n^{\sigma_n}[Y] \overset{(8)}{\equiv} W_{b,\sigma_1,\ldots,\sigma_n}[\boldsymbol{Z}]$, which establishes the first part of (9). The other parts are similar. Next we note that, by (7), $\boldsymbol{L}, \boldsymbol{W}_1, \cdots, \boldsymbol{W}_p$ are expanded, and therefore $\boldsymbol{L}[\boldsymbol{Z}_1 \vee \boldsymbol{Z}_2, \boldsymbol{X}] \equiv \boldsymbol{L}[\boldsymbol{Z}_1,\boldsymbol{X}] \vee \boldsymbol{L}[\boldsymbol{Z}_2,\boldsymbol{X}]$ and $\boldsymbol{W}_a[\boldsymbol{Z}_1 \vee \boldsymbol{Z}_2] \equiv \boldsymbol{W}_a[\boldsymbol{Z}_1] \vee \boldsymbol{W}_a[\boldsymbol{Z}_2]$, for $a = 1, \cdots, p$. Using these formulas and the fact that every $\boldsymbol{Z}$ can be expressed in the form $\boldsymbol{Z} = \boldsymbol{Z}_1 \vee \cdots \vee \boldsymbol{Z}_r$, whereby each of the $\boldsymbol{Z}_1,\cdots,\boldsymbol{Z}_r$ is of form $\boldsymbol{Y}_1^{\nu_1}\cdots \boldsymbol{Y}_n^{\nu_n}$, one sees that the assumption $Z_{\nu_1\ldots\nu_n} \equiv Y_1^{\nu_1}\cdots Y_n^{\nu_n}$ can be dropped in (9). Consequently $\boldsymbol{G}, \boldsymbol{L}, \boldsymbol{W}_1, \cdots, \boldsymbol{W}_p$ is a special recursion i.e. $\mathfrak{B}$ is a p-automaton. It is clearly in expanded form, see (7). Thus, it only remains to show that the behaviors $\mathrm{beh}_{\mathfrak{A}}^{\alpha}$ and $\mathrm{beh}_{\mathfrak{B}}^{\alpha}$ are equal, for any α.

Let $\boldsymbol{r} = \zeta_{\mathfrak{A}}(\boldsymbol{i})$ and let $s = \zeta_{\mathfrak{B}}(\boldsymbol{i})$ (see Lemma 1). Using the first two formulas of (8) one shows, by induction over α, that $s_{\sigma_1,\ldots,\sigma_n}t \equiv (r_1t)^{\sigma_1}\cdots(r_nt)^{\sigma_n}$, for all $t \leqq \alpha$. Therefore, by the third formula of (8), $Q[s\alpha] \equiv O[r\alpha]$. Thus by Definition 2, $\mathrm{beh}^{\alpha}_{\mathfrak{A}} = \mathrm{beh}^{\alpha}_{\mathfrak{B}}$. Q.e.d.

Based on Lemma 3 we now prove that the class of α-behaviors is closed under finite-predicate quantification. This is the core of our argument; it will be used in Section 3 to eliminate predicate quantifiers in WST$[\alpha, <]$. We suggest that the reader carry out the proof of Lemma 4 in the special case $p = 1$. This will show clearly why the order p of a recursion is stepped up by a quantification of part of the input predicates.

LEMMA 4 (*projection lemma*): *To every p-recursion* $\mathfrak{R} = \boldsymbol{E}, \boldsymbol{H}[\boldsymbol{Y}, \boldsymbol{X}, \boldsymbol{Z}]$, $\boldsymbol{U}_1[\boldsymbol{Y}], \cdots, \boldsymbol{U}_p[\boldsymbol{Y}]$ *one can construct a* $(p+1)$*-recursion* $\mathfrak{S} = \boldsymbol{E}, \boldsymbol{L}[\boldsymbol{Y}, \boldsymbol{X}]$, $\boldsymbol{W}_1[\boldsymbol{Y}], \cdots, \boldsymbol{W}_{p+1}[\boldsymbol{Y}]$ *such that for any output* $O[\boldsymbol{Y}]$ *and any ordinal* α, $(\exists \boldsymbol{j})\mathrm{beh}^{\alpha}_{\mathfrak{R},O}(\boldsymbol{i}, \boldsymbol{j}) . \equiv . \mathrm{beh}^{\alpha}_{\mathfrak{S},O}(\boldsymbol{i})$.

PROOF. Given $\mathfrak{R}$ we let $\boldsymbol{U}_0[\boldsymbol{Y}]$ stand for $\boldsymbol{H}[\boldsymbol{Y}, \boldsymbol{F}, \boldsymbol{F}]$. By assumption on $\mathfrak{R}$ we have

$$(10) \qquad \boldsymbol{U}_p\boldsymbol{U}_p = \boldsymbol{U}_p, \ \boldsymbol{U}_a\boldsymbol{U}_b = \boldsymbol{U}_a, \quad \text{for } 0 \leqq b < a \leqq p.$$

By Lemma 3 we may assume that $\mathfrak{R}$ is in expanded form, from which it follows that

$$(11) \qquad \begin{aligned} &\boldsymbol{Y} \supset \boldsymbol{Y}' . \supset . \ \boldsymbol{U}_a[\boldsymbol{Y} \supset \boldsymbol{U}_a[\boldsymbol{Y}'], \quad \text{for } 0 \leqq a \leqq p; \\ &\boldsymbol{Y} \supset \boldsymbol{Y}' . \supset . \ \boldsymbol{H}[\boldsymbol{Y}, \boldsymbol{X}, \boldsymbol{Z}] \supset \boldsymbol{H}[\boldsymbol{Y}', \boldsymbol{X}, \boldsymbol{Z}]. \end{aligned}$$

Next we define $\boldsymbol{L}$ by

$$(12) \qquad \boldsymbol{L}[\boldsymbol{Y}, \boldsymbol{X}] . \equiv . \bigvee_{\boldsymbol{Z}} \boldsymbol{H}[\boldsymbol{Y}, \boldsymbol{X}, \boldsymbol{Z}]$$

and we let $\boldsymbol{W}_0[\boldsymbol{Y}]$ stand for $\boldsymbol{L}[\boldsymbol{Y}, \boldsymbol{F}]$. Then, clearly,

$$(\mathrm{u}_0) \qquad \boldsymbol{U}_0[\boldsymbol{Y}] \supset \boldsymbol{W}_0[\boldsymbol{Y}].$$

By (11) and (u_0) it follows that $\boldsymbol{U}_1\boldsymbol{U}_0[\boldsymbol{Y}] \supset \boldsymbol{U}_1\boldsymbol{W}_0[\boldsymbol{Y}]$. Therefore, by (10), $\boldsymbol{U}_1[\boldsymbol{Y}] \supset \boldsymbol{U}_1\boldsymbol{W}_0[\boldsymbol{Y}]$. Hence,

$$\boldsymbol{U}_1[\boldsymbol{Y}] \supset \boldsymbol{U}_1\boldsymbol{W}_0[\boldsymbol{Y}] \supset \boldsymbol{U}_1\boldsymbol{W}_0^2[\boldsymbol{Y}] \supset \cdots \supset \boldsymbol{U}_1\boldsymbol{W}_0^e[\boldsymbol{Y}] \supset \ldots..$$

Because there are but a finite number of propositional expressions in $\boldsymbol{Y}$, this sequence must break up at some place e_0 with $\boldsymbol{U}_1\boldsymbol{W}_0^{e_0} = \boldsymbol{U}_1\boldsymbol{W}_0^{e_0+1}$. Note that such an e_0 can be effectively found. We now define $\boldsymbol{W}_1$ by

$$(\mathrm{w}_1) \qquad \boldsymbol{W}_1[\boldsymbol{Y}] \equiv \boldsymbol{U}_1[\boldsymbol{W}_0^{e_0}[\boldsymbol{Y}]]$$

and we note that

$$(\mathrm{u}_1) \qquad \boldsymbol{U}_1[\boldsymbol{Y}] \supset \boldsymbol{W}_1[\boldsymbol{Y}] \qquad\qquad (\mathrm{v}_1) \qquad \boldsymbol{W}_1\boldsymbol{W}_0 = \boldsymbol{W}_1$$

Iterating the argument which leads from (u_0) to the definition (w_1) and the formulas (u_1, v_1) we obtain numbers $e_0, \cdots, e_{p-1}$ and expressions $W_1, \cdots, W_p$ such that

(w) $$W_{a+1}[Y] \equiv U_{a+1}[W_a^{e_a}[Y]], \quad a = 0, \cdots, p-1;$$

(u) $$U_a[Y] \supset W_a[Y], \quad a = 0, \cdots, p;$$

(v) $$W_{a+1} W_a = W_{a+1}, \quad a = 0, \cdots, p-1.$$

From (w) and (v) one concludes

($\bar{\text{v}}$) $$W_a W_b = W_a, \quad 0 \leqq b < a \leqq p.$$

Using (10) and (w) one finds $W_p = U_p W_p$. This, together with (u), yields,

$$U_p[Y] \supset W_p[Y] \equiv U_p W_p[Y] \supset W_p^2[Y] \equiv U_p W_p^2[Y] \supset W_p^3[Y] \equiv \ldots..$$

It follows that one can find e_p such that $W_p^{e_p} = W_p^{e_p+1} = U_p W_p^{e_p}$. We define W_{p+1} by

(13) $$W_{p+1}[Y] \equiv U_p[W_p^{e_p}[Y]]$$

and we obtain

(14) $$U_p[Y] \supset W_{p+1}[Y], \quad W_{p+1} W_p = W_{p+1}, \; W_{p+1} W_{p+1} = W_{p+1}.$$

By (14), ($\bar{\text{v}}$) it follows that $W_{p+1} W_b = W_{p+1}$, for $0 \leqq b \leqq p+1$. This, together with ($\bar{\text{v}}$), means that $\mathfrak{S} = E, L, W_1, \cdots, W_{p+1}$ is a special recursion. It remains to be shown that, for any expanded output $O[Y]$ and any α,

(a) $$\text{beh}_{\mathfrak{R},O}^{\alpha}(i,j) \supset \text{beh}_{\mathfrak{S},O}^{\alpha}(i), \quad \text{for any finite } i,j;$$

(b) $$\text{beh}_{\mathfrak{S},O}^{\alpha}(i) \supset (\exists j)\, \text{beh}_{\mathfrak{R},O}^{\alpha}(i,j), \quad \text{for any finite } i.$$

PROOF OF (a). Let i,j be any sequences of finite predicates on α, let $r = \zeta_{\mathfrak{R}}(i,j)$ and $s = \zeta_{\mathfrak{S}}(i)$ (see Lemma 1). Then

(15)
$$\begin{aligned}
r0 &\equiv E; \\
rx' &\equiv H[rx, ix, jx]; \\
rx &\equiv \lim_{t<x} U_1[rt], \text{ if } \mathrm{lm}_1 x; \\
&\vdots \\
rx &\equiv \lim_{t<x} U_{p-1}[rt], \text{ if } \mathrm{lm}_{p-1} x; \\
rx &\equiv \lim_{t<x} U_p[rt], \text{ if } \mathrm{Lm}_p x;
\end{aligned}$$

(16)
$$\begin{aligned}
s0 &\equiv E; \\
sx' &\equiv L[sx, ix]; \\
sx &\equiv \lim_{t<x} W_1[st], \text{ if } \mathrm{lm}_1 x; \\
&\vdots \\
sx &\equiv \lim_{t<x} W_{p-1}[st], \text{ if } \mathrm{lm}_{p-1} x; \\
sx &\equiv \lim_{t<x} W_p[st], \text{ if } \mathrm{lm}_p x; \\
sx &\equiv \lim_{t<x} W_{p+1}[st], \text{ if } \mathrm{Lm}_{p+1} x.
\end{aligned}$$

By transfinite induction on $x \leqq \alpha$, and using (15), (16), (12), (u), (14), one shows $\boldsymbol{r}x \supset \boldsymbol{s}x$, for all $x \leqq \alpha$. In particular $\boldsymbol{r}\alpha \supset \boldsymbol{s}\alpha$. Because $O[\boldsymbol{Y}]$ is expanded it follows that $O[\boldsymbol{r}\alpha] \supset O[\boldsymbol{s}\alpha]$. Thus, if $(\boldsymbol{i},\boldsymbol{j})$ belongs to $\mathrm{beh}^{\alpha}_{\mathfrak{R},O}$, then $\boldsymbol{i}$ belongs to $\mathrm{beh}^{\alpha}_{\mathfrak{S},O}$, which establishes (a).

We use notations like $\mathbf{j}[y,z)$ to refer to predicates defined on the interval $[y,z) = \{t; y \leqq t < z\}$ of ordinals. $\boldsymbol{r}[y,z) = \zeta_C \boldsymbol{i}, \boldsymbol{j}[yz)$ means that $\boldsymbol{r}[y,z)$ is defined from $\boldsymbol{i}[y,z)$ and $\boldsymbol{j}[y,z)$ by the recursion (15), modified to start at y with $\boldsymbol{r}y \equiv \boldsymbol{C}$. To prove (b) we need the following lemma.

(c) *Let* $\boldsymbol{i}$ *be a vector of finite predicates on* α, *and let* $\boldsymbol{s} = \zeta_{\mathfrak{S}}(\boldsymbol{i})$ *be given by the recursion* (16). *If* $z \neq 0$ *and* $s_h z$, *then there exist* $g, y < z, \boldsymbol{j}[y,z)$ *such that* $s_g y$, *and if* $C_g \equiv T$ *and* $\boldsymbol{r}[y,z) = \zeta_C \boldsymbol{i}, \boldsymbol{j}[y,z)$ *then* $r_h z$, *and each component of* $\boldsymbol{j}[y,z)$ *is finite.*

The proof splits into 3 cases.

Case $z = t'$: By (16), $\boldsymbol{s}z \equiv \boldsymbol{L}[\boldsymbol{s}t, \boldsymbol{i}t]$, and because $s_h z$ it follows that $L_h[\boldsymbol{s}t, \boldsymbol{i}t]$. By (12) there is a $\boldsymbol{Z}$ such that $H_h[\boldsymbol{s}t, \boldsymbol{i}t, \boldsymbol{Z}]$, and because H_h is expanded there must be a g such that $s_g t$ and $H_g[\boldsymbol{C}, \boldsymbol{i}t, \boldsymbol{Z}]$, for arbitrary $\boldsymbol{C}$ with $C_g \equiv T$. Thus $g, y = t, \boldsymbol{j}[y,z) = \boldsymbol{Z}$ satisfy the requirements of (c).

Case $\mathrm{lm}_a z, 1 \leqq a \leqq p$: Since $\boldsymbol{i}$ consists of finite predicates there is an $x < z$ such that $(\forall t)^z_x(\boldsymbol{i}t \equiv \boldsymbol{F})$. We may also assume that $(\forall t)^z_x \sim \mathrm{Lm}_a t$. Then, by (16) and $(\bar{\mathrm{v}})$, $\boldsymbol{s}z \equiv \boldsymbol{W}_a[\boldsymbol{s}x]$, and by (w), $\boldsymbol{s}z \equiv \boldsymbol{U}_a \boldsymbol{W}^{e_{a-1}}_{a-1}[\boldsymbol{s}x]$. But, by (16) and $(\bar{\mathrm{v}})$, $\boldsymbol{W}^{e_{a-1}}_{a-1}[\boldsymbol{s}x] \equiv \boldsymbol{s}(x + \omega^{a-1} e_{a-1})$. Therefore, if $y = x + \omega^{a-1} e_{a-1}$, we have $\boldsymbol{s}z \equiv \boldsymbol{U}_a[\boldsymbol{s}y]$. Because $s_h z$ it follows that $U_{a,h}[\boldsymbol{s}y]$, and because $U_{a,h}$ is expanded there must be a g, such that $s_g y$ and $U_{a,g}[\boldsymbol{C}]$ in case $C_g \equiv T$. Now let $\boldsymbol{j}[y,z) = \boldsymbol{F}\boldsymbol{F}\cdots$. Then clearly $g, y, \boldsymbol{j}[y,z)$ satisfy the requirements of (c).

Case $\mathrm{Lm}_{p+1} z$: Since $\boldsymbol{i}$ consists of finite predicates there is an $x < z$ such that $(\forall t)^z_x(\boldsymbol{i}t \equiv \boldsymbol{F})$. Then, by (16) and $\boldsymbol{W}_{p+1}\boldsymbol{W}_a = \boldsymbol{W}_{p+1}$, $\boldsymbol{s}z \equiv \boldsymbol{W}_{p+1}[\boldsymbol{s}x]$, and by (13), $\boldsymbol{s}z \equiv \boldsymbol{U}_p \boldsymbol{W}^{e_p}_p[\boldsymbol{s}x]$. But, by (16) and $(\bar{\mathrm{v}})$, $\boldsymbol{W}^{e_p}_p[\boldsymbol{s}x] \equiv \boldsymbol{s}(x + \omega^p e_p)$. Therefore, if $y = x + \omega^p e_p$, we have $\boldsymbol{s}z \equiv \boldsymbol{U}_p[\boldsymbol{s}y]$. Because $s_h z$ it follows that $U_{p,h}[\boldsymbol{s}y]$, and because $U_{p,h}$ is expanded there must be a g, such that $s_g y$ and $U_{p,h}[\boldsymbol{C}]$ in case $C_g \equiv T$. Now let $\boldsymbol{j}[y,z) = \boldsymbol{F}\boldsymbol{F}\cdots$. Then clearly $g, y, \boldsymbol{j}[y,z)$ satisfy the requirements of (c).

This establishes (c) and there remains only to prove (b). Let $\boldsymbol{i}$ be any vector of finite predicates on α. Let $\boldsymbol{s} = \zeta_{\mathfrak{S}}(\boldsymbol{i})$ be given by (16) and assume $\mathrm{beh}_{\mathfrak{S},O}(\boldsymbol{i})$. Then $O[\boldsymbol{s}\alpha]$, and because O is expanded, there is an h_0 such that $s_{h_0}\alpha$ and

$$Y_{h_0} \supset O[\boldsymbol{Y}]. \tag{17}$$

By $s_{h_0}\alpha$ and (c) there exist $h_1, y_1, \boldsymbol{j}[y_1,\alpha)$ such that $y_1 < \alpha, s_{h_1} y_1$, and $r_{h_0}\alpha$ if $\boldsymbol{r}[y_1, \alpha) = \zeta_C \boldsymbol{i}, \boldsymbol{j}[y_1,\alpha)$ and $C_{h_1} \equiv T$. Because $s_{h_1} y_1$ we can use (c) again, and iterating this procedure we find the existence of $h_1, h_2, \cdots$; $\alpha = y_0 > y_1 > y_2 > \cdots$; $j[y_1, y_0), j[y_2, y_1), \cdots$, such that, for $v = 0, 1, 2 \cdots$,

(18) $\quad s_{h_v} y_v$, every component of $j[y_{v+1}, y_v)$ is finite;

(19) $\quad$ if $r[y_{v+1}, y_v) = \zeta_C i, j[y_{v+1}, y_v)$ then $C_{h_v} \supset r_{h_v} y_v$.

Because α is an ordinal the sequence $y_0 > y_1 > y_2 > \cdots$ must end, say with $y_m = 0$. Now $j[y_m, y_{m-1}), \cdots, j[y_1, y_0)$ make up a j defined on $[y_m, y_0) = [0, \alpha) = \alpha$, and this j is finite, by (18). Let $r = \zeta_{\mathfrak{R}}(i, j)$ be defined by (15). Then we have $r0 \equiv s0 \equiv E$. Because $y_m = 0$, and (18), it follows that $r_{h_m} y_m$. But r clearly satisfies $r[y_{v+1}, y_v) = \zeta_{r y_{v+1}} i, j[y_{v+1}, y_v)$, for $v + 1 = m, \cdots, 1$. Therefore, starting from $r_{h_m} y_m$ and using (19) we successively obtain $r_{h_{m-1}} y_{m-1}, \cdots, r_{h_0} y_0$. Put $y_0 = \alpha$, thus $r_{h_0}\alpha$, and by (17) it follows that $O[r\alpha]$, i.e., $\mathrm{beh}^{\alpha}_{\mathfrak{R},0}(i, j)$. Thus, for any finite i in $\mathrm{beh}^{\alpha}_{\mathfrak{S},0}$ we have proved the existence of a finite j such that i, j belongs to $\mathrm{beh}^{\alpha}_{\mathfrak{R},0}$. This establishes (b). Q.e.d.

The crucial step in this proof is the proper construction of the W's from the U's; the step from (u_0) to (w_1), and from $(\bar{v})$ to (13). A simple form of this idea was used in [1], and goes back to Church's synthesis algorithm for finite automata.

In analogy to $\alpha = \omega$, one might define a set X of vectors i of finite predicates on α to be an *α-regular event of order p*, in case X is the α-behavior of a p-automaton. In this terminology we have established the closure of the class of α-regular events under Boolean operations (not changing the order) and projection (stepping up the order by 1). We will now show that an α-event X is regular if and only if it is definable in WST$[\alpha, <]$.

3. Definability in weak second order theory of $[\alpha, <]$. Note that all the interpreted theories WST$[\alpha, <]$ have the same formulas; we will call them WS-formulas or just formulas. We will say that formulas Σ and Γ are *α-equivalent* in case $\Sigma \equiv \Gamma$ is valid in WST$[\alpha, <]$, i.e., in case Σ and Γ define the same relation (or truth value) in $[\alpha, <]$. In case Σ and Γ are α-equivalent, for all ordinals α, we will also say that Σ and Γ are equivalent. Note that the bounded quantifier $(\forall x)_0^y \cdots$ may be introduced in WST as an abbreviation for $(\forall x)[x < y \supset \cdots]$. We introduce the following special sorts of formulas.

Matrix. A formula in which $\forall$, $\exists$, and $<$ do not occur. A notation like $M[ix_1, \cdots, ix_n]$ stands for a matrix in which only the indicated atomic formulas may occur.

Kernel. A formula of form $(\forall x_1)(\forall x_2)_0^{x_1} \cdots (\forall x_n)_0^{x_{n-1}} M[ix_1, \cdots, ix_n] \wedge (\exists x) Q_1[ix] \wedge \cdots \wedge (\exists x) Q_m[ix]$, where M and the Q's are matrices.

Prenex formulas. A formula $(j)\Gamma(i,j)$, where Γ is a kernel and (j) is a prefix consisting of predicate quantifiers only, and where the right-most quantifier is existential. Thus a prenex formula might look thus $(\forall j_1 j_2)(\exists j_3).(\forall x)(\forall y)_0^x M[ix,jx,iy,jy] \wedge (\exists x)j_3 x \wedge (\exists x)j_1 x$, where M is a matrix. Note that in prenex formulas $<$ does not occur, except in the form $(\forall x)_0^y$. The *height* of prenex formulas is defined as follows. If Γ is a kernel, then the height of $(\exists j)\Gamma$ is 1. If Σ is of odd (even) height h, then $(\forall j)\Sigma$ (respectively $(\exists j)\Sigma$) is of height $h+1$. Thus, odd (even) height means that the prefix of Σ starts with $\exists$ (with $\forall$).

The predicate quantifiers $\overset{*}{\exists}$ and $\overset{*}{\forall}$ are defined as follows:

$$(\overset{*}{\exists} i)\Sigma:\ (\exists i)[(\exists x)ix \wedge \Sigma] \qquad (\overset{*}{\forall} i)\Sigma:(\forall i)\ [(\forall x)\widetilde{ix} \vee \Sigma]\,.$$

We make the following remarks.

LEMMA 5. *Let A be a quantifier free formula and let (prefix) be any prefix in which $\mathbf{x}$ does not occur. Then one can find a quantifier free formula B such that $(\exists \mathbf{x})(prefix)A. \equiv .(\overset{*}{\exists}\mathbf{i})(\forall \mathbf{x})(prefix)B$ is a logical equivalence, if $\mathbf{i}$ does not occur in (prefix)A.*

PROOF. Say, for example, $\mathbf{x} = x_1,x_2$. Then the following are all equivalent. $(\exists x_1 x_2)(\text{prefix})A$ and $(\overset{*}{\exists} i_1 i_2)(\forall x_1 x_2).[ix_1 \wedge ix_2] \supset (\text{prefix})A$ and $(\overset{*}{\exists} i_1 i_2)(\forall x_1 x_2)(\text{prefix})\cdot[ix_1 \wedge ix_2] \supset A$. Thus B can be taken to be $[ix_1 \wedge ix_2)] \supset A$.

LEMMA 6. *If $A(\mathbf{i},x_1,\cdots,x_n)$ is a quantifier free formula then one can find a matrix $D[\mathbf{i}x_1,\cdots,\mathbf{i}x_n]$ such that $(\forall x_1 \cdots x_n)A$ is α-equivalent to $(\forall x_1)(\forall x_2)_0^{x_1}\cdots(\forall x_n)_0^{x_{n-1}}D$, for every α.*

PROOF. Given the quantifier free formula $A(\mathbf{i},x_1,\cdots,x_n)$. Let $B(\mathbf{i},x_1,\cdots,x_n)$ be the conjunction of all formulas $A(\mathbf{i},y_1,\cdots,y_n)$, where $(y_1,\cdots,y_n)$ is a combination (with or without repetitions) of the variables $(x_1,\cdots,x_n)$. Because $[\alpha,<]$ is a linear order it follows that $(\forall x_1 \cdots x_n)A$ is equivalent to $(\forall x_1 \cdots x_n)\cdot x_n < x_{n-1} < \cdots < x_1 \supset B$. Note that B is a propositional expression in the atomic parts $\mathbf{i}x_1,\cdots,\mathbf{i}x_n$, and $x_\nu < x_\mu$ with $1 \leqq \nu,\mu \leqq n$. Let D be obtained from B by replacing the atomic parts $x_\nu < x_\mu$ by T or F according to whether $\mu < \nu$ or $\nu \leqq \mu$. Then clearly $(\forall x_1 \cdots x_n).x_n < \cdots < x_1 \supset B$ is equivalent to $(\forall x_1 \cdots x_n).x_n < \cdots < x_1 \supset D$. Thus we have found D such that $(\forall x_1 \cdots x_n)A$ is equivalent to $(\forall x_1)(\forall x_2)_0^{x_1}\cdots(\forall x_n)_0^{x_{n-1}}D$. But D is a propositional expression in the atomic parts $\mathbf{i}x_1,\cdots,\mathbf{i}x_n$ only. In other words D is a matrix.

LEMMA 7 (PRENEX FORM LEMMA): *To every WS-formula $\Sigma(\mathbf{i})$ one can construct a prenex formula $\Gamma(\mathbf{i})$ which is α-equivalent to Σ, for every α.*

PROOF. Starting from $\Sigma(\boldsymbol{i})$ we construct successively equivalent formulas $\Sigma_1(\boldsymbol{i})$, $\Sigma_2(\boldsymbol{i})$, $\Sigma_3(\boldsymbol{i})$, $\Sigma_4(\boldsymbol{i})$, $\Gamma(\boldsymbol{i})$ which are of the following forms

Σ_1: (ind-pred)A
Σ_2: (pred)*(ind)B
Σ_3: (pred)*$(\forall x_1 \cdots x_n)C$
Σ_4: (pred)$.(\forall x_1 \cdots x_n)D \wedge (\exists x)Q_1(x) \wedge \cdots \wedge (\exists x)Q_m(x)$
Γ: prenex form

Here A, B, C, D are quantifier-free formulas, $Q_1,\cdots,Q_m$ are matrices, (ind) signifies a string of individual quantifiers, (pred) signifies a string of predicate quantifiers and (pred)* indicates that $\overset{*}{\forall}$ and $\overset{*}{\exists}$ may also occur in the string. We now indicate how $\Sigma_1,\cdots,\Gamma$ are obtained, each from its predecessor.

Step 1: In the usual manner move quantifiers, occurring in Σ, to the front, to obtain Σ_1.

Step 2: Use Lemma 5 and its dual to successively move the left-most individual quantifiers to the right, to eventually obtain Σ_2 starting from Σ_1. A single step in this procedure may look like this. (pred 1)*$(\exists \boldsymbol{x})(\forall \boldsymbol{s})$(pred-ind)$G$, using Lemma 5 pass to (pred 1)*$(\overset{*}{\exists}\boldsymbol{i})(\forall \boldsymbol{x})(\forall \boldsymbol{s})$(pred-ind)$H$, interchange $(\forall \boldsymbol{x})(\forall \boldsymbol{s})$, (pred 2)*$(\forall \boldsymbol{x})$(pred-ind)$H$.

Step 3: Use Lemma 5 and its dual to successively decrease the number of changes (from $\forall$ to $\exists$ and $\exists$ to $\forall$) in the part (ind) of Σ_2, to eventually arrive at Σ_3. A single step in this procedure may look like this (pred 1)*$(\exists \boldsymbol{x})(\forall \boldsymbol{y})(ind)G$, using Lemma 5 pass to (pred 2)*$(\overset{*}{\forall}\boldsymbol{xy})(ind)H$.

Step 4: Let $(\overset{*}{\forall} j_1),\cdots,(\overset{*}{\forall} j_p)$ be all the quantifiers of form $(\overset{*}{\forall} j)$ occurring in Σ_3. Replace them by $(\forall j_1),\cdots,(\forall j_p)$ and replace C by $[C \vee \widetilde{j_1 x_1} \vee \cdots \vee \widetilde{j_p x_1}]$ in Σ_3 to obtain an equivalent Σ_3' of form (pred)*$(\forall x_1 \cdots x_n)C'$. Note that $\overset{*}{\forall}$ does not occur in Σ_3'. Let $(\overset{*}{\exists} s_1),\cdots,(\overset{*}{\exists} s_m)$ be all the quantifiers of form $(\overset{*}{\exists} s)$ occurring in Σ_3'. Replace them by $(\exists s_1),\cdots,(\exists s_m)$ and conjoin $(\exists x)s_1 x \wedge \cdots \wedge (\exists x)s_m x$ to $(\forall x_1 \cdots x_m)C'$ in Σ_3'. This yields the formula Σ_4 equivalent to Σ_3.

Step 5: Using Lemma 6 replace $(\forall x_1 \cdots x_n)D$ by $(\forall x_1)(\forall x_2)_0^{x_1} \cdots (\forall x_n)_0^{x_{n-1}} M$ in Σ_4. The resulting formula Γ is prenex, because M, $Q_1,\cdots,Q_m$ are matrices, q.e.d.

Next consider the system of formulas

$$\text{(a)} \quad \begin{aligned} \boldsymbol{r}x &\equiv (\forall t)_0^x \boldsymbol{A}[\boldsymbol{r}t, \boldsymbol{s}t, \boldsymbol{i}t] \\ \boldsymbol{s}x &\equiv (\exists t)_0^x \boldsymbol{B}[\boldsymbol{r}t, \boldsymbol{s}t, \boldsymbol{i}t] \end{aligned}$$

It determines recursively an operator ζ, defined for vectors $\boldsymbol{i}$ of predicates on ordinals and taking as values vectors $\boldsymbol{r}, \boldsymbol{s}$.

Lemma 8. *From A, B one can construct a special recursion $\mathfrak{R}$ of order 1, such that ζ given by* (a) *and $\zeta_{\mathfrak{R}}$ are identical for vectors $\mathbf{i}$ of finite predicates.*

Proof. It is sufficient to deal with a recursion of form

$$\text{(b)} \qquad rx \equiv (\exists t)_0^x A[rt, it],$$

because (a) can be reduced to this form by replacing in (a) rx by $\sim rx$. Now (b) is equivalent to

$$\text{(c)} \qquad \begin{array}{ll} r0 \equiv F & r(x+1) .\equiv. rx \vee A[rx, ix] \\ rx \equiv (\exists t)_0^x rt, & \text{if } \mathrm{Lm}_1 x. \end{array}$$

Let $H[Y, X]$ stand for $Y \vee A[Y, X]$, and let $U_0[Y]$ stand for $H[Y, F]$. Then (c) can be restated in the following form,

$$\text{(c}'\text{)} \qquad \begin{array}{ll} r0 \equiv F & r(x+1) \equiv H[rx, ix] \\ rx \equiv \lim_{t<x} rt, & \text{if } \mathrm{Lm}_1 x. \end{array}$$

This is not a special recursion. But note that $Y \supset U_0[Y]$, and therefore $Y \supset U_0[Y] \supset U_0^2[Y] \supset \cdots$. It follows that one can find e such that $U_0^e[Y] \equiv U_0^{e+1}[Y]$. If we now define U_1 by

$$\text{(d)} \qquad U_1[Y] \equiv U_0^e[Y]$$

we have

$$\text{(e)} \qquad U_1 U_0 = U_1, \quad U_1^2 = U_1, \quad Y \supset U_1[Y].$$

Therefore $\mathfrak{R} = F, H, U_1$ is a special recursion of order 1. Its operator $\zeta_{\mathfrak{R}}$ is given by

$$\text{(c}''\text{)} \qquad \begin{array}{ll} r0 \equiv F & r(x+1) \equiv H[rx, ix] \\ rx \equiv \lim_{t<x} U_1[rt] & \text{if } \mathrm{Lm}_1 x. \end{array}$$

It remains to show that (c′) and (c″) are equivalent for all vectors $\mathbf{i}$ of finite predicates. This is done by transfinite induction and comes down to showing that $\lim_{t<x} rt \equiv \lim_{t<x} U_1[rt]$, if $\mathbf{i}$ is finite, $\mathrm{Lm}_1 x$, and $r[0, x)$ is given by (c″).

Suppose therefore that $\lim_{t<x} U_1[rt] = C$. Then, because $\mathbf{i}$ is finite, there is a $y < x$ such that $it \equiv F, U_1[rt] = C$ for all $y \leqq t < x$. By (d) it follows $U_0^e[rt] = C$. Because $it \equiv F$ and (c″) we have $U_0^e[rt] \equiv r(t+e)$, and therefore $r(t+e) \equiv C$, for all $y \leqq t < x$. Thus, $\lim_{t<x} rt = C$. Q.e.d.

LEMMA 9. *To every kernel* $\Gamma(\boldsymbol{i})$ *one can construct an automaton* $\mathfrak{A}$ *of order* 1 *such that for every* α, *the behavior* $\mathrm{beh}_{\mathfrak{A}}^{\alpha}$ *is equal to the set* $\hat{\boldsymbol{i}}\Gamma(\boldsymbol{i})$ *defined by* Γ *in* $\mathrm{WST}[\alpha, <]$.

PROOF. For typographical reasons let us assume that the kernel $\Gamma(\boldsymbol{i})$ is of form $(\forall x_1)(\forall x_2)_0^{x_1}(\forall x_3)_0^{x_2}M[\boldsymbol{i}x_1, \boldsymbol{i}x_2, \boldsymbol{i}x_3] \wedge (\exists x)B[\boldsymbol{i}x]$, where M and B are matrices. (Longer kernels can be handled quite analogously.) Because M contains only monadic atomic parts, its conjunctive normal form looks as follows:

$$M[\boldsymbol{i}x_1, \boldsymbol{i}x_2, \boldsymbol{i}x_3] \,.\!\equiv\!.\, A[\boldsymbol{i}x_1, \boldsymbol{i}x_2, \boldsymbol{i}x_3] \wedge \cdots \wedge A'[\boldsymbol{i}x_1, \boldsymbol{i}x_2, \boldsymbol{i}x_3]$$

where each A is of the form $A_1[\boldsymbol{i}x_1] \vee A_2[\boldsymbol{i}x_2] \vee A_3[\boldsymbol{i}x_3]$. We now distribute $(\forall x_1)\ (\forall x_2)_0^{x_1}(\forall x_3)_0^{x_2}$ in $\Gamma(\boldsymbol{i})$ over the conjuncts of M to get

$$(1)\ \Gamma(\boldsymbol{i}).\!\equiv\!.(\forall x_1)(\forall x_2)_0^{x_1}(\forall x_3)_0^{x_2}[A_1[\boldsymbol{i}x_1] \vee A_2[\boldsymbol{i}x_2] \vee A_3[\boldsymbol{i}x_3]] \wedge \cdots \wedge (\exists x)B[\boldsymbol{i}x].$$

The $\cdots$ stands for zero or more conjuncts of the same form as the first. For typographical reasons, let us assume that no more occur; the general case can be handled analogously. Because the variables x_1, x_2, x_3 are isolated, we can move the universal quantifiers in (1) inside to get

$$(2)\quad \Gamma(\boldsymbol{i}) \equiv (\forall x_1)[A_1(x_1) \vee (\forall x_2)_0^{x_1}[A_2(x_2) \vee (\forall x_3)_0^{x_2}A_3(x_3)]] \wedge (\exists x)B(x).$$

Now consider the following system of formulas,

$$(3)\qquad \begin{aligned} r_3x &\equiv (\forall t)_0^x A_3[\boldsymbol{i}t] \\ r_2x &\equiv (\forall t)_0^x \,.\, A_2[\boldsymbol{i}t] \vee r_3t \\ r_1x &\equiv (\forall t)_0^x \,.\, A_1[\boldsymbol{i}t] \vee r_2t \\ r_0x &\equiv (\exists t)_0^x B[\boldsymbol{i}t]. \end{aligned}$$

It determines recursively an operator ζ, defined for vectors $\boldsymbol{i}$ of finite predicates on α and taking as values vectors $\boldsymbol{r} = \zeta\boldsymbol{i}$ of predicates r_0, r_1, r_2, r_3 on $\alpha + 1$. Furthermore one easily verifies that the right side of (2) is equivalent to $(\forall \boldsymbol{r}).\boldsymbol{r} = \zeta\boldsymbol{i} \supset [r_1\alpha \wedge r_0\alpha]$, for any particular α. Thus, for any α,

$$(4)\qquad \Gamma(\boldsymbol{i}) \equiv (\forall \boldsymbol{r}).\boldsymbol{r} = \zeta\boldsymbol{i} \supset [r_1\alpha \wedge r_0\alpha].$$

But, by Lemma 8, the recursive operator ζ defined by (3) is also definable by a special recursion $\mathfrak{R}$ of order 1. Now, by Definition 2, the right side of (4) just says that $\boldsymbol{i}$ belongs to the behavior of $\mathfrak{R}, O$ where the output $O[\boldsymbol{r}\alpha]$ is given by $[r_1\alpha \wedge r_0\alpha]$. Thus, $\Gamma(\boldsymbol{i})$ defines $\mathrm{beh}_{\mathfrak{R},O}^{\alpha}$, for any α. Furthermore we have actually given a procedure for constructing the 1-automaton $\mathfrak{R}, O$. Q.e.d.

The reader will find two ideas of Skolem's in the previous proofs. Namely "replacing bounded quantifiers by primitive recursions" in the proof of Lemma 8, and "isolation of individual quantifiers" in formulas containing monadic predicates only.

It is now easy to prove our first theorem.

THEOREM 1 (SYNTHESIS THEOREM): *To every* WS-*formula* $\Sigma(\boldsymbol{i})$ *one can construct a special automaton* $\mathfrak{A}$ *such that, for any ordinal* α, *the set* $\mathfrak{t}\Sigma(\boldsymbol{i})$ *defined in* WST$[\alpha, <]$ *by* Σ *is just the* α-*behavior of* $\mathfrak{A}$. *Moreover, if* Σ *is a prenex formula of height* h *then* $\mathfrak{A}$ *is of order* $(h+1)$.

PROOF. By Lemma 7 we may assume that $\Sigma(\boldsymbol{i})$ is a prenex formula, say for example of height $h = 3$. Then $\Sigma(\boldsymbol{i}) . \equiv . (\exists \boldsymbol{j}_3) \sim (\exists \boldsymbol{j}_2) \sim (\exists \boldsymbol{j}_1) \Gamma(\boldsymbol{i}, \boldsymbol{j}_1, \boldsymbol{j}_2, \boldsymbol{j}_3)$ where Γ is a kernel. By Lemma 9 one can construct a 1-automaton $\mathfrak{A}_1$ whose α-behavior satisfies

$$\mathrm{beh}^{\alpha}_{\mathfrak{A}}(\boldsymbol{i}, \boldsymbol{j}_1, \boldsymbol{j}_2, \boldsymbol{j}_3) \equiv \Gamma(\boldsymbol{i}, \boldsymbol{j}_1, \boldsymbol{j}_2, \boldsymbol{j}_3).$$

Using the Projection Lemma 4 three times, and Lemma 2 (twice) one obtains from $\mathfrak{A}_1$ an automaton $\mathfrak{A}$ of order $h + 1 = 4$, as required. Q.e.d.

The converse to Theorem 1 is rather trivial in comparison. None of the previous lemmas is needed for its proof.

THEOREM 2 (ANALYSIS THEOREM): *For every special automaton* $\mathfrak{A}$ *one can construct a formula* $\Sigma(\boldsymbol{i})$ *such that, for every ordinal* α, Σ *defines in* WST$[\alpha, <]$, *the* α-*behavior of* $\mathfrak{A}$.

PROOF. Let $\mathfrak{A} = \boldsymbol{E}, \boldsymbol{H}, \boldsymbol{U}_1, \cdots, \boldsymbol{U}_p$ be the given p-automaton. By Definition 3, the α-behavior of $\mathfrak{A}$ is given by

$$\text{(a)} \qquad \mathrm{beh}^{\alpha}(\boldsymbol{i}) . \equiv . (\exists \boldsymbol{r})[(\forall x) R(x, \boldsymbol{i}, \boldsymbol{r}) \wedge O[\boldsymbol{r}\alpha]].$$

Here, $R(x, \boldsymbol{i}, \boldsymbol{r})$ is the conjunction of the formulas occurring in (3) of Section 2. Thus the formula $\Gamma(\boldsymbol{i})$ on the right side of (a) defines $\mathrm{beh}^{\alpha}(\boldsymbol{i})$. Note however that $\Gamma(\boldsymbol{i})$ is not a WS-formula because (1) R and $O[\boldsymbol{r}\alpha]$ are not WS-formulas (this could easily be corrected), and (2) the quantifiers $\exists \boldsymbol{r}$ in (a) must be interpreted to range also over non-finite predicates. To remedy this second point, we note that in calculating $\boldsymbol{r}\alpha$ from $\boldsymbol{i}$ (finite!) we actually make use of $\boldsymbol{r}x$ at finitely many places $x < \alpha$, only. These "significant places" make up a finite predicate j. For places x outside of j we then may assign the value $\boldsymbol{F}$ to $\boldsymbol{r}x$. With these hints in mind consider the conjunction $S(x, y, \boldsymbol{i}, \boldsymbol{j}, \boldsymbol{r})$ of the following formulas,

$\boldsymbol{i}x \not\equiv \boldsymbol{F} \supset j(x+1)$

$\mathrm{Lm}_a x \wedge j(x+\omega^a). \supset .\ jx$, for $0 \leqq a \leqq (p-1)$

$\mathrm{Lm}_p y \wedge \boldsymbol{j}y\ . \supset . (\exists x)_0^y [jx \wedge \mathrm{lm}_{p-1} x \wedge (\forall t)_{x+1}^y \widetilde{jt}]$

$\boldsymbol{r}0 \equiv \boldsymbol{E}$

$jx\ . \supset .\ \boldsymbol{r}(x+1) \equiv \boldsymbol{H}[\boldsymbol{r}x, \boldsymbol{i}x]$

$\tilde{j}(x+\omega^{a-1}) \wedge \cdots \wedge \tilde{j}(x+1) \wedge jx\ . \subset .\ \boldsymbol{r}(x+\omega^a) \equiv \boldsymbol{U}_a[\boldsymbol{r}x]$, for $1 \leqq a \leqq (p-1)$

$x < y \wedge \mathrm{Lm}_p y \wedge (\forall t)_{x+1}^y \widetilde{jt} \wedge jx\ . \supset .\ \boldsymbol{r}y \equiv \boldsymbol{U}_p[\boldsymbol{r}x]$.

Using $\boldsymbol{U}_a \boldsymbol{U}_b = \boldsymbol{U}_a$, for $0 \leqq b \leqq a \leqq p$ and $\boldsymbol{U}_p^2 = \boldsymbol{U}_p$, it is easy to verify that for "significant places" $x \leqq \alpha$ (i.e. if jx) these formulas assign the same value to $\boldsymbol{r}x$ as the formulas (3) of Section 2. Therefore we have

(b) $$\mathrm{beh}^\alpha(\boldsymbol{i})\ . \equiv .\ (\exists \boldsymbol{r}j) . (\forall xy)\, S(x,y,\boldsymbol{i},j,\boldsymbol{r}) \wedge j\alpha \wedge O[\boldsymbol{r}\alpha].$$

Furthermore, in (b), the quantifiers $(\exists \boldsymbol{r}j)$ may be interpreted to range over finite predicates on $\alpha+1$. Next we note that in (b) one may equivalently replace $j\alpha \wedge O[\boldsymbol{r}\alpha]$ by the disjunction $Q(\boldsymbol{i},j,\boldsymbol{r})$ of the following formulas

$$(\exists x).(\forall t)[t \leqq x] \wedge jx \wedge O[\boldsymbol{H}[\boldsymbol{r}x, \boldsymbol{i}x]]$$

$$(\forall x)(\exists t)_{x+1} \mathrm{Lm}_{a-1} t \wedge (\exists x).(\forall t)_x \widetilde{\mathrm{Lm}_a t} \wedge jx \wedge \widetilde{j(x+1)} \wedge \cdots \wedge \widetilde{j(x+\omega^a)} \wedge O[\boldsymbol{U}_a[\boldsymbol{r}x]],\ 1 \leqq a \leqq p-1$$

$$(\forall x)(\exists t)_{x+1} \mathrm{Lm}_{p-1} t \wedge (\exists x).jx \wedge (\forall t)_{x+1} \widetilde{jt} \wedge O[\boldsymbol{U}_p[\boldsymbol{r}x]].$$

(Note that the $p+1$ disjuncts of Q correspond to the cases $[\mathrm{lm}_0 \alpha \vee \alpha = 0]$, $\mathrm{lm}_1 \alpha, \cdots, \mathrm{lm}_{p-1}\alpha, \mathrm{Lm}_p \alpha$, describing various terminal characteristics of α.) Finally we remark that S and Q can be expressed by WS-formulas, because for any $0 \leqq a \leqq p-1$ the expression $y = x + \omega^a$ is WS-definable by $x < y \wedge \mathrm{Lm}_a y \wedge (\forall t)[x<t<y \supset \sim \mathrm{Lm}_a t]$. Therefore the formula $(\exists \boldsymbol{r}j).(\forall xy) S \wedge Q$ defines beh^α in WST$[\alpha, <]$, for any α. Q.e.d.

Remark 1. Theorems 1 and 2 provide a clear understanding of those relations $R(i_1, \cdots, i_k)$ between finite subsets of α, which are definable in WST$[\alpha, <]$. Namely, they are just the α-behaviors of p-automata, and thus can be surveyed in a satisfactory manner. We will not engage here in a detailed study of α-behaviors; the interested reader will find it easy to generalize from the discussion of ω-behaviors given in [1]. Note, that for a p-recursion $\mathfrak{R}$, the equivalence relation $\boldsymbol{i} \equiv \boldsymbol{j}(\mathfrak{R})$ given by $\zeta_{\mathfrak{R}}\boldsymbol{i} = \zeta_{\mathfrak{R}}\boldsymbol{j}$ is of finite index and has the congruence property $\boldsymbol{i} \equiv \boldsymbol{j}(\mathfrak{R}) . \supset . \boldsymbol{i}\hat{\ }\boldsymbol{s} \equiv \boldsymbol{j}\hat{\ }\boldsymbol{s}(\mathfrak{R})$.

The behavior of an automaton $\mathfrak{R}, O$ is a (finite) union of congruence classes of $\equiv(\mathfrak{R})$.

Remark 2. An element x of α can be represented by the finite predicate $it \equiv [x = t]$. Therefore, we could also survey the relations of type $R(i_1, \cdots, i_n, x_1, \cdots, x_m)$, definable in WST$[\alpha, <]$, and thus obtain a complete picture of definability in these theories. In the next section we will study, in some detail, the particular case $R(x)$.

Remark 3. The formula $\Sigma(\boldsymbol{i})$ constructed in the proof of Theorem 2 is of the form $(\exists \boldsymbol{j})(\forall x)\Gamma(\boldsymbol{i}, \boldsymbol{j}, x)$, where Γ is a "generalized kernel" containing only propositional connectives, atomic parts $\boldsymbol{i}t$ and $\boldsymbol{j}t$, and bounded individual quantifiers. Note also the uniformity in α of both Theorem 1 and 2. Thus, to every WS-formula $\Sigma(\boldsymbol{i})$ we can construct a formula $\Sigma'(\boldsymbol{i})$ of form $(\exists \boldsymbol{j})(\forall x)$(generalized kernel), such that $\Sigma(\boldsymbol{i}) \equiv \Sigma'(\boldsymbol{i})$ holds in all WST$[\alpha, <]$. Furthermore, such a $\Sigma'(\boldsymbol{i})$ can also be constructed in the dual form $(\forall \boldsymbol{j})(\exists x)$(generalized kernel). Just note that the right side of (a), in the proof of Theorem 2, can be replaced by $(\forall \boldsymbol{r})[(\forall x)R(x, \boldsymbol{i}, \boldsymbol{r}) \supset O[\boldsymbol{r}\alpha]]$.

4. Input free special recursions. For any $p < \omega$ the ordinal x can be represented in the form $x = y + \omega^{p-1}c_{p-1} + \cdots + \omega^0 c_0$, where $\mathrm{Lm}_p y$ (note that y may be 0), and $0 \leqq c_0, \cdots, c_{p-1} < \omega$. We set $c = 0$ if $y = 0$, and $c = 1$ if $\omega^p \leqq y$, and call $(c, c_{p-1}, \cdots, c_0)$ the *p-terminal character* of x. Suppose now that $x = y + \omega^n c_n + \cdots + \omega^0 c_0$, where $\mathrm{Lm}_\omega y$, $c_n \neq 0$, $0 \leqq c_0, \cdots, c_n < \omega$. Set $c = 0$ if $y = 0$ and $c = 1$ if $y \neq 0$, and call $(c, c_n, \cdots, c_0)$ the *ω-terminal character* of x, $\omega / x = y$ the *ω-head* of x, and $x / \omega = \omega^n c_n + \cdots + \omega^0 c_0$ the *ω-tail* of x. Note that n may be 0, namely in case $\mathrm{Lm}_\omega x$.

Suppose now that $\mathfrak{R} = \boldsymbol{E}, \boldsymbol{U}_0[\boldsymbol{Y}], \cdots, \boldsymbol{U}_p[\boldsymbol{Y}]$ is an input free p-recursion, i.e., $\boldsymbol{U}_a \boldsymbol{U}_b = \boldsymbol{U}_a$ for $0 \leqq b < a \leqq p$ and $\boldsymbol{U}_p^2 = \boldsymbol{U}_p$. Thus the predicate vector $\boldsymbol{r} = \zeta_{\mathfrak{R}}$ defined by the recursion $\mathfrak{R}$ is given by

(1) $\quad \boldsymbol{r}x \equiv \boldsymbol{U}_0^{c_0} \cdots \boldsymbol{U}_{p-1}^{c_{p-1}} \boldsymbol{U}_p^c[\boldsymbol{E}]$ if $(c, c_{p-1}, \cdots, c_0)$ is the p-character of x.

Here $\boldsymbol{U}^0[\boldsymbol{Y}]$ stands for $\boldsymbol{Y}$, $\boldsymbol{U}^{n+1}[\boldsymbol{Y}]$ stands for $\boldsymbol{U}[\boldsymbol{U}^n[\boldsymbol{Y}]]$. It is now easy to prove the following theorem.

Definition 4. $[\alpha, <]$ and $[\beta, <]$ are called WS-*equivalent* in case the same sentences Σ are true in WST$[\alpha, <]$ and WST$[\beta, <]$.

Theorem 3. *$[\alpha, <]$ and $[\beta, <]$ are* WS-*equivalent just in case their ω-terminal characters are equal. For any α, the theory* WST$[\alpha, <]$ *is decidable. In fact there is a method M which, for a given sequence of* *numbers* $(c, c_n, \cdots, c_0)$ *and a given* WS-*sentence* Σ, *decides whether or not* Σ *is true in* $[\alpha, <]$ *if* α *has* ω-*terminal character* $(c, c_n, \cdots, c_0)$.

PROOF. Suppose first that $[\alpha, <]$ and $[\beta, <]$ are WS-equivalent and that $(c, c_n, \cdots, c_0)$ is the ω-terminal character of α. We show, by cases, that $(c, c_n, \cdots, c_0)$ must also be the ω-terminal character of β.

Case 1. $c = 0$. Then $\alpha = \omega^n c_n + \cdots + \omega^0 c_0$. It is easy to write a WS-sentence Σ which holds in $[\gamma, <]$ if and only if $\gamma = \omega^n c_n + \cdots + \omega^0 c_0$. Because $[\alpha, <]$ and $[\beta, <]$ are WS-equivalent, it follows that Σ also holds in $[\beta, <]$ and therefore $\beta = \alpha$. Thus $(0, c_n, \cdots, c_0)$ is the ω-terminal character of β.

Case 2. $c = 1$, $n = 0$. This means $\mathrm{Lm}_\omega \alpha$, i.e., the WS-sentences $(\forall x)(\exists y)_x \mathrm{Lm}_k y$, $k = 1, 2, 3, \cdots$ all hold in $[\alpha, <]$. Because $[\alpha, <]$ and $[\beta, <]$ are WS-equivalent, all these sentences also hold in $[\beta, <]$. Therefore (1) is also the ω-terminal character of β.

Case 3. $c = 1$, $n \neq 0$. It is easy to write WS-sentences $\Sigma_k, k = 1, 2, 3, \cdots$ such that Σ_k holds in $[\gamma, <]$ just in case $(\exists y)_1^\gamma [\mathrm{Lm}_k y \wedge \gamma = y + \omega^n c_n + \cdots + \omega^0 c_0]$. Because $(1, c_n, \cdots, c_0)$ is the ω-terminal character of α, all sentences Σ_k hold in $[\alpha, <]$. Because $[\beta, <]$ is WS-equivalent to $[\alpha, <]$, these sentences also hold in $[\beta, <]$. It follows that $(1, c_n, \cdots, c_0)$ must also be the ω-terminal character of β.

Suppose next that α and β have the same ω-terminal character. Note that α and β then have the same p-terminal character for any p. Now, let Σ be any WS-sentence. By Theorem 1 there is an input free p-recursion $\mathfrak{R}$ and an output O such that, for $\boldsymbol{r} = \zeta_{\mathfrak{R}}$, Σ holds in $[\gamma, <]$ if and only if $O[\boldsymbol{r}\gamma]$. But α and β have the same p-terminal character. Therefore, by (1), $\boldsymbol{r}\alpha \equiv \boldsymbol{r}\beta$. Thus $O[\boldsymbol{r}\alpha] \equiv O[\boldsymbol{r}\beta]$, i.e., Σ holds in $[\alpha, <]$ if and only if it holds in $[\beta, <]$. Thus $[\alpha, <]$ and $[\beta, <]$ are WS-equivalent.

This establishes the first part of Theorem 3. It remains to describe a decision method for $\mathrm{WST}[\alpha, <]$, if the ω-terminal character of α is $(c, c_n, \cdots, c_0)$. Therefore let Σ be any WS-sentence. Using Theorem 1 we construct an input free p-recursion $\mathfrak{R} = \boldsymbol{E}, \boldsymbol{U}_0[\boldsymbol{Y}], \cdots, \boldsymbol{U}_p[\boldsymbol{Y}]$ and an output $O[\boldsymbol{Y}]$ such that, for $\boldsymbol{r} = \zeta_{\mathfrak{R}}$, Σ holds in $[\alpha, <]$ if and only if $O[\boldsymbol{r}\alpha]$. We may assume that $p \geqq n$, else modify $\mathfrak{R}$ letting $\boldsymbol{U}_{p+1} = \cdots = \boldsymbol{U}_n = \boldsymbol{U}_p$. Then the p-terminal character of α is $(c, 0, \cdots, 0, c_n, \cdots, c_0)$, and by (1), $\boldsymbol{r}\alpha \equiv \boldsymbol{U}_0^{c_0} \cdots \boldsymbol{U}_n^{c_n} \boldsymbol{U}_p^{c}[\boldsymbol{E}]$. Thus, Σ holds in $[\alpha, <]$ if and only if $\boldsymbol{O}\boldsymbol{U}_0^{c_0} \cdots \boldsymbol{U}_n^{c_n} \boldsymbol{U}_p^{c}[\boldsymbol{E}]$. Thus the truth of Σ in WST $[\alpha, <]$ can be decided by evaluating the truth value of the propositional expression $\boldsymbol{O}\boldsymbol{U}_0^{c_0} \cdots \boldsymbol{U}_n^{c_n} \boldsymbol{U}_p^{c}[\boldsymbol{E}]$. Q.e.d.

DEFINITION 5. Let h be an isomorphism of $[\alpha, <]$ into $[\beta, <]$. h is called a WS-*embedding* of $[\alpha, <]$ into $[\beta, <]$, if for any WS-formula $\Sigma(\boldsymbol{i})$, and for any vector $\boldsymbol{i}$ of finite subsets of α, $\Sigma(\boldsymbol{i})$ holds in $[\alpha, <]$ if and only if $\Sigma(h\boldsymbol{i})$ holds in $[\beta, <]$. If there is a WS-embedding of $[\alpha, <]$ into $[\beta, <]$, we call $[\beta, <]$ a WS-*extension* of $[\alpha, <]$.

THEOREM 4. *For any* $\alpha < \beta$, $[\beta, <]$ *is a* WS-*extension of* $[\alpha, <]$ *if and only if* α *and* β *have the same* ω-*terminal character (i.e., if and only if* $[\alpha, <]$ *and* $[\beta, <]$ *are* WS-*equivalent). If this is the case, the* WS-*embedding* h *of* $[\alpha, <]$ *into* $[\beta, <]$ *is given by*

$$(2) \qquad hx = \begin{cases} x, & \text{if } 0 \leqq x < (\omega/\alpha) \\ (\omega/\beta) + y, & \text{if } x = (\omega/\alpha) + y,\ y < (\alpha/\omega) \end{cases}$$

In particular, if $\omega \leqq \alpha < \beta$ *then* $[\omega^\beta, <]$ *is a* WS-*extension of* $[\omega^\alpha, <]$, *and the embedding is* $hx = x$.

PROOF. Suppose $[\beta, <]$ is a WS-extension of $[\alpha, <]$. Then $[\beta, <]$ and $[\alpha, <]$ are trivially WS-equivalent, and by Theorem 3, α and β have the same ω-terminal character.

Suppose next that $\alpha < \beta$ have the same ω-terminal character $(c, c_n, \cdots, c_0)$, so that $\alpha/\omega = \beta/\omega = \omega^n c_n + \cdots + \omega^0 c_0$. Note that $c = 1$ (else $\alpha = \beta$), thus $0 < (\omega/\alpha) < (\omega/\beta)$ are in Lm_ω. We are to show that (2) defines a WS-embedding h of $[\alpha, <]$ into $[\beta, <]$. Let therefore $\Sigma(\boldsymbol{i})$ be any WS-formula, and let $\boldsymbol{i}$ be any vector of finite subsets of α. By Theorem 1 there is a p-recursion $\mathfrak{R} = \boldsymbol{E}, \boldsymbol{H}[\boldsymbol{Y}, \boldsymbol{X}], \boldsymbol{U}_1[\boldsymbol{Y}], \cdots, \boldsymbol{U}_p[\boldsymbol{Y}]$ and an output $O[\boldsymbol{Y}]$ such that for $\boldsymbol{r} = \zeta_{\mathfrak{R}}\boldsymbol{i}, \boldsymbol{s} = \zeta_{\mathfrak{R}}(h\boldsymbol{i})$,

$$(3) \qquad \begin{array}{l} \Sigma(\boldsymbol{i}) \text{ holds in } [\alpha, <] \text{ if and only if } O[\boldsymbol{r}\alpha] \\ \Sigma(h\boldsymbol{i}) \text{ holds in } [\beta, <] \text{ if and only if } O[\boldsymbol{s}\beta]. \end{array}$$

Let $\boldsymbol{Y} \equiv \boldsymbol{r}(\omega/\alpha)$. Because $\mathrm{Lm}_p(\omega/\alpha)$ and $\omega/\alpha \neq 0$ it follows that $\boldsymbol{U}_p[\boldsymbol{Y}] \equiv \boldsymbol{Y}$. By (2) we have $(\forall t)_0^{\omega/\alpha}[\boldsymbol{i}t \equiv (h\boldsymbol{i})t]$, therefore $\boldsymbol{r}(\omega/\alpha) \equiv \boldsymbol{s}(\omega/\alpha) \equiv \boldsymbol{Y}$. By (2) we have $(\forall t)_{\omega/\alpha}^{\omega/\beta}[(h\boldsymbol{i})t \equiv \boldsymbol{F}]$, therefore $\boldsymbol{s}(\omega/\beta) \equiv \boldsymbol{U}_p[\boldsymbol{s}(\omega/\alpha)]$, thus $\boldsymbol{s}(\omega/\beta) \equiv \boldsymbol{Y}$. By (2) we have $(\forall t)_0^\gamma[\boldsymbol{i}(\omega/\alpha + t) \equiv (h\boldsymbol{i})(\omega/\beta + t)]$ if $\gamma = \alpha/\omega = \beta/\omega$; furthermore $\boldsymbol{r}(\omega/\alpha) \equiv \boldsymbol{s}(\omega/\beta) \equiv \boldsymbol{Y}$. Therefore $\boldsymbol{r}\alpha \equiv \boldsymbol{s}\beta$. By (3) it now follows that $\Sigma(\boldsymbol{i})$ holds in $[\alpha, <]$ if and only if $\Sigma(h\boldsymbol{i})$ holds in $[\beta, <]$. Q.e.d.

If $\alpha < \omega^\omega$, it is easy to see that every element of α is definable in $\mathrm{WST}[\alpha, <]$, and therefore, every finite predicate $\boldsymbol{i} \subseteq \alpha$ is definable in $\mathrm{WST}[\alpha, <]$. Also, if $\beta < \omega^\omega$, all finite predicates on $\omega^\omega + \beta$ are definable in $\mathrm{WST}[\omega^\omega + \beta, <]$. Consider now $[\alpha + \beta, <]$ where $\mathrm{Lm}_\omega \alpha, \alpha \neq 0, \beta < \omega^\omega$. By Theorem 4, this is a WS-extension of $[\omega^\omega + \beta, <]$. It follows that x is definable in WST $[\alpha+\beta,<]$ if and only if either $x<\omega^\omega$ or $\alpha \leqq x<\alpha+\beta$. Therefore,

COROLLARY 1. *If* $\mathrm{Lm}_\omega \alpha, \alpha \neq 0, \beta<\omega^\omega$ *then the finite predicate i is definable in* WST$[\alpha+\beta,<]$ *if and only if* $i \subseteq \{x ; x<\omega^\omega\} \cup\{x ; \alpha \leqq x<\alpha+\beta\}$, *i.e., i does not enter into* $[\omega^\omega, \alpha]$.

To further test the strength of our definability criterion (Theorems 1 and 2) we discuss definability of subsets of α by WS-formulas $\Sigma(x)$. Consider the formula $\Sigma'(i):(\exists x) . \Sigma(x) \wedge(\forall t)[i t \equiv(t=x)]$. By Theorem 1 there is a p-recursion $\mathfrak{R}=\boldsymbol{E}, \boldsymbol{H}[Y, X], \boldsymbol{U}_1[Y], \cdots, \boldsymbol{U}_p[Y]$ and an output O such that for any finite i on α, $\Sigma'(i)$ holds in $[\alpha,<]$ if and only if $O[\boldsymbol{r} \alpha]$, if $\boldsymbol{r}=\zeta_{\mathfrak{R}} i$. Let $U_0[Y]$ stand for $\boldsymbol{H}[Y, F], \boldsymbol{V}[Y]$ for $\boldsymbol{H}[Y, T]$. By (1) and the relationship between $\Sigma(x)$ and $\Sigma'(i)$ it follows that, for any $x<\alpha$, $\Sigma(x)$ holds in $[\alpha,<]$ if and only if

$$
O U_0^{a_0} \cdots U_{p-1}^{a_{p-1}} U_p^a V U_0^{b_0} \cdots U_{p-1}^{b_{p-1}} U_p^b[E] \tag{4}
$$

where $(b, b_{p-1}, \cdots, b_0)$ is the p-terminal character of x and $(a, a_{p-1}, \cdots, a_0)$ is the p-terminal character of $-(x+1)+\alpha$. Thus,

COROLLARY 2. *Let S be a subset of* α *which is definable in* WST$[\alpha,<]$. *There are propositional expressions* $\boldsymbol{E}, \boldsymbol{U}_0[Y], \cdots, \boldsymbol{U}_p[Y], \boldsymbol{V}[Y], \boldsymbol{O}[Y]$, $\boldsymbol{U}_c \boldsymbol{U}_d=\boldsymbol{U}_c$ *for* $c<d$, $\boldsymbol{U}_d \boldsymbol{U}_d=\boldsymbol{U}_d$, *such that* $x \in S$ *if and only if* (4) *holds with* $(b, b_{p-1}, \cdots, b_0)=$ *the p-terminal character of* x, *and* $(a, a_{p-1}, \cdots, a_0)$ $=$ *the p-terminal character of* $-(x+1)+\alpha$.

Such a set S might well be called *ultimately periodic of order* p. Let us for example take the case $\alpha=\omega^\omega$. Then any $x<\omega^\omega$ is of the form $\omega^n c_n+\cdots+\omega^0 c_0$, $c_n \neq 0$. The p-terminal character of $-(x+1)+\omega^\omega$ is (1). The p-terminal character of x is $(0,0, \cdots, 0, c_n, \cdots, c_0)$ if $n<p$, and $(1, c_{p-1}, \cdots, c_0)$ if $p \leqq n$. If we let $Q[Y]$ stand for $O[V[Y]]$ we therefore have that $x=\omega^n c_n+\cdots+\omega^0 c_0\left(c_n \neq 0\right)$ belongs to S just in case

$$
\begin{aligned}
& Q U_0^{c_0} \cdots U_{n-1}^{c_{n-1}} U_n^{c_n}[E], \quad \text{if } n<p \\
& Q U_0^{c_0} \cdots U_{p-1}^{c_{p-1}} U_p[E], \quad \text{if } p \leqq n .
\end{aligned} \tag{4'}
$$

It follows that, from ω^p on, S repeats itself with period ω^p; i.e., $S[\omega^p, \omega^p+\omega^p)=S[y, y+\omega^p)$ for all $0<y<\omega^\omega$, $\mathrm{lm}_p y$. Note furthermore that for each $0 \leqq d<p-1$ there are numbers e, q such that $\boldsymbol{U}_d^e=\boldsymbol{U}_d^q \boldsymbol{U}_d^e$. It follows that S also becomes ultimately periodic in each segment $[y, y+\omega^{d+1}]$, $\mathrm{Lm}_d y$, for any $0 \leqq d<p-1$.

The subsets of α, definable in WST$[\alpha,<]$, thus are closely related to certain right-congruences of finite index on $[\alpha,+]$. Similarly, the study of α-behaviors of p-automata (i.e. sets S of vectors of finite predicates on α, definable in WST[α, $<$]) is related to certain right-congruences of finite index on well-ordered sequences of letters from a finite alphabet. We do not want to elaborate here on this extension of automata theory to the transfinite.

5. Elementary theory of addition of ordinals. Every ordinal x can be uniquely represented in the form $x = 2^{u_1} + \cdots + 2^{u_n}$, where $u_1 > \cdots > u_n$. The finite predicate $\phi x = \{u_n, \cdots, u_1\}$ is called the *binary expansion* of x. For any α, ϕ is a one-to-one map of 2^α onto the class Fin_α, consisting of all finite subsets of α. In fact ϕ is an isomorphism between the systems

$$(1) \qquad \begin{array}{l} 2^\alpha, \quad P, E, \prec \\ \mathrm{Fin}_\alpha, \ \alpha, \ A, \ < \end{array}$$

where $P = \{2^x; x < \alpha\}$, Exy stands for "x is a power of 2 which occurs in the binary expansion of y" (i.e. $Exy . \equiv . (\exists u)[x = 2^u \wedge u \in \phi y]$), $\prec$ stands for the ordering relation on P, Axi stands for ix, and $<$ is the ordering relation on α. Thus

(2) The first order theory FT[$2^\alpha, P, E, \prec$] is identical with the weak second order theory WST[α, $<$].

Let $i \oplus j = \phi(\phi^{-1} i + \phi^{-1} j)$, for any finite predicates i,j on ordinals. We note next that the algorithm for addition of natural numbers in binary expansion can be extended to ordinals. Namely $i \oplus j = s$ holds if and only if there exist an ordinal y and a (finite) predicate r such that

$$(3) \qquad \begin{array}{l} y \text{ is the largest } \mathrm{Lm}_1 \text{ which is surpassed by members of } j. \\ rx \equiv F, \text{ for all } \mathrm{Lm}_1 x \\ r(x+1) . \equiv . [ix \wedge jx] \vee [ix \wedge rx] \vee [jx \wedge rx], \text{ for all } x \\ sx . \equiv . ix \,\overline{\vee}\, jx \,\overline{\vee}\, rx, \quad \text{for all } x \geqq y \ (\overline{\vee} = \text{exclusive or}) \\ sx \equiv jx, \text{ for all } x < y \end{array}$$

(to see this, note that $2^u + 2^v = 2^v$ if $u + \omega \leqq v$, $2^u + 2^v = 2^v + 2^u$ if $u - v$ is finite, $2^u + 2^u = 2^{u+1}$). It is easy to express (3) as a WS-formula $B(i,j,s,r,y)$. Thus, $i \oplus j = s$ is definable by a WS-formula $(\exists ry)B$. On the other hand P is definable from E, and $\prec$ is definable from $+$ and P, by elementary formulas. Thus, the theory WST[α, $<$] is but a reformulation of FT[2^α, $+$, E] differing only in the choice of primitives. In particular, we have shown

LEMMA 10. a) *For every* WS-*formula* $\Sigma_<(i_1, \cdots, i_k)$ *one can effectively construct an elementary formula* $\Sigma'_{+,E}(x_1,\cdots,x_n)$ *such that, for any finite predicates* $i_1,\cdots,i_k$ *on* α, $\Sigma(i_1,\cdots,i_k)$ *holds in* WST $[\alpha,<]$ *if and only if* $\Sigma'(\phi^{-1}i_1,\cdots,\phi^{-1}i_k)$ *holds in* FT$[2^\alpha,+,E]$.

b) *For every elementary formula* $\Sigma_{+,E}(x_1,\cdots,x_k)$ *one can construct a* WS-*formula* $\Sigma^*_<(i_1,\cdots,i_k)$ *such that for any* $x_1,\cdots,x_n<2^\alpha$, $\Sigma(x_1,\cdots,x_k)$ *holds in* FT$(2^\alpha,+,E]$ *if and only if* $\Sigma^*(\phi x_1,\cdots,\phi x_k)$ *holds in* WST$[\alpha,<]$.

This lemma together with Theorems 1 and 2 yield complete information on those sets of ordinals which are elementarily definable in $[2^\alpha,+,E]$. In particular, we can restate Theorems 3, 4, and 5 in the following form.

THEOREM 3′. $[2^\alpha,+,E]$ *and* $[2^\beta,+,E]$ *are elementarily equivalent just in case* α *and* β *have the same* ω-*terminal character. For any* α, *the elementary theory* FT$[2^\alpha,+,E]$ *is decidable.*

THEOREM 4′. *For any* $\alpha<\beta$, $[2^\alpha,+,E]$ *is elementarily embeddable into* $[2^\beta,+,E]$ *if and only if* α *and* β *have the same* ω-*terminal character. If this is the case, the embedding h is given by*

$$h(2^{\omega/\alpha}x_1+x_2) = 2^{\omega/\beta}x_1+x_2$$

for $x_1<2^\gamma, x_2<2^{\omega/\alpha}$, $\gamma=\alpha/\omega=\beta/\omega$. *In particular, if* $\omega\leqq\alpha<\beta$ *then* $[\omega^{(\omega^\beta)},+,E]$ *is an elementary extension of* $[\omega^{(\omega^\alpha)},+,E]$; *the embedding is* $hx=x$.

In particular, if Od is the class of all ordinals, $[Od,+,E]$ is an elementary extension of $[\omega^{(\omega^\omega)},+,E]$. As a corollary to this we get Ehrenfeucht's result [6]: $[Od,+]$ is an elementary extension of $[\omega^{(\omega^\omega)},+]$. (Note that $2^{(\omega^\omega)}=\omega^{(\omega^\omega)}$.)

COROLLARY 1′. *If* $\mathrm{Lm}_\omega\alpha$, $\alpha\neq 0$ $\beta<\omega^\omega$ *then x is elementarily definable in* $[2^{\alpha+\beta},+,E]$ *if and only if* $x=2^\alpha x_1+x_2$ *for* $x_1<2^\beta$ *and* $x_2<2^{(\omega^\omega)}$.

REFERENCES

[1] J. R. BÜCHI, Weak second order arithmetic and finite automata. *Zeitschrift für Math. Log. und Grundl. der Math.* **6** (1960), pp. 66–92.

[2] ———, On a decision method in restricted second order arithmetic. *Logic, Method. and Phil. of Sc., Proc.* 1960 *Int. Congress*, Stanford Univ. Press, 1962.

[3] S. FEFERMAN, Some recent work of Ehrenfeucht and Fraïssé. Summer Institute for Symbolic Logic, Cornell Univ. 1957, Commun. Research Div., Institute for Defense Analysis, 1960, pp. 201–209.

[4] S. FEFERMAN and R. L. VAUGHT, The first order properties of products of algebraic systems. *Fund. Math.* **47** (1959), pp. 57–103.

[5] R. MCNAUGHTON, Review of [1, 2], *Journ. Symb. Logic* **28** (1963), pp. 100–102.

[6] A. EHRENFEUCHT, Application of games to some problems of mathematical logic. *Bull. de l'Acad. Pol. Sci.* **5** (1957), pp. 35–37.

THE INTRINSIC COMPUTATIONAL DIFFICULTY OF FUNCTIONS

ALAN COBHAM
I.B.M. Research Center, Yorktown Heights, N. Y., U.S.A.

The subject of my talk is perhaps most directly indicated by simply asking two questions: first, is it harder to multiply than to add? and second, why? I grant I have put the first of these questions rather loosely; nevertheless, I think the answer, ought to be: *yes*. It is the second, which asks for a justification of this answer which provides the challenge.

The difficulty does not stem from the fact that the first question has been imprecisely formulated. There seems to be no substantial problem in showing that using the standard algorithms it is in general harder—in the sense that it takes more time or more scratch paper—to multiply two decimal numbers than to add them. But this does not answer the question, which is concerned with the computational difficulty of these two operations without reference to specific computational methods; in other words, with properties intrinsic to the functions themselves and not with properties of particular related algorithms. Thus to complete the argument, I would have to show that there is no algorithm for multiplication computationally as simple as that for addition, and this proves something of a stumbling block. Of course, as I have implied, I feel this must be the case and that multiplication is in this absolute sense harder than addition, but I am forced to admit that at present my reasons for feeling so are of an essentially extra-mathematical character.

Questions of this sort belong to a field which I think we might well call *metanumerical-analysis*. The name seems appropriate not only because it is suggestive of the subject matter, namely, the methodology of computation, but also because the relationship of the field to computation is closely analogous to that of metamathematics to mathematics. In metamathematics, we encounter problems concerned with specific proof systems; e.g., the existence of proofs having a certain form, or the adequacy of a system in a given context. We also encounter problems concerned with provability but independently of any particular proof system; e.g., the undecidability of mathematical theories. So in metanumerical-analysis we encounter problems related to specific computational systems or categories of computing machines as well as problems such as those mentioned above which, though concerned with computation, are independent of any particular method of computation. It is this latter segment of metanumerical-analysis I would like to look at more closely.

Let me begin by stating two of the very few results which fall squarely within this area. (Others appear in references [3], [4] and [5].) Both are drawn from Ritchie's work on predictably computable functions [6], the first being an almost immediate generalization of one part of that work, the second an incomplete rendering of another part. These relate, perhaps not in the most happy fashion, the computational complexity of a function with its location in the Grzegorczyk hierarchy [2]. Recall that this hierarchy is composed of a sequence of classes:

$$\mathscr{E}^0 \subset \mathscr{E}^1 \subset \mathscr{E}^2 \subset \ldots,$$

each properly contained in the next, and having as union the class of all primitive recursive functions. $\mathscr{E}^2$ can be characterized as the smallest class containing the successor and multiplication functions, and closed under the operations of explicit transformation, composition and limited recursion. The classes $\mathscr{E}^3$ (which is the familiar class of Kalmar elementary functions) and $\mathscr{E}^4$ have similar characterizations, with exponentiation and the function $\underbrace{x^{x^{\cdot^{\cdot^{\cdot^{x}}}}}}_{y}$ respectively, replacing multiplication in the characterization of $\mathscr{E}^2$. The higher classes can be characterized inductively in like manner.

With each single-tape Turing machine Z which computes a function of one variable we may associate two functions, σ_Z and τ_Z. Assuming some standard encoding of natural numbers—we might take decimal notation to be specific—we define $\sigma_Z(n)$, where n is a natural number, to be the numbers of steps (instruction executions) in the computation on Z starting with n encoded on its tape, and define $\tau_Z(n)$ to be the number of distinct tape squares scanned during the course of this computation. Restricting attention to functions of one variable, we have the following.

THEOREM. *For each $k \geqq 3$ the following five statements are equivalent:*

1. $f \in \mathscr{E}^k$;

2. *there exists a Turing machine Z which computes f and such that $\sigma_Z \in \mathscr{E}^k$;*

3. *there exists a Turing machine Z which computes f and a function $g \in \mathscr{E}^k$ such that, for all n, $\sigma_Z(n) \leqq g(n)$;*

4. – 5. *Same as 2 – 3 with τ_Z in place of σ_Z.*

This theorem has an immediate generalization to functions of several variables. From it we can infer that (in effect, but not quite precisely) if one function is simpler to compute than another, in the very strong sense that for any value of the argument the computation can be done in fewer steps or with less tape, then that function lies no higher than the other in the Grzegorczyk hierarchy. We cannot conclude the converse however: that if one function lies lower than another, it is necessarily simpler to compute. As a matter of fact, it appears that a function in the lower part of the hierarchy may actually be *on average* harder to compute than one higher up, though, of course, it cannot be harder for all values of the argument.

A word is needed as to why I have included a theorem involving Turing machines in a discussion which I said was going to be about method-independent aspects of computation. The fact is the theorem remains correct even if one considers far wider classes of computing machines. In particular, it holds for Turing machines with more than one tape or with multi-dimensional tapes providing the cells of the latter are arranged in reasonably orderly fashion. It also holds if the set of possible instructions is extended to include, e.g., erasure of an entire tape or resetting of a scanning head to its initial position (although I doubt such operations should be considered *steps* since it does not appear that they can be executed in a bounded amount of time).

The reason for such general applicability can be found on examination of the proof of this theorem. There we find that the fact that we are dealing with a particular class of Turing machines is quite incidental: it is the form of their arithmetization which counts. The geometry and basic operations of a Turing machine are of a sort which admit an arithmetization in which the functions which describe the step-by-step course of a computation on it are of a very simple nature, lying, in particular, well within the class $\mathscr{E}^2$. This is all that is needed to obtain the preceding theorem (as well as the one which follows). Now the class $\mathscr{E}^2$ is so rich in functions that it is almost inconceivable to me that there could exist *real* computers not having mathematical models whose arithmetization could be carried out in such a way that these associated functions would fall within it. Thus I suspect this theorem does indeed say something about the absolute computational properties of functions, and so fits properly in the discussion.

The five equivalences of the preceding theorem do not hold for $k < 3$. Ritchie has obtained a hierarchy which decomposes the range between $\mathscr{E}^2$ and $\mathscr{E}^3$ into classes of functions of varying degrees of computational difficulty; however, rather than go into this, I would like now to turn to the problem of classifying the functions within $\mathscr{E}^2$, where many of the functions most frequently encountered in computational work, addition and multiplication in particular, are located. First, concerning $\mathscr{E}^2$ itself, we have [6] the following.

THEOREM. *A function f belongs to* $\mathscr{E}^2$ *if and only if there exists a Turing machine Z which computes f and constants* c_1 *and* c_2 *such that* $\tau_Z(n) \leqq c_1 l(n) + c_2$, *for all n.*

Here $l(n)$ is the length of n, that is, the number of digits in its decimal representation. Machines which compute in the fashion described are equivalent to those which Myhill has called linear bounded automata [4]. Since merely writing n requires $l(n)$ tape squares, we must have $c_1 \geqq 1$. As a matter of fact, if we consider machines with arbitrarily large alphabets, then c_1 need only be enough larger than this to permit writing of the answer on the tape; e.g., if $f(n) = n^2$ we can take $c_1 = 2$; if $f(n) \leqq n$ for all n we can take $c_1 = 1$. In other words, if we have enough space to write the larger of the value and the argument of a function in $\mathscr{E}^2$ then we have enough space to carry out the entire computation. Consequently, the function τ is not a suitable tool for making fine distinctions concerning the computational difficulty of functions within $\mathscr{E}^2$. We might attempt to redefine what we mean by the amount of tape used during a computation by distinguishing between those locations used for writing input and output and those used in the actual computation. But the artificiality of such a seemingly ad hoc distinction would seem to be trending away from our goal of obtaining a natural analysis independent of the method or type of machine used in the computation.

This may be a good point to mention that, although I have so far been tacitly equating computational difficulty with time and storage requirements, I don't mean to commit myself to either of these measures. It may turn out that some measure related to the physical notion of work will lead to the most satisfactory analysis; or we may ultimately find that no single measure adequately reflects our intuitive concept of difficulty. In any case, for the present, I see no harm in restricting the discussion somewhat and, having discarded τ as a tool for reasons just stated, confining further attention to the analysis of computation time.

This leaves us some latitude for differentiating among functions in $\mathscr{E}^2$. The closest analog of the foregoing theorem concerning σ, rather than τ, that I know of states that for any f in $\mathscr{E}^2$ there exists a Turing machine Z which computes it and such that σ_Z is bounded by a polynomial in its argument. f itself must also be bounded by a polynomial in its argument, but I don't know whether these two conditions in turn imply that f is in $\mathscr{E}^2$.

To obtain some idea as to how we might go about the further classification of relatively simple functions, we might take a look at how we ordinarily set about computing some of the more common of them. Suppose, for example, that m and n are two numbers given in decimal notation with one written above the other and their right ends aligned. Then to add m and n we start at the right and proceed digit-by-digit to the left writing down the sum. No matter how large m and n, this process terminates with the answer after a number of steps equal at most to one greater than the larger of $l(m)$ and $l(n)$. Thus the process of adding m and n can be carried out in a number of steps which is bounded by a linear polynomial in $l(m)$ and $l(n)$. Similarly, we can multiply m and n in a number of steps bounded by a quadratic polynomial in $l(m)$ and $l(n)$. So, too, the number of steps involved in the extraction of square roots, calculation of quotients, etc., can be bounded by polynomials in the lengths of the numbers involved, and this seems to be a property of simple functions in general. This suggests that we consider the class, which I will call $\mathscr{L}$, of all functions having this property.

For several reasons the class $\mathscr{L}$ seems a natural one to consider. For one thing, if we formalize the above definition relative to various general classes of computing machines we seem always to end up with the same well-defined class of functions. Thus we can give a mathematical characterization of $\mathscr{L}$ having some confidence that it characterizes correctly our informally defined class. This class then turns out to have several natural closure properties, being closed in particular under explicit transformation, composition and limited recursion on notation (digit-by-digit recursion). To be more explicit concerning the latter operation, which incidentally seems quite appropriate to computational work, we say that a function f is defined from functions $g, h_0,\ldots,h_9$, and k by limited recursion on notation (assuming decimal notation) if

$$\begin{aligned} f(\mathbf{x}, 0) &= g(\mathbf{x}) \\ f(\mathbf{x}, s_i(y)) &= h_i(\mathbf{x}, y, f(\mathbf{x}, y)) \qquad (i = 0,\ldots, 9; \qquad i \neq 0 \text{ if } y = 0) \\ f(\mathbf{x}, y) &\leqq k(\mathbf{x}, y)\,, \end{aligned}$$

where s_i is the generalized successor: $s_i(y) = 10y + i$. $\mathscr{L}$ is in fact the smallest class closed under these operations and containing the functions s_i and $x^{l(y)}$. It is closely related to, perhaps identical with, the class of what Bennett has called the extended rudimentary functions [1]. Since $\mathscr{L}$ contains $x^{l(y)}$, which cannot, by the second of the theorems mentioned earlier, belong to $\mathscr{E}^2$, $\mathscr{L}$ is not a subclass of $\mathscr{E}^2$. On the other hand, I strongly suspect that the function $f(n)$ = the nth prime, which is known to be in $\mathscr{E}^2$, does not belong to $\mathscr{L}$. If this is the case then $\mathscr{E}^2$ and $\mathscr{L}$ are incomparable and we have the unsurprising result that the categorization of the simpler functions as to computational difficulty yields divergent classifications according to the criterion of difficulty selected—in this case time and storage requirements. Concerning functions which are relatively simple under both criteria, that is, those in both $\mathscr{E}^2$ and $\mathscr{L}$, I can only offer further conjecture, namely that $\mathscr{E}^2 \cap \mathscr{L}$ ir a subclass of the constructive arithmetic functions, probably even a proper subclass. (The function $f(n) = 1$ or 0, according as n is or is not prime, is constructive arithmetic but seemingly not in $\mathscr{L}$.)

An attempt to construct a natural computational hierarchy within $\mathscr{L}$ now brings out quite sharply one of the basic problems entailed in the study of absolute or intrinsic computational properties of functions. Suppose we start out in the obvious way and define, for each k, a subclass $\mathscr{L}^k$ of $\mathscr{L}$ consisting of all functions which can be computed in such a way that the number of steps in the computation is bounded by a polynomial of degree k in the lengths of the arguments. So defined, the classes $\mathscr{L}^k$ form an increasing sequence whose union is $\mathscr{L}$. Clearly, almost as a matter of definition, the analog of the theorem concerning the Grzegorczyk hierarchy I mentioned earlier will hold for this hierarchy: a function in the upper part of the hierarchy cannot be simpler to compute for every argument than one further down.

If we are to make any application of this theorem, we need a precise, mathematical characterization of the classes $\mathscr{L}^k$. Unlike the foregoing situation, however, we find that it makes a definite difference what class of computational methods and devices we consider in our attempt to formalize the definition. Thus, if we restrict attention to single-tape Turing machines, we find that addition does not belong to $\mathscr{L}^1$, whereas it does if we permit our machines to have several tapes. Similarly, multiplication gets into $\mathscr{L}^2$ only if we permit multi-tape machines. This certainly does not mean that there is no reasonable formalization of the classes of this hierarchy, but it does suggest that there may be some difficulty both in finding this formalization and, once found, in convincing oneself that it correctly captures all relevant aspects of the intuitive model.

The problem is reminiscent of, and obviously closely related to, that of the formalization of the notion of effectiveness. But the emphasis is different in that the physical aspects of the computation process are here of predominant concern. The question of what may legitimately be considered to constitute a step of a computation is quite unlike that of what constitutes an effective operation. I did not dwell particularly on what I consider to be the properties of legitimate step when I was discussing the classification of functions outside of $\mathscr{E}^2$ because, as I pointed out, one could admit all sorts of questionable operations as steps and, so long as they could be represented by functions in $\mathscr{E}^2$, the results obtained would remain unaltered. Quite similar remarks can be made concerning permissible geometric arrangements of the working area of a computation, and even concerning the types of notation used for representing natural numbers. If, however, we are to make fine distinctions, say between functions in $\mathscr{L}^1$ and functions in $\mathscr{L}^2$, then we must have an equally fine analysis of all phases of the computational process. It is no longer a problem of finding convincing arguments that every conceivable computing method can be arithmetized within $\mathscr{E}^2$ but rather of finding convincing arguments that these can somehow be arithmetized within whatever presumably more restricted class we settle upon as a formalization for $\mathscr{L}^1$. Of course, at the same time, we must be prepared to argue that we haven't taken too broad a class for $\mathscr{L}^1$, and thus admitted to it functions not in actuality computable in a number of steps linearly bounded by the lengths of its arguments. I think this is one of the fundamental problems of metanumerical-analysis and one whose resolution may well call for considerable patience and discrimination, but until it, and several related problems, have received more intensive treatment, I doubt we can find any really satisfying proof that multiplication is indeed harder than addition.

REFERENCES

[1] J. H. BENNETT, On Spectra, doctoral dissertation, Princeton University, 1962.

[2] A. GRZEGORCZYK, Some Classes of Recursive Functions, *Rozprawy Matematyczne*, 1953.

[3] J. HARTMANIS and R. E. STEARNS, On the Computational Complexity of Algorithms, Notes for the University of Michigan Summer Conference on Automata Theory, 1963, 59–79.

[4] J. MYHILL, Linear Bounded Automata, WADD Tech. Notes 60–165, University of Pennsylvania, Report No. 60–22, 1960.

[5] M. O. RABIN, Degree of Difficulty of Computing a Function and a Partial Ordering of Recursive Sets, Applied Logic Branch, Hebrew University, Jerusalem, Technical Report No. 2, 1960.

[6] R. W. RITCHIE, Classes of Predictably Computable Functions, *Trans. Amer. Math. Soc.* **106,** 1963, 139–173.

CLASSICAL EXTENSIONS OF INTUITIONISTIC MATHEMATICS[1]

S. C. KLEENE
University of Wisconsin, Madison, Wisconsin, U.S.A.

§ **1. Introduction.** Between the oral presentation of this paper and its publication, the speaker's and R. E. Vesley's monograph *The Foundations of Intuitionistic Mathematics* [10] (cited as "FIM") will have appeared. The original formalization of intuitionistic mathematics was by Heyting in 1930 [1, 2], using some specifically intuitionistic symbolism. In contrast, the symbolism in FIM is the same as that of a system of classical mathematics. This is accomplished by the use in FIM of type-1 function variables $\alpha, \beta, \gamma, \cdots$ (i.e. variables for one-place number-theoretic functions) for Brouwer's choice sequences, as well as for other uses of functions (i.e. particular functions or laws for Brouwer, and the functions of classical mathematics). There are also type-0 variables $a, b, c, \cdots, x, y, z, \cdots$ (ranging over the natural numbers $0, 1, 2, \cdots$), the logical symbols of a two-sorted predicate calculus, the equality symbol $=$, symbols for some particular primitive recursive functions and functionals (including 0), and Church's λ-operator (with number variables).[2]

The use of a common symbolism facilitates the study of relationships between intuitionistic and classical systems. Indeed, in FIM postulates are given for three formal systems: a *basic system B*, and divergent classical and intuitionistic systems. The *classical system C* arises from the basic system *B* by adding the law of double negation $\neg\neg A \supset A$ (or the law of the excluded middle $A \vee \neg A$) as an additional axiom schema, or in place of the intuitionistic negation-elimination postulate $\neg A \supset (A \supset B)$. The *intuitionistic system I* arises from the basic system by adding a postulate which expresses Brouwer's principle that, if to each (one-place number-theoretic) function α a number b is correlated, the correlated number b must be determined (under the operation of an algorithm) by some initial segment $\alpha(0), \cdots, \alpha(y-1)$ of the values of α. (Actually, we have postulated the generalization of this to the case that to each α a function β is correlated; cf. FIM § 7.)

---

[1] Part of the work reported herein was done in 1963-64 under Grant GP1624 from the National Science Foundation of the U.S.A.

[2] Many details must be left out in the half-hour talk. The reader of the published version will be able to consult FIM. The formalization in FIM was forecasted in [4, 6].

The divergence between the classical and intuitionistic systems is manifested by outright contradictions. Thus $\vdash_C \forall\alpha(\forall x\alpha(x) = 0 \vee \neg \forall x\alpha(x) = 0)$ and $\vdash_I \neg\forall\alpha(\forall x\alpha(x) = 0 \vee \neg \forall x\alpha(x) = 0)$ (FIM *27.17), where $\vdash_C$ ($\vdash_I$) asserts provability in C (in I).

On the other hand, by no means every classically provable but intuitionistically unprovable sentence (i.e. closed formula) is refutable intuitionistically. One theorem on this situation is given in FIM (Theorem 9.14). In the present paper we shall chart the territory of compatibility between the classical and intuitionistic systems from another viewpoint. We shall show that *any classically provable sentence* E *which does not include both a universal function quantifier* $\forall\alpha$ *and an* $\exists$- *or* $\vee$-*symbol* (in the scope of an $\forall\alpha$, and not applied directly to a prime formula) *can be consistently adjoined to the intuitionistic system* (provided C is consistent).

§2. Realizing and realizability. A notion of when a number *e realizes* a number-theoretic sentence E was given by us in [3] or in our *Introduction to Metamathematics* [5] (cited as "IM") §82. In FIM §8 we introduce a corresponding notion of when a function ε *realizes*-Ψ a formula E in the symbolism of FIM, where Ψ is a specification of values of the (number and function) variables Ψ including all that occur free in E. (In [3] and IM, we avoided this "Ψ" by first substituting numerals for the free variables.) An equivalent notion was first introduced in [6], and we shall not repeat the discussion given there and in FIM to motivate the notion. The notion is not in general equivalent for number-theoretic formulas to the notion of [3] and IM.

We shall now formalize the notion ε *realizes*-Ψ E in the symbolism of FIM (as *e realizes* E for number-theory was formalized in [3] §12). We depart slightly from a direct translation of the definition in FIM 8.5, so as to arrange that the result $\varepsilon\,\mathbf{r}\,E$ contain no $\exists$, except in parts $\exists x P(x)$ with P(x) prime, and no $\vee$.[3] Accordingly, here we let "$\exists! x A(x)$" abbreviate $\exists x A(x) \,\&\, \forall x \forall y(A(x) \,\&\, A(y) \supset x = y)$, which is equivalent to the expression so abbreviated in IM (cf. p. 199, bottom p. 200). The definition of $\varepsilon\,\mathbf{r}\,E$, for any formula E, is by recursion on the number of (occurrences of) logical symbols in E, as follows. The variable ε shall be distinct from the free variables of E; in Clauses 4–6, 8 when $\alpha\,\mathbf{r}\,F$ (or $\beta\,\mathbf{r}\,F$) appears for F a part of E, the variable α (or β) shall be distinct from the free variables of F; and in Clause 4 α shall not occur free in B. Besides, all variables differently designated below shall be distinct from each other (cf. FIM 3.2 ¶s 3, 4).

---

[3] As is the case, in systems providing $T_1(z,x,y)$ as a prime formula, for one of the formalizations $e\,\mathbf{r}_1\,E$ of the earlier number-theoretic notion, [9] Theorem 1.

DEFINITION 1. 1. $\varepsilon \mathbf{r} P$ is P, for P a prime formula.

2. $\varepsilon \mathbf{r}(A \& B)$ is $(\varepsilon)_0 \mathbf{r} A \& (\varepsilon)_1 \mathbf{r} B$.

3. $\varepsilon \mathbf{r}(A \vee B)$ is $[(\varepsilon(0))_0 = 0 \supset (\varepsilon)_1 \mathbf{r} A] \& [(\varepsilon(0))_0 \neq 0 \supset (\varepsilon)_1 \mathbf{r} B]$. This is equivalent in B to $[(\varepsilon(0))_0 = 0 \& (\varepsilon)_1 \mathbf{r} A] \vee [(\varepsilon(0))_0 \neq 0 \& (\varepsilon)_1 \mathbf{r} B]$, which is the direct translation of FIM 8.5 Clause 3.

4. $\varepsilon \mathbf{r}(A \supset B)$ is $\forall \alpha \{\alpha \mathbf{r} A \supset \forall t \exists ! y\, \varepsilon(2^{t+1} * \bar{\alpha}(y)) > 0$ & $\forall \beta [\forall t \exists y\, \varepsilon(2^{t+1} * \bar{\alpha}(y)) = \beta(t) + 1 \supset \beta \mathbf{r} B]\}$. (Motivation follows.)

5. $\varepsilon \mathbf{r} \neg A$ is $\forall \alpha \neg \alpha \mathbf{r} A$.

6. $\varepsilon \mathbf{r} \forall x A$ is $\forall x \{\forall t \exists ! y\, \varepsilon(2^{t+1} * \overline{\lambda s x}(y)) > 0$ & $\forall \beta [\forall t \exists y\, \varepsilon(2^{t+1} * \overline{\lambda s x}(y)) = \beta(t) + 1 \supset \beta \mathbf{r} A]\}$.

7. $\varepsilon \mathbf{r} \exists x A$ is $\forall x [(\varepsilon(0))_0 = x \supset (\varepsilon)_1 \mathbf{r} A]$. Writing A(x) for A with ε or the bound variables in A so chosen that ε is free for x in A(x) (cf. IM §18), this is equivalent in B to $(\varepsilon)_1 \mathbf{r} A((\varepsilon(0))_0)$.

8. $\varepsilon \mathbf{r} \forall \alpha A$ is $\forall \alpha \{\forall t \exists ! y\, \varepsilon(2^{t+1} * \bar{\alpha}(y)) > 0$ & $\forall \beta [\forall t \exists y\, \varepsilon(2^{t+1} * \bar{\alpha}(y)) = \beta(t) + 1 \supset \beta \mathbf{r} A]\}$.

9. $\varepsilon \mathbf{r} \exists \alpha A$ is $\forall t \exists ! y (\varepsilon)_0(2^{t+1} * \overline{\lambda s 0}(y)) > 0$ & $\forall \alpha [\forall t \exists y (\varepsilon)_0(2^{t+1} * \overline{\lambda s 0}(y)) = \alpha(t) + 1 \supset (\varepsilon)_1 \mathbf{r} A]$.

In $\varepsilon \mathbf{r} E$, the prime-formula parts of E are intact (but other formula-parts of E may be altered). The free variables of $\varepsilon \mathbf{r} E$ are always those of E (occurring only in the prime-formula parts) and sometimes ε (not so occurring). Different allowed choices of the bound variables introduced in $\varepsilon \mathbf{r} E$ lead from a given E (or a pair of congruent E's, IM p. 153) to congruent formulas. Given a term t, by choosing appropriately the bound variables introduced in $\varepsilon \mathbf{r} E$ (not the x in Clauses 6, 7, nor the α in Clauses 8, 9), we can arrange that, if E is E(z) with t free for z, then $\varepsilon \mathbf{r} E$ is D(z) with t free for z; and similarly with a functor u and function variable γ. This remark will justify free substitutions, and applications of Axiom Schemata 10N, 11N, 10F, 11F involving E in $\varepsilon \mathbf{r} E$. Similarly we can justify free replacements under equality. (Both are used in the remark in Clause 7.) Cf. FIM 4.1, 4.4, 4.5.

In Clause 4, we want a function β realizing-Ψ B to be determined by ε from any function α which realizes-Ψ A. We accomplish this by letting ε determine each value $\beta(t)$ of β from the function α', where $\alpha'(0) = t$ and $\alpha'(x+1) = \alpha(x)$. This ε does by serving as associate of a countable type-2 functional [8] 1.2; indeed here $\varepsilon(\overline{\alpha'}(x)) > 0$ for a unique $x = y + 1$, for which $\varepsilon(\overline{\alpha'}(x)) = \beta(t) + 1$. The number $\overline{\alpha'}(x) = \Pi_{i<x} p_i^{\alpha'(i)+1} = 2^{t+1} * \bar{\alpha}(y)$ represents the initial segment $t, \alpha(0), \cdots, \alpha(y-1)$ of values of α'. Similar considerations explain Clauses 6 ($\lambda s x$ replacing α), 8, and 9 ($\lambda s 0$ replacing α).[4] In Clauses 2, 3, etc. $(\varepsilon)_i$ is $\lambda t (\varepsilon(t))_i$ (cf. IM p. 230, FIM 5.7 ¶1).

[4] In FIM this representation by a function ε of functionals of α, $\lambda s x$ or $\lambda s\, 0$ is incorporated into notations analogous to IM §65 ¶ 1, which facilitate the proofs in FIM.

A sentence E is *realizable*, if a general recursive function ε realizes E. An open formula E is *realizable*, if its closure $\forall$E is.

It is shown in FIM (Theorem 9.3(a)) that every provable sentence of I is realizable. But classically a sentence E may be realized by a function ε which is not general recursive (FIM 8.6 ¶s 6, 7). (If then ε is general recursive in a class or list T of number-theoretic functions, we say E is *realizable*/T.) Furthermore, a sentence E is realizable, i.e. a general recursive function realizes $\neg$E (under Clause 5 of the definition), exactly if no function ε whatsoever realizes E. So only in the latter case can $\neg$E be provable in I. Thus the sentences E such that $(E\varepsilon)[\varepsilon$ realizes E] are ones which can be consistently adjoined to the intuitionistic system I.

For any sentence E, "$(E\varepsilon)[\varepsilon$ realizes E]" is expressed in the formal symbolism by $\exists\varepsilon\varepsilon\,\mathbf{r}\,\mathrm{E}$, while "E is realizable" is expressed by $\exists\varepsilon_{\mathrm{GR}(\varepsilon)}\varepsilon\,\mathbf{r}\,\mathrm{E}$ abbreviated "$\mathbf{r}\,\mathrm{E}$", where GR(ε) expresses "ε is general recursive".[5]

We shall now study $\exists\varepsilon\varepsilon\,\mathbf{r}\,\mathrm{E}$, rather than $\mathbf{r}\,\mathrm{E}$, because the first of these is appropriate to our present problem (end §1 above), and because in the present situation we do not wish to get seriously involved in formalizing the theory of general and partial recursive functionals, as we would have to to study $\mathbf{r}\,\mathrm{E}$.[6]

In the sequel, we shall (now formally) abbreviate $\lambda t\langle a_1, \cdots, a_k, \alpha_1(t), \cdots, \alpha_l(t)\rangle$ as "$\langle a_1, \cdots, a_k, \alpha_1, \cdots, \alpha_l\rangle^1$" or "$\langle\Psi\rangle^1$" where Ψ is any permutation of $a_1, \cdots, a_k, \alpha_1, \cdots, \alpha_l$ preserving the order within each type, omitting the superscript "1" when the context makes it clear that $l > 0$ or that the result is a functor (FIM preceding (8.1a) and 5.7 ¶1; [7] 2.4).

§ 3. Evaluation of $\exists\varepsilon\varepsilon\,\mathbf{r}\,\mathrm{E}$ from outside.

LEMMA 2. Under the same stipulations as in Definition 1:

1. $\vdash_B \exists\varepsilon\varepsilon\mathbf{r}\mathrm{P} \sim \mathrm{P}$.
2. $\vdash_B \exists\varepsilon\varepsilon\mathbf{r}(\mathrm{A} \,\&\, \mathrm{B}) \sim \exists\varepsilon\varepsilon\mathbf{r}\mathrm{A} \,\&\, \exists\varepsilon\varepsilon\mathbf{r}\mathrm{B}$.
3. $\vdash_B \exists\varepsilon\varepsilon\mathbf{r}(\mathrm{A} \vee \mathrm{B}) \sim \exists\varepsilon\varepsilon\mathbf{r}\mathrm{A} \vee \exists\varepsilon\varepsilon\mathbf{r}\mathrm{B}$.

4a. $\vdash_B \exists\varepsilon\varepsilon\mathbf{r}(\mathrm{A} \supset \mathrm{B}) \supset (\exists\varepsilon\varepsilon\mathbf{r}\mathrm{A} \supset \exists\varepsilon\varepsilon\mathbf{r}\mathrm{B})$.

4b. $\vdash_C (\exists\varepsilon\varepsilon\mathbf{r}\mathrm{A} \supset \exists\varepsilon\varepsilon\mathbf{r}\mathrm{B}) \supset \exists\varepsilon\varepsilon\,\mathbf{r}(\mathrm{A} \supset \mathrm{B})$.

5. $\vdash_B \exists\varepsilon\varepsilon\mathbf{r}\neg\mathrm{A} \sim \neg\exists\varepsilon\varepsilon\mathbf{r}\mathrm{A}$.
6. $\vdash_B \exists\varepsilon\varepsilon\mathbf{r}\forall x\mathrm{A} \sim \forall x\exists\varepsilon\varepsilon\mathbf{r}\mathrm{A}$.
7. $\vdash_B \exists\varepsilon\varepsilon\mathbf{r}\exists x\mathrm{A} \sim \exists x\exists\varepsilon\varepsilon\mathbf{r}\mathrm{A}$.

8a. $\vdash_B \exists\varepsilon\varepsilon\mathbf{r}\forall\alpha\mathrm{A} \supset \forall\alpha\exists\varepsilon\varepsilon\mathbf{r}\mathrm{A}$.

---

5 Such formulas GR(ε) are clearly available, but a really convenient choice awaits the result of study now in progress.

6 As in Nelson [11] the theory of general and partial recursive functions was formalized to study the formulas ®E expressing the earlier realizability notion for number theory [3] § 12.

8b. $\vdash_C \neg\{\forall\alpha\exists\varepsilon\varepsilon\,\mathbf{r}A \supset \exists\varepsilon\varepsilon\,\mathbf{r}\,\forall\alpha A\}$ *when* A *is* $\forall x\,\alpha(x)=0 \vee \neg\forall x\,\alpha(x)=0$.
9. $\vdash_B \exists\varepsilon\varepsilon\,\mathbf{r}\,\exists\alpha A \sim \exists\alpha\exists\varepsilon\varepsilon\,\mathbf{r}A$.

PROOFS. 1. By IM p. 162 *76 (with Clause 1 of Definition 1 above).

3. I. Assume **(a)** $\exists\varepsilon\varepsilon\,\mathbf{r}(A \vee B)$. Assume preparatory to $\exists\varepsilon$-elim. (IM § 23): **(b)** $\varepsilon\,\mathbf{r}(A \vee B)$, whence equivalently (cf. 3 Definition 1) $[(\varepsilon(0))_0=0\,\&\,(\varepsilon)_1\,\mathbf{r}A] \vee [(\varepsilon(0))_0 \neq 0\,\&\,(\varepsilon)_1\,\mathbf{r}B]$, whence by cases ($\vee$-elim.), &-elim., $\exists$-introd. (since $(\varepsilon)_1$ is a functor) and $\vee$-introd., $\exists\varepsilon\varepsilon\,\mathbf{r}A \vee \exists\varepsilon\varepsilon\,\mathbf{r}B$. Now complete the $\exists\varepsilon$-elim. (discharging (b)), and use $\supset$-introd. (discharging (a)). II. Assume $\exists\varepsilon\varepsilon\,\mathbf{r}A \vee \exists\varepsilon\varepsilon\,\mathbf{r}B$. CASE 1: $\exists\varepsilon\varepsilon\,\mathbf{r}A$. Assume (for $\exists\varepsilon$-elim.) $\varepsilon\,\mathbf{r}A$. But $(\langle 0,\varepsilon\rangle)_1 = (\lambda t\langle 0,\varepsilon(t)\rangle)_1$ [unabbreviating] $= \lambda x((\lambda t\langle 0,\varepsilon(t)\rangle)(x))_1$ [unabbreviating] $= \lambda x(\langle 0,\varepsilon(x)\rangle)_1$ [FIM 4.3 *0.1, 4.5 Lemma 4.2] $= \lambda x\varepsilon(x)$ [FIM *25.1] $= \varepsilon$ [*0.4]. So $(\langle 0,\varepsilon\rangle)_1\,\mathbf{r}A$. Also easily $(\langle 0,\varepsilon\rangle(0))_0=0$. So, using the equivalent in 3 Def. 1, by &-, $\vee$- and $\exists$-introd. with $\langle 0,\varepsilon\rangle$ as the functor (and completing the $\exists\varepsilon$-elim.), $\exists\varepsilon\varepsilon\,\mathbf{r}(A \vee B)$. CASE 2: $\exists\varepsilon\varepsilon\,\mathbf{r}B$. Similarly, with $\langle 1,\varepsilon\rangle$ as the functor.

4b. Assume **(a)** $\exists\varepsilon\varepsilon\,\mathbf{r}A \supset \exists\varepsilon\varepsilon\,\mathbf{r}B$. We use classical cases (by IM *51). CASE 1: $\exists\alpha\alpha\,\mathbf{r}A$. By (a), $\exists\beta\beta\,\mathbf{r}B$. Assume prior to $\exists\beta$-elim.: **(b)** $\beta\,\mathbf{r}B$. Using FIM 5.7 Lemma 5.5 (a) (with FIM Remark 4.1, and IM *50), assume prior to $\exists\varepsilon$-elim.

$$\forall x\varepsilon(x) = \begin{cases} \beta((x)_0 \dot{-} 1)+1 & \text{if } \mathrm{Seq}(x)\,\&\,\mathrm{lh}(x)=1, \\ 0 & \text{if } \neg(\mathrm{Seq}(x)\,\&\,\mathrm{lh}(x)=1). \end{cases}$$

Then, using *22.8, *22.5, *20.3 (with $p_0 = 2$), *23.5, *22.6 (with $\bar{\alpha}(0) = 1$), *19.9, *6.3: **(c)** $\varepsilon(2^{t+1} * \bar{\alpha}(0)) = \beta(t)+1$ and **(d)** $y>0 \supset \varepsilon(2^{t+1} * \bar{\alpha}(y))=0$. So, using *170 with 0 as the t: **(e)** $\forall t\exists!y\varepsilon(2^{t+1} * \bar{\alpha}(y)) > 0$. Toward (f), assume $\forall t\exists y\varepsilon(2^{t+1} * \bar{\alpha}(y)) = \gamma(t)+1$. After using $\forall$-elim., assume prior to $\exists y$-elim. $\varepsilon(2^{t+1} * \bar{\alpha}(y)) = \gamma(t)+1$. Thence by *172 with (c) and (e), $y=0$; so $\gamma(t)+1 = \varepsilon(2^{t+1} * \bar{\alpha}(0)) = \beta(t)+1$, whence by *132, $\gamma(t)=\beta(t)$. Completing the $\exists y$-elim., and using $\forall$-introd. (cf. FIM 4.5 ¶ 1), $\gamma=\beta$. So by (b), $\gamma\,\mathbf{r}B$. By $\supset$- and $\forall$-introd.: **(f)** $\forall\gamma[\forall t\exists y\varepsilon(2^{t+1} * \bar{\alpha}(y)) = \gamma(t)+1 \supset \gamma\,\mathbf{r}B]$. Combining (e) and (f) by &-introd., and using *11 and $\forall\alpha$-introd., $\varepsilon\,\mathbf{r}(A \supset B)$, whence $\exists\varepsilon\varepsilon\,\mathbf{r}(A \supset B)$. Now complete the $\exists\varepsilon$- and $\exists\beta$-elim. CASE 2: $\neg\exists\alpha\alpha\,\mathbf{r}A$. Then by *86, $\forall\alpha\neg\alpha\,\mathbf{r}A$, whence by $\forall$-elim., *10a, and $\forall$-introd., $\varepsilon\,\mathbf{r}(A \supset B)$, whence $\exists\varepsilon\varepsilon\,\mathbf{r}(A \supset B)$.

5. $\exists\varepsilon\varepsilon\,\mathbf{r}\neg A \sim \exists\varepsilon\forall\alpha\neg\alpha\,\mathbf{r}A$ [5 Def. 1] $\sim \forall\alpha\neg\alpha\,\mathbf{r}A$ [*76] $\sim \neg\exists\varepsilon\varepsilon\,\mathbf{r}A$ [*86].

6. I. Assume $\exists\varepsilon\varepsilon\,\mathbf{r}\,\forall xA$. Omitting $\exists\varepsilon$ prior to $\exists$-elim., and using $\forall$- and &-elim. and *87: **(a)** $\forall t\exists y\varepsilon(2^{t+1} * \lambda s x(y)) > 0$ and

(b) $\forall\beta[\forall t\exists y\varepsilon(2^{t+1} * \overline{\lambda sx}(y)) = \beta(t)+1 \supset \beta\,\mathbf{r}A]$. Applying *2.2 to (a), and omitting $\exists\upsilon$: **(c)** $\forall t\varepsilon(2^{t+1} * \overline{\lambda sx}(\upsilon(t))) > 0$. Using FIM 5.6 Lemma 5.3 (a), assume prior to $\exists\beta$-elim. $\forall t\beta(t) = \varepsilon(2^{t+1} * \overline{\lambda sx}(\upsilon(t))) \dot{-} 1$. Using (c) and *6.7, $\varepsilon(2^{t+1} * \overline{\lambda sx}(\upsilon(t))) = \beta(t)+1$, whence $\forall t\exists y\varepsilon(2^{t+1} * \overline{\lambda sx}(y)) = \beta(t)+1$, so by (b) $\beta\,\mathbf{r}A$, whence $\exists\varepsilon\varepsilon\,\mathbf{r}A$. Completing the $\exists\beta$- and $\exists\upsilon$-elim., and using $\forall$-introd., $\forall x\exists\varepsilon\varepsilon\,\mathbf{r}A$. Complete the $\exists\varepsilon$-elim. II. Assume $\forall x\exists\varepsilon\varepsilon\,\mathbf{r}A$, whence by $^{\times}$2.1 $\exists\zeta\forall x\,\lambda t\zeta(\langle x,t\rangle)\,\mathbf{r}A$. Assume $\forall x\,\lambda t\zeta(\langle x,t\rangle)\,\mathbf{r}A$. Assume

$$\forall u\varepsilon(u) = \begin{cases} \zeta(\langle (u)_1 \dot{-} 1,\ (u)_0 \dot{-} 1\rangle) + 1 & \text{if } \mathrm{Seq}(u)\,\&\,\mathrm{lh}(u) = 2, \\ 0 & \text{if } \neg(\mathrm{Seq}(u)\,\&\,\mathrm{lh}(u) = 2), \end{cases}$$

so that (using *22.8, *22.5, *20.3, *23.5; $^{\times}$21.1, *B4, *B3, *23.2, $^{\times}$0.1; *19.11 with *18.5, *3.9, *129; *19.9, *19.10, *6.3) $\varepsilon(2^{t+1} * \overline{\lambda sx}(1)) = \zeta(\langle x,t\rangle)+1$ and $y \neq 1 \supset \varepsilon(2^{t+1} * \overline{\lambda sx}(y)) = 0$. Now $\forall t\exists! y\varepsilon(2^{t+1} * \overline{\lambda sx}(y)) > 0$, and like (f) in 4b (via $\gamma = \lambda t\zeta(\langle x,t\rangle)$) $\forall\gamma[\forall t\exists y\varepsilon(2^{t+1} * \overline{\lambda sx}(y)) = \gamma(t)+1 \supset \gamma\,\mathbf{r}A]$. So, using &-, $\forall$- and $\exists$-introd. (and completing the $\exists\varepsilon$- and $\exists\zeta$-elims.), $\exists\varepsilon\varepsilon\,\mathbf{r}\,\forall xA$.

7. Write A as A(x), supposing the variables chosen so that ε is free for x. I. Assume $\exists\varepsilon\varepsilon\,\mathbf{r}\,\exists xA(x)$. Assume $\varepsilon\,\mathbf{r}\,\exists xA(x)$. Via the equivalent in 7 Def. 1 $(\varepsilon)_1\,\mathbf{r}A((\varepsilon(0))_0)$, whence $\exists x\exists\varepsilon\varepsilon\,\mathbf{r}A(x)$ (cf. remarks following Def.1). II. Assume $\exists x\exists\varepsilon\varepsilon\,\mathbf{r}A(x)$. Assume $\varepsilon\,\mathbf{r}A(x)$. But $(\langle x,\varepsilon\rangle(0))_0 = x$ and $(\langle x,\varepsilon\rangle)_1 = \varepsilon$, so $(\langle x,\varepsilon\rangle)_1\,\mathbf{r}A((\langle x,\varepsilon\rangle(0))_0)$, so via the equivalent $\langle x,\varepsilon\rangle\,\mathbf{r}\,\exists xA(x)$, whence $\exists\varepsilon\varepsilon\,\mathbf{r}\,\exists xA(x)$.

8b. By *51, $\vdash_C A$. But $\vdash_B \exists\varepsilon\varepsilon\,\mathbf{r}A \sim \forall x\exists\varepsilon\varepsilon\,\mathbf{r}\alpha(x) = 0 \vee \neg\forall x\exists\varepsilon\varepsilon\,\mathbf{r}\alpha(x) = 0$ [3, 5, 6 of this lemma] $\sim \forall x\alpha(x) = 0 \vee \neg\forall x\alpha(x) = 0$ [1] $\sim A$. So $\vdash_C \exists\varepsilon\varepsilon\,\mathbf{r}A$, and by $\forall$-introd. $\vdash_C \forall\alpha\exists\varepsilon\varepsilon\,\mathbf{r}A$. So by *41 it will suffice to show that $\vdash_B \neg\exists\varepsilon\varepsilon\,\mathbf{r}\,\forall\alpha A$. Let A($\alpha$), B($\alpha$) be $\forall x\alpha(x) = 0$, $\neg\forall x\alpha(x) = 0$. Assume $\exists\varepsilon\varepsilon\,\mathbf{r}\,\forall\alpha A$. Assume $\varepsilon\,\mathbf{r}\,\forall\alpha A$, whence by $\forall$-elim., *87 and &-elim.: **(i)** $\forall t\exists y\varepsilon(2^{t+1} * \bar{\alpha}(y)) > 0$, **(ii)** $\forall t\forall y\forall z(\varepsilon(2^{t+1} * \bar{\alpha}(y)) > 0\ \&\ \varepsilon(2^{t+1} * \bar{\alpha}(z)) > 0 \supset y = z)$ and **(iii)** $\forall\beta[\forall t\exists y\varepsilon(2^{t+1} * \bar{\alpha}(y)) = \beta(t)+1 \supset \beta\,\mathbf{r}(A(\alpha) \vee B(\alpha))]$. Assume **(iv)** $\forall s\tau(s) = \min(2, (\varepsilon(2^{0+1} * s) \dot{-} 1)_0 + 1)\,\mathrm{sg}\,\varepsilon(2^{0+1} * s)$. From (i) by *2.2, and omitting $\exists\upsilon$: **(v)** $\forall t\varepsilon(2^{t+1} * \bar{\alpha}(\upsilon(t))) > 0$. Assume **(vi)** $\forall t\beta(t) = \varepsilon(2^{t+1} * \bar{\alpha}(\upsilon(t))) \dot{-} 1$. Using (v) and *6.7, $\varepsilon(2^{t+1} * \bar{\alpha}(\upsilon(t))) = \beta(t)+1$, so $\forall t\exists y\varepsilon(2^{t+1} * \bar{\alpha}(y)) = \beta(t)+1$. By (iii), $\beta\,\mathbf{r}(A(\alpha) \vee B(\alpha))$. So $((\beta(0))_0 = 0\ \&\ (\beta)_1\,\mathbf{r}A(\alpha)) \vee ((\beta(0))_0 \neq 0\ \&\ (\beta)_1\,\mathbf{r}B(\alpha))$. But $(\beta)_1\,\mathbf{r}A(\alpha) \vdash \exists\varepsilon\varepsilon\,\mathbf{r}A(\alpha) \vdash A(\alpha)$ [6, 1], and similarly $(\beta)_1\,\mathbf{r}B(\alpha) \vdash B(\alpha)$. So **(vii)** $((\beta(0))_0 = 0\ \&\ A(\alpha)) \vee ((\beta(0))_0 \neq 0\ \&\ B(\alpha))$. Using (iv), *10.3 etc., $\tau(\bar{\alpha}(x)) > 0 \supset \varepsilon(2^{0+1} * \bar{\alpha}(x)) > 0$; so using () and (ii): **(viii)** $\forall x[\tau(\bar{\alpha}(x)) > 0 \supset \upsilon(0) = x]$. Also, using (iv)-(vi), $\tau(\bar{\alpha}(\upsilon(0))) = \min(2, (\beta(0))_0 + 1)$; so using (vii) and *7.2: **(ix)** $(A(\alpha)\ \&\ \tau(\bar{\alpha}(\upsilon(0))) = 1) \vee (B(\alpha)\ \&\ \tau(\bar{\alpha}(\upsilon(0))) = 2)$. From (viii) and (ix), using &- and $\exists y$-introd., completing the $\exists\beta$- and

$\exists\upsilon$-elims., and using $\forall$-introd., we obtain (a) of the proof of *27.1 . We continue as in that proof (with $\exists\tau$- and $\exists\varepsilon$-elim. to be completed at the end).

REMARK 3. In Part II for 6 we have our first significant use of $^{\times}$2.1 not via *2.2 (cf. FIM 7.15).

REMARK 4. We do not know whether the $\vdash_C$ in 4b can be strengthened to $\vdash_B$.

§ **4. Evaluation of $\exists\varepsilon\varepsilon$ r E from inside.** LEMMA 5.[7] *To each term* t *containing free only* $a_1, \cdots, a_k$, $\alpha_1, \cdots, \alpha_l$, *there is a prime formula* $P(a_1, \cdots, a_k, s_1, \cdots, s_l, w)$ (*containing free only the variables shown*) *such that* $\vdash_B t = w \sim \exists x P(a_1, \cdots, a_k, \bar{\alpha}_1(x), \cdots, \bar{\alpha}_l(x), w)$; *and dually.*

PROOF. After we prove the first statement, the dual will follow by the usual duality argument, in view of FIM 5.5 #D, and *158, *86. By FIM Lemma 4.3 and its proof, we may first reduce t to a normal term, which has only the same variables free. (Cf. Remark 4.6. By Remark 4.5, actually this is true of any normal form of t.) Now, with t assumed normal, we use induction on the subscript h of the function symbol f_h in t of highest subscript (with $h=0$ if no function symbol occurs in t). Within the induction on h, we use induction on the sum g of the numbers of (occurrences of) function symbols (including 0), of function variables not as argument of a function symbol, and of λ's, in t. The cases arise from IM 3.3. CASE 1: t is a number variable or 0. Trivial. CASE 2: t is $f_i(t_1, \cdots, t_{k_i}, u_1, \cdots, u_{l_i})$ $(i > 0)$. SUBCASE 2.1: $t_1, \cdots, t_{k_i}, u_1, \cdots, u_{l_i}$ are distinct variables. Then $i = h$. SUB^{2}CASE 2.1.1: $i = 1$ (f_i is $'$). Trivial. SUB^{2}CASE 2.1.2: f_i is introduced by **(b)** in FIM 5.1 $(i > 1)$. To simplify notation, say the variables are y, a, α. By hyp. ind. on h: **(a)** $q(a, \alpha) = w \sim \exists x Q(a, \bar{\alpha}(x), w)$ and **(c)** $r(y, z, a, \alpha) = w \sim \exists x R(y, z, a, \bar{\alpha}(x), w)$. Simplifying IM p. 243 by using $(c)_i$ in place of $\beta(c, d, i)$, let $P_0(y, a, \alpha, w)$ be $\exists c\{q(a, \alpha) = (c)_0$ & $\forall i_{i<y} r(i, (c)_i, a, \alpha) = (c)_{i'}$ & $(c)_y = w\}$. Adapting IM p. 244 Remark 1, and using ind. on b and (b): **(d)** $f_i(y, a, \alpha) = w \sim P_0(y, a, \alpha, w)$. Using (a) and (c) to replace $q(a, \alpha) = (c)_0$ and $r(i, (c)_i, a, \alpha) = (c)_{i'}$ in $P_0(y, a, \alpha, w)$, and advancing and contracting quantifiers (by *25.9, *91 and *25.4): **(e)** $P_0(y, a, \alpha, w) \sim \exists x\{Q(a, \bar{\alpha}((x)_1), (x)_{0,0})$ & $\forall i_{i<y} R(i, (x)_{0,i}, a, \bar{\alpha}((x)_{2,i}), (x)_{0,i'})$ & $(x)_{0,y} = w\}$. Using *23.4 with *19.2, *19.6: **(f)** $\bar{\alpha}((x)_1) = \Pi_{j<(x)_1} p_j^{(\bar{\alpha}(x))_j}$, $\bar{\alpha}((x)_{2,1}) = \Pi_{j<(x)_{2,1}} p_j^{(\bar{\alpha}(x))_j}$. By *23.5: **(g)** $x = lh(\bar{\alpha}(x))$. Using (f) and (g) in (e), then #D and #E can be used to express the resulting scope of $\exists x$ in the form $P(y, a, \bar{\alpha}(x), w)$ with $P(y, a, s, w)$ prime. SUB^{2}CASE 2.1.3: f_i is introduced by **(a)** in FIM 5.1 $(i > 7)$. Then $f_i(y, a, \alpha) = w \sim p(y, a, \alpha) = w$ [(a)]

[7] In FIM *25.4 and *25.3, $\exists a$ and $\forall a$ can be replaced by $\exists a_{a>0}$ and $\forall a_{a>0}$ resp. Hence (using *23.4 with *19.2 and *23.5), in Lemmas 5, 6 and 17 here, $\exists x$ and $\forall x$ can be replaced by $\exists x_{x>0}$ and $\forall x_{x>0}$ resp.

$\sim \exists x P(y, a, \bar{\alpha}(x), w)$ [hyp. ind. on h]. SUBCASE 2.2: $t_1, \cdots, t_{k_i}, u_1, \cdots, u_{l_i}$ are variables not all distinct. Again $i = h$. Apply Subcase 2.1 to f_i with distinct variables, and identify variables in the result. SUBCASE 2.3: otherwise. Say e.g. t is $f_i(t_1, u_1)$ with t_1, u_1 not both variables. Say e.g. they contain free at most a, α. If t_1 is not a variable, the g of $f_i(b, \beta)$ is less than that of $f_i(t_1, u_1)$ by at least the g of t_1; and if u_1 is not a variable, by at least the g of u_1. So by hyp. ind. on g or h: **(a)** $f_i(b, \beta) = w \sim \exists x Q(b, \bar{\beta}(x), w)$. Now **(b)** $f_i(t_1, u_1) = w \sim \exists x Q(t_1, \bar{u}_1(x), w)$ [subst. into (a)] $\sim$ $\exists b \exists c \exists x[Q(b, c, w) \ \& \ t_1 = b \ \& \ \mathrm{Seq}(c) \ \& \ \mathrm{lh}(c) = x \ \& \ \forall i_{i<x} u_1(i) = (c)_1 \dot{-} 1]$ [using *23.5, *23.2, *6.3; and *22.3, ˣ23.1, ˣ22.1, *B19, *6.7]. If u_1 is a variable, then t_1 is not, and $u_1(i)$ is normal and has a g less than the g of $f_i(t_1, u_1)$ by the g of t_1. By hyp. ind. on g or h: **(c)** $t_1 = b \sim \exists x R(a, \bar{\alpha}(x), b)$, and **(d)** $u_1(i)' = d \sim \exists x S(a, \bar{\alpha}(x), d)$ where $u_1(i)'$ is a normal form of $u_1(i)$, so **(e)** $u_1(i) = u_1(i)'$. Using (c)–(e) in (b), and advancing and contracting quantifiers, etc., we obtain as before the desired form. CASE 3: t is (u)(s) for u a functor, s a term. SUBCASE 3.1: u is a function variable α. Say e.g. s contains free only a, α. Now $\alpha(s) = w \sim \exists b[\alpha(b) = w \ \& \ s = b] \sim \exists b[\alpha(b) = w \ \&$ $\exists x Q(a, \bar{\alpha}(x), b]$ [hyp. ind. on g] $\sim \exists x_{x>0}[\alpha((x)_0) = w \ \& \ Q(a, \bar{\alpha}((x)_1), (x)_0)]$ [*91, *25.4[7]], etc. using $x > 0 \supset \alpha((x)_0) = (\bar{\alpha}(x))_{(x)_0} \dot{-} 1$ by *23.2 with *19.2, *6.3. SUBCASE 3.2: u is an f_i. Comes under Case 2. SUBCASE 3.3: u is a λ-functor. Excluded by t being normal.

LEMMA 6.[7] (a) *To each term* $t(\Psi)$ *containing free only* Ψ, *there are prime formulas* $R(s, w)$ *and* $S(s, w)$ *(containing free only* s, w*) such that*

$$\vdash_B t(\Psi) = w \sim \exists x R(\overline{\langle \Psi \rangle}(x), w) \sim \forall x S(\overline{\langle \Psi \rangle}(x), w).$$

(b) *To each prime formula* $P(\Psi)$ *containing free only* Ψ, *there are prime formulas* $R(s)$ *and* $S(s)$ *(containing free only* s*) such that*

$$\vdash_B P(\Psi) \sim \exists x R(\overline{\langle \Psi \rangle}(x)) \sim \forall x S(\overline{\langle \Psi \rangle}(x)).$$

PROOF. (For "$\langle \Psi \rangle$" cf. end § 2.) (a) From Lemma 5, using $a_j = ((\overline{\langle \Psi \rangle}(x'))_0 \dot{-} 1)_m$ $(m = j - 1)$ and $\overline{\alpha_j}(x) = \Pi_{i<x} p_i \exp(((\overline{\langle \Psi \rangle}(x'))_i \dot{-} 1)_n + 1)$ $(n = k + j - 1)$ (by *23.2, *6.3, ˣ0.1, *25.1; and ˣ23.1, *B19). (b) Say $P(\Psi)$ is $t_1(\Psi) = t_2(\Psi)$. Apply (a) to $|t_1(\Psi) - t_2(\Psi)|$ and replace w by 0 in the result (using *11.2).

LEMMA 7. *To each formula* E, *containing free only* Ψ, *and containing no* $\exists$, *other than in parts of the form* $\exists x P(x)$ *with* $P(x)$ *prime or* $\exists \alpha P(\alpha)$ *with* $P(\alpha)$ *prime, and no* $\vee$, *there is a functor* p *such that*

(i) $\vdash_B \exists \varepsilon \, \varepsilon \, \mathbf{r} \, E \supset E$,

(ii) $\vdash_B E \supset \forall t \exists! y \, p(2^{t+1} * \overline{\langle \Psi \rangle}(y)) > 0 \ \&$
$\forall \varepsilon [\forall t \exists y \, p(2^{t+1} * \overline{\langle \Psi \rangle}(y)) = \varepsilon(t) + 1 \supset \varepsilon \, \mathbf{r} \, E]$,

and hence

(iii) $\vdash_B \exists\varepsilon\varepsilon\,\mathbf{r}\,E \sim E$.

PROOF of (i) and (ii) follows, by ind. on the number of logical symbols in E, with eight cases (numbered as in Def. 1). The implication $E \supset \exists\varepsilon\varepsilon\,\mathbf{r}\,E$ for (iii) follows from (ii) for the same E; indeed, essentially as in the proof of *27.2 (1)–(5), from (ii) and E we obtain $\lambda t p(2^{t+1} * \overline{\langle\Psi\rangle}(\upsilon(t))) \dot{-} 1\ \mathbf{r}\,E$.

CASE 1. E is a prime formula P. Then (i) holds by *1 and *76, and (ii) holds upon taking p to be $\lambda s p(s)$ where p(s) is chosen by FIM 5.5 #F with $m = 1$ so that $Seq(s) \& lh(s) = 1 \supset p(s) = 1, \neg(Seq(s) \& lh(s) = 1) \supset p(s) = 0$.

CASE 4: E is $A \supset B$. (i) Assume $\exists\varepsilon\varepsilon\,\mathbf{r}(A \supset B)$. Assume A. By hyp. ind. (iii), $\exists\varepsilon\varepsilon\,\mathbf{r}A$. So using 4a Lemma 2, $\exists\varepsilon\varepsilon\,\mathbf{r}B$. By hyp. ind. (i), B (ii) By hyp. ind. (ii): **(a)** $B \supset \forall t \exists ! y q(2^{t+1} * \overline{\langle\Psi\rangle}(y)) > 0\ \&$ $\forall\beta[\forall t \exists y\, q(2^{t+1} * \overline{\langle\Psi\rangle}(y)) = \beta(t) + 1 \supset \beta\,\mathbf{r}\,B]$. We shall construct a term p(s) so that

(b) $p(2^{2^{t+1} * \bar{\alpha}(y)+1} * \overline{\langle\Psi\rangle}(y)) = q(2^{t+1} * \overline{\langle\Psi\rangle}(y)) + 1$,
(c) $y \neq z \supset p(2^{2^{t+1} * \bar{\alpha}(y)+1} * \overline{\langle\Psi\rangle}(z)) = 0$,
(d) $\neg(Seq(t) \& lh(t) > 0) \supset p(2^{t+1} * 1) = 1$,
(e) $\neg(Seq(t) \& lh(t) > 0) \& z > 0 \supset p(2^{t+1} * \overline{\langle\Psi\rangle}(z)) = 0$.

To do so, apply #F with $m = 2$ so that $Seq(s) \& lh(s) > 0 \& Seq((s)_0 \dot{-} 1) \&$ $lh((s)_0 \dot{-} 1) = lh(s) \supset p(s) = q(2^{((s)_0 \dot{-} 1)_0} * \prod_{i < lh(s) \dot{-} 1} p_i^{(s)_{i+1}}) + 1$ and $Seq(s) \& lh(s) = 1 \& \neg(Seq((s)_0 \dot{-} 1) \& lh((s)_0 \dot{-} 1) > 0) \supset p(s) = 1$. Using *22.8, *22.5, *20.3, *23.5, ˣ21.1, *19.11, *19.9, *19.13, *6.3, ˣ23.1, *23.2, etc., (b)–(d) are deducible. Now let p be $\lambda s p(s)$. Using (b)–(e): **(f)** $\forall t \exists ! y\, p(2^{t+1} * \overline{\langle\Psi\rangle}(y)) > 0$. Assume **(g)** $A \supset B$. Toward (h), assume **(A)** $\forall t \exists y\, p(2^{t+1} * \overline{\langle\Psi\rangle}(y)) = \varepsilon(t) + 1$. After using *2.2, assume **(B)** $\forall t\, p(2^{t+1} * \overline{\langle\Psi\rangle}(\upsilon(t))) = \varepsilon(t) + 1$. Now $0 < \varepsilon(2^{t+1} * \bar{\alpha}(y)) + 1 =$ $p(2^{2^{t+1} * \bar{\alpha}(y)+1} * \overline{\langle\Psi\rangle}(\upsilon(2^{t+1} * \bar{\alpha}(y))))$, so by (c) $\upsilon(2^{t+1} * \bar{\alpha}(y)) = y$; thus $\varepsilon(2^{t+1} * \bar{\alpha}(y)) + 1 = p(2^{2^{t+1} * \bar{\alpha}(y)+1} * \overline{\langle\Psi\rangle}(y)) = q(2^{t+1} * \overline{\langle\Psi\rangle}(y)) + 1$ [(b)], whence by *132 and $\forall$-introd.: **(C)** $\forall t \forall y\, \varepsilon(2^{t+1} * \bar{\alpha}(y)) = q(2^{t+1} * \overline{\langle\Psi\rangle}(y))$. Assume **(D)** $\alpha\,\mathbf{r}\,A$, whence by $\exists$-introd. and hyp. ind. (i), A, whence by (g): **(E)** B. So by (a) and (C): **(F)** $\forall t \exists ! y\, \varepsilon(2^{t+1} * \bar{\alpha}(y)) > 0$. Assuming $\forall t \exists y\, \varepsilon(2^{t+1} * \bar{\alpha}(y)) = \beta(t) + 1$, then (E), (a) and (C) give $\beta\,\mathbf{r}\,B$; by $\supset$- and $\forall$-introd.: **(G)** $\forall\beta[\forall t \exists y\, \varepsilon(2^{t+1} * \bar{\alpha}(y)) = \beta(t) + 1 \supset \beta\,\mathbf{r}\,B]$. Using (F) and (G) with &- and $\supset$-introd. (discharging (D)), and $\forall\alpha$-introd., completing the $\exists\upsilon$-elim. (discharging (B)), and using $\supset$-introd. (discharging (A)) and $\forall\varepsilon$-introd.: **(h)** $\forall\varepsilon[\forall t \exists y\, p(2^{t+1} * \overline{\langle\Psi\rangle}(y)) = \varepsilon(t) + 1 \supset \varepsilon\,\mathbf{r}(A \supset B)]$.

CASES 6 AND 8: E is $\forall x A(x)$ or $\forall\alpha A(\alpha)$, resp. Similar to Case 4. In

Case 6 (Case 8), $\langle\Psi\rangle$ in the right side of (b) is replaced by $\langle\Psi, x\rangle$ (by $\langle\Psi,\alpha\rangle$); and in Case 6 α is replaced in the left of (b) and (c) by λsx, and $\&\forall i_{0<i<lh(t)}(t)_i = x+1$ is inserted into the negated conjunction in the left of (d) and (e).

CASE 9. E is $\exists\alpha P(\alpha)$ with $P(\alpha)$ prime. (i) Assume $\exists\varepsilon\, \varepsilon\, \mathbf{r}\, \exists\alpha P(\alpha)$. By 9 Lemma 2, $\exists\alpha\exists\varepsilon\,\varepsilon\,\mathbf{r} P(\alpha)$. By Case 1 (i) and *70, $\exists\alpha\exists\varepsilon\,\varepsilon\,\mathbf{r}P(\alpha) \supset \exists\alpha P(\alpha)$. (ii) By Lemma 6 (b): **(a)** $P(\alpha) \sim \exists x R(\overline{\langle\Psi,\alpha\rangle}(x))$. We construct a prime formula $S(s)$ so that (using *23.5): **(b)** $S(\overline{\langle\Psi\rangle}(y)) \sim \mathrm{Seq}(y)\ \&$ $R(\Pi_{i<lh(y)}\, p_i \exp(((\overline{\langle\Psi\rangle}(y)) \dot{-} 1)p_{k+l}{}^{(y)_i \dot{-} 1} + 1))$, where Ψ consists of $k+l$ variables. Then (using *23.1, *20.5, *22.1, etc.) **(c)** $S(\overline{\langle\Psi\rangle}(y)) \supset R(\overline{\langle\Psi, \lambda t(y)_t \dot{-} 1\rangle}(lh(y)))$. Now construct a term $p(s)$ so that

$$\textbf{(d)}\quad S(\overline{\langle\Psi\rangle}(y))\ \&\ \forall v_{v<y} \neg S(\overline{\langle\Psi\rangle}(v)) \supset$$

$$p(2^{2^{t+1}*\overline{\lambda s0}(y)+1} * \overline{\langle\Psi\rangle}(y)) = \langle((y)_t \dot{-} 1)+1, 0\rangle + 1,$$

$$\textbf{(e)}\quad S(\overline{\langle\Psi\rangle}(y))\ \&\ \forall v_{v<y} \neg S(\overline{\langle\Psi\rangle}(v))\ \&\ z \neq y \supset p(2^{2^{t+1}*\overline{\lambda s0}(z)+1} * \overline{\langle\Psi\rangle}(y)) = 1,$$

$$\textbf{(f)}\quad \neg(S(\overline{\langle\Psi\rangle}(y))\ \&\ \forall v_{v<y} \neg S(\overline{\langle\Psi\rangle}(v))) \supset p(2^{2^{t+1}*\overline{\lambda s0}(z)+1} * \overline{\langle\Psi\rangle}(y)) = 0,$$

$$\textbf{(g)}\quad \neg(\mathrm{Seq}(t)\ \&\ lh(t)>0\ \&\ \forall i_{0<i<lh(t)}(t)_i = 1) \supset p(2^{t+1} * 1) = 1,$$

$$\textbf{(h)}\quad \neg(\mathrm{Seq}(t)\ \&\ lh(t)>0\ \&\ \forall i_{0<i<lh(t)}(t)_i = 1)\ \&\ z>0 \supset p(2^{t+1} * \overline{\langle\Psi\rangle}(z)) = 0$$

Let p be $\lambda s p(s)$. Assume $\exists\alpha P(\alpha)$, whence by (a), $\exists\alpha\exists x R(\overline{\langle\Psi,\alpha\rangle}(x))$. Assume $R(\overline{\langle\Psi,\alpha\rangle}(x))$, whence by (b) $S(\overline{\langle\Psi\rangle}(\bar\alpha(x)))$, whence by $\exists$-introd. and *149a (completing the $\exists\alpha\exists x$-elim.) $\exists y[S(\overline{\langle\Psi\rangle}(y))\ \&\ \forall v_{v<y}\neg S(\overline{\langle\Psi\rangle}(v))]$. Assume **(i)** $S(\overline{\langle\Psi\rangle}(y_0))\ \&\ \forall v_{v<y_0} \neg S(\overline{\langle\Psi\rangle}(v))$. Thence, using *174a and (d)–(h): **(j)** $\forall t \exists! y\, p(2^{t+1} * \overline{\langle\Psi\rangle}(y)) > 0$. By (i) and (c), $R(\overline{\langle\Psi, \lambda t(y_0)_t \dot{-} 1\rangle}(lh(y_0)))$, whence by $\exists$-introd. $\exists x R(\overline{\langle\Psi, \lambda t(y_0)_t \dot{-} 1\rangle}(x))$, whence by (a): **(k)** $P(\lambda t(y_0)_t \dot{-} 1)$. Toward (ł), assume **(A)** $\forall t \exists y\, p(2^{t+1} * \overline{\langle\Psi\rangle}(y)) = \varepsilon(t)+1$. After using *2.2, assume **(B)** $\forall t\, p(2^{t+1} * \overline{\langle\Psi\rangle}(\upsilon(t))) = \varepsilon(t) + 1$. Now $0 < \varepsilon(2^{t+1} * \overline{\lambda s0}(z)) + 1 =$ $p(2^{2^{t+1}*\overline{\lambda s0}(z)+1} * \overline{\langle\Psi\rangle}(\upsilon(2^{t+1} * \overline{\lambda s0}(z))))$ [(B)], so by (f) (and FIM Remark 4.1), $S(\overline{\langle\Psi\rangle}(\upsilon(2^{t+1} * \overline{\lambda s0}(z))))\ \&\ \forall v_{v<\upsilon(2^{t+1}*\overline{\lambda s0}(z))} \neg S(\overline{\langle\Psi\rangle}(v))$, whence by (i), *174a, *172 and $\forall$-introd.: **(C)** $\forall z \upsilon(2^{t+1} * \overline{\lambda s0}(z)) = y_0$. So by (B), $\varepsilon(2^{t+1} * \overline{\lambda s0}(y_0)) + 1 = p(2^{2^{t+1}*\overline{\lambda s0}(y_0)+1} * \overline{\langle\Psi\rangle}(y_0))$, whence by (d), (i) and *132, $\varepsilon(2^{t+1} * \overline{\lambda s0}(y_0)) = \langle((y_0)_t \dot{-} 1)+1, 0\rangle$, whence (completing the $\exists\upsilon$-elim.): **(D)** $(\varepsilon)_0(2^{t+1} * \overline{\lambda s0}(y_0)) = ((y_0)_t \dot{-} 1) + 1 > 0$. By (B), (C), (i), (e) and *132, $z \neq y_0 \supset \varepsilon(2^{t+1} * \overline{\lambda s0}(z)) = 0$, whence **(E)** $z \neq y_0 \supset (\varepsilon)_0(2^{t+1} * \overline{\lambda s0}(z)) = 0$. By (D) and (E): **(F)** $\forall t \exists! y (\varepsilon)_0(2^{t+1} * \overline{\lambda s0}(y)) > 0$. Assume **(G)** $\forall t \exists y (\varepsilon)_0(2^{t+1} * \overline{\lambda s0}(y)) = \alpha(t)+1$. Using *2.2, assume $\forall t (\varepsilon)_0(2^{t+1} * \overline{\lambda s0}(\tau(t))) = \alpha(t) + 1$. Using $\forall$-elim., then (F), (D) and *172 give $\tau(t) = y_0$ and $\alpha(t)+1 = (\varepsilon)_0(2^{t+1} * \overline{\lambda s0}(y_0)) = ((y_0)_t \dot{-} 1)+1$, whence by *132 and $\forall t$-introd. (completing the $\exists\tau$-elim.), $\alpha = \lambda t(y_0)_t \dot{-} 1$, whence by (k), $P(\alpha)$, whence by

1 Def. 1: **(H)** $(\varepsilon)_1 \mathbf{r} P(\alpha)$. Now by $\supset$-introd. (discharging (G)), $\forall\alpha$-introd., &-introd. (with (F)), $\supset$-introd. (discharging (A)), and $\forall\varepsilon$-introd.:
(I) $\forall\varepsilon[\forall t \exists y\, p(2^{t+1} * \overline{\langle\Psi\rangle}(y)) = \varepsilon(t)+1 \supset \varepsilon\, \mathbf{r}\, \exists\alpha P(\alpha)]$.

§ 5. Consistent extensions of the intuitionistic system by classically provable formulas.[8] THEOREM 8. *For each formula* E *in which no* $\exists$, *other than in parts of the form* $\exists x P(x)$ *with* $P(x)$ *prime or* $\exists\alpha P(\alpha)$ *with* $P(\alpha)$ *prime, and no* $\vee$ *lies in the scope of a universal function quantifier* $\forall\alpha$ [*an* $\forall\alpha$ *or* $\supset$]:

$$\vdash_C \exists\varepsilon\,\varepsilon\,\mathbf{r}\,E \sim E \qquad [\vdash_B \exists\varepsilon\,\varepsilon\,\mathbf{r}\,E \sim E].$$

PROOF. Consider the minimal components $C_1, \cdots, C_l$ of E which stand inside no $\forall\alpha$ [no $\forall\alpha$ or $\supset$]. These components are prime, or have an $\forall\alpha$ [an $\forall\alpha$ or $\supset$] outermost. So they contain no $\exists$, other than in parts $\exists x P(x)$ or $\exists\alpha P(\alpha)$, and no $\vee$. So by Lemma 7 (iii), $\vdash_B \exists\varepsilon\,\varepsilon\,\mathbf{r}\,C_i \sim C_i$ $(i = 1, \cdots, l)$. By Lemma 2, $\vdash_C \exists\varepsilon\,\varepsilon\,\mathbf{r}\,E \sim E'$ [$\vdash_B \exists\varepsilon\,\varepsilon\,\mathbf{r}\,E \sim E'$] where E' is composed from $\exists\varepsilon\,\varepsilon\,\mathbf{r}\,C_1, \cdots, \exists\varepsilon\,\varepsilon\,\mathbf{r}\,C_l$ in the same manner as E from $C_1, \cdots, C_l$.

THEOREM 9. *Assuming the validity of the classical reasoning formalized in C: Each formula* E *of either type described in Theorem 8 which is provable in C (or any class of such formulas) can be consistently adjoined to the intuitionistic system I.*

PROOF. SINGLE FORMULA E. It will suffice to show that not $\vdash_I \neg E$ (cf. IM p. 212 lines 12–17). By hyp., $\vdash_C E$. By Theorem 8, $\vdash_C \exists\varepsilon\,\varepsilon\,\mathbf{r}\,E \sim E$. So $\vdash_C \exists\varepsilon\,\varepsilon\,\mathbf{r}\,E$. By our assumption of the validity of reasoning in C, for each choice Ψ of numbers and functions as values of the variables Ψ in E, there is a function ε which realizes-Ψ E. Then $\neg E$ is not realizable; for under FIM § 8, or (the informal analogs of) Clauses 5, 6 and 8 of Definition 1 here, if $\neg E$ were realizable, no α could realize-Ψ E for any Ψ. Hence by FIM Theorem 9.3(a), not $\vdash_I \neg E$.

CLASS OF FORMULAS $E_0, E_1, E_2, \cdots$. It will suffice to show that, for each finite conjunction $E_{i_1} \& \cdots \& E_{i_n}$, not $\vdash_I \neg(E_{i_1} \& \cdots \& E_{i_n})$ (cf. IM p. 425 lines 4–7). But $E_{i_1} \& \cdots \& E_{i_n}$ is itself of the type in question.

---

[8] Theorem 8 for the prefix $\exists\varepsilon\,\varepsilon\,\mathbf{r}$ is analogous to [9] Theorem 2 with Footnotes 9 and 10 there for the prefix $\mathbf{r}$ (written "$\mathbf{r}'$" in [9]; the "$\mathbf{r}$" of [9] is for the number-theoretic realizability notion Ⓡ of [3] §12). Like the latter, it depends on two lemmas, one for evaluating from outside and one for evaluating from inside. The "outside" lemma is different for $\exists\varepsilon\,\varepsilon\,\mathbf{r}$ (Lemma 2 here) than for $\mathbf{r}$ ([9] Lemma 2.2 with Footnote 10); i.e. different operators are excepted ($\supset$, $\forall\alpha$ vs. $\supset, \neg, \forall x, \forall\alpha$)., $\exists\alpha$ The inside lemma is the same for $\exists\varepsilon\,\varepsilon\,\mathbf{r}$ as for $\mathbf{r}$, and indeed is FIM Lemma 8.4b. In the present Lemma 7 we have accomplished a quick formalization of this, detached from the theory of partial recursive functionals. The further discussion of $\mathbf{r}$ belongs in the context of a formalization of that theory.

§ 6. Further results proved metamathematically under assumption (*). In Theorem 15 we shall give a metamathematical version of this result, subject to the assumption that the proof of FIM Theorem 9.3(a) can be formalized in B to establish:[9]

$$(*) \qquad \vdash_I \mathrm{E} \rightarrow \vdash_B \exists\varepsilon\,\varepsilon\,\mathbf{r}\,\mathrm{E}.$$

We are highly confident of this hypothesis, although not all of the detailed work necessary to confirm it has yet been carried out. The results under this hypothesis will be starred.

* LEMMA 10. *If* $\vdash_I \mathrm{E} \sim \mathrm{E}'$, *then* $\vdash_B \exists\varepsilon\,\varepsilon\,\mathbf{r}\,\mathrm{E} \sim \exists\varepsilon\,\varepsilon\,\mathbf{r}\,\mathrm{E}'$.

PROOF. Assume $\vdash_I \mathrm{E} \sim \mathrm{E}'$. By (*), $\vdash_B \exists\varepsilon\,\varepsilon\,\mathbf{r}(\mathrm{E} \sim \mathrm{E}')$, whence using 2, 4a Lemma 2, $\vdash_B \exists\varepsilon\,\varepsilon\,\mathbf{r}\,\mathrm{E} \sim \exists\varepsilon\,\varepsilon\,\mathbf{r}\,\mathrm{E}'$.

* THEOREM 11. *If* E *is prenex*: $\vdash_B \exists\varepsilon\,\varepsilon\,\mathbf{r}\,\mathrm{E} \supset \mathrm{E}$, *hence* (*12; 5 Lemma 2) $\vdash_B \neg\mathrm{E} \supset \exists\varepsilon\,\varepsilon\,\mathbf{r}\,\neg\mathrm{E}$, *hence* $\vdash_B \exists\varepsilon\,\varepsilon\,\mathbf{r}\,\neg\neg\mathrm{E} \supset \neg\neg\,\mathrm{E}$.

PROOF. Using FIM # D, the scope F of the prefix $Q\Theta$ in E is equivalent in B to a prime formula F′. By 6, 7, 8a, 9 Lemma 2 (with *69, *70, *2), $\vdash_B \exists\varepsilon\,\varepsilon\,\mathbf{r}\,\mathrm{E} \supset Q\Theta\,\exists\varepsilon\,\varepsilon\,\mathbf{r}\,\mathrm{F}$. Also $\vdash_B \exists\varepsilon\,\varepsilon\,\mathbf{r}\,\mathrm{F} \sim \exists\varepsilon\,\varepsilon\,\mathbf{r}\,\mathrm{F}'$ [Lemma 10] $\sim \mathrm{F}'$ [1 Lemma 2] $\sim \mathrm{F}$.

* THEOREM 12. (a) *For each formula* E *which is of the first type described in Theorem* 8 [*of the second type in Theorem* 8 *or is prenex or the double negation of a prenex formula*]: *If* $\vdash_I \mathrm{E}$, *then* $\vdash_C \mathrm{E}$ [$\vdash_B \mathrm{E}$].

(b) *Likewise, when* E *is* $\neg(\mathrm{E}_1 \,\&\cdots\&\, \mathrm{E}_n)$ *and each* E_i *is of the first type described in Theorem* 8 [*of the second type described in Theorem* 8 *or is the negation of a prenex formula*].

PROOF. (a) Assume $\vdash_I \mathrm{E}$. By (*), $\vdash_B \exists\varepsilon\,\varepsilon\,\mathbf{r}\,\mathrm{E}$. By Theorem 8 or 11, $\vdash_C \exists\varepsilon\,\varepsilon\,\mathbf{r}\,\mathrm{E} \supset \mathrm{E}$ [$\vdash_B \exists\varepsilon\,\varepsilon\,\mathbf{r}\,\mathrm{E} \supset \mathrm{E}$]. Hence $\vdash_C \mathrm{E}$ [$\vdash_B \mathrm{E}$].

(b) E.g. take $n = 2$. In C [In B], $\mathrm{E}_i \supset \exists\varepsilon\,\varepsilon\,\mathbf{r}\,\mathrm{E}_i$, whence $\mathrm{E}_1 \,\&\, \mathrm{E}_2 \supset \exists\varepsilon\,\varepsilon\,\mathbf{r}\,\mathrm{E}_1 \,\&\, \exists\varepsilon\,\varepsilon\,\mathbf{r}\,\mathrm{E}_2$ [Theorema praeclarum], whence $\neg(\exists\varepsilon\,\varepsilon\,\mathbf{r}\,\mathrm{E}_1 \,\&\, \exists\varepsilon\,\varepsilon\,\mathbf{r}\,\mathrm{E}_2) \supset \neg(\mathrm{E}_1 \,\&\, \mathrm{E}_2)$ [*12], whence $\exists\varepsilon\,\varepsilon\,\mathbf{r}\,\mathrm{E} \supset \mathrm{E}$ [2, 5 Lemma 2].

* COROLLARY 13. *For each formula* E: *If* $\vdash_I \exists\varepsilon\,\varepsilon\,\mathbf{r}\,\mathrm{E}$, *then* $\vdash_B \exists\varepsilon\,\varepsilon\,\mathbf{r}\,\mathrm{E}$.

REMARK 14. The proof of FIM Theorem 9.14 provides examples of formulas E such that $\vdash_C \exists\varepsilon\,\varepsilon\,\mathbf{r}\,\mathrm{E}$ but not $\vdash_I \exists\varepsilon\,\varepsilon\,\mathbf{r}\,\mathrm{E}$. For, when $P(a)$ is not recursive: $\vdash_C \mathrm{E}$, not $\vdash_I \mathrm{E}$, but by Theorem 8 $\vdash_B \exists\varepsilon\,\varepsilon\,\mathbf{r}\,\mathrm{E} \sim \mathrm{E}$.

* THEOREM 15. *If* C *is simply consistent*: *Each formula* E, *which is*

[9] Cf. FIM 9.2 ¶ 5. Formalization of the proof of Theorem 9.3 (a) should give: (**) $\vdash_I \mathrm{E} \rightarrow \vdash_B \mathbf{r}\,\forall\mathrm{E}$. But a fortiori (i.e. &-elim. with *70), $\vdash_B \mathbf{r}\,\forall\mathrm{E} \supset \exists\varepsilon\,\varepsilon\,\mathbf{r}\,\forall\mathrm{E}$; and by 6, 8a Lemma 2 (with *69, *2) and $\forall$-elim., $\vdash_B \exists\varepsilon\,\varepsilon\,\mathbf{r}\,\forall\mathrm{E} \supset \exists\varepsilon\,\varepsilon\,\mathbf{r}\,\mathrm{E}$. Thus if (**), then (*). Very likely a simplified argument would give (*) directly; but (**) is of interest itself.

of either type described in Theorem 8 or is the negation of a prenex formula, such that $\vdash_C$ E *(or any class of such formulas) can be consistently adjoined to I.*

PROOF. SINGLE FORMULA. By Theorem 12(a), if $\vdash_I \neg$ E, then $\vdash_C \neg$ E. So by $\vdash_C$E and the consistency of C, not $\vdash_I \neg$E. CLASS OF FORMULAS. Similarly, using Theorem 12(b).

REMARK 16. Theorem 11 illustrates how more detailed consideration of the way $\exists$, $\vee$, $\forall\alpha$, $\supset$ enter into a formula E may give results beyond those formulated in Theorem 8. The scope of Theorems 8, 9, 11, 12, 15 can be extended by using equivalences in I with Lemma 10. For example, each formula consisting of only existential quantifiers and bounded existential and universal number quantifiers (in any sequence) applied to a quantifier-free scope is equivalent in B to a formula of the form $\exists\alpha$P(α) with P(α) prime (and, if there are no function quantifiers, to one of the form $\exists$xP(x) with P(x) prime), by using first *25.9, *25.10, *91 and *78, then *0.6 and *25.6 (or *25.4), and finally # D and # E. For P prime, $\vdash_B$ P $\vee$ A $\sim \neg$P $\supset$ A.

LEMMA 17.[7] *Let* P(Ψ, w) *be the formula provided by the proofs of* IM *Theorems* 27 *and* 32(a) *to numeralwise represent a (primitive or) general recursive function* $\phi(\Psi)$. *There are prime formulas* R(s, w) *and* S(s, w) *(containing free only* s, w) *such that*

$$\vdash_B \mathrm{P}(\Psi, \mathrm{w}) \sim \exists \mathrm{x} \mathrm{R}(\overline{\langle\Psi\rangle}(\mathrm{x}), \mathrm{w}) \sim \forall \mathrm{x} \mathrm{S}(\overline{\langle\Psi\rangle}(\mathrm{x}), \mathrm{w}).$$

Similarly, for the formula P(Ψ) *provided by the proofs of* IM *Corollaries to Theorems* 27 *and* 32 *to numeralwise express a general recursive predicate* $P(\Psi)$. (Also cf. FIM Lemma 8.7.)

PROOF. $\exists$X-FORM. For $\phi(\Psi)$ primitive recursive, we use induction on k with cases as on IM p. 243, and advance quantifiers etc. as in the proofs of Lemmas 5 and 6. (In Case 5, we may replace B(c,d,i,w) by rm(c,(i$'\cdot$d)$'$)=w, using *180b, *13.5, *13.4, *12.3.) For $\phi(\Psi)$ general recursive, we use on IM p. 296 line 3 the fact that S(e, Ψ, z) is S(e, Ψ, z, 0) where $\exists$!wS(e, Ψ, z, w) (by the proof of Corollary Theorem 27, and Remark 1 p. 244 with FIM Remark 8.6), so $\neg$S(e, Ψ, z) $\sim$ $\exists$wS(e, Ψ, z, w$'$).

$\forall$X-FORM. For $\phi(\Psi)$ primitive recursive, $\neg$P(Ψ, w) $\sim$ $\exists$u[P(Ψ, u) & u $\neq$ w] [using $\exists$!wP(Ψ, w), from Remarks 1 and 8.6] $\sim$ $\exists$u[$\exists$xR($\overline{\langle\Psi\rangle}$(x), u) & u $\neq$ w] $\sim$ $\exists$xQ($\overline{\langle\Psi\rangle}$(x), w), so P($\Psi$, w) $\sim \neg\neg$ P(Ψ, w) [FIM Lemma 8.8] $\sim$ $\neg$ $\exists$xQ($\overline{\langle\Psi\rangle}$(x), w) $\sim$ $\forall$x$\neg$ Q($\overline{\langle\Psi\rangle}$(x), w) [*86] $\sim$ $\forall$xS($\overline{\langle\Psi\rangle}$(x), w). For $\phi(\Psi)$ general recursive, we use the $\exists$x-form (and then *86) in replacing S(e, Ψ, z) on IM p. 296 line 3.

REFERENCES

[1] AREND HEYTING, *Die formalen Regeln der intuitionistischen Logik*, **Sitzungsberichte der Preussischen Akademie der Wissenschaften, Physikalisch-mathematische Klasse,** 1930, pp. 42–56.

[2] AREND HEYTING, *Die formalen Regeln der intuitionistischen Mathematik*, ibid., pp. 57–71, 158–169.

[3] S. C. KLEENE, *On the interpretation of intuitionistic number theory*, **The journal of symbolic logic,** vol. 10 (1945), pp. 109–124.

[4] S. C. KLEENE, *Recursive functions and intuitionistic mathematics*, **Proceedings of the International Congress of Mathematicians (Cambridge, Mass., U.S.A., Aug. 30-Sept. 6, 1950),** 1952, vol. 1, pp. 679–685.

[5] S. C. KLEENE, **Introduction to metamathematics,** Amsterdam (North-Holland Pub. Co.), Groningen (Noordhoff), New York and Toronto (Van Nostrand), 1952, X + 550 pp.

[6] S. C. KLEENE, *Realizability*, **Summaries of talks presented at the Summer Institute of Symbolic Logic in 1957 at Cornell University,** vol. 1, pp. 100–104 (2nd. ed., Princeton, N.J. (Communications Research Division, Institute for Defense Analyses) 1960), reprinted in **Constructivity in mathematics,** Amsterdam (North-Holland Pub. Co.), 1959, pp. 285–289.

[7] S. C. KLEENE, *Recursive functionals and quantifiers of finite types* I, **Transactions of the American Mathematical Society,** vol. 91 (1959), pp. 1–52.

[8] S. C. KLEENE, *Countable functionals*, **Constructivity in mathematics,** Amsterdam (North Holland Pub. Co.), 1959, pp. 81–100.

[9] S. C. KLEENE, *Realizability and Shanin's algorithm for the constructive deciphering of mathematical sentences*, **Logique et analyse,** 3e Année, Oct. 1960, 11–12, pp. 154–165.

[10] S. C. KLEENE and R. E. VESLEY, **The foundations of intuitionistic mathematics, especially in relation to recursive functions,** Amsterdam (North-Holland Pub. Co.), 1965, VIII + 206 pp.

[11] DAVID NELSON, *Recursive functions and intuitionistic number theory*. **Transactions of the American Mathematical Society,** vol. 61 (1947), pp. 307–368.

THE CALCULUS OF PARTIAL PROPOSITIONAL FUNCTIONS

SIMON KOCHEN
Cornell University, Ithaca, N.Y., U.S.A.
and
ERNST P. SPECKER
Eidg. Technische Hochschule, Zurich, Switzerland

1. The calculus of partial propositional functions has been introduced in [2]. It is a variant of the classical propositional calculus, a variant which takes into account that pairs of propositions may be "incompatible" and cannot therefore be connected. As is well known, such pairs are considered in Quantum Theory; but they may also be said to occur in natural languages. Difficulties arising from propositions of the type "If two times two are five, then there exist centaurs" seem to be due as much to incompatibility as to material implication.

The calculus has been based in [2] on the connectives $\neg$, $\vee$ and a relation $\circlearrowleft$ (called "commeasurability"). A method of eliminating $\circlearrowleft$ has been sketched; if carried out, this elimination leads to a rather complicated system.

The choice of connectives being as free in the new calculus as in the classical one, we choose falsity (f) and implication ($\rightarrow$) as new basic connectives. For commeasurability of ϕ, ψ is most naturally expressed as $f \rightarrow (\phi \rightarrow \psi)$. "If ϕ, ψ are commeasurable, then $\phi \rightarrow \psi$ makes sense; whatever makes sense is implied by f. Conversely, what is implied by anything makes sense; if $\phi \rightarrow \psi$ makes then sense, ϕ, ψ are commeasurable."

Presented this way, the calculus $\boldsymbol{PP}_1$ of partial propositional functions has the same set of formulas as the classical propositional calculus; it differs from it by the notion of validity. The formal notion of validity in $\boldsymbol{PP}_1$ (called "Q-validity" in [2]) is based on the notion of partial Boolean algebra. In order to make this paper somewhat independent from [2], we assume familiarity with this notion only in the last section. Whenever partial Boolean algebras are mentioned in earlier sections, the reader may think of the partial algebra of linear subspaces of the 3-dimensional orthogonal space (as defined in section 3, example 2) or of the partial algebra of closed linear subspaces of Hilbert space. These algebras are the most interesting examples and are at the origin of the notion of partial Boolean algebras. Their relation to Quantum Theory has been considered in [2], [3].

The notion of validity in $\boldsymbol{PP}_1$ may also be explained somewhat informally.

Let S be a set of propositions. Assume that there is defined on S a binary relation $\mathfrak{c}$ of commeasurability and a partial function $\rightarrow$ from $S \times S$ to S, $s_1 \rightarrow s_2$ being defined if and only if $\mathfrak{c}(s_1, s_2)$; assume furthermore that the set S_1 of true sentences of S is given. (In general, propositions depend on parameters and are therefore neither true nor false.) Let ϕ be a formula of $\boldsymbol{PP}_1$, e.g. $x_1 \rightarrow (x_2 \rightarrow x_1)$. ϕ can be evaluated for a pair if and only if $\mathfrak{c}(s_2, s_1)$ and $\mathfrak{c}(s_1, s_2 \rightarrow s_1)$ hold: $s_2 \rightarrow s_1$ has to be defined and putting $s_3 = s_2 \rightarrow s_1$ also $s_1 \rightarrow s_3$. If these conditions are satisfied, the value assigned to ϕ for $< s_1, s_2 >$ is $s_1 \rightarrow (s_2 \rightarrow s_1)$. The formula ϕ holds in the structure $< S, \mathfrak{c}, \rightarrow, S_1 >$ if the assigned value is an element of S_1 for all such pairs $< s_1, s_2 >$: "ϕ holds iff it is true whenever it makes sense." Our axiom system is based on the assumption that ϕ makes sense if and only if $f \rightarrow \phi$ holds. The question of validity of ϕ is thereby reduced to the question whether ϕ is derivable from $f \rightarrow \phi$. The notion of derivation will be formalized by rules of inference $R_1, \cdots, R_7$ (given in 7.1). The rules are adopted from a system of Wajsberg [4] for the propositional calculus. We have learnt from Wajsberg's work also in an other respect; indeed, the main idea behind the series of derived rules in section 8 is due to him. We shall prove completeness, i.e. we show that ϕ holds in all partial Boolean algebras iff ϕ is derivable from $f \rightarrow \phi$. The formulas in such a derivation all make sense provided ϕ does; a proof of ϕ in the system $\boldsymbol{PP}_1$ is therefore essentially a proof based on subformulas of ϕ.

As pointed out in [2], most formulas of *Principia Mathematica* hold in the calculus of partial propositional functions. Contrary to a conjecture of [2], it is not true for all formulas, the "praeclarum theorema" of Leibniz (PM 3.47) being a counter-example. The formula PM 3.47 holds however in the partial Boolean algebra $\boldsymbol{B}(E^\omega)$ associated with Hilbert space E^ω and a fortiori in $\boldsymbol{B}(E^3)$. Axiomatizations of the sets of formulas holding in $\boldsymbol{B}(E^\alpha)$ $(\alpha = 3, \cdots, \omega)$ and relations between these sets will be given in another paper.

In some of the following sections, we write $\phi\psi$ instead of $\phi \rightarrow \psi$, $\phi\psi\chi$ instead of $\phi(\psi\chi)$. There is no danger of misunderstanding since conjunction does nor occur explicitly.

2. Let $\boldsymbol{P}_1$ be the system of the classical propositional calculus as defined e.g. in [1]: Symbols of $\boldsymbol{P}_1$ are

$$(\rightarrow) \; f x_0 \; x_1 \; x_2 \; \cdots$$

f and $x_0, x_1, x_2, \cdots$ are formulas; if ϕ, ψ are formulas, then $(\phi \rightarrow \psi)$ is a formula. Outermost parentheses in formulas may be omitted.

3. A structure $\boldsymbol{B} = < B, \mathfrak{c}, 0, 1, \neg, \vee >$ is of type PB (partial Boolean) if it satisfies the following conditions

a) B is a non-empty set;
b) δ is a binary relation on B ($\delta(a,b)$ is read: "a and b are commeasurable");
c) 0 and 1 are elements of B;
d) $\neg$ is a unary function from B to B;
e) $\vee$ is a binary function. The domain of $\vee$ is the set of those ordered pairs $\langle a,b \rangle$ of $B \times B$ for which $\delta(a,b)$; the co-domain of $\vee$ is the set **B.**

The notion of partial Boolean algebra is defined by imposing restrictions on structures of type *PB*. An example of such a restriction is: $\neg\neg a = a$ for all $a \in B$.

We define two structures of type *PB* which are partial Boolean algebras:

1° The Boolean algebra of two elements. B is the set (0,1); $0 \neq 1$. $\delta(a,b)$ holds for all a, b in B. $\neg 0 = 1, \neg 1 = 0$. $0 \vee 0 = 0$, $0 \vee 1 = 1$, $1 \vee 0 = 1$, $1 \vee 1 = 1$.

2° The partial algebra $\boldsymbol{B}(E^3)$ of linear subspaces of E^3 (3-dimensional orthogonal spacce).

a) B is the set of linear subspaces of E^3;
b) $\delta(a,b)$ for subspaces a, b iff a and b are orthogonal in the sense of elementary geometry, i.e. if there exists a basis of E^3 containing a basis of a and of b; (if a is a subspace of b, $\delta(a,b)$ holds.)
c) 0 is the 0-dimensional, 1 is the 3-dimensional subspace of E^3;
d) $\neg a$ is the orthogonal complement of a;
e) $a \vee b$ is the union (span) of a and b, defined only for those pairs $\langle a,b \rangle$ for which $\delta(a,b)$ holds.

4. We state some properties of the structure $\boldsymbol{B}(E^3)$ defined in example 2° of section 3. These properties hold in all partial Boolean algebras as defined in [2]; it will follow from the completeness theorem in section 10 that they form an axiom system for partial Boolean algebras.

For all elements a, b, c of B:

4.1 $\neg 0 = 1, \neg 1 = 0$
4.2 $\neg\neg a = a$
4.3 $\delta(1,a)$
4.4 *If* $\delta(a,b)$, *then* $\delta(b,a)$
4.5 *If* $\delta(\neg a,b)$, *then* $\delta(a,b)$
4.6 $1 \vee a = 1$, $a \vee 1 = 1$
4.7 $0 \vee a = a$, $a \vee 0 = a$ ($\delta(0,a)$ *holds by* 4.1, 4.3, 4.5)
4.8 *If* $\neg a \vee b = 1$ *and* $\neg b \vee a = 1$, *then* $a = b$
4.9 *If* $\delta(a,b)$, *then* $\delta(\neg b,a)$, $\delta(\neg a, \neg b \vee a)$ *and* $\neg a \vee (\neg b \vee a) = 1$
4.10 *If* $\delta(a,\neg b)$, $\delta(a,c)$, $\delta(b,c)$ *then* $\delta(\neg a,b)$, $\delta(\neg a,c)$, $\delta(\neg b,c)$,

$\mathfrak{Z}(\neg a, \neg b \vee c)$, $\mathfrak{Z}(\neg(\neg a \vee b), \neg a \vee c)$, $\mathfrak{Z}(\neg(\neg a \vee (\neg b \vee c)),$ $\neg(\neg a \vee b) \vee (\neg a \vee c))$, *and* $\neg(\neg a \vee (\neg b \vee c)) \vee (\neg(\neg a \vee b)$ $\vee (\neg a \vee c)) = 1$. (All operations are defined by the hypotheses.)

The theorems 4.9, 4.10 are special cases of the following: *If* $\mathfrak{Z}(a,b)$, $\mathfrak{Z}(a,c)$, $\mathfrak{Z}(b,c)$, *then all Boolean identities in a, b, c hold.*

5. Let $\boldsymbol{B} = < B, \mathfrak{Z}, 0, 1, \neg, \vee >$ be a structure of type PB as defined in section 3 and let N be the set of natural numbers. We associate functions with formulas ϕ of $\boldsymbol{P}_1$ (defined in section 2). The domain D_ϕ of the function $[\phi]$ associated with ϕ is a subset of B^N(the set of functions from N to B), the codomain of $[\phi]$ is B. The functions $[\phi]$ and their domains D_ϕ are defined simultaneously by recursion (with respect to the length of ϕ).

1° $D_f = B^N$ and $[f](q) = 0$ for all $q \in D$. ($[f]$ is the constant function 0 defined for all sequences.)

2° $D_{x_0} = D_{x_1} = D_{x_2} = \cdots = B^N$ and $[x_0](q) = q(0)$, $[x_1](q) = q(1)$, $[x_2](q) = q(2)$, $\cdots$ ($[x_1]$ *is the* projection of B^N on its coordinate 1.)

3° $q \in D_{\phi \to \psi}$ if and only if $q \in D_\phi$ and $q \in D_\psi$ and $\mathfrak{Z}(\neg [\phi](q), [\psi](q))$.

The function

$$[\phi \to \psi] : D_{\phi \to \psi} \Rightarrow B$$

is defined as follows:

$$[\phi \to \psi](q) = \neg([\phi](q)) \vee [\psi](q) \text{ for } q \in D_{\phi \to \psi}.$$

EXAMPLE: The set $D_{x_0 \to x_1}$ consists of those sequences $< q(0), q(1), \cdots >$ for which $\mathfrak{Z}(\neg q(0), q(1))$ and $[x_0 \to x_1](q) = \neg q(0) \vee q(1)$. Roughly speaking, D_ϕ is the set of those sequences in B^N for which ϕ can be evaluated and $[\phi](q)$ is the result of the evaluation.

DEFINITION of validity in a structure of type PB: A formula ϕ of $\boldsymbol{P}_1$ holds (is valid) in the structure $\langle B, \mathfrak{Z}, 0, 1, \neg, \vee \rangle$ of type PB if and only if $[\phi](q) = 1$ for all $q \in D_\phi$.

REMARKS.

1) If $\langle B, \mathfrak{Z}, 0, 1, \neg, \vee \rangle$ is the two element Boolean algebra defined in example 1° of section 3, D_ϕ is equal to B^N for all formulas ϕ and the above construction is the one given by Tarski for the notion of satisfaction.

2) A formula valid in all partial Boolean algebras has been called "Q-valid" in [2].

DEFINITION of (semantic) consequence in the structure of type PB: The formula ψ of $\boldsymbol{P}$ is a semantic consequence of the formulas $\phi_1, \phi_2, \cdots, \phi_n$

of $\boldsymbol{P}_1$ in the structure $\langle B, \mathfrak{C}, 0, 1, \neg, \vee\rangle$ of type PB if and only if the following condition is satisfied for all q in B^N:
If $q \in D_{\phi_i}$ and $[\phi_i](q) = 1$ for all i, $1 \leqq i \leqq n$, then $q \in D_\psi$ and $[\psi](q) = 1$. Semantic consequence is expressed as follows:

$$\phi_1, \cdots, \phi_n \Vdash \psi$$

6. We introduce a shorter notation: Instead of $\phi \to \psi$ we write $\phi\psi$; association is to the right, i.e. $\phi\psi\chi$ is $\phi(\psi\chi)$. The formula $(\phi \to (\psi \to \chi)) \to ((\phi \to \psi) \to (\phi \to \chi))$ is therefore written $(\phi\psi\chi)(\phi\psi)(\phi\chi)$. Throughout this section, validity and semantic consequence is with respect to a fixed structure $\boldsymbol{B} = \langle B, \mathfrak{C}, 0, 1, \neg, \vee\rangle$ of type PB satisfying 4.1–4.10 (i.e. a partial Boolean algebra). Formulas are formulas of $\boldsymbol{P}_1$.

6.1 *$q \in D_{f\phi}$ iff $q \in D_\phi$*

PROOF. $q \in D_{f\phi}$ iff $q \in D_f$, $q \in D_\phi$ and $\mathfrak{C}(\neg [f](q), [\phi](q))$. Therefore, if $q \in D_{f\phi}$ then $q \in D_\phi$. Assume $q \in D_\phi$. By definition, $D_f = B^N$, $[f](q) = 0$; by 4.1, $\neg 0 = 1$; by 4.3 $\mathfrak{C}(1, [\phi](q))$, *i.e.* $q \in D_{f\phi}$.

6.2 *$f\phi$ is valid for all formulas ϕ of $\boldsymbol{P}_1$.*

PROOF. Assume $q \in D_{f\phi}$; then $[f\phi](q) = \neg [f](q) \vee [\phi](q) = 1 \vee a = 1$ (*by* 4.1, 4.6).

6.3 *ϕ is valid iff $f\phi \Vdash \phi$.*

PROOF. 1° Assume ϕ valid and $q \in D_{f\phi}$; then $q \in D_\phi$ by 6.1 and $[\phi](q) = 1$ by validity. 2° Assume $f\phi \Vdash \phi$ and $q \in D_\phi$; then $q \in D_{f\phi}$ by 6.1, $[f\phi](q) = 1$ by 6.2, and therefore $[\phi](q) = 1$ by $f\phi \Vdash \phi$.

6.4 *$\phi \Vdash f\phi$*

PROOF. If $q \in D_\phi$, then $q \in D_{f\phi}$ by 6.1; $[f\phi](q) = 1$ by 6.2.

6.5 *$f\phi\psi \Vdash f\psi$*

PROOF. Assume $q \in D_{f\phi\psi}$; then $q \in D_{\phi\psi}$, $q \in D_\psi$; by 6.1, $q \in D_{f\psi}$; by 6.2 $[f\psi](q) = 1$.

6.6 *$(\phi f) f \Vdash \phi$*

PROOF. Assume $q \in D_{(\phi f) f}$; then $q \in D_\phi$. Putting $[\phi](q) = a$ and assuming $[(\phi f) f](q) = 1$, we have $\neg(\neg a \vee 0) \vee 0 = 1$. By 4.7, $b \vee 0 = b$ for all b; therefore $\neg\neg a = 1$; by 4.2, $\neg\neg a = a$, i.e. $[\phi](q) = 1$.

6.7 *$f\phi\psi \Vdash \phi\psi\phi$*

PROOF. Assume $q \in D_{f\phi\psi}$; then $q \in D_{\phi\psi}$, $q \in D_\phi$, $q \in D_\psi$. Putting $[\phi](q) = a$ $[\psi](q) = b$, we have $\mathfrak{C}(\neg a, b)$. Therefore by 4.5, $\mathfrak{C}(a, b)$; by 4.9,

♂$(\neg b, a)$, ♂$(\neg a, \neg b \vee a)$ and $\neg a \vee (\neg b \vee a) = 1$. Hence $q \in D_{\psi\phi}$, $q \in D_{\phi\psi\phi}$, and $[\phi\psi\phi](q) = 1$.

6.8 $f\psi\chi, f\phi\psi, f\phi\chi \Vdash (\phi\psi\chi)(\phi\psi)(\phi\chi)$

PROOF. Assume $q \in D_{f\psi\chi}, q \in D_{f\phi\psi}$, and $q \in D_{f\phi\chi}$; then $q \in D_{\psi\chi}$, $q \in D_{\phi\psi}$, $q \in D_{\phi\chi}, q \in D_{\phi}, q \in D_{\psi}, q \in D_{\chi}$. Putting $[\phi](q) = a, [\psi](q) = b, [\chi](q) = c$, we have ♂$(\neg b, c)$, ♂$(\neg a, b)$, ♂$(\neg a, c)$. Therefore, by 4.5, ♂$(a, b)$, ♂$(b, c)$, ♂$(a,c)$ and, by 4.10, ♂$(\neg(\neg a \vee b), \neg a \vee c)$; hence $q \in D_{(\phi\psi)(\phi\chi)}$. Furthermore by 4.10, ♂$(\neg(\neg a \vee (\neg b \vee c)), \neg(\neg a \vee b) \vee (\neg a \vee c))$, i.e. $q \in D_{(\phi\psi\chi)(\phi\psi)(\phi\chi)}$. Again by 4.10, $\neg(\neg a \vee (\neg b \vee c)) \vee (\neg(\neg a \vee b) \vee (\neg a \vee c)) = 1$, i.e. $[(\phi\psi\chi)(\phi\psi)(\phi\chi)](q) = 1$.

6.9 $\phi, \phi\psi \Vdash \psi$

PROOF. Assume $q \in D_{\phi\psi}$; then $q \in D_{\psi}$. Assuming $[\phi](q) = 1$ and putting $[\psi](q) = a$, we have $[\phi\psi](q) = \neg[\phi](q) \vee [\psi](q) = \neg 1 \vee a = 0 \vee a = a$ (by 4.1, 4.7). Assuming $[\phi\psi](q) = 1$, we have $1 = a$, i.e. $[\psi](q) = 1$.

6.10 $f\phi\psi, \psi\chi, \chi\psi \Vdash f\phi\chi$

PROOF. Assume $q \in D_{f\phi\psi}$, $q \in D_{\psi\chi}$, $q \in D_{\chi\psi}$. Then $q \in D_{\phi}$, $q \in D_{\psi}$, $q \in D_{\chi}$. Putting $[\phi](q) = a$, $[\psi](q) = b$, $[\chi](q) = c$, we have ♂$(\neg a, b)$; $[\psi\chi](q) = \neg b \vee c$, $[\chi\psi](q) = \neg c \vee b$. Assuming $[\psi\chi](q) = 1$ and $[\chi\psi](q) = 1$, we have $\neg b \vee c = 1$ and $\neg c \vee b = 1$. Therefore by 4.8, $b = c$. Hence ♂$(\neg a, c)$, $q \in D_{\phi\chi}$; by 6.1, 6.2, $q \in D_{f\phi\chi}$, $[f\phi\chi](q) = 1$.

6.11 REMARK. All rules $\phi_1, \cdots, \phi_m \Vdash \psi$ in 6.1—6.9 have the property that $q \in D_{\psi}$ provided $q \in D_{\phi_i}$, $i = 1, \cdots, m$. The rule 6.10 does not have this property as can be shown by an example in $\boldsymbol{B}(E^3)$.

7. 7.1 DEFINITION of the calculus $\boldsymbol{PP}_1$ of partial propositional functions.

1° Formulas of $\boldsymbol{PP}_1$ are the formulas of $\boldsymbol{P}_1$.

2° $\boldsymbol{PP}_1$ has the following rules of inference

$R_1: \phi \vdash f \to \phi$

$R_2: f \to (\phi \to \psi) \vdash f \to \psi$

$R_3: (\phi \to f) \to f \vdash \phi$

$R_4: f \to (\phi \to \psi) \vdash \phi \to (\psi \to \phi)$

$R_5: f \to (\psi \to \chi), f \to (\phi \to \psi), f \to (\phi \to \chi) \vdash (\phi \to (\psi \to \chi)) \to ((\phi \to \psi) \to (\phi \to \chi))$

$R_6: \phi,\ \phi \to \psi \vdash \psi$

$R_7: f \to (\phi \to \psi),\ \psi \to \chi,\ \chi \to \psi \vdash f \to (\phi \to \chi)$

3° A rule

$$\phi_1, \cdots, \phi_m \vdash \gamma_n$$

is a derivable rule of $\boldsymbol{PP}_1$ iff there exists a sequence $\gamma_1 \cdots \gamma_n$ of formulas of

$\boldsymbol{PP}_1$ such that each $\gamma_i (i \leqq n)$ is either one of the formulas $\phi_1, \cdots, \phi_m$ or follows from formulas $\gamma_{i_1}, \cdots, \gamma_{i_k}$ $(i_j < i, j = 1, \cdots, k)$ by one of the rules $R_1, \cdots, R_7$.

4° A formula ϕ of $\boldsymbol{PP}_1$ is provable in $\boldsymbol{PP}_1$ iff $f \to \phi \vdash \phi$ is a derivable rule of $\boldsymbol{PP}_1$.

7.2 THEOREM. *If* $\phi_1, \cdots, \phi_m \vdash \psi$ *is a derivable rule of* $\boldsymbol{PP}_1$, *then* $\phi_1, \cdots, \phi_m \Vdash \psi$ *holds in every partial Boolean algebra.*

Proof. Let $\langle \gamma_1, \cdots, \gamma_n \rangle$ be a sequence as defined in 3° of 7.1 and assume $q \in D_{\phi i}$, $[\phi_i](q) = 1$ for $i = 1, \cdots, m$. We prove by induction with respect to $j : q \in D_{\gamma_j}$ and $[\gamma_j](q) = 1$. The inductive step is provided for each of the rules R_i by $6.3 + i\,(i = 1, \cdots, 7)$.

THEOREM. *A provable formula of* $\boldsymbol{PP}_1$ *holds in all partial Boolean algebras (is "Q-valid").*

PROOF. Assume $f \to \phi \vdash \phi$; then $f \to \phi \Vdash \phi$ by the preceding theorem. By 6.1, ϕ is valid iff $f \to \phi \Vdash \phi$ holds.

7.3 The rest of the paper is devoted to the proof of the converse: If ϕ holds in all partial Boolean algebras, then ϕ is provable in $\boldsymbol{PP}_1$. By 6.1, it suffices to show: If $f \to \phi \Vdash \phi$ holds in all partial Boolean algebras, then $f \to \phi \vdash \phi$ is a derivable rule of $\boldsymbol{PP}_1$.

7.4 It might be suspected that $\phi_1, \cdots, \phi_m \vdash \psi$ follows generally from $\phi_1, \cdots, \phi_m \Vdash \psi$. This is not so as shown by the following counterexample. Clearly $f \Vdash x_0$ holds in all partial Boolean algebras as there is no q such that $[f](q) = 1$. However, $f \vdash x_0$ is not a derivable rule. For, if the variable x_0 does not occur in the premise of the rules $R_1, \cdots, R_7$, neither does it occur in the conclusion. $f \vdash \phi$ is therefore derivable only for formulas ϕ not containing x_0. The system $\boldsymbol{PP}_1$ can be made complete in the above strong sense by adjoining the infinite list of axioms $fx_0, fx_1, \cdots$.

7.5 We shall state a series of derivable rules, numbered $S_1, S_2, \cdots$. For clarity, they will be included in brackets:

$$S_i : [\phi_1, \cdots, \phi_m \vdash \psi].$$

Proofs of such rules will be given in the following form

$$[\gamma_1; \gamma_2; \cdots; R_2 : \gamma_5; \cdots; S_2 : \gamma_7; \cdots; D : \gamma_9; \cdots; \gamma_n]$$

$\gamma_1, \cdots, \gamma_n$ will be formulas of $\boldsymbol{P}_1$, γ_n the formula ψ. A formula γ_k $(1 \leqq k \leqq n)$ not preceded by some R_i, S_i or D is one of the formulas $\phi_1, \cdots, \phi_m, \gamma_1, \cdots, \gamma_{k-1}$. If γ_k is preceded by R_i (or: by S_i), it follows from $\gamma_{k-t_i}, \cdots, \gamma_{k-1}$ by the rule R_i (or: S_i), where t_i is the number of formulas in the premise of R_i (or: S_i).

If γ_k is preceded by D, it is obtained from γ_{k-1} by substituting w (truth) for the subformula ff or by substituting ff for w.

8. Derivable rules:

S_1: $[f\phi\psi \vdash f\psi\phi]$
  $[f\phi\psi;\ R_4 : \phi\psi\phi;\ R_1 : f\phi\psi\phi;\ R_2 : f\psi\phi$

S_2: $[f\phi \vdash f\phi f]$
  $[f\phi;\ R_1 : ff\phi;\ S_1 : f\phi f]$

S_3: $[f\phi\psi \vdash f\phi]$
  $[f\phi\psi;\ S_1 : f\psi\phi;\ R_2 : f\phi]$

S_4: $[f\phi\psi\chi \vdash f\chi]$
  $[f\phi\psi\chi;\ R_2 : f\psi\chi;\ R_2 : f\chi]$

S_5: $[f\phi\psi\chi \vdash f\psi]$
  $[f\phi\psi\chi;\ R_2 : f\psi\chi;\ S_3 : f\psi]$

S_6:$[\phi \vdash ff]$
  $[\phi;\ R_1 : f\phi;\ R_1 : ff\phi;\ S_3 .\ ff]$

DEFINITION: w is ff

S_7: $[\phi \vdash w]$
  $[\phi;\ S_6 : ff;\ D : w]$

S_8: $[f\phi \vdash f\phi w]$
  $[f\phi;\ S_7 : w;\ S_6 : ff;\ R_1 : fff; f\phi;\ S_2 : f\phi f; fff; f\phi f; f\phi f;\ R_5 : (\phi ff)(\phi f)(\phi f);$
  $R_1 : f(\phi ff)(\phi f)(\phi f);\ S_3 : f\phi ff;\ D : f\phi w]$

S_9: $[f\phi \vdash \phi w]$
  $[f\phi;\ S_8 : f\phi w;\ S_1 : fw\phi;\ R_4 : w\phi w;\ S_7 : w;\ w\phi w;\ R_6 : \phi w]$

S_{10}: $[f\phi \vdash \phi w\phi]$
  $[f\phi;\ S_8 : f\phi w;\ R_4 : \phi w\phi]$

S_{11}: $[f\phi\psi \vdash f\phi\psi\phi]$
  $[f\phi\psi;\ R_4 : \phi\psi\phi;\ R_1 : f\phi\psi\phi]$

S_{12}: $[f\phi\phi, f\phi\psi \vdash f\phi\phi\psi]$
  $[f\phi\psi; f\phi\phi; f\phi\psi;\ R_5 : (\phi\phi\psi)(\phi\phi)(\phi\psi);\ R_1 : f(\phi\phi\psi)(\phi\phi)(\phi\psi);\ S_3 : f\phi\phi\psi]$

S_{13}: $[f\phi\psi, f\phi\chi, f\psi\chi \vdash f\phi\psi\chi]$
  $[f\psi\chi;\ f\phi\psi;\ f\phi\chi;\ R_5 : (\phi\psi\chi)(\phi\psi)(\phi\chi);\ R_1 : f(\phi\psi\chi)(\phi\psi)(\phi\chi);\ S_3 : f\phi\psi\chi]$

S_{14}: $[f\phi\psi, f\phi\chi, f\psi\chi \vdash f(\phi\psi)\chi]$
  $[f\psi\chi;\ S_1 : f\chi\psi; f\phi\chi;\ S_1 : f\chi\phi; f\chi\psi; f\phi\psi;\ S_{13} : f\chi\phi\psi;\ S_1 : f(\phi\psi)\chi]$

S_{15}: $[f\phi\psi, f\phi\chi, \psi\chi \vdash \phi\psi\chi]$
  $[f\phi\psi;\ S_1 : f\psi\phi;\ f\phi\chi;\ S_1 : f\chi\phi;\ \psi\chi;\ R_1 : f\psi\chi;\ f\psi\phi;\ f\chi\phi;\ S_{14} : f(\psi\chi)\phi;$
  $R_4 : (\psi\chi)\phi(\psi\chi);\ \psi\chi;\ (\psi\chi)\phi(\psi\chi);\ R_6 : \phi(\psi\chi)]$

S_{16}: $[f\phi\psi, f\phi\chi, \psi\chi \vdash (\phi\psi)(\phi\chi)]$
  $[f\psi\chi;\ f\phi\psi;\ f\phi\chi;\ R_5 : (\phi\psi\chi)(\phi\psi)(\phi\chi);\ f\phi\psi : f\phi\chi;\ \psi\chi;\ S_{15} : \phi\psi\chi;\ (\phi\psi\chi)$
  $(\phi\psi)(\phi\chi) : R_6 : (\phi\psi)(\phi\chi)]$

S_{17}: $[f\phi\phi, f\phi\psi, f\phi\chi, \phi\psi\chi \vdash \psi\phi\chi]$

$[\phi\psi\chi;\ R_1: f\phi\psi\chi;\ R_2: f\psi\chi; f\phi\psi;\ f\phi\chi;\ R_5: (\phi\psi\chi)(\phi\psi)(\phi\chi);\ \phi\psi\chi;\ (\phi\psi\chi)(\phi\psi)(\phi\chi);\ R_6: (\phi\psi)(\phi\chi);\ R_1: f(\phi\psi)(\phi\chi);\ \phi\psi\chi;\ R_1: f\phi\psi\chi;\ R_2: f\psi\chi; f\phi\psi;\ S_1: f\psi\phi; f\psi\chi; f\phi\chi;\ S_{13}: f\psi\phi\chi; f\psi\phi;\ R_4: \psi\phi\psi;\ R_1: f\psi\phi\psi; f\psi\phi\chi; (\phi\psi)(\phi\chi);\ S_{16}: (\psi\phi\psi)(\psi\phi\chi);\ \psi\phi\psi;\ (\psi\phi\psi)(\psi\phi\chi);\ R_6: \psi\phi\chi]$

S_{18}: $[f\phi\phi, \phi\psi\phi \vdash \psi\phi\phi]$

$[\phi\psi\phi;\ R_1: f\phi\psi\phi;\ R_2: f\psi\phi;\ S_1: f\phi\psi; f\phi\phi; f\phi\psi; f\phi\phi;\ \phi\psi\phi;\ S_{17}: \psi\phi\phi]$

S_{19}: $[f\phi \vdash f\phi w]$

$[f\phi;\ S_9: \phi w;\ R_1: f\phi w]$

S_{20}: $[f\phi \vdash fw\phi]$

$[f\phi;\ S_{19}: f\phi w;\ S_1: fw\phi]$

S_{21}: $[f\phi \vdash f(w\phi)w]$

$[f\phi;\ S_{20}: fw\phi;\ S_{20}: fww\phi;\ S_{21}: f(w\phi)\,w]$

S_{22}: $[f\phi \vdash f\phi w\phi]$

$[f\phi;\ S_{10}: \phi w\phi;\ R_1: f\phi w\phi]$

S_{23}: $[f\phi \vdash f(w\phi)\,\phi]$

$[f\phi;\ S_{22}: f\phi w\phi;\ S_1: f(w\phi)\,\phi]$

S_{24}: $[f\phi \vdash f(w\phi)(w\phi)]$

$[f\phi;\ S_{20}: fw\phi; f\phi;\ S_{21}: f(w\phi)\,w; f\phi;\ S_{23}: f(w\phi)\,\phi;\ f(w\phi)\,w; fw\phi;\ S_{13}: f(w\phi)(w\phi)]$

S_{25}: $[f\phi \vdash (w\phi)(w\phi)]$

$[f\phi;\ S_{21}: f(w\phi)\,w;\ R_4: (w\phi)\,w\,(w\phi);\ f\phi;\ S_{24}: f(w\phi)(w\phi);\ (w\phi)\,w\,(w\phi);\ S_{18}: w(w\phi)(w\phi);\ S_7: w;\ w(w\phi)(w\phi):\ R_6: (w\phi)(w\phi)]$

S_{26}: $[f\phi \vdash (w\phi)\,\phi]$

$[f\phi;\ S_{25}: (w\phi)(w\phi); f\phi;\ S_{23}: f(w\phi)\,\phi;\ f\phi;\ S_{21}: f(w\phi)\,w;\ f\phi;\ S_{24}: f(w\phi)(w\phi); f(w\phi)\,w; f(w\phi)\,\phi;\ (w\phi)(w\phi);\ S_{17}: w(w\phi)\phi;\ S_7: w;\ w(w\phi)\,\phi;\ R_6: (w\phi)\,\phi]$

S_{27}: $[f\phi \vdash f\phi\phi]$

$[f\phi;\ S_{10}: \phi w\phi; f\phi;\ S_{26}: (w\phi)\phi; f\phi;\ S_{22}: f\phi w\phi; (w\phi)\,\phi;\ \phi w\phi;\ R_7: f\phi\phi]$

S_{28}: $[f\phi\psi \vdash f\phi\phi]$

$[f\phi\psi;\ S_3: f\phi;\ S_{27}: f\phi\phi]$

S_{29} $[f\phi\psi \vdash f\psi\psi]$

$[f\phi\psi;\ S_1: f\psi\phi;\ S_{28}: f\psi\psi]$

S_{30}: $[f\phi\psi \vdash f\phi\phi\psi]$

$[f\phi\psi;\ S_{28}: f\phi\phi; f\phi\psi;\ S_{12}: f\phi\phi\psi]$

S_{31}: $[f\phi\psi, f\phi\chi, \phi\psi\chi \vdash \psi\phi\chi]$

$[f\phi\psi;\ S_{28}: f\phi\phi; f\phi\psi; f\phi\chi;\ \phi\psi\chi;\ S_{17}: \psi\phi\chi]$

S_{32}: $[\phi\psi\phi \vdash \psi\phi\phi]$

$[\phi\psi\phi;\ R_1: f\phi\psi\phi;\ S_{28}: f\phi\phi;\ \phi\psi\phi;\ S_{18}: \psi\phi\phi]$

S_{33}: $[f\phi \vdash \phi\phi]$

$[f\phi;\ S_{10}: \phi w\phi;\ S_{32}: w\phi\phi;\ S_7: w;\ w\phi\phi;\ R_6: \phi\phi]$

S_{34}: $[f\phi\psi \vdash f(\phi f)\psi]$
$[f\phi\psi;\ R_2: f\psi;\ R_1: ff\psi;\ f\phi\psi;\ S_3: f\phi;\ S_2: f\phi f;\ f\phi\psi;\ ff\psi;\ S_{13}: f(\phi f)\psi)]$

S_{35}: $[f\phi\psi \vdash f\phi\psi f]$
$[f\phi\psi;\ S_1: f\psi\phi;\ S_{34}: f(\psi f)\phi;\ S_1: f\phi\psi f]$

S_{36}: $[f\phi\psi, f\phi\chi, f\psi\chi \vdash f((\phi f)\psi)\chi]$
$[f\phi\chi;\ S_{34}: f(\phi f)\chi;\ f\phi\psi;\ S_{34}: f(\phi f)\psi;\ f(\phi f)\chi;\ f\psi\chi;\ S_{13}: f((\phi f)\psi)\chi]$

9. 9.1 Let $\boldsymbol{P}_1^n(y)$ be the system $\boldsymbol{P}_1$ introduced in section 2 in which the series $x_0, x_1, \cdots$ is replaced by $y_0, y_1, \cdots$ and where no other variables than $y_0, \cdots, y_n$ occur. If γ is a formula of $\boldsymbol{P}_1(y)$ and $*$ is an n-sequence $\langle\phi_0, \cdots, \phi_n\rangle$ of formulas of $\boldsymbol{P}_1$ then γ^* is the result of substituting ϕ_i for y_i $(i = 1, \cdots, n)$. We have $f^* = f$, $y_0^* = \phi_0, \cdots; (\gamma_1\gamma_2)^* = \gamma_1^*\gamma_2^*$. If $*$ is the sequence $\langle\phi_0, \cdots, \phi_n\rangle$ of formulas of $\boldsymbol{P}_1$, then f^{**} is the following sequence of formulas: It is $f\phi_0\phi_0$ in case $n = 0$; *it is* $f\phi_0\phi_1, \cdots, f\phi_i\phi_j (i < j), \cdots, f\phi_{n-1}\phi_n$ in case $n \geqq 1$. We state a metarule

M_1: *If* γ_1, γ_2 *are formulas of* $\boldsymbol{P}_1^n(y)$ *and if* $*$ *is an n-sequence of formulas of* $\boldsymbol{P}_1$, *then* $[f^{**} \vdash f\gamma_1^*\gamma_2^*]$ *is a derivable rule.*

The proof (by induction) follows from the rules R_2, S_3, S_{13}, S_{27}.

9.2 M_2: *If the formula* γ *of* $\boldsymbol{P}_1^n(y)$ *is an identity of the classical propositional calculus and if* $*$ *is an n-sequence of formulas of* $\boldsymbol{P}_1$, *then*

$$[f^{**} \vdash \gamma^*]$$

is a derivable rule.

PROOF. γ being an identity, there exists by Wajsberg [4], p. 138, a sequence $\langle\gamma_1, \cdots, \gamma_m\rangle$, $\gamma_m = \gamma$, of formulas of $\boldsymbol{P}_1^n(y)$ having the following property:

For each i, $1 \leqq i \leqq m$, one of the following alternatives hold:

(a) there exist formulas ϕ, ψ, χ of $\boldsymbol{P}_1^n(y)$ such that γ_i is one of the following formulas ("γ_i is an axiom")

(a_1) $f\phi$
(a_2) $\phi\psi\phi$
(a_3) $(\phi\psi\chi)(\phi\psi)(\phi\chi)$.

(b) There exists j, $j < i$, such that γ_j is $(\gamma_i f)f$.

(c) There exist j, k, $j < i$, $j < k$ such that γ_k is $\gamma_j\gamma_i$.

We describe a modification of the sequence $\langle\gamma_1^*, \cdots, \gamma_m^*\rangle$ which transforms it into a proof of $[f^{**} \vdash \gamma_m^*]$. The formula γ_i^* will be replaced by one of the following sequences (the last formula being γ_i^* itself):

(a_1) M_1: $f\phi^*\phi^*$; R_2: $f\phi^*$
(a_2) M_1: $f\phi^*\psi^*$; R_4: $\phi^*\psi^*\phi^*$
(a_3) M_1: $f\psi^*\chi^*$; M_1: $f\phi^*\psi^*$; M_1: $f\phi^*\chi^*$; R_5: $(\phi^*\psi^*\chi^*)(\phi^*\psi^*)(\phi^*\chi^*)$
(b) $(\gamma_i^* f)f$; R_3: γ_i^*
(c) γ_j^*; $\gamma_j^*\gamma_i^*$; R_6: γ_i^*

9.3 The following rules S_{37}, S_{38}, S_{39} are special cases of the metarule M_2:

S_{37}: $[f\phi\phi \vdash ((\phi f)f)\phi]$
S_{38}: $[f\phi\phi \vdash \phi(\phi f)f]$
S_{39}: $[f\phi_1\phi_2, f\phi_1\psi_1, f\phi_1\psi_2, f\phi_2\psi_1, f\phi_2\psi_2, f\psi_1\psi_2 \vdash (\psi_1\psi_2)(\phi_2\phi_1)(\phi_1\psi_1)(\phi_2\psi_2)]$

Rule S_{40} follows easily from R_6, R_7, S_1, S_{39}.

S_{40}: $[f\phi_1\psi_1, \phi_1\phi_2, \phi_2\phi_1, \psi_1\psi_2, \psi_2\psi_1 \vdash (\phi_1\psi_1)(\phi_2\psi_2)]$

9.4 We proceed to prove the substitutivity property of equivalence. Let γ be a formula of $\boldsymbol{P}_1''(y)$, let $\langle\phi_0', \phi_1, \cdots, \phi_n\rangle$ and $\langle\phi_0'', \phi_1, \cdots, \phi_n\rangle$ be n-sequences of formulas of $\boldsymbol{P}_1^*$; let γ_1^* be the formula corresponding to the first, γ_2^* the formula corresponding to the second sequence. We then have the two following metarules:

M_3: $[f\gamma_1^*, \phi_0'\phi_0'', \phi_0''\phi_0', \vdash \gamma_1^*\gamma_2^*]$
M_4: $[\gamma_1^*, \phi_0'\phi_0'', \phi_0''\phi_0' \vdash \gamma_2^*]$

The proof of M_3 is by induction with respect to the length of γ; the inductive step is provided by S_{40}.

Proof of M_4: $[\gamma_1^*;\ R_1: f\gamma_1^*;\ \phi_0'\phi_0'';\ \phi_0''\phi_0';\ M_3: \gamma_1^*\gamma_2^*;\ \gamma_1^*;\ \gamma_1^*\gamma_2^*;\ R_6: \gamma_2^*]$.

10. Theorem. *If the formula ϕ of $\boldsymbol{PP}_1$ holds in all partial Boolean algebras* (as defined in [2]), *then $f \to \phi \vdash \phi$ is a derivable rule of $\boldsymbol{PP}_1$.*

Instead of giving the (rather tedious) reduction of this completeness theorem to the one given in [2], we outline the adaptation of the proof in [2] to the present case.

10.1 With each formula ϕ of $\boldsymbol{P}_1$ we associate a partial Boolean algebra $\boldsymbol{B}_\phi$ such that $f \to \phi \vdash \phi$ is a derivable rule of $\boldsymbol{PP}_1$, if ϕ holds in $\boldsymbol{B}_\phi$. Let Ω be the set of formulas ψ of $\boldsymbol{P}_1$ such that $[f\phi \vdash f\psi]$ is a derivable rule and let the relation $\simeq$ on $\Omega \times \Omega$ be defined as follows: $\psi_1 \simeq \psi_2$ iff the rules $[f\phi \vdash \psi_1\psi_2]$ and $[f\phi \vdash \psi_2\psi_1]$ are derivable. Ω and $\simeq$ have the following properties:

1° $f \in \Omega$, $w \in \Omega$, $\phi \in \Omega$ (S_6, S_7, R_1).
2° *If $\psi_1\psi_2 \in \Omega$, then $\psi_i \in \Omega$ ($i = 1,2$; by R_2, S_3)*
3° *$\simeq$ is an equivalence relation (S_{33}, M_4).*
4° *If $\psi \simeq \psi'$, then $\psi f \simeq \psi' f (S_{40})$.*
5° *If $\psi_1\psi_2 \in \Omega$, then $((\psi_1 f)f)\psi_2 \in \Omega$ and $((\psi_1 f)f)\psi_2 \simeq \psi_1\psi_2 (S_{37}, S_{38}$*
6° *If $\psi_1 \simeq \psi_1'$ and $\psi_2 \simeq \psi_2'$, then $\psi_1\psi_2 \in \Omega$ iff $\psi_1'\psi_2' \in \Omega (M_4)$.*
7° *If $\psi_1 \simeq \psi_1'$, $\psi_2 \simeq \psi_2'$ and $\psi_1\psi_2 \in \Omega$, then $(\psi_1 f)\psi_2 \simeq (\psi_1' f)\psi_2' (M_4)$.*
8° *$\psi \simeq w$ iff $[f\phi \vdash \psi]$ is derivable.*

Proof. Assume $[f\phi \vdash \psi]$; then $[f\phi \vdash \psi w]$ by S_9, $[f\phi \vdash w\psi]$ by S_{10}, R_6. If $[f\phi \vdash w\psi]$, then $[f\phi \vdash \psi]$ by R_6.

10.2 We define a structure $\boldsymbol{B}_\phi = \langle B,$ ♂$, 0, 1, \neg, \vee\rangle$ of type PB (cf. section 3):

a) B is the set of equivalence classes of the relation $\simeq$ on Ω.

b) ♂(a_1, a_2) holds for $a_i \in B (i = 1, 2)$ iff there exist formulas $\psi_1 \in a_i (i = 1, 2)$ such that $[f\phi \vdash f\psi_1\psi_2]$ is derivable. By 6°, ♂(a_1, a_2) iff $[f\phi \vdash f\psi_1\psi_2]$ is derivable for all formulas $\psi_i \in a_i (i = 1, 2)$.

c) 0 is the class of f, 1 is the class of $w (f, w \in \Omega$ by 1°).

d) By 4°, there exists for every class $a \in B$ a class $b \in B$ such that $(\psi f) \in b$ if $\psi \in a$; let this class b be $\neg a$.

e) Assume $a_i \in B$, $\psi_i \in a_i$, $\psi_i' \in a_i (i = 1, 2)$ and ♂(a_1, a_2). Then $\psi_1\psi_2 \in \Omega$, $\psi_1'\psi_2' \in \Omega$ and the formulas $(\psi_1 f)\psi_2, (\psi_1' f)\psi_2'$ belong to the same class b: let $a_1 \vee a_2$ be this class b.

10.3 *The structure* $\boldsymbol{B}_\phi$ *defined in* 10.2 *is a partial Boolean algebra,* i.e. it satisfies the following 5 axioms of [2]:

A1) The relation ♂ is symmetric and reflexive (symmetry by S_1, reflexivity by S_{28}).

A2) For all $b \in B$: ♂$(b, 1)$, ♂$(b, 0) (S_2$ and S_8).

A3) The partial function $\vee$ is defined exactly for those pairs $\langle b_1, b_2\rangle$ for which ♂(b_1, b_2) (by definition).

A4) If ♂(b_1, b_2), ♂(b_1, b_3) and ♂(b_2, b_3), then ♂$(b_1 \vee b_2, b_3)$, ♂$(\neg b_1, b_2)$ (the first conclusion by S_{36}, the second by S_{34}).

A5) For all $b_0, b_1, b_2 \in B$: If ♂(b_0, b_1), ♂(b_0, b_2) and ♂(b_1, b_2), then the Boolean polynomials in b_0, b_1, b_2 form a Boolean algebra.

Proof. By 4.8, it suffices to show: If P is a Boolean polynomial such that $P(y_0, y_1, y_2) = 1$ in the Boolean sense, then $P(b_0, b_1, b_2) = 1$ in $\boldsymbol{B}_\phi$. Let γ be the formula of $\boldsymbol{P}_1^2(y)$ translating the polynomial $\boldsymbol{P}$ (the translation of $y_0 \vee y_1$ being $(y_0 f) y_1$ etc.); $P = 1$ being a Boolean identity, γ is an identity of the classical propositional calculus. Assume $\psi_i \in b_i (i = 0, 1, 2)$, $* = \langle\psi_0, \psi_1, \psi_2\rangle$ and ♂(b_0, b_1), ♂(b_0, b_2), ♂(b_1, b_2). We then have $f**$ and by M_4: $[f** \vdash \gamma*]$, i.e. $[f\phi \vdash \gamma*]$; by 8° of 10.1 therefore $\gamma* \simeq w$, i.e. $\gamma* \in 1$. The formula $\gamma*$ is an element of $P(b_0, b_1, b_2)$: The class of $(\phi_0 f)\phi_1$ is by definition $\{(\psi_0 f)f\} \vee \{\psi_1\}$ which is the same as $\{\psi_0\} \vee \{\psi_1\}$, i.e. $b_0 \vee b_1$. We therefore have $\gamma* \in 1$, $\gamma* \in P(b_0, b_1, P_2)$, i.e. $P(b_0, b_1, b_2) = 1$.

10.4 *There exists a sequence* $q \in B^N$ *such that for all formulas* ψ *of* Ω, $q \in D_\psi$ *and* $[\psi](q) = \{\psi\}$ (equivalence class of ψ).

Proof. q is defined as follows: If the variable x_n is an element of Ω, then $q(n) = \{x_n\}$; otherwise $q(n) = 0$. The theorem is then proved by induction with respect to the length of ψ. If ψ is a variable or f, it holds by definition. Assume therefore $\psi = \psi_1\psi_2$; then $\psi_1, \psi_2 \in \Omega (2°$ of 10.1) and $q \in D_{\psi_i}$, $[\psi_i](q) = \{\psi_i\} (i = 1, 2)$ by the hypothesis of the induction. In order to prove $q \in D\psi_1\psi_2$, we have to show: ♂$(\neg[\psi_1](q), [\psi_2](q))$. We have

$\neg[\psi_1](q) = \neg\{\psi_1\} = \{\psi_1 f\}$; $[\psi_2](q) = \{\psi_2\}$; we therefore have to show $[f\phi \vdash f(\psi_1 f)\psi_2]$, which follows immediately from S_{34}. Furthermore $[\psi_1\psi_2]\,(q) = \neg[\psi_1]\,(q) \vee [\psi_2]\,(q) = \neg\{\psi_1\} \vee \{\psi_2\} = \{\psi_1 f\} \vee \{\psi_2\} = \{((\psi_1 f)f)\psi_2\}$; by 5° of 10.1, $((\psi_1 f)f)\psi_2 \simeq \psi_1\psi_2$ and therefore $[\psi_1\psi_2](q) = \{\psi_1\psi_2\}$.

10.5 *If the formula ψ of Ω holds in the partial Boolean algebra $\boldsymbol{B}_\phi$, then $[f\phi \vdash \psi]$ is a derivable rule of $\boldsymbol{PP}_1$.*

PROOF. Let q be the sequence defined in 10.4; then $[\psi](q) = \{\psi\}$. If ψ holds in $\boldsymbol{B}_\phi$, then $[\psi](q) = 1$, i.e. $\{\psi\} = 1$, $\psi \simeq w$. By 8° of 10.1, $\psi \simeq w$ iff the rule $[f\phi \vdash \psi]$ is derivable.

By 1° of 10.1, ϕ is a formula of Ω. Therefore:

ϕ holds in the partial Boolean algebra $\boldsymbol{B}_\phi$ iff $[f \to \phi \vdash \phi]$ is a derivable rule of $\boldsymbol{PP}_1$.

REFERENCES

[1] A CHURCH, *Introduction to Mathematical Logic*, Volume 1, Princeton University Press, Princeton 1956.

[2] S. KOCHEN and E. P. SPECKER, Logical Structures Arising in Quantum Theory, to appear in the *Proc. of the Model Theory Symp.* held in Berkeley, June–July 1963.

[3] E. SPECKER, Die Logik nicht gleichzeitig entscheidbarer Aussagen, *Dialectica* **14** (1960), 239–246.

[4] M. WAJSBERG, Metalogische Beiträge II, *Wiadomości Matematyczne* **47** (1939), 119–139.

A SIMPLE METHOD FOR UNDECIDABILITY PROOFS AND SOME APPLICATIONS*

MICHAEL O. RABIN
The Hebrew University, Jerusalem, Israel

In the following note we present a new principle for establishing the undecidability of theories. Whereas the usual method for proving that a theory T_1 is undecidable employs some finitely axiomatizable and essentially undecidable theory T_0 (with the undecidability of T_1 being demonstrated by interpreting T_0 in T_1), our method can use any undecidable theory T. Roughly speaking the principle may be stated as follows: If T is an undecidable theory and T_1 is a theory such that by using appropriate formulas of T_1 to represent the universe of T and the non-logical constants of T, every model of T is obtained from some model of T_1, then T_1 is also undecidable.

It turns out that the semantical form of the principle is particularly convenient for applications. In this way one can get very quickly most of the undecidability results in the literature as well as several new results.

In Section 1 we state and prove the main theorems on which the method is based. Theorem 5 is an elegant generalization, due to D. Scott, of the principle. This theorem is expressed in syntactical terms.

By way of illustration of the principle, we apply it to give short new proofs of the known results concerning the undecidability of the theory of a symmetric irreflexive relation, theory of distributive atomistic lattices and the theory of groups (for which theory we actually have a stronger, new, result). We conclude with a proof that the theory of finite commutative rings is undecidable. This last result seems to be new and its proof by our method is very natural though somewhat harder.

I. The General Theorems and Some Examples

0. Notation and terminology.

Even though the general theorem is extendable to higher order calculi, we shall restrict our attention to the first-order logic case. We shall adopt the terminology of [3]. We also assume as known some of the

* Work on most of the material in Chapter I was done under NSF Grant G-14006 at the University of California, Berkeley. Subsequent work was supported by ONR Contract No. 62558-3882, NR 049-130, at the Hebrew University. A fuller discussion of the method and many further applications will appear in a forthcoming joint paper with D. Scott.

basic concepts of model theory such as *structure* (or *relational system*), *satisfaction* of a formula by elements of a structure, etc.

A *language* L will be an applied first-order predicate calculus with equality. We shall restrict ourselves to languages with a finite number of non-logical constants. A *theory* T based on L is a set of sentences of L which is closed under logical deduction in L. Thus, for example, when speaking about the (elementary) theory of groups we mean the following. L will be the language having the non-logical constant $\bullet$ to denote a binary operation and the individual constant $\mathbf{1}$ (to denote the unit element). T will be the set of all sentences true in all structures $G = \langle A, \cdot, i \rangle$ which are groups, where $\cdot$ is the group operation and $i \in A$ is the unit element. Alternatively, T is the set of all logical consequences in L of the usual axioms of group theory written in terms of $\bullet$ and $\mathbf{1}$.

A similar definition can be given for the theory of finite commutative rings. In this case, however, we have only the semantical definition since this theory will turn out to be non-axiomatizable.

Let T be a theory based on a language L. An *inessential extension* T' of T is obtained by adding to L a number of individual constants and forming the closure T' of T with respect to logical deduction in the augmented language. It is well known that T' is undecidable if and only if T is undecidable [3, p. 16].

1. General theorems. Let L_1 be a language and let L_2 be an inessential extension obtained from L_1 by addition of a finite number of individual constants $c_1, \cdots, c_n$. If M_1 is a structure of L_1 and $a_1, \cdots, a_n$ belong to the domain of M_1, then $\langle M_1, a_1, \cdots, a_n \rangle$ will denote the structure of L_2 obtained from M_1 by addition of the a_i as distinguished elements, where it is understood that c_i is interpreted as a_i, $1 \leqq i \leqq n$.

From now on L will denote a fixed language and we shall assume for simplicity that L has just one binary predicate constant $\boldsymbol{P}$.

DEFINITION. Let $D(x)$ and $F(x, y)$ be formulas of L_2 (all notations retain their previous meanings). The structure $M = M_1(D, F, a_1, \cdots, a_n)$ of L *induced by D and F* will have, by definition, the domain A consisting of all elements a satisfying $D(x)$ in $\langle M_1, a_1, \cdots, a_n \rangle$ and the binary relation R consisting of all pairs $\langle a, b \rangle$ of elements satisfying $F(x, y)$ in $\langle M_1, a_1, \cdots, a_n \rangle$ i.e., $M = \langle A, R \rangle$.

We can now state the theorem which is the basis of our method of undecidability proofs.

THEOREM 1. *Let T be an undecidable theory based on the language L*

and let T_1 *be a theory based on* L_1. *Assume that there exists an inessential extension* L_2 *of* L_1 *and formulas* $D(x)$ *and* $F(x, y)$ *of* L_2 *such that* (1) *for every model* N *of* T *there exists a model* M_1 *of* T_1 *and elements* $a_1, \cdots, a_n$ *of* M_1 *such that the induced structure* $M = M_1(D, F, a_1, \cdots, a_n)$ *is isomorphic to* N; (2) *for every model* M_1 *of* T_1 *and every* $a_1, \cdots, a_n$ *in* M_1, *the induced structure is a model of* T. *Under these conditions the theory* T_1 *is undecidable.*

PROOF. It will suffice to prove that the logical closure T_2 of T_1 in L_2 is undecidable (T_2 is an inessential extension of T_1).

Let σ be a sentence of L. Let $\sigma(D, F)$ be the sentence of L_2 obtained from σ by relativizing all quantifiers of σ to $D(x)$ (see [3, p. 24]) and replacing all occurrences of $P(u, v)$ in σ by $F(u, v)$ taking care at the same time, by changing bound variables of F, that the free variables, u and v, of the predicate $P(u, v)$, which is being replaced, do not become bound in $F(u, v)$.

It is clear that for a structure $M_2 = \langle M_1, a_1, \cdots, a_n \rangle$ of L_2, $\sigma(D, F)$ is true in M_2 if and only if σ is true in the induced structure $M_1(D, F, a_1, \cdots, a_n)$. Conditions (1) and (2) ensure that σ is a theorem of T if and only if $\sigma(D, F)$ is a theorem of T_2. This proves the theorem.

THEOREM 2. *If* T *is finitely axiomatizable then condition* (2) *in the previous theorem can be dropped and the conclusion retained.*

PROOF. Let α be an axiom for T. It is readily seen that $\alpha(D, F) \rightarrow \sigma(D, F)$ is a theorem of T_2, if and only if σ is a theorem of T.

Two further generalizations of Theorem 1 will be useful in applications.

Call a class K of models of T *characteristic* if for every σ which is not in T there exists a model $M \in K$ such that $\sim \sigma$ is true in M. An example of a characteristic class is the class of all countable models of a theory T.

THEOREM 3. *If* K *is a characteristic class of models of* T *then we can replace condition* (1) *of Theorem* 1 *by: For every* $N \in K$ *there exists a model* M_1 *of* T_1 *and elements* $a_1, \cdots, a_n$ *of* M_1 *such that the induced structure* $M_1(D, F, a_1, \cdots, a_n)$ *is isomorphic to* N. *Also, if* T *is finitely axiomatizable then* (2) *can be dropped.*

The proof is clear.

Up to now we constructed models of T by using as domains (definable) subsets of models of T_2. We shall now generalize further and allow the domains of the models of T to be sets of equivalence classes of elements of a model of T_2.

Let D be a set and E an equivalence relation on D. We shall denote by D/E the set of equivalence classes of elements of D with respect to E. Let furthermore R be a binary relation on D which is *invariant* with respect to E

(i.e. if $E(x, x_1)$ and $E(y, y_1)$, then $R(x, y)$ iff $R(x_1, y_1)$). We can define in a natural way a *quotient structure* $\langle D/E, \bar{R} \rangle$ where two classes $\bar{x}, \bar{y} \in D/E$ are in the relation $\bar{R}$ iff some representatives $x \in \bar{x}$, $y \in \bar{y}$, are in the relation R.

We retain the notations and terminology introduced in the paragraph preceding Theorem 1. Let $D(x)$, $E(x, y)$, and $F(x, y)$ be formulas of L_2. Let M_1 be a model of T_1 and $a_1, \cdots, a_n$ be elements of M_1. Denote by D the subset of M_1 defined by $D(x)$ in $\langle M_1, a_1, \cdots, a_n \rangle$ and by E and R the binary relations defined by $E(x, y)$ and $F(x, y)$ in this structure. If E is an equivalence relation on D and R is invariant with respect to E, then $M_1(D, E, F, a_1, \cdots, a_n) = \langle D/E, \bar{R} \rangle$ will be the *quotient structure induced by the formulas* $D(x)$, $E(x, y)$, $F(x, y)$.

THEOREM 4. *Let T, T_1, L_1, and L_2 be as in Theorem* 1. *Assume that there exist formulas $D(x)$, $E(x, y)$, $F(x, y)$ of L_2 such that* (1) *for every model M_1 of T_1 and elements $a_1, \cdots, a_n$ of M_1, the relation E corresponding to $E(x, y)$ is an equivalence relation on the set D corresponding to $D(x)$, and the relation R corresponding to $F(x, y)$ is invariant with respect to E*; (2) *for every model N of T there exist a model M_1 of T_1 and elements $a_1, \cdots, a_n \in M_1$ such that the induced quotient structure $M_1(D, E, F, a_1, \cdots, a_n)$ is isomorphic to N*; (3) *every induced quotient structure $M_1(D, E, F, a_1, \cdots, a_n)$ is a model of N. Under these conditions the theory T_1 is undecidable.*

The proof is similar to the proof of Theorem 1 and need not be given. We can again generalize Theorem 4 in an obvious way following the patterns of Theorems 2–3.

We shall now give Scott's generalization of Theorems 1–3 which applies, however, only to finitely axiomatizable theories.

THEOREM 5. *If T is finitely axiomatizable and undecidable and every finite extension of T is relatively weakly interpretable in an inessential extension of T_1, then T_1 is undecidable.*

The very simple proof of this principle will be omitted and a full discussion of this theorem and its relation to the classical method in undecidability proofs will be given elsewhere.

In actual applications the only way known to us of establishing the relation of Theorem 5 between theories T and T_1 is by the explicit semantical method of Theorems 1–3. These are therefore the results to be applied when proving a theory undecidable along these lines.

2. Symmetric relations.

THEOREM 6 [1]. *The theory T_1 of an irreflexive, symmetric, binary (i.s.b.) relation is undecidable.*

PROOF. (Due to D. Scott.) Let T be the theory of a single binary relation $\boldsymbol{P}$. We shall write $x\boldsymbol{P}y$ for $\boldsymbol{P}(x, y)$. Consider the formulas

$$D(x) \quad = \quad \exists t \forall s[s\boldsymbol{P}t \rightarrow s = x]$$

$$F(x, y) \quad = \quad \exists u \exists v \exists w[x\boldsymbol{P}u\boldsymbol{P}v\boldsymbol{P}x \wedge u\boldsymbol{P}w\boldsymbol{P}y \wedge v \neq w]\,.$$

For any structure $M_1 = \langle A, S \rangle$ where S is an i.s.b. relation, the induced structure $M_1(D, F)$ is a model of the theory of the general binary relation.

Let now $M = \langle B, R \rangle$ be a structure where $R \subseteq B \times B$. Add to B two elements $t_1(x)$, $t_2(x)$ for each $x \in B$ and three elements $u(x, y)$, $v(x, y)$, $w(x, y)$ for each $\langle x, y \rangle \in R$. Let A be the resulting set

$$A = B \cup \{t_1(x), t_2(x), u(x, y), v(x, y), w(x, y) \mid x \in B,\ \langle x, y \rangle \in R\}.$$

Define on S an i.s.b. relation S by

$$S_1 = \{\langle t_i, x \rangle, \langle x, u \rangle, \langle x, v \rangle, \langle v, u \rangle, \langle u, w \rangle, \langle w, y \rangle \mid i = 1, 2, x \in B, \langle x, y \rangle \in R\},$$

and $S = S_1 \cup \breve{S}_1$, where t abbreviates $t(x)$, u abbreviates $u(x, y)$, etc., and $\breve{S}_1$ is the converse of the relation S_1, so that S is indeed symmetric and irreflexive. The following diagram where $x, y \in B$, $\langle x, y \rangle \in R$, and every two points joined by a line are in the relation S, explains the construction of S.

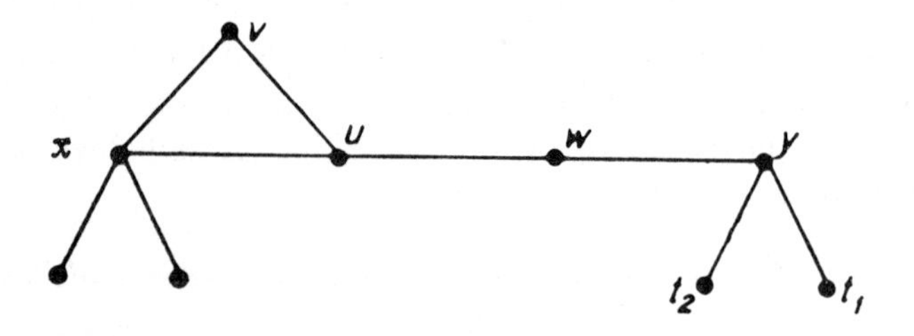

It is readily seen that in the structure $M_1 = \langle B, S \rangle$ an element x satisfies $D(x)$ if and only if $x \in A$. Elements x, y, satisfy $F(x, y)$ if and only if $\langle x, y \rangle \in R$. Thus the induced structure $M_1(D, F)$ is $\langle A, R \rangle$.

REMARK. If we start from the fact that the theory of a single binary *finite* relation is undecidable then the same proof will show that the theory of a finite i.s.b. relation is undecidable.

3. Theory of distributive lattices.

THEOREM 7 [2]. *The theory T_1 of atomistic distributive lattices is undecidable.*

PROOF. Let $A(x)$ be the formula $\forall u[u \leqq x \rightarrow u = \mathbf{0} \vee u = x] \wedge x \neq \mathbf{0}$. An element in a lattice satisfies $A(x)$ if and only if it is an *atom*. Let Ind(x) be the formula $\sim \exists u \exists v[u \neq \mathbf{0} \wedge v \neq \mathbf{0} \wedge u \cap v = \mathbf{0} \wedge u \cup v = x]$. If an element satisfies Ind(x) we shall say that it is *join-indecomposable*.We adjoin to the first-order language of lattice theory an individual constant $\boldsymbol{c}$ and define

$$D(x) \quad = \quad A(x) \wedge x \leqq \boldsymbol{c}$$

$$F(x, y) = \quad x \neq y \wedge \exists u[x \cup y = u \cap \boldsymbol{c} \wedge \operatorname{Ind}(u)].$$

We shall show that for every model $M = \langle C, S \rangle$ of the theory of an i.s.b. relation there exist a distributive atomistic lattice M_1 and an element $C \in M_1$ such that the structure $M_1(D, F, C) \approx M$.

Let B be an infinite set and for each $\langle x, y \rangle \in S$ let $B_{yx} = B_{xy} \subseteq B$ be an infinite set such that if $\{x, y\} \neq \{u, v\}$ then $B_{xy} \cap B_{uv} = \emptyset$.

Let M_1 be the closure under unions and intersections of the collection of subsets of $B \cup C$ which consists of (1) all finite subsets, (2) C, (3) all sets $\{x, y\} \cup B_{xy}$ where $\langle x, y \rangle \in S$.

Note that the only subsets of B_{xy} which are in M_1 are the finite sets. It follows that the $\{x, y\} \cup B_{xy}$ are join-indecomposable in M_1. These are, with the possible exception of C, the singeltons $\{x\}$, and $\emptyset$, the only indecomposable elements of M_1. Now, an element $a \in M_1$ satisfies $D(x)$ if and only if $a = \{x\}$ for $x \in C$. Elements $a, b \in M_1$ satisfy $F(x, y)$ if and only if $a = \{x\}$, $b = \{y\}$, and $\langle x, y \rangle \in S$.

4. Theory of Groups. In this section we prove a certain generalization of Tarski's theorem [3] to the effect that the theory of groups is undecidable. The same proof will yield Tarski's result.

THEOREM 8. *The theory of groups G which are the free product of two free groups with an amalgamated subgroup is undecidable.*

PROOF. We shall reproduce in groups of the above type the general denumerable binary relation and thus reach the conclusion by virtue of Theorem 3.

Let us denote by $\mathbf{1}$ the group unit element and let us consider the inessential extension of the language of group theory by the individual constants $\boldsymbol{a}, \boldsymbol{b}, \boldsymbol{c}_1, \boldsymbol{c}_2, \boldsymbol{d}_1, \boldsymbol{d}_2$.

Consider the formulas

$$D(x) \quad = \quad x\boldsymbol{a} = \boldsymbol{a}x \wedge x \neq \mathbf{1} \wedge \exists y[x = y^2],$$

$$F(x, y) \quad = \quad \exists u_1 \exists u_2 \exists v_1 \exists v_2[u_1 \neq \mathbf{1} \wedge u_2 \neq \mathbf{1} \wedge u_1 \boldsymbol{c}_1 = \boldsymbol{c}_1 u_1 \wedge$$

$$_2\boldsymbol{c}_2 = \boldsymbol{c}_2 u_2 \wedge v_1 \boldsymbol{d}_1 = \boldsymbol{d}_1 v_1 \wedge v_2 \boldsymbol{d}_2 = \boldsymbol{d}_2 v_2 \wedge u_1 x u_2 = v_1 y v_2].$$

For any group G and elements a, b, c_1, c_2, d_1, d_2 in G, the induced structure $G(D, F, a, b, c_1, c_2, d_1, d_2)$ is just a structure $M = \langle A, R \rangle$ where R is a binary relation on A.

Let $M = \langle A, R \rangle$ be a denumerable structure where R is a binary relation. We may assume that A is the set of all non-zero even integers. Let f be a pairing function from $A \times A$ into positive even integers. Consider the following presentation of a finitely generated group G

$$(a, b, c_1, c_2, d_1, d_2 : a^2 = b^2, c_1^k a^{2n} c_2^k = d_1^k b^{2m} d_2^k, \text{for} \langle 2n, 2m \rangle \in R, k = f(2n, 2m))$$

We shall need some algebraic properties of the group G. Let F_1 be the free group on the generators a, c_1, c_2, let $h_0 = a^2$, $h_i = c_i^{k_i} a^{2n_i} c_2^{k_i}, \cdots$ be an enumeration of the left hand sides of the relations of G, let H_1 be the subgroup of F_1 generated by $h_0, h_1, \cdots$. Similarly denote by F_2 the free group on the generators b, d_1, d_2, let $h_0', h_1', \cdots$ be an enumeration of the right hand sides of the relations of G, and let H_2 be the subgroup generated by these elements.

Consider a *reduced* product

$$p = h_{i_1}^{e_1} \cdots h_i^{e_n}, \; e_j = \pm 1, \tag{1}$$

of the generators of H_1. Note that if $i \neq j$ and $e, g = \pm 1$, then $k_i \neq k_j$ and in a product

$$h_i^e h_j^g = (c_1^{k_i} a^{2n_i} c_2^{k_i})^e (c_1^{k_j} a^{2n_j} c_2^{k_j})^g,$$

a^{2n_i} and a^{2n_j} will be separated by a non-unit product of c_1 and c_2. This implies that (a) $p \neq 1$ and the generators $h_0, h_1, \cdots$, of H_1 are free generators of this subgroup; (b) $p = c_1^m a^l c_2^k$, $m \neq 0$, implies that in (1), $n = 1$, $e_1 = 1$ and hence $p = h_{i_1}$.

Similar statements hold for F_2 and its subgroup H_2. The group G is therefore the free product of the free groups F_1 and F_2 with the amalgamation of subgroups H_1 and H_2 given by the relations $h_i = h_i'$, $0 \leqq i < \infty$. If an element $p = c_1^m a^n c_2^k$, with $m \neq 0$, equals an element $d_1^{m'} b^{n'} d_2^{k'}$ then $p \in F_1 \cap F_2$ in G, hence $p \in H_1$, hence, by (b), $p = h_i$ for some i, $0 < i < \infty$ and $d_1^{m'} b^{n'} d_2^{k'} = h_i'$.

It can be verified that if in the free product with amalgamated subgroups

$$G = (F_1 * F_2)_{H_1 = H_2}$$

an element $x \in F_1$ commutes with an element $y \notin F_1$ then x is conjugate in F_1 to an element $h \in H_1$, i.e. for some $t \in F_1$, $x = tht^{-1}$. Now $a \in F_1$ is not conjugate to any $h \in H_1$ because all conjugates of elements of H_1 contain an *even* number of factors a. Hence the only elements of G commuting with a are in F_1 and therefore of the form a^n. Similar results hold for c_1, c_2, d_1, d_2.

It follows from the previous two paragraphs that (1) the only elements $x \in G$ satisfying $D(x)$ in G are the even non-unit powers a^{2n} of a; (2) two elements $x, y \in G$ such $D(x)$ and $D(y)$ hold, satisfy $F(x, y)$ if and only if $x = a^{2n_i}$, $y = a^{2m_i} = b^{2m}$ where $\langle 2n_i, 2m_i \rangle \in R$. Thus the induced structure $G(D, F, a, b, c_1, c_2, d_1, d_2)$ is isomorphic to $\langle A, R \rangle$.

REMARK. The class of groups for which we established the undecidability result has, unlike the general case, a solvable word problem.

II. UNDECIDABILITY OF THE THEORY OF FINITE COMMUTATIVE RINGS

In the application of the general method to the case of commutative rings we shall use a class of special (finite) rings. We shall star: / defining and studying some algebraic properties of these rings.

Denote by Z_p the field of p elements (p is a prime) which is taken as the set of integers i, $0 \leqq i \leqq p-1$ with addition and multiplication mod p. Let $R_k \subset Z_p[x_1, \cdots, x_k]$ denote the ring of all polynomials $f(x_1, \cdots, x_k)$ with coefficients in Z_p such that $f(0, \cdots, 0) = 0$ (i.e. with constant term zero).

DEFINITION. Let n be a positive integer and let I_n be the ideal generated in R_k by all products $f_1 f_2 \cdots f_n$ of n polynomials. Define $R(k, n, p) = R_k / I_n$.

5. Algebraic properties.

It is readily seen that each element $f \in R(k, n, p)$ has a unique representation of the form

$$f = \Sigma a_{i_1 \ldots i_k} x_1^{i_1} \cdots x_k^{i_k} \tag{2}$$

where $0 \leqq a_{i_1 \ldots i_k} < p$ and $0 < \Sigma_1^k i_j < n$. In fact, $R(k, n, p)$ can be viewed as the set of all formal sums (2) with addition and multiplication performed subject to the identities

$$py = 0, \ y_1 y_2 \cdots y_n = 0. \tag{3}$$

In particular it follows that *$R(k, n, p)$ is finite.*

LEMMA 9. *Let R be a commutative ring satisfying the identities* (3) *and let $r_1, \cdots, r_k \in R$. There exists a unique homomorphism ϕ: $R(k,n,p) \rightarrow R$ such that $\phi(x_i) = r_i$, $1 \leqq i \leqq k$.*

PROOF. For $f \in R(k, n, p)$ as in (2), define $\phi(f) = \Sigma a_{i_1 \ldots i_k} r_1^{i_1} \cdots r_k^{i_k}$. Because of the uniqueness of the representation (2) of f, ϕ is a well defined mapping.

Since R satisfies the identities (3) it follows from the remark preceding the Lemma that ϕ is a homomorphism.

We shall establish for special elements of $R(k,n,p)$ certain propositions concerning divisibility and factors which somewhat resemble properties of unique factorization rings.

Call an element $l \in R(k,n,p)$ *linear* if $l = a_1x_1 + \cdots + a_kx_k$, $a_i \in Z_p$. We observe that *if $m < n$ then the product of any m non-zero linear elements is not* 0.

LEMMA 10. *Let $m < n$. If $q = l_1 \cdots l_m$ where $l_i \neq 0$ is linear, $1 \leqq i \leqq m$, and l is a linear element such that $q = lr$ for some $r \in R(k,n,p)$ then for some i, $1 \leqq i \leqq m$, and some $c \in Z_p$ we have $l = cl_i$.*

PROOF. Let $l = a_1x_1 + \cdots + a_kx_k$; we may assume $a_1 \neq 0$. Consider the element $l' = a_1^{-1}l = x_1 + b_2x_2 + \cdots + b_kx_k$ (a_1^{-1} is the inverse of a_1 in Z_p); for $r' = a_1r$ we have $q = l'r'$. Consider the homomorphism $\phi: R(k,n,p) \to R(k,n,p)$ such that $\phi(x_1) = -b_2x_2 - \cdots - b_kx_k$, $\phi(x_i) = x_i$, $2 \leqq i \leqq k$. We have

$$\phi(q) = \phi(l')\phi(r') = 0 = \phi(l_1) \cdots \phi(l_m).$$

Each $\phi(l_i)$ is a linear element. Since $m < n$, the observation preceding our Lemma implies that for some i, $1 \leqq i \leqq m$, $\phi(l_i) = 0$. This immediately implies that for some $c \in Z_p$, $l = cl_i$.

Define a weight function w on $R(k,n,p)$ by

$$w(f) = \min_{a_{i_1 \ldots i_k} \neq 0} (i_1 + \cdots + i_k)$$

where f is again as in (2).

DEFINITION. $x, y \in R(k,n,p)$ are called *equivalent* (notation: $x \equiv y$) if $x = l_1 + f$, $y = l_2 + g$ where l_1 and l_2 are linear, $2 \leqq w(f)$, $2 \leqq w(g)$, and for some $c \in Z_p$, $c \neq 0$, we have $l_1 = cl_2$.

LEMMA 11. *Let $q = l_1l_2 \cdots l_{n-1}$ where $l_i \neq 0$ is linear, $1 \leqq i \leqq n-1$. For $x \in R(k,n,p)$ there exists an r such that $n-2 \leqq w(r)$ and $xr = q$, if and only if for some i, $1 \leqq i \leqq n-1$, x is equivalent to l_i.*

PROOF. Write $x = l + f$ where $2 \leqq w(f)$. If x is equivalent to (say) l_1 then $l = cl_1$ for some $c \in Z_p$, $c \neq 0$. Put $r = c^{-1}l_2l_3 \cdots l_{n-1}$. We have $w(r) = n-2$ and $lr = l_1l_2 \cdots l_{n-1} + fc^{-1}l_2l_3 \cdots l_{n-1} = q$, since $2 \leqq w(f)$ implies $fl_2l_3 \cdots l_{n-1} = 0$.

Assume now that for some r we have $n-2 \leqq w(r)$ and $xr = q$. Now

$(l+f)r = lr + fr = lr = q$ (we have again $fr = 0$). It follows now from Lemma 10 that for some i, $1 \leqq i \leqq n-1$, and some $c \neq 0$ we have $l = cl_i$. Hence $x \equiv l_i$.

6. The ring corresponding to a relation.

Let S be an i.s.b. relation over the finite domain $\{1, \cdots, k\}$. Choose an integer n such that

$$\max(k, \bar{\bar{S}}) < n, \tag{4}$$

and let p be a prime. In the ring $R(k,n,p)$, denote by d and s elements such that

$$d = \prod_{1 \leqq i \leqq k} x_i^{n_i}, \quad \Sigma n_i = n - 1, \tag{5}$$

$$s = \prod_{\langle i,j \rangle \in S} (x_i + x_j)^{n_{ij}}, \quad \Sigma n_{ij} = n - 1. \tag{6}$$

The elements d, s, satisfy the condition on q in Lemma 11 so that we immediately have the following three propositions.

PROPOSITION 1. *The element* $x \in R(k,n,p)$ *satisfies*

$$\exists r[n - 2 \leqq w(r) \wedge xr = d] \tag{7}$$

if and only if $x \equiv x_i$ *for some* i, $1 \leqq i \leqq k$.

PROPOSITION 2. *For elements* u *and* v *such that* $u \equiv x_i$, $v \equiv x_j$, *we have* $u \equiv v$ *(i.e.* $i = j$*) if and only if* $2 \leqq w(u-v)$, *or* $x = u - v$ *satisfies* (7).

PROPOSITION 3. *There exist elements* $u \equiv x_i$ *and* $v \equiv x_j$ *such that*

$$\exists r[n - 2 \leqq w(r) \wedge (u+v)r = s]$$

if and only if $\langle i,j \rangle \in S$.

7. The undecidability result.

THEOREM 12. *The theory of finite commutative rings is undecidable.*

PROOF. We shall use the undecidable theory T of a finitary i.s.b. (irreflexive symmetric binary) relation S. Consider the inessential extension of the language of rings by the individual constants $\boldsymbol{d}$, $\boldsymbol{s}$, $\boldsymbol{c}$, $\boldsymbol{e}$. Define

$$D(x) = \exists r[rc = 0 \wedge rx = \boldsymbol{d}] \tag{8}$$

$$E(x,y) = D(x) \wedge D(y) \wedge [(x-y)\boldsymbol{e} = 0 \vee D(x-y)] \tag{9}$$

$$F(x,y) = D(x) \wedge D(y) \wedge \sim E(x,y) \wedge \exists u \exists v \exists r[E(x,u) \wedge E(y,v) \wedge rc = 0 \wedge (u+v)r = \boldsymbol{s}]. \tag{10}$$

For every finite (commutative) ring R and every assignment of values

$d, s, c, e \in R$ to the corresponding individual constants, the above formulas determine a set $D \subseteq R$ and binary relations $F, E \subseteq D \times D$ in the usual way (e.g., D is the set of all $x \in R$ satisfying $D(x)$ in $R(d,s,c,e)$, etc.). It is readily seen that E is an equivalence relation on D and that F induces an i.s.b. relation on the set D/E of equivalence classes of elements of D.

Our proof of undecidability will be completed by showing that every finitary i.s.b. relation S is obtained in the above manner. Assume that the domain of S is $\{1, \cdots, k\}$. Let p be a prime and let n satisfy (4). Choose $d, s \in R(k,n,p)$ as in (5) and (6). Finally let $c = x_1^2$, $e = x_1^{n-2}$. We see that for $r \in R(k,n,p)$, $rc = 0$ if and only if $n-2 \leqq w(r)$; $re = 0$ if and only if $2 \leqq w(r)$. Propositions 1–3 now imply that $x \in R(k,n,p)$ satisfies $D(x)$ if and only if $x \equiv x_i$ for some i, $1 \leqq i \leqq k$; i.e. the set D (here D, E and F are understood as in the previous paragraph) is the union of the equivalence classes of $x_1, \cdots, x_k$. Two elements x, $y \in D$ satisfy $E(x,y)$ if and only if $x \equiv y$; i.e. E is the relation $\equiv$ restricted to D. The elements x, $y \in D$ satisfy $F(x,y)$ if and only if for their representatives $x_j \equiv x$ and $x_i \equiv y$ we have $\langle i,j \rangle \in S$. Thus the relation F on the quotient set D/E is isomorphic to S so that every finitary i.s.b. relation is reproduced by the formulas (8), (9), (10), which completes the proof by Theorem 4.

We note that the rings $R(k,n,p)$ used in the previous proof do not have a unit element. However, the result also holds for rings with unit.

COROLLARY. *The theory of finite commutative rings with unit is undecidable.*

PROOF. Let U be the finite ring obtained from $R(k,n,p)$ by adjoining a unit without changing the characteristic p. The elements of U are of the form $x = c - f$ where $c \in Z_p$ and $f \in R(k,n,p)$. The original elements $f \in R(k,n,p)$ are nilpotent ($f^n = 0$) and hence divisors of zero. On the other hand, if $c \neq 0$ then

$$(c-f)(c^{n-1}+c^{n-2}f+ \cdots +f^{n-1}) = c^n - f^n = c^n \neq 0.$$

Since $c \in Z_p$, it is invertible in U so that x is not a divisor of zero. Thus the formula $R(x) = \exists y[y \neq 0 \wedge yx = 0]$ is satisfied in U just by the elements $f \in R(k,n,p)$. We can now use the formula $R(x)$ to relativize all quantifiers in the formulas (8), (9), (10) and define the general finitary relation S in rings with unit so that the result follows.

REFERENCES

[1] A. CHURCH and W. V. QUINE, Some theorems on definability and decidability, *J. Symb. Logic*, vol. **17** (1952), pp. 179–187.

[2] A. GRZEGORCZYK, Undecidability of some topological theories, *Fund. Math.*, vol. **38** (1951), pp. 137–152.

[3] A. TARSKI, A. MOSTOWSKI, and R. M. ROBINSON, *Undecidable Theories*, North-Holland Publishing Co., Amsterdam, 1953.

MACHINE CONFIGURATION AND WORD PROBLEMS OF GIVEN DEGREE OF UNSOLVABILITY*

J. C. SHEPHERDSON
The University, Bristol, England

1. Introduction. This is a variation on the earliest of all applications of the theory of mathematical machines to mathematical logic. For the halting problem for Turing machines was one of the first problems to be shown recursively unsolvable, by Turing himself in his original paper [1]; and a similar configuration problem was used by him to show the undecidability of the first order predicate calculus. Post's proof [2] of the unsolvability of the word problem for Thue systems also starts from a Turing machine configuration problem.

The sort of variation I consider is also by now fairly common and consists of replacing 'unsolvable' by 'of any given recursively enumerable degree of unsolvability'. My reason for presenting it in this symposium is that I think it provides an example of a mathematical result which is most easily established by thinking in terms of machines, programs and sub-routines as long as possible.

2. Thue systems. The mathematical result I refer to concerns the degrees of unsolvability of word problems of Thue systems. A Thue system T is simply a finitely presented associative system, i.e. one given by a finite alphabet of generators and a finite set of defining relations consisting of equations between pairs of words in this alphabet:

$T.$ $$A_1 = B_1, A_2 = B_2, \cdots, A_k = B_k$$

The *word problem* for T is that of deciding, for arbitrarily given words W_1, W_2 of the given alphabet, whether $W_1 = W_2$ in the associative structure defined by the above set of relations, i.e. whether W_1 can be sent into W_2 by means of a finite number of applications of the 'two-way productions':

T_2: $$PA_1Q \Leftrightarrow PB_1Q, \cdots, PA_kQ \Leftrightarrow PB_kQ$$

We shall write

$$W_1 \leftrightarrow W_2(T_2)$$

---

* Since limitations of space did not allow the inclusion of full statements and proofs of theorems I have given here only the text of my lecture; the full details will appear in a paper in the *Zeitschrift für Mathematische Logik und Grundlagen der Mathematik.*

when this is true, i.e. when there exist words $W_{11}, \cdots, W_{1n}$ such that

$$W_1 = W_{11} \Leftrightarrow W_{12} \Leftrightarrow W_{13} \Leftrightarrow \cdots \Leftrightarrow W_{1(n-1)} \Leftrightarrow W_{1n} = W_2$$

where $X \Leftrightarrow Y$ here means as above that, for some i ($i = 1, \cdots, k$) and words P, Q, $X = PA_iQ$ and $Y = PB_iQ$, or vice-versa.

Post [2] and Markov [3] independently showed in 1947 that there exist Thue systems for which this problem is recursively unsolvable. Turing [4] did the same for cancellation semigroups and ultimately Novikov [5] did it for groups. Boone [6] and also, apparently, Ceitin [7] have shown that there are Thue systems whose word problem is of any given recursively enumerable degree of unsolvability and recently Clapham [8] has shown this for groups. What I shall do here is outline a simple proof of Boone's result, in which most of the work is done with machines, which have a unique next move, leaving a minimum to be done with systems with non-unique and two-way productions. In support of my claim that this is a simple approach I shall show that one can also get results about the relation between the degrees of the *special word problems*,—of deciding for fixed W_0 but arbitrarily given W, whether $W = W_0$—and the degree of the overall or *general word problem* of the system, i.e. the problem of deciding for arbitrarily given W_1, W_2 whether $W_1 = W_2$. In the case of groups all the special word problems and the general one are recursively equivalent for

$$W_1 = W_2 \text{ iff } W_1W_2^{-1} = 1 \text{ iff } W_1W_2^{-1}W_0 = W_0,$$

so Clapham's example shows there exist associative systems in which all special word problems and the general word problem are of the same given r.e.d.u. (recursively enumerable degree of unsolvability). However this is obviously not a general phenomenon for associative systems; usually many of the special problems are solvable (e.g. if W_0 is shorter than any of the A's or B's). The best result one might hope to prove is that there exists a system with special problems ranging over any r.e. (recursive enumerable) set of r.e. degrees and general w.p. (world problem) of any r.e.d.u. greater than or equal to all the former. I don't know whether this is true but I can arrange for the special w.p.'s to *include* any r.e. set of r.e. degrees. More precisely, if $A(x, n)$ is any r.e. predicate, $\mathbf{d}_n$ = degree of $\lambda x A(x, n)$, $\mathbf{d}$ = degree of $\lambda x n A(x, n)$ then there exists a Thue system whose degrees of special w.p.s range over the set $\{\mathbf{d}_1, \mathbf{d}_2, \cdots\}$ together with the set of joins $\mathbf{d}_{i_1} \cup \mathbf{d}_{i_2} \cup \cdots \cup \mathbf{d}_{i_n}$ $(n = 0, 1, 2, \cdots)$ of a finite number of $\mathbf{d}_i$'s and whose general w.p. is of any r.e. degree $\geqq \mathbf{d}$.

Interesting special cases of this are: 1) $A(x, n)$ recursive. Here all special w.p.s are solvable: the general w.p. can be taken to be of anv r.e.d.u.

2) $A(x, n)$ a universal r.e. predicate (e.g. $(Ey)\, T_1(n, x, y)$). Here there are special w.p.s of all r.e.d.u. and the general w.p. is naturally of highest r.e.d.u. $\mathbf{0}'$.

3. Method of approach to Thue systems. As usual we try to establish a connection with corresponding problems for the one-way or semi-Thue system with productions:

$$T_1: \qquad PA_1Q \Rightarrow PB_1Q, \cdots, PA_kQ \Rightarrow PB_kQ.$$

The connection one would hope for is that

$$E: \qquad W_1 \leftrightarrow W_2(T_2) \text{ iff } (EW)(W_1 \to W(T_1) \,\&\, W_2 \to W(T_1))$$
$$\text{i.e. iff } W_1, W_2 \textit{ conflue} \text{ in } T_1.$$

(Here $W_1 \to W(T_1)$ means that W is obtainable from W_1 by a string of productions of T_1.)

Newman [9] considered this question for combinatorial systems in a more general setting and one of the simplest sufficient conditions he gave for E to hold was:

M: *If two moves of T_1 are applicable to a word they can be performed one after the other and give the same result in either order.*

For the particular case of Thue systems this is certainly true provided:

$$O: \qquad A_1, \cdots, A_n \textit{ are non overlapping.}$$

It is easy to set up the productions of a semi-Thue system T_1 so that they imitate, on words of a special form, the moves of a Turing Machine. But it is very tedious to arrange that condition M is satisfied, for if a word contains *two* letters corresponding to heads of the Turing machine they may 'compete' for letters lying between them. This suggests it would be easier to use a form of machine where action takes place only at the ends of a tape, not in the middle. It turns out that if you use a FIFO store, which has been shown to be a universal machine [10], you can easily get the stronger condition O satisfied. The moves of this machine are of the types:

Print 0 *on end of tape.*
Print 1 *on end of tape.*
Scan and delete first letter of tape: *if* 0, *proceed to next instruction*; *if* 1, *go to instruction* k.

This gives a fairly quick reduction of the general w.p. of the Thue system to the *confluence problem* of such a machine M:

Do C_1, C_2 *conflue in* M? i.e. *does there exist* C *such that*

$$C_1 \to C(M),\ C_2 \to C(M)?$$

Here C, C_1, C_2 stand for (complete) configurations of M, and $C_1 \rightarrow C(M)$ means that configuration C_1 leads to C in M, i.e. that when started in C_1, M passes through C at some later time. Special w.p.s of the Thue system correspond to joins of a finite number of special confluence problems of M, i.e. problems of the form 'do C, C_0 conflue in M?' where C_0 is fixed.

4. Machine configuration problems. So there are at least two machine configuration problems of interest, namely the halting problem for Turing machines—because it is the original one and one of the simplest to arise—and the confluence problem for FIFO stores—because that provides the easiest way of getting results on word problems of Thue systems. By configuration, here, I mean complete configuration; the combination of internal state (or number of program line), position of reading/writing head, complete state of marking of the tape, which together uniquely determine the next move. This suggests it would be profitable to develop a general method of attack which will enable us to deal with a variety of configuration problems for a variety of machines. We therefore start with a fairly general definition, regarding a *machine* M simply as any combination of a denumerable set Γ_M (called the set of configurations of M) and a singulary operation (immediate successor, next move) defined on a subset of Γ_M. We write C^1 for the immediate successor of C; we also write $C \Rightarrow C^1(M)$. We define $C^0 = C$, $C^{n+1} = (C^n)^1$ and as above we use $\rightarrow$ for the transitive closure of $\Rightarrow$, i.e. we write $C \rightarrow C_1(M)$ (and say C leads to C_1 or C_1 is a successor of C) when $C_1 = C^n$ for some $n \geqq 0$. If C^1 is not defined we say C is a *terminal configuration;* if, for some n, C^n is terminal we say C leads to a halt or, in short, C halts. The *halting problem* of such a machine M is that of determining for arbitrarily given C in Γ_M whether C leads to a halt. This does not become a definite problem until we have, in addition to M, a particular method of giving configurations, i.e. a particular one-one mapping of the set Γ_M of configurations of M onto the set of natural numbers or, more generally, into a set of words on a finite alphabet. This is quite separate from M and there are of course infinitely many recursively inequivalent such mappings, giving rise to halting problems of infinitely many different degrees of unsolvability. But for the particular kinds of machine we consider there is in each case a natural mapping or several recursively equivalent ones, which we shall indicate in the usual way by showing how we represent configurations by combinations of tuples of integers and words on some finite alphabet and assuming one of the standard natural mappings of these objects onto the natural numbers is used. For our general theorems about machines we need to assume only two properties of this mapping, which are obviously satisfied in all our applications · namely that a) *the set of* *terminal configurations is recursive*, b) *the relation* $C_1 \Rightarrow C_2$ *is recursive*. In other words you can tell whether there exists a next move and, if so, find it. There are all sorts of problems about configurations which one might consider and which have been used in applications, e.g. does C lead to a cycle? or to a C_1 of a certain form? (e.g. one indicating the printing of a certain letter). The general method of attack I shall use would be applicable to many of these problems but I shall confine myself to the simplest ones which seem to be the *halting* and *confluence* problems already mentioned and also the *derivability* problem: does $C_1 \rightarrow C_2(M)$? In all of these we consider fixed M. We are not interested in the single problem you get when M also varies over a class of machines (e.g. Turing machines)—a problem which is usually of the highest r.e.d.u. $\mathbf{0}'$—but in the range of d.u.s of these problems for fixed M as M varies over such a class. However, as already mentioned, we are interested in the further specialisation obtained in the case of the last two problems by fixing one of C_1, C_2.

5. Large scale machines. Our plan is first to produce large scale machines with configuration problems of given r.e.d.u. and then to reduce these systematically to Turing and other simple forms of machine. The large scale machines will have complicated basic moves and the reductions will mainly be by the usual procedure of expansion of these by subroutines of simpler instructions.

We start with the halting problem. As always in attempting to construct problems of a given d.u. there is no difficulty in ensuring that the halting problem is of degree $\geqq \mathbf{a}$; viz: take an r.e. predicate $A(x)$ of degree $\mathbf{a}$ and take ϕ recursive so that

$$A(x) \text{ iff } (Ey)(\phi(x, y) = 0)$$

and construct a machine for computing the function:

$$f(x) = \mu y(\phi(x, y) = 0)$$

e.g. one with three registers holding numbers x, y, z and whose program of instructions is:

1. Put $y = 0$
2. Put $z = \phi(x, y)$
3. If $z = 0$ go to 5
4. Add 1 to y; go to 2
5. Stop

This can be thought of as a machine in the above sense; e.g. if configurations are 'given' by quadruples (i, x, y, z) the first move is

$$(1, x, y, z) \Rightarrow (2, x, 0, z).$$

Clearly the set of terminal configurations—those of the form $(5, x, y, z)$—is recursive; so is the next move relation. The halting problem is of degree at least **a** for

$$(1, x_0, y_0, z_0) \text{ halts iff } (Ey)(\phi(x_0, y) = 0) \text{ i.e. iff } A(x_0).$$

But we have other types of configuration to consider, e.g.

$$(2, x_0, y_0, z_0) \text{ halts iff } (Ey)(y \geqq y_0 \ \& \ \phi(x_0, y) = 0),$$

and this predicate is not in general reducible to $A(x_0)$. Indeed even if $f(x)$ is defined for all x, so that $A(x)$ is decidable, this may be of any r.e.d.u. Davis [11] has shown how to avoid this difficulty and produce a machine which in this case always halts. He showed that *every recursive function is strongly computable by a Turing machine, i.e. by one which always halts, whatever configuration it is started in.* Essentially what we need is to extend this to: *every partial recursive function is computable by a Turing machine whose halting problem is equivalent to the 'definition problem' of* f (i.e. *of deciding for arbitrary* x *whether* $f(x)$ *is defined*).

At present of course we are not concerned with doing this by a Turing machine but by a large scale machine and we are not interested in whether the machine computes f but only in the degree of its halting problem. Davis's trick is to include in the loop not merely the evaluation of $\phi(x, y)$ for one value of y but for all earlier values as well. The following machine $M_1(\phi)$ does this:

1. Put $y = \mu t \leqq z \ (\phi(x, t) = 0)$; if $\phi(x, y) \neq 0$ add 1 to z and go to 1
2. Stop

This is easily seen as above to be a machine in our sense. Its halting problem is of the same degree as $A(x)$, for $(2, x, y, z)$ is terminal, so always halts; and $(1, x, y, z)$ halts iff $(Ey)(\phi(x, y) = 0)$, i.e. iff $A(x)$. The derivability problem of $M_1(\phi)$ is however solvable, for z increases by 1 every time the loop 1—1 is traversed and is otherwise unaltered, so that

$$(1, x_1, y_1, z_1) \rightarrow (1, x_2, y_2, z_2)$$

only if $z_1 \leqq z_2$ and it $\rightarrow$ in $z_2 - z_1$ steps which is decidable. Similarly the case $(1, x_1, y_1, z_1) \rightarrow (2, x_2, y_2, z_2)$ can be decided; the remaining cases are trivial. The confluence problem of $M_1(\phi)$ is also seen to be solvable; e.g. $(1, x_1, y_1, z_1)$, $(1, x_2, y_2, z_2)$ conflue only if $x_1 = x_2$ (since x is unaltered in the program) and will then conflue in $(1, x_1, z, z - 1)$ where $z = \max(z_1, z_2)$ unless one of them has already led to line 2; this happens only if

$(Et) \leqq z\ (\phi(x, t) = 0)$ and then *both* lead to terminal forms which can be checked for identity.

It is very easy to work with these large scale machines; if we alter line 2 to

2. Put $y = z = 0$ and go to 1

we get a machine $M_2(\phi)$ with a solvable halting problem—it never halts —, a derivability problem of the same degree as $A(x)$ ($(1, x, 0, 1) \to (1, x, 0,0)$ iff $A(x)$, since z can be changed only by going through line 2), and a solvable confluence problem. Similarly, by adding a 4th register to hold a number u and changing line 2 to

2. Put $u = 0$, go to 2

we get a machine $M_3(\phi)$ with a *confluence* problem of the same degree as $A(x)$, the other two problems solvable. Now take any 3 r.e. predicates $A_1(x)$, $A_2(x)$, $A_3(x)$, with corresponding primitive recursive functions ϕ_1, ϕ_2, ϕ_3, and take a machine with 4 registers holding numbers x, y, z, u and a 6 line program consisting of program $M_1(\phi_1)$ followed by $M_2(\phi_2)$ followed by $M_3(\phi_3)$, viz:

1. Put $y = \mu t \leqq z(\phi_1(x, t) = 0)$; if $\phi_1(x, y) \neq 0$ add 1 to z and go to 1
2. Stop
3. Put $y = \mu t \leqq z(\phi_2(x, t) = 0)$; if $\phi_2(x, y) \neq 0$ add 1 to z and go to 3
4. Put $y = z = 0$ and go to 3
5. Put $y = \mu t \leqq z(\phi_3(x, t) = 0)$; if $\phi_3(x, y) \neq 0$ add 1 to z and go to 5
6. Put $u = 0$ and go to 6

This gives a *machine whose halting, derivability and confluence problems are respectively of the same degrees as* $A_1(x)$, $A_2(x)$, $A_3(x)$.

Similarly, for any r.e. predicate $A(x, n)$ one may construct a machine M which for each n has a configuration C_n which is a recursive function of n such that the predicates $A(x, n)$, $Cf(C, C_n)$ (C, C_n *conflue in* M) are recursive in each other recursively in n. Furthermore the degree of each special confluence predicate $Cf(C, C_0)$ is either $\mathbf{0}$ or is the degree of one of $A(x, 0)$, $A(x, 1), \cdots$. In addition the halting, derivability and confluence problems of M can be chosen to be of any 3 r.e. degrees $\mathbf{a}_1, \mathbf{a}_2, \mathbf{a}_3$ such that $\mathbf{a}_3 \geqq$ degree $A(x, n)$. This is the large scale machine from which one starts in order to get the above mentioned results on Thue systems. Similarly one can produce a machine which satisfies the conditions of the generalization of Davis's result mentioned above.

6. Reductions to simpler machines. The next problem is to reduce these large scale machines to Turing machines, FIFO stores etc. This is done by replacing their complicated single instructions by subroutines of simpler ones. For example the first step in the reduction of $M_1(\phi)$ would be to $M_1'(\phi)$:

1. Introduce u, v registers; put $u = z, y = 0$
2. Put $v = \phi(x, y)$; if $v = 0$ go to 5
3. Add 1 to y; if $u \neq 0$ subtract 1 from u and go to 2
4. Add 1 to z, remove u, v registers, subtract 1 from y and go to 1
5. Remove u, v registers
6. Stop

If we consider only lines 1 and 6 of $M_1'(\phi)$ and follow the intermediate lines until we get back to 1 or 6 again we have a machine $M_1''(\phi)$ which is a 'restriction' of $M_1'(\phi)$ and is virtually the same as $M_1(\phi)$. What we need is a general theorem to the effect that this procedure of restriction doesn't alter the degrees of the configuration problems we are interested in. It turns out (for further details see the Appendix) that this goes through alright provided that the subroutines involved are 'strong', i.e. lead back to the main program after a finite number of steps. This is usually a simple matter to check; indeed it is only necessary to verify what happens when you start on a program line to which a backward jump is made, for a sequence of instructions cannot fail to lead to a halt unless it goes through such a line.

In this way one gets down very easily to what Sturgis and I [10] have called a LRM (Limited Register Machine). This has at any time a finite number N $(= 1, 2, 3, \cdots)$ of registers each of which stores a natural number. The content of the nth register, i.e. the number stored in it, will be denoted by $\langle n \rangle$. The basic instructions are

$P_N(n)$: Add 1 to $\langle n \rangle$
$SD_N(n)[k]$: If $\langle n \rangle \neq 0$ subtract 1 from it and go to k
$N \to N + 1$: Bring in new (empty register) register number $N + 1$
$N \to N - 1$: Remove register N

The next step, also a straightforward one, is to a machine with a single register which holds the contents of these N registers in the form of a single word W:

$$W = 1^{<1>} 0 \, 1^{<2>} 0 \cdots 1^{<N>}$$

and has instructions of the form:

$P_N^{(0)}$: Print 0 on the end of W
$P_N^{(1)}$: Print 1 on the end of W
$SD_N[k]$: Scan and delete the first letter of W; if it is 1 go to k

where N is one more than the number of 0's in W and a pair (i, W) is considered to be a configuration only when W is a word on 0, 1 with $N_i - 1$ zeros in it, where N_i is the subscript of line i of the program of the machine in question. This is not quite the FIFO store referred to above whose instructions were independent of N but it is near enough (see Appendix) for the application to Thue systems to go through easily; one simply has to add a marker which goes through the word checking on the number of 0's in it before it allows any move to take place.

But in attempting reductions to even simpler forms of machines we begin to encounter other difficulties. This is only to be expected. We saw at the beginning that it was easy to construct machines whose configuration problems were of the right degree if you only consider 'intended' configurations, i.e. those which correspond to starting on the first line of the program with properly coded numerical data on the tape. By our introduction of large scale machines and their subsequent 'sub-routine' reduction to simpler ones we have shown how to ensure that the introduction of configurations corresponding to intermediate lines does not change the degrees of the problems we are interested in. But now we have to face the other problem and ensure that the machine's behaviour on non-numerical or 'nonsense' tapes does not change the degrees. This is not something one can hope to prove very general theorems about, for different types of basic machine do yield different results about the solvability of configuration problems. For example Rabin and Wang [12] have shown that for the non-erasing Turing machines the derivability problem is (uniformly) solvable. So in reducing the single-register programs above to more basic machines one must expect to have to consider each type of basic machine separately. This turns out to be quite easy for Turing machines; all the above mentioned results about the existence of machines with configuration problems of given degrees hold if the word 'Turing' is inserted before 'machine'. For the non-erasing Turing machines introduced by Wang [13] this cannot be true since the derivability problem is solvable. I haven't investigated confluence problems for them but as far as halting problems go the above results hold, i.e. for each r.e. degree **a** there is such a machine whose halting problem is of degree **a**; every recursive function is computable by such a machine which always halts; every partial recursive function f by one whose halting problem is of the same degree as the definition problem of f. Ironically, for the FIFO store itself which (see section 3) started the investigation, I was unable to prove any result, even the strong computability of all recursive functions. However for the slightly modified form of the last paragraph all the results hold as for the Turing machines.

I haven't space to go into the details of this last stage of the reduction. Like the first stage it calls for some modification of the usual programs.

This will be obvious if you think of the usual basic Turing machine programs for moving words about, moving the head to a particular place etc. Most of these will fail to stop on tapes which are not of the expected form, e.g. the head may go off to the right forever looking for a 1 which is not there. The rough idea is to modify these programs so that on 'nonsense' or abnormal tapes (those not of the form .... (head) $01^{x_1} 01^{x_2} \cdots 01^{x_N} 00 \cdots$ with the correct N) the program either leads to a halt or ultimately sends these into normal tapes.

In conclusion it is worth pointing out that a given Turing machine can be regarded in two distinct ways as a machine in our sense. If we think of the tape as fixed in space with the head moving along it, it is natural to think of the tape as having a fixed square recognisable by us (but not by the head) as the origin or middle square, and to consider two tape configurations to be identical only if the same squares are marked and the head is in the same position on both. The natural description of configurations in this sense is by means of a word W obtained by taking the smallest portion of the tape which includes all marked squares, the square under scan and the middle square, and writing out in order the symbols on these squares, prefixing the middle square and the scanned square by symbols m and q_i where i is the number of the program line (e.g. $10m\,01\,q_1 01$). This gives a machine we call a TM_1. However if we think of the tape flowing past a fixed head it is natural to think of the tape as uniform and tape squares to be distinguishable only by their position relative to the head. This amounts to identifying tape configurations of the TM_1 which are translations of each other. This, the more usual way of thinking of Turing machine configurations, gives what we call a TM_2; its configurations are described in the same way as those of a TM_1 except that all reference to the middle square and m is deleted (e.g. tape configurations $10m01q_101, m001001q_101$ of the TM_1 are considered to be the same configuration $1001q_101$ of the TM_2). The TM_1, TM_2 corresponding to a given Turing machine are different machines with different configuration problems. In proving the above results one can produce programs which simultaneously give the right results for both types but it is by no means generally true that corresponding TM_1, TM_2 have the same degrees for a given configuration problem. This is actually the case for the halting problem but e.g. the derivability problem for a TM_1 can be solvable and that for the corresponding TM_2 unsolvable.

7. Appendix. The general reduction results referred to in section 5 are:

DEFINITION A. A representation of a machine M_b by a machine M_a is a recursive map ϕ of Γ_{M_a} into Γ_{M_b} together with a recursive map

ϕ^{-1} of Γ_{M_b} into Γ_{M_a} which is inverse to ϕ^{-1} in the sense that $\phi(\phi^{-1}(C)) = C$ for all C in Γ_{M_b} and such that for all C, C_1, C_2:

(i) If $C_1 \Rightarrow C_2(M_a)$ then $\phi(C_1) \rightarrow \phi(C_2)\,(M_b)$

(ii) If $C_1 \Rightarrow C_2(M_b)$ then $\phi^{-1}(C_1) \rightarrow \phi^{-1}(C_2)\,(M_a)$

(iii) If $\phi\,(C)$ is terminal (M_b) then $C \rightarrow \text{halt}(M_a)$; if not then there exists $n > 0$ such that $\phi(C^n) = \phi(C)^1$

(iv) For C_1, C_2 of M_a such that $\phi(C_1) = \phi(C_2)$ the derivability problem $C_1 \rightarrow C_2(M_a)$ is reducible to the derivability problem of M_b, and the confluence problem $Cf(C_1, C_2)$ is solvable.

THEOREM A. *If there is a representation of M_b by M_a then the halting, derivability and confluence problems of M_a, M_b are equivalent and M_a always halts iff M_b does. Also the special predicates*

$$Cf\,(C, C_0)(M_b),\; Cf(C, \phi^{-1}(C_0))\,(M_a)$$

are recursive in each other recursively in C_0.

The usual way of obtaining a representation will be by restricting a given program to those lines which are terminal or to which a backward jump is made. The general definition and theorem which covers this is:

DEFINITION B. If N is a recursive set of configurations of a machine M such that every configuration of M leads to a member of N then the restriction $M|N$ is a machine defined thus: $\Gamma_{M|N} = N$, $C_1 \Rightarrow C_2\ (M|N)$ iff C_2 is the first element of N in the sequence $C_1, C_1^2, \cdots$.

THEOREM B. *A restriction $(M|N)$ has halting, derivability and confluence problems of the same degree as those of M. It always halts iff M does. There is a recursive function ϕ^{-1} such that for all C_0 the predicates $Cf(C, C_0)(M|N)$, $Cf(C, \phi^{-1}(C_0))(M)$ are recursive in each other recursively in C_0. The sets of degrees of special confluence predicates of $M|N$ and M coincide.*

The transition from a program P of instructions of the form $P_N^{(0)}, P_N^{(1)}$ —print 0, 1 on the end of W, $SD_N[k, l]$—scan and delete the first letter of W; if it is 0 or if $W = \Lambda$ go to k; if it is 1 go to l (the further reduction to $SD_N[l]$ above is not worth making here) — to a semi-Thue system $T_1(P)$ whose confluence problems are of the same degree as the corresponding ones for P is made as follows:

If P has s lines and I_i is the ith instruction of P and N_i the subscript of I_i then the alphabet of $T_1(P)$ consists of the letters $0, 1, \bar{0}, \bar{1}, b, e, q_i, \bar{q}_i$ (for $i = 1, \cdots, s$), $q_{i,j}^{(r)}$ (for $i = 1, \cdots, s-1$; $j = 1, \cdots, s$; $r = 1, \cdots, N_i$). A configuration (i, W) of P will be represented by a word bq_iWe of T_1 and the ith move of P will correspond to a sequence of moves of T_1 in which the symbol q_i carries out the required operation (if any) at the beginning of W, and then, in the form $q_{i,j}^{(r)}$ passes through the word registering the next program line j and the number $r-1$ of the zeroes it has already passed, changing 0 into $\bar{0}$, $\bar{1}$ into 1 as it goes; if it arrives at the end e with $r = N_i$, i.e. having passed the correct number of zeroes, it carries out the required operation (if any) and changes itself into $\bar{q}_j$. This then goes back to the beginning changing, $\bar{0},\bar{1}$ back into $0,1$ and when it finally gets back to b it changes into q_j to complete the operation. This is accomplished by the following productions ($i = 1, \cdots, s-1$ for $1-3$; $i = 1, \cdots, s$ for 4,5);

$$
\begin{array}{llcll}
1\ \mathrm{a,b} & bq_i & \Rightarrow & bq^{(1)}_{i,i+1} & \text{if } I_i = P^{(0)}_{N_i} \text{ or } P^{(1)}_{N_i} \\
1\ \mathrm{c.1} & bq_i 0 & \Rightarrow & bq^{(2)}_{\imath,j} & \\
\quad 2 & bq_i e & \Rightarrow & bq^{(1)}_{i,j} e & \text{if } I_i = SD_{N_i}[j,k] \\
\quad 3. & bq_i 1 & \Rightarrow & bq^{(1)}_{i,k} & \\
2.1 & q^{(r)}_{i,j} 0 & \Rightarrow & \bar{0} q^{(r+1)}_{i,j} & (r = 1, \cdots, N_i - 1) \\
2.2 & q^{(r)}_{i,j} 1 & \Rightarrow & \bar{1} q^{(r)}_{i,j} & (r = 1, \cdots, N_i) \\
3.\mathrm{a} & q^{(N^i)}_{i,i+1} e & \Rightarrow & \bar{q}_{i+1} 0\, e & \text{if } I_i = P^{(0)}_{N} \\
\quad \mathrm{b} & q^{(N_i)}_{i,i+1} e & \Rightarrow & \bar{q}_{\imath+1} 1 e & \text{if } I_i = P^{(1)}_{N_i} \\
\quad \mathrm{c} & q^{(N_i)}_{i,j} e & \Rightarrow & \bar{q}_j e & \text{if } I_i = SD_{N_i}[k,1] \text{ and } j = k \text{ or } 1 \\
4. & \bar{x}\bar{q}_i & \Rightarrow & \bar{q}_i x & (x = 0,1) \\
5. & b\bar{q}_i & \Rightarrow & bq_i &
\end{array}
$$

Since the left hand sides of the above productions are non-overlapping we can as above proceed from the confluence problem for $T_1(P)$ to the word problem for the corresponding two-way Thue System.

REFERENCES

[1] TURING, A. M., On computable numbers, with an application to the Entscheidungsproblem, *Proc. London Math. Soc.*, ser. 2, **42** (1936–7), 230–265. A correction, *ibid.* **43** (1937), 544–546.

[2] POST, E. L., Recursive unsolvability of a problem of Thue, *Journal of Symbolic Logic* **12** (1947), 1–11.

[3] MARKOV, A. A., On the impossibility of certain algorithms in the theory of associative systems (Russian), *Doklady AN SSSR* **55** (1947), 587–590.

[4] TURING, A. M., The word problem in semigroups with cancellation, *Annals of Math.* **52** (1950), 491–505.

[5] NOVIKOV, P. S., On the algorithmic unsolvability of the word problem in group theory (Russian), *AN SSSR Mat. Inst. Trudy.* No. 44, Moscow 1955.

[6] BOONE, W. W., Partial results regarding word problems and recursively enumerable degrees of unsolvability, *Bull. Amer. Math. Soc.* **68** (1962), 616–623.

[7] CEITIN, Verbal reference by P. S. Novikov in half-hour lecture at International Congress of Mathematicians, Stockholm, 1962.

[8] CLAPHAM, C. R. J., Finitely presented groups with word problems of arbitrary degrees of insolubility, *Proc. London Math. Soc.* **56** (1964), 633–676.

[9] NEWMAN, M. H. A., On theories with a combinatorial definition of equivalence, *Annals of Math.* **43** (1942), 223–243.

[10] SHEPHERDSON, J. C. and STURGIS, H. E., The computability of partial recursive functions, *J. Assoc. Comp. Mach.* **10** (1963), 217–255.

[11] DAVIS, M., A note on universal Turing machines, *Automata Studies*, Princeton, 1956, pp. 172–175.

[12] WANG, H. and RABIN, M., Words in the history of a Turing Machine with a fixed input. *J. Assoc. Comp. Mach.* **10** (1963), 526–527.

[13] WANG, H., A variant to Turing's theory of computing machines, *J. Assoc. Comp. Mach.* **10** (1963), 217–255.

THE LÖWENHEIM-SKOLEM THEOREM*

R. L. VAUGHT
University of California, Berkeley, California, U.S.A.

In its simplest form, the Löwenheim-Skolem Theorem states that:

1) *If a countable set Σ of E-sentences has a model $\mathfrak{A}$ of infinite power then Σ has a model $\mathfrak{B}$ of power ω* (where by E we mean the elementary, or first-order, language).

The earliest proofs of (1) showed that, moreover, $\mathfrak{B}$ can be taken to be a subsystem of $\mathfrak{A}$. A later proof by Skolem and the familiar proof using Gödel's completeness theorem (or its proof) do not yield a subsystem $\mathfrak{B}$, but can be directly generalized to show that in (1) 'ω' can be replaced by 'κ', denoting an arbitrary infinite cardinal. It is customary to call these two different improvements of (1) the Downward and the Upward Löwenheim-Skolem Theorems, respectively.[1]

The proof of the Downward Löwenheim-Skolem Theorem is very straightforward; indeed, its essence is the familiar argument which shows that countably many elements of, say, a group generate a countable subgroup. Using basically the same argument, various authors have extended the Downward Theorem, in suitable form, to the case where Σ has arbitrary power and to a number of languages richer than E (see Tarski-Vaught [1], Mostowski [1], Scott-Tarski [1], Hanf [1], Karp [1]). Despite (or perhaps because of) the naturalness and simplicity of their proofs, the Downward Theorem and its generalizations have many extremely important applications. For example: the Downward Theorem is used both in the work of Gödel and in that of Cohen concerning the Continuum Hypothesis; and the Downward Theorem for the weak second-order language was shown by Scott-Tarski [1] to yield a precise result embodying almost exactly the so-called "Lefschetz Principle" in algebraic geometry.

In the last few years a number of other generalizations of the Löwenheim-Skolem Theorem have been established by various authors, and still

* Work supported by a National Science Foundation fellowship at the University of California at Los Angeles 1963–4, and by a National Science Foundation grant.

1 The Upward Theorem was first proved by Tarski. For a history of the Löwenheim-Skolem Theorem and references, see Tarski-Vaught [1]. (References to the bibliography will be made in this style or in the shorter style [TV1].)

others conjectured. These may be called generalizations of the Upward Theorem, just in order to indicate that they are not obtained by using only the basic downward argument. The purpose of this talk will be to summarize and discuss these proved or proposed generalizations of the Upward Löwenheim-Skolem Theorem. With the exception of a result of Helling (3.4 below), the principal results to be described are already in various papers either in print or soon to be in print. What we shall do here is summarize in one place these results and the many related open problems; in addition, we make a number of small remarks bearing upon these results or problems.

The generalizations of the Löwenheim-Skolem Theorem to be considered fall into three different forms, which will be considered in Sections 1, 2, and 3, respectively.

1. Problems concerning two cardinals

The letters 'α' and 'β' always denote ordinals; 'δ' a limit ordinal; 'n' a natural number; 'κ', 'λ', and 'μ' infinite cardinals. $2^{\kappa,\alpha}$ is 2 to κ α times, i.e., $2^{\kappa,0} = \kappa$, $2^{\kappa,\alpha+1} = 2^{2^{\kappa,\alpha}}$, and $2^{\kappa,\delta} = \Sigma\{2^{\kappa,\alpha} : \alpha < \delta\}$. $\beth_\alpha = 2^{\omega,\alpha}$.

A relational structure $\mathfrak{A} = \langle A, U, R_\beta \rangle_{\beta<\alpha}$, where $U \subseteq A$, is said to have *type* $\langle \kappa, \lambda \rangle$ if $\overline{\overline{A}} = \kappa$ and $\overline{\overline{U}} = \lambda$. (Thus, in discussing a type $\langle \kappa, \lambda \rangle$, it is assumed without mention that $\lambda \leqq \kappa$.) One may consider possible generalizations of (1) of the form:

(2) *For any countable set Σ of E-sentences, if Σ has a model of type $\langle \kappa, \lambda \rangle$, then Σ has a model of type $\langle \kappa', \lambda' \rangle$.*

The statement (2) will be abbreviated: $\kappa, \lambda \Rightarrow \kappa', \lambda'$. The following list of theorems and conjectures gives a complete picture of what is known concerning statements of this form, at least if one assumes the Generalized Continuum Hypothesis (GCH). (Of course, the ordinary Upward Theorem implies that $\kappa, \kappa \Rightarrow \lambda, \lambda$.)

THEOREM 1.1. *For each n there is an E-sentence σ_n whose models have just those types $\langle \kappa, \lambda \rangle$ for which $\kappa \leqq 2^{\lambda,n}$.* (R. Robinson, see [MV1])

THEOREM 1.2. (GCH if $\lambda \neq \omega$) $\kappa^+, \kappa \Rightarrow \lambda^+, \lambda$ *if λ is regular.* ([MV1] for $\lambda = \omega$, Chang [1] for $\lambda > \omega$.)

THEOREM 1.3. (GCH) $\kappa, \lambda \Rightarrow \kappa', \lambda'$ *if* $\kappa \geqq \kappa' \geqq \lambda' \geqq \lambda$. (Chang-Keisler [1], using ultraproducts.)

THEOREM 1.4. $2^{\kappa,\omega}, \kappa \Rightarrow \lambda, \mu$. ([V3])

CONJECTURE 1.5 (GCH). $2^{\kappa,n}, \kappa \Rightarrow 2^{\lambda,n}, \lambda$.

Concerning these results and problems a number of remarks will now be made.

(a) By 1.2, if $\kappa > \kappa'$ then $\kappa, \kappa' \Rightarrow \omega_1, \omega$. Chang raised the question (see [VI]) whether this (as well as other parts of 1.2) can be improved in Downward Theorem style to read: *Any structure* $\mathfrak{A}$ *of type* $\langle \kappa, \kappa' \rangle$ (where $\kappa > \kappa'$ and $\mathfrak{A}$ has countably many specified relations) *has an elementary substructure of type* $\langle \omega_1, \omega \rangle$. This, being a downward-type question, is outside of our announced topic. But it should at least be mentioned that F. Rowbottom has recently obtained some remarkable results concerning this problem which he has used to obtain a considerable strengthening of Scott's theorem: if the Axiom of Constructibility holds, then there are no measurable cardinals.

(b) 1.2 is proved using homogeneous models when $\lambda = \omega$, and saturated models when $\lambda > \omega$. In some proofs using saturated models one can get by instead with the so-called special models (of Morley-Vaught [1]), which in singular powers are the closest things to saturated models which exist. However, all attempts so far to extend 1.2 to singular λ by this method have failed. This may suggest that a new proof even for regular λ is what is needed.

(c) For the sake of simplicity we considered only countable Σ in the notation $\Rightarrow$, but it is natural to consider other Σ as well. Chang's proof shows that 1.2, for regular $\lambda > \omega$, is correct even if we allow all Σ such that $\overline{\overline{\Sigma}} < \lambda$. Recently I found that this can be improved to allow $\overline{\overline{\Sigma}} \leqq \lambda$ (clearly the best possible). The proof uses an extremely simple technique which might be useful in other places, so we sketch it here.

Chang's proof actually shows that (GCH):

(3) *If* Σ, *of power* $< \lambda$, *has a model* $\mathfrak{A}$ *of type* $\langle \kappa^+, \kappa \rangle^2$ *then* Σ *has a model* $\mathfrak{B}$ *of type* $\langle \lambda^+, \lambda \rangle$; *moreover* $\mathfrak{B}$ *can be found as an elementary extension of an arbitrary given structure* $\mathfrak{C}$, *of power* λ, *elementarily equivalent to* $\mathfrak{A}$.[2]

Now suppose $\overline{\overline{\Sigma}} = \lambda$ and Σ has a model $\mathfrak{A}$ of type $\langle \kappa^+, \kappa \rangle$.[2] We can assure that $\kappa > \lambda$ by applying (3). Let $\mathfrak{A} = \langle A, U, R_\xi \rangle_{\xi < \lambda}$. To simplify the notation, suppose each R_ξ is singulary. Let $a_0, \cdots, a_\xi, \cdots (\xi < \lambda)$ be distinct elements of A; put $P = \{\langle a_\xi, x \rangle / R_\xi x \text{ and } \xi < \lambda\}$; let $\mathfrak{A}' = \langle A, U, P \rangle$; and let $\mathfrak{C}'$ be an elementary subsystem of $\mathfrak{A}'$ of power λ containing all a_ξ's. By (3), $\mathfrak{C}'$ has an elementary extension $\mathfrak{B}' = \langle B, V, Q \rangle$ of type $\langle \lambda^+, \lambda \rangle$. For $\xi < \lambda$, put $S_\xi = \{y / Q a_\xi y\}$. Then clearly $\langle B, V, S_\xi \rangle_{\xi < \lambda}$ is a model of Σ of type $\langle \lambda^+, \lambda \rangle$.

---

[2] It is to be understood here that $\mathfrak{A}$ has no relations not named in Σ.

(d) Chang asked if 1.2 remains valid with no restriction at all on $\overline{\overline{\Sigma}}$ provided $\kappa < \lambda$ (and assuming GCH). He observed that this is indeed the case if $\kappa^+ < \lambda$; for (as A. Robinson first remarked in connection with the ordinary Upward Theorem) a structure $\mathfrak{A}$ of power κ^+ has at most $2^{(\kappa^+)} = \kappa^{++}$ distinct relations so one may assume $\overline{\overline{\Sigma}} \leqq \kappa^{++}$ with no loss of generality. There remains, however, the case $\kappa^+ = \lambda$, i.e., the question whether the complete structure (having all possible relations) on κ^+ of type $\langle \kappa^+, \kappa \rangle$ is elementarily equivalent to a structure of type $\langle \kappa^{++}, \kappa^+ \rangle$. By the fundamental result of Keisler [1], this amounts to the question whether $\langle \kappa^+, \kappa \rangle$ has a limit ultrapower of type $\langle \kappa^{++}, \kappa^+ \rangle$. This problem is open and so is the question whether $\langle \kappa^+, \kappa \rangle$ has an (ordinary) ultrapower of type $\langle \kappa^{++}, \kappa^+ \rangle$. Here we have again made contact with a whole series of problems about the cardinality of ultraproducts, which have been raised and discussed in Frayne-Morel-Scott [1]. Most of these very interesting problems remain open, though some important progress was made in Chang-Keisler [1] and Keisler [1], [3].

(e) The proof of 1.4 given in [V3] also yields, as is shown there, various theorems concerning three or more cardinals. (A structure $\langle A, U, V, R_\xi \rangle_{\xi < \alpha}$, where $\overline{\overline{A}} = \kappa$, $\overline{\overline{U}} = \lambda$, and $\overline{\overline{V}} = \mu$ is said to have type $\langle \kappa, \lambda, \mu \rangle$; and then $\kappa, \lambda, \mu \Rightarrow \kappa', \lambda', \mu'$ has the obvious meaning.) For example, it is shown there that:

THEOREM 1.6. *$\kappa, \lambda, \mu \Rightarrow \kappa', \lambda', \mu'$ if $\kappa \geqq 2^{\lambda, \omega}, \lambda \geqq 2^{\mu, \omega}$, and $\kappa' \geqq \lambda' \geqq \mu'$.*

Recently, J. Silver [1] has used 1.6 in answering negatively the question raised in Addison [1], whether the reduction principle for $\bigwedge_1^1$ holds.

(f) With the exception of 1.2 for singular λ, the simplest open part of Conjecture 1.5 is:

(4) CONJECTURE: (GCH) $\kappa^{++}, \kappa \Rightarrow \lambda^{++}, \lambda$.

A simple argument (noticed independently by Silver and the author) shows that (4) would imply the "three-cardinal theorem":

(4') (GCH) $\kappa^{++}, \kappa^+, \kappa \Rightarrow \lambda^{++}, \lambda^+, \lambda$.

This is an immediate consequence of the following consideration: If $A \supseteq U \supseteq V$, $\overline{\overline{A}} = \kappa^{++}$ and $\overline{\overline{V}} = \kappa$, then a necessary and sufficient condition for $\overline{\overline{U}} = \kappa^+$ is that U can be represented as an increasing union of sets equinumerous to V and A as an increasing union of sets equinumerous to U. But this condition is clearly equivalent to one of the form: there exist additional relations $R, S, \cdots$ such that $\langle A, U, V, R, S, \cdots \rangle$ is a model of σ—for a certain E-sentence σ.

This remark perhaps increases the estimated difficulty of proving (4). Our final remark adds both to the estimated difficulty and to the importance of proving (4):

(g) A well-known set-theoretical conjecture of Kurepa is as follows:

$K(\kappa)$: (GCH) *There is a family* $\mathfrak{F}$ *of subsets of* κ^+ *such that* $\overline{\overline{\mathfrak{F}}} = \kappa^{++}$ *but, for each* $X \subseteq \kappa^+$, *if* $\overline{\overline{X}} \leqq \kappa$ *then* $\{X \cap Y \mid Y \in \mathfrak{F}\}$ *has power* $\leqq \kappa$.

$K(\kappa)$ is unresolved for any κ (though recently Rowbottom and Lévy (independently) have obtained results concerning its consistency). By methods like that used in (f) above, it is easily seen that $K(\kappa)$ can be expressed equivalently in the form: σ has a model of type $\langle \kappa^{++}, \kappa \rangle$—for a certain E-sentence σ. Thus our conjecture (4) would include as a special case the result: $K(\kappa) \to K(\lambda)$.

2. Transfer principles between languages

Let Q_κ be the language obtained from E by adding a new quantifier symbol Q and requiring that a formula $Qx\phi$ be interpreted as saying "there are at least κ x such that ϕ." Thus there is only one Q-syntax but the semantics of Q_κ depends on κ. The study of these languages was initiated in Mostowski [1]. G. Fuhrken ([1], [2]) has obtained several results of the form:

(5) *For any countable set* Σ *of* Q*-sentences, if* Σ *has a* Q_κ*-model, then* Σ *has a* Q_λ*-model.*

Let us abbreviate (5) by writing $\kappa \Rightarrow \lambda$. The following list of theorems gives a complete picture, assuming GCH, of what is known concerning propositions of this form.

THEOREM 2.1. (GCH *if* $\lambda > \omega$) $\kappa \Rightarrow \lambda^+$ *if* κ *and* λ *are regular.* (Fuhrken [1])

THEOREM 2.2. $\omega \Rightarrow \kappa$ (Fuhrken [1]).

THEOREM 2.3. *If* κ *is weakly inaccessible then, for some singular* $\lambda, \kappa \Rightarrow \lambda$. (D. Scott, see Fuhrken [1])

THEOREM 2.4. $\kappa \not\Leftarrow \lambda$ *if* κ *is singular,* λ *not; or if* κ *is a successor cardinal,* λ *not*; or if $\kappa \neq \omega$, $\lambda = \omega$. (Fuhrken [1])

CONJECTURE 2.5. (GCH) $\kappa \Rightarrow \lambda$ *if* κ *and* λ *are singular, or if* κ *and* λ *are inaccessible. In* 2.1, λ *need not be regular*[3]. (Fuhrken [1], [2])

[3] This last assertion would follow from 1.2 for λ singular. (See Fuhrken [1], [2])

To obtain these results, Fuhrken developed a method (or two methods) for reducing questions about Q_κ to questions in the model theory of E. The latter, in turn, were answered by using either 1.2 above, or a result of MacDowell and Specker [1], or various new arguments.

We want to mention briefly now two important aspects of the proof of some of the results of this section and §1. Just as the ordinary Upward Theorem is closely related to the Compactness Theorem, so to each of 2.1, 1.2, 1.3, and 1.4 there corresponds a certain 'compactness theorem'. Moreover, also in analogy to the situation for the ordinary Upward Theorem, to each of 2.1, 1.2, and 1.4 there corresponds a certain 'completeness theorem'. For some of the open problems mentioned above there also are corresponding open 'compactness problems' and 'completeness problems'. For full details the papers Fuhrken [1], [2] and Vaught [2], [3] should be consulted. We will state here one example of each kind of theorem or problem mentioned.

Denote by $\mathscr{C}_\omega$ the class of all κ such that either $\omega = cf\kappa$ (the confinality character of κ) or else $\kappa = \lambda^+$ where $\mathrm{cf}\lambda = \omega$. Call Q_κ *complete* if its set of universally valid sentences is recursively enumerable. Call Q_κ *ω-compact* if, whenever every finite subset of a countable set Σ of Q-sentences has a Q_κ-model, so has Σ. Q_ω is not ω-complete (Mostowski [1]); and clearly Q_ω is not ω-compact.

THEOREM 2.6. (GCH) *If* $\kappa \notin \mathscr{C}_\omega$ *or* $\kappa = \omega_1$, *then* Q_κ *is ω-compact* (Fuhrken [1]). Q_{ω_1} *is complete* (Vaught [2]); *hence* (GCH), *by* 2.1, *so is* Q_{λ^+} *for any regular* λ.

PROBLEM 2.7. Are all Q_κ, for $\kappa \neq \omega$, ω-compact? complete?

3. Hanf numbers.

For each language L we consider, the Hanf number νL (or $\nu_\kappa L$) is defined as the least μ such that any sentence σ of L (resp., any set Σ of L-sentences of power $\leqq \kappa$), which has a model of power at least μ, has arbitrarily large models. Hanf [1] observed that these numbers exist for almost any language (roughly speaking, as long as the class of all L-sentences is a set). In this section we consider, as a third type of generalization of the Upward Theorem, the problem of exactly evaluating these numbers for various languages.

THEOREM 3.1. $\nu Q_{\omega_1} = \nu_\omega Q_{\omega_1} = \beth_\omega$. ([V3])

THEOREM 3.2. (GCH) If $\kappa \notin \mathscr{C}_\omega$, $\nu Q_\kappa = \nu_\kappa Q_\kappa = 2^{\kappa,\omega}$.

PROOF. In [V3], it was shown that $vQ_\kappa = v_\omega Q_\kappa = 2^{\kappa,\omega}$, when $\kappa \notin \mathscr{C}_\omega$. At the end of the same paper, some (too loose!) remarks are made about the situation when uncountably many symbols are allowed. But the author only realized recently that the full 3.2, above, can be proved if one proceeds with more care. To accomplish this, the following two facts must be noted:

Firstly, when uncountably many symbols are allowed, the main theorem, 4.1, of [V3] should be restated as follows:

For each signature ρ there are a signature $\rho' \supseteq \rho$ and sets $T_0 \subseteq T_1 \subseteq \cdots \subseteq T_n \subseteq \cdots$ of E-sentences of signature ρ' such that:

(.1) Every model of $\bigcup_n T_n$ has a proper elementary extension with the same U; and

(6) *(.2) If Σ is any set of E-sentences of signature ρ and Σ has for each n a model $\mathfrak{D}_n$ of some type $\langle \kappa_n, \lambda_n \rangle$ such that $\kappa_n \geqq 2^{\lambda,n}$, then for each n $\Sigma \cup T_n$ has a model $\mathfrak{A}_n$ such that, for some m, $\mathfrak{A}_n \prec^{*}_{U} \mathfrak{D}_m$.*

The proof of 4.1 in [V3], with only some minor changes, establishes (6).

The second fact needed is a strengthening of part 2.6, above:

(7) *For $\kappa \notin \mathscr{C}_\omega$, Q_κ is "strongly ω-compact," i.e., if $\Sigma_0 \subseteq \Sigma_1 \subseteq \cdots \subseteq \Sigma_n \subseteq \cdots$ are arbitrary sets of Q-sentences such that, for each n, Σ_n has a Q_κ-model, then so has $\bigcup_n \Sigma_n$.*

(7) follows at once from the result of Fuhrken [2] that, when $\kappa \notin \mathscr{C}_\omega$, Q_κ-sentences are preserved by denumerable ultraproducts.

With the aid of (6) and (7), the proof in [V3] that $v_\omega Q_\kappa = 2^{\kappa,\omega}$ (if $\kappa \notin \mathscr{C}_\omega$) can be extended in an obvious way to yield 3.2.

By the same argument, one can even show, assuming still that $\kappa \notin \mathscr{C}_\omega$, that $v_\lambda Q_\kappa = 2^{\kappa,\omega}$, as long as $\lambda < 2^{\kappa,\omega}$, and also that $v_\lambda Q_\kappa \leqq 2^{\lambda,\omega}$ if $\kappa \leqq \lambda$. *But the exact value of $v_\lambda Q_\kappa$ is unknown when $\kappa \notin \mathscr{C}_\omega$ and $\lambda \geqq 2^{\kappa,\omega}$.*

Morley [1] developed an ingenious method which allows the computation of Hanf numbers for some quite rich languages. His method is related to the work of Ehrenfeucht and Mostowski [1]; but while they apply the ordinary Ramsey's Theorem to model theory, Morley applies a generalization of Ramsey's Theorem due to Erdös and Rado. Following Morley, let μ_κ be the least μ such that, for any set Σ of E-sentences and any set Γ of E-formulas with one (fixed) free variable, if $\bar{\bar{\Sigma}}, \bar{\bar{\Gamma}} \leqq \kappa$ and there is a model of Σ of power $\geqq \mu$ which has no element satisfying all members of Γ, then there are arbitrarily large such models. What Morley showed is:

THEOREM 3.3. $\mu_\omega = \beth_{\omega_1}$ *and, in general,* $\beth_{\kappa^+} \leqq \mu_\kappa \leqq \beth_{(2^\kappa)^+}$. (Morley [1])

Recently, M. Helling has discovered that by refining Morley's proof he can establish:

THEOREM 3.4. $\mu_\kappa = \beth_{\kappa^+}$ *if* $\kappa = \beth_\delta$ *and* $\mathrm{cf}\,\delta = \omega$. (Helling [1])

In all other cases, however, the exact location of μ_κ in the interval given in 3.3 remains an open problem:

CONJECTURE 3.5. (GCH?) *In general* $\mu_\kappa = \beth_{\kappa+}$. (Morley [1])

It turns out that several Hanf number problems can be reduced to the determination of μ_κ. Let W be the weak second-order language of Scott-Tarski [1], in which, roughly speaking, the notion of finite sequence is available. It is seen easily that $\nu W = \nu Q_\omega$ and $\nu_\omega W = \nu_\omega Q_\omega$. Helling pointed out to Morley that 3.3 allows the computation of $\nu_\omega W$ $(= \nu_\omega Q_\omega)$ since, as is easily proved:

THEOREM 3.6. $\mu_\omega = \nu_\omega W$.

Morley noticed that similarly, for example, $\nu_{\omega_1} Q_{\omega_1}$ is related to μ_{ω_1}. Recently Helling has verified that, from various results of Craig-Hanf [1], Fuhrken [2], and Keisler [1], one can easily infer that, in general:

THEOREM 3.7 (GCH). *If* $\kappa \in \mathscr{C}_\omega$ *and* κ *is less than the first measurable cardinal* m, *then* $\mu_\kappa = \nu_\kappa Q_\kappa$. *Indeed, if* $\kappa \in \mathscr{C}_\omega$, $\kappa \leqq \lambda < m$, *and* λ *is not inaccessible, then* $\mu_\lambda = \nu_\lambda Q_\kappa = \nu_\lambda Q_\omega$.[4]

Let L_κ be the language whose rules of formation differ from those of E in allowing conjunctions and disjunctions of length $\leqq \kappa$ (but only ordinary quantification). Lopez-Escobar showed, by a direct but rather involved process, that $\mu_\omega = \nu L_\omega$. Recently C. C. Chang found that the same thing holds for all κ, i.e.,

THEOREM 3.8. $\mu_\kappa = \nu L_\kappa$.

An interesting case not discussed above is the evaluation of νW (which is the same as νQ_ω). This problem was raised by Scott several years ago. He observed, by means of simple examples, that $\nu W \geqq \beth_{\omega_1^c}$, where ω_1^c is the 'constructible ω_1', i.e., the first ordinal not the order type of some recursive well-ordering.

CONJECTURE 3.9. $\nu W = \beth_{\omega_1^c}$. (Scott)

---

[4] Keisler pointed out to the author in Jerusalem that from a result in Keisler [2] it follows that the hypothesis 'λ is not inaccessible' in 3.7 can be considerably weakened.

PROBLEM 3.10. *In general, what are* vQ_κ *and* $v_\lambda Q_\kappa$, if $\kappa \in \mathscr{C}_\omega$ *and* $\lambda < \kappa$?

There is some connection between Problems 3.10 and 2.7. In fact, the method of proof of 3.2, above, shows that $vQ_\kappa = v_\omega Q_\kappa = 2^{\kappa,\omega}$ if Q_κ is ω-compact.

REFERENCES

J. ADDISON [1], The theory of hierarchies, *Logic, methodology, and philosophy of science* Proc. 1960 Int. Congress, Stanford, 1962, pp. 26–37.

C. C. CHANG [1], A note on the two cardinal problem. Submitted to *Bull. Amer. Math. Soc.*

C. C. CHANG and H. J. KEISLER [1], Applications of ultraproducts of pairs of cardinals to the theory of models, *Pacific J. Math.* vol. **12** (1962), pp. 835–845.

W. CRAIG and W. HANF [1], On relative characterizability in a language, *Notices Amer. Math. Soc.* vol. **9** (1962), pp. 152–153.

A. EHRENFEUCHT and A. MOSTOWSKI [1], Models of axiomatic theories admitting automorphisms, *Fund. Math.* vol. **43** (1956), pp. 50–68.

T. FRAYNE, A. MOREL, and D. SCOTT [1], Reduced direct products, *Fund. Math.* vol. 51 (1962), pp. 195–228.

G. FUHRKEN [1], Skolem-type normal forms for first-order languages with a generalized quantifier, *Fund. Math.* vol. **64** (1964), pp. 291–302.

G. FUHRKEN [2], Languages with added quantifier 'there exist at least $\aleph_\alpha$', *Proc.* 1963 *Berkeley Symposium on Theory of Models*, to appear.

W. HANF [1], *Some fundamental problems concerning languages with infinitely long expressions*, Doctoral dissertation, University of Calif., Berkeley, 1962.

M. HELLING [1], Hanf numbers for some generalizations of first-order language, *Notices Amer. Math. Soc.* vol. **11** (1964), p. 679.

C. KARP [1], *Languages with formulas of infinite length*, Doctoral dissertation, University of Calif., Berkeley, 1959.

H. J. KEISLER [1], Limit ultrapowers, *Trans. Amer. Math. Soc.* vol. **107** (1963), pp. 382–408.

H. J. KEISLER [2], The equivalence of certain problems in set theory with problems in the theory of models, *Notices Amer. Math. Soc.* vol. **9** (1962), pp. 339–340.

H. J. KEISLER [3], On cardinalities of ultraproducts, *Bull. Amer. Math. Soc.*, vol. **70** (1964), pp. 644–647.

R. MACDOWELL and E. SPECKER [1], Modelle der Arithmetik, *Infinitistic Methods, Proc.* 1959 *Warsaw Symposium on Foundations of Mathematics*, 1961, pp. 257–263.

M. MORLEY [1], Omitting classes of elements, *Proc.* 1963 *Berkeley Symposium on Theory of Models*, to appear.

M. MORLEY and R. VAUGHT [1], Homogeneous universal models, *Math. Scand.* vol. **11** (1962), pp. 37–57.

A. MOSTOWSKI [1], On a generalization of quantifiers, *Fund. Math.* vol. **44** (1957), pp. 12–36.

D. SCOTT and A. TARSKI [1], Extension principles for algebraically closed fields, *Notices Amer. Math. Soc.* vol. **5** (1958), pp. 778–779.

J. SILVER [1], Second separation property for existential second-order classes, *Notices Amer. Math. Soc.*, vol. **12** (1965), p. 241.

A. TARSKI and R. VAUGHT [1], Arithmetical extensions of relational systems, *Compositio Math.* vol. **13** (1957), pp. 81–102.

R. VAUGHT [1], Models of complete theories, *Bull. Amer. Math. Soc.* vol. **69** (1963), pp. 299–313.

R. VAUGHT [2], The completeness of logic with the added quantifier "there are uncountably many," *Fund. Math.* vol. **64** (1964), pp. 301–304.

R. VAUGHT [3], A Löwenheim-Skolem theorem for cardinals far apart, *Proc.* 1963 *Berkeley Symposium on the Theory of Models*, to appear.

Section II

Foundations of Mathematical Theories

INFINITE VALUED LOGIC AS A BASIS FOR SET THEORY*

C. C. CHANG

University of California, Los Angeles, California, U.S.A.

This talk is a brief survey of some results and problems in a study of set theory with infinite valued logic as a basis. The main problem is to consider the consistency of the axiom schema of comprehension in infinite valued logic. We shall start at the very beginning of the subject and finish with some open problems. Our discussion is informal throughout the paper and no proofs will be given. However, we supply the reader with a liberal selection of bibliographical references, where he may pursue some particular areas of interest to him.

On a quick trip like this one, it is impossible to give full historical accounts of the development of the theory. We again rely on appropriate references where the reader will find fuller, more complete, and more accurate treatments.

The language L. Consider the usual first order language L with identity based on the following symbols:

binary predicate constant symbols $\in$ and $\equiv$;
individual variables $v_0, v_1, \cdots$;
propositional connectives $\neg$ (unary), $\rightarrow$, $\wedge$, $\vee$, $\leftrightarrow$ (binary);
quantifiers $\exists$ and $\forall$;
improper symbols (,) .

Formulas of L are constructed in the usual way. We write $U(x_1, \cdots, x_m)$ to indicate that the free variables of the formula U are among the variables $x_1, \cdots, x_m$. A sentence is a formula with no free variables.

We shall henceforth regard the language L as fixed, independent of the number or kind of truth values.

Interpretations of the logical connectives. Let $X = [0, 1]$ be the closed real unit interval. The following interpretations of the propositional connectives are due to Łukasiewicz [11]. The function $\neg$ is defined on X to X as follows: for all x in X,

* This work was partly supported by a NSF research grant GP-220.

$$\neg x = 1 - x .$$

The functions $\rightarrow$, $\wedge$, $\vee$, $\leftrightarrow$ are defined on $X \times X$ to X as follows: for all x, y in X,

$$x \rightarrow y = \min(1, 1 - x + y),$$

$$x \wedge y = \min(x, y),$$

$$x \vee y = \max(x, y),$$

$$x \leftrightarrow y = 1 - |x - y| .$$

The functions $\exists$ and $\forall$ are natural generalizations of the two-valued quantifiers and they are defined on the set of all non-empty subsets of X to X as follows: for all non-empty $Y \subset X$,

$$\exists Y = \sup Y \quad \text{and} \quad \forall Y = \inf Y .$$

If we endow X with the natural topology, then each of the functions $\neg$, $\rightarrow$, $\wedge$, $\vee$, $\leftrightarrow$ is continuous. Furthermore, the functions $\exists$ and $\forall$ are continuous with respect to a natural topology on the set of all non-empty subsets of X. A little inspection shows that the seven functions defined above can be defined from $\neg$, $\rightarrow$, and $\exists$. For further details on these functions see, e.g., [14], [4], and [1].

The set X is referred to as the set of truth values of the infinite valued logic. In order to discuss finite valued logics, we consider the following finite subsets of X. For each $n \geqq 2$, let

$$X_n = \left\{0, \frac{1}{n-1}, \frac{2}{n-1}, \cdots, 1\right\} .$$

Each set X_n is referred to as the set of truth values of n-valued logic. Each one of the seven functions introduced above, when restricted to X_n, or $X_n \times X_n$, or the set of all non-empty subsets of X_n, yields values in X_n. Observe that in case $n = 2$, all functions have their standard 2-valued meanings.

Models. A model M for L is a triple $M = \langle A, E, I \rangle$ where A is a non-empty set, E is a mapping of $A \times A$ into X, and I is the mapping of $A \times A$ into X such that for all a, b in A

$$I(a, b) = 0 \quad \text{if} \quad a \neq b ,$$

$$I(a, b) = 1 \quad \text{if} \quad a = b .$$

Let $\mathcal{M}$ be the class of all such models. To obtain the finite valued models, for each $n \geqq 2$, let $\mathscr{M}_n$ be the class of all models $M = \langle A, E, I \rangle$ in $\mathscr{M}$ such that the range of E is a subset of X_n. Members of $\mathscr{M}_n$ are referred to as n-valued models for L. Clearly, for all m, $n \geqq 2$,

$$\mathscr{M}_n \subset \mathscr{M}$$

and

$$\mathscr{M}_n \subset \mathscr{M}_m \text{ if and only if } (n-1) \text{ divides } (m-1).$$

The class $\mathscr{M}_2$ is simply the familiar class of all 2-valued models for L.

The value function U_M. For each formula $U(x_1, \cdots, x_m)$, model $M = \langle A, E, I \rangle$, and interpretation of $x_1, \cdots, x_m$ as elements $a_1, \cdots, a_m$ in A, we define a real number $U_M(a_1, \cdots, a_m)$ in X by induction on the formulas. The definition of $U_M(a_1, \cdots, a_m)$ is quite simple and straightforward. We give the complete definition for those who may not be familiar with it. (When the model M is understood, we agree to drop the subscript M and write $U(a_1, \cdots, a_m)$ for $U_M(a_1, \cdots, a_m)$.)

(i) If U is the atomic formula $x_i \equiv x_j$, where $1 \leqq i, j \leqq m$, define $U(a_1, \cdots, a_m) = I(a_i, a_j)$.

(ii) If U is the atomic formula $x_i \in x_j$, where $1 \leqq i, j \leqq m$, define $U(a_1, \cdots, a_m) = E(a_i, a_j)$.

(iii) Suppose U, V are formulas such that $U(a_1, \cdots, a_m)$ and $V(a_1, \cdots, a_m)$ are defined, then define

$$(\neg U)(a_1, \cdots, a_m) = \neg(U(a_1, \cdots, a_m)),$$

$$(U \to V)(a_1, \cdots, a_m) = U(a_1, \cdots, a_m) \to V(a_1, \cdots, a_m),$$

$$(U \wedge V)(a_1, \cdots, a_m) = U(a_1, \cdots, a_m) \wedge V(a_1, \cdots, a_m),$$

$$(U \vee V)(a_1, \cdots, a_m) = U(a_1, \cdots, a_m) \vee V(a_1, \cdots, a_m),$$

$$(U \leftrightarrow V)(a_1, \cdots, a_m) = U(a_1, \cdots, a_m) \leftrightarrow V(a_1, \cdots, a_m).$$

(iv) Suppose $U(x_1, \cdots, x_{m+1})$ is a formula such that $U(a_1, \cdots, a_{m+1})$ is defined for all $a_1, \cdots, a_{m+1}$ in A. Then define

$$(\exists x_{m+1} U)(a_1, \cdots, a_m) = \exists\{U(a_1, \cdots, a_{m+1}) : a_{m+1} \text{ in } A\},$$

$$(\forall x_{m+1} U)(a_1, \cdots, a_m) = \forall\{U(a_1, \cdots, a_{m+1}) : a_{m+1} \text{ in } A\}.$$

By (i)–(iv), $U_M(a_1, \cdots, a_m)$ is a uniquely defined real number in X. Furthermore, for each $n \geqq 2$,

$$\text{if } M \text{ is in } \mathscr{M}_n, \text{ then } U_M(a_1, \cdots, a_m) \text{ is in } X_n.$$

In case U is a sentence, the value of $U_M(a_1, \cdots, a_m)$ is independent of the elements $a_1, \cdots, a_m$ in A and we simply write U_M for $U_M(a_1, \cdots, a_m)$. The real number U_M, in case U is a sentence, is referred to as the value of the sentence U on the model M. Clearly U_M may be considered as a function mapping the class

$$\text{(the set of all sentences)} \times \mathscr{M}$$

into the set X. Also, if M is in $\mathscr{M}_n$, then U_M is in X_n for every sentence U. It should be evident that in case $n = 2$, we have the usual value function U_M as M ranges in $\mathscr{M}_2$ and U ranges over all sentences.

The operations Mod **and** Th. Let Σ be a set of sentences of L, K be a class of models in $\mathscr{M}$, and Y be a subset of X. Define

$$\mathrm{Mod}(\Sigma, Y) = \{M \text{ in } \mathscr{M}_n : U_M \text{ is in } Y \text{ for every } U \text{ in } \Sigma\},$$

$$\mathrm{Th}(K, Y) = \{U \text{ sentences of } L : U_M \text{ is in } Y \text{ for every } M \text{ in } K\}.$$

For each $n \geqq 2$, define

$$\mathrm{Mod}_n(\Sigma, Y) = \{M \text{ in } \mathscr{M}_n : U_M \text{ is in } Y \text{ for every } U \text{ in } \Sigma\}.$$

Notice that $\mathrm{Mod}_2(\Sigma, \{1\})$ is an *elementary class* in 2-valued model theory, and

$$\mathrm{Th}(\mathrm{Mod}_2(\Sigma, \{1\}), \{1\})$$

is the set of all *semantical consequences* of Σ in 2-valued logic.

Using the notation we have defined so far, we may formulate and prove many model theoretical results in infinite valued logic. For instance, 2-valued theorems such as the compactness theorem, the Löwenheim-Skolem theorems all have their natural infinite valued generalizations. We refer the reader to [6] for statements of some of these results. Model theoretical theorems in even more general settings can be found in the summary [8] and in a forthcoming monograph by Chang and Keisler.

The axiom schema of comprehension in many valued logics. Let Σ_0 be the set of all sentences of L of the form

$$(*) \qquad \forall x_1 \cdots x_m \, \exists y \, \forall t (t \in y \leftrightarrow U(t, x_1, \cdots, x_m)),$$

where the formula U does not contain y free. The free variables of U are called the parameters of the formula U. We refer to the set Σ_0 as the axiom schema of comprehension.

The very first result about Σ_0 due to Russell is the following (we use 0 to denote the empty set):

(A) $$\mathrm{Mod}_2(\Sigma_0, \{1\}) = 0.$$

A natural extension of Russell's result (A) is

(B) $$\mathrm{Mod}_n(\Sigma_0, \{1\}) = 0 \text{ for each } n \geqq 2.$$

The result (B) has been noticed by several authors, in particular, by Skolem in [17] and Tarski (by oral communication). (A) and (B) can be easily sharpened to the following.

(C) If n is odd, then $\mathrm{Mod}_n(\Sigma_0, (\frac{1}{2}, 1]) = 0$ and $\mathrm{Mod}_n(\Sigma_0, [\frac{1}{2}, 1]) \neq 0$. If n is even, let p be the rational $n-2/(2n-2)$. Then $\mathrm{Mod}_n(\Sigma_0, (p, 1]) = 0$ and $\mathrm{Mod}_n(\Sigma_0, [p, 1]) \neq 0$.

The problem of the consistency of the axiom schema of comprehension in infinite valued logic is to decide whether or not $\mathrm{Mod}(\Sigma_0, \{1\}) \neq 0$. As far as we know this problem is still open. In the next few paragraphs we describe some partial positive answers.

Let Σ_1 be the set of all sentences of L of the form (*) where $U(t, x_1, \cdots, x_m)$ contains no bound variables. Skolem [17] proved in 1957 that

(D) $$\mathrm{Mod}(\Sigma_1, \{1\}) \neq 0.$$

Notice that this is already a drastic departure from the finite valued results (A), (B), and (C), since each of these results holds with Σ_0 replaced by Σ_1.

In 1963, the author proved the following two theorems, see [7]. Let Σ_2 be the set of all sentences of L of the form (*) where $U(t, x_1, \cdots, x_m)$ does not contain any parameters $x_1, \cdots, x_m$, but may contain arbitrary bound variables.

(E) $$\mathrm{Mod}(\Sigma_2, \{1\}) \neq 0.$$

Let Σ_3 be the set of all sentences of L of the form (*) where $U(t, x_1, \cdots, x_m)$ may contain parameters $x_1, \cdots, x_m$, but each bound variable u of U is restricted to occur only in the second place in atomic formulas of the form $v \in w$.

(F) $$\mathrm{Mod}(\Sigma_3, \{1\}) \neq 0.$$

The result (F) may be considered as an improvement of Skolem's result (D), since $\Sigma_1 \subset \Sigma_3$.

Just recently, Fenstad in [9] obtained another result in this direction. Let Σ_4 be the set of all sentences of L of the form (*) where $U(t, x_1, \cdots, x_m)$ carries the single restriction that the free variable t is allowed to occur only in the first place in atomic formulas of the form $v \in w$.

(G) $$\mathrm{Mod}(\Sigma_4, \{1\}) \neq 0.$$

The proofs of all of the results (D)–(G) are based on the original method of Skolem using the Brouwer fixed point theorem for the space X^n. In addition, the result (E) requires also an application of the compactness theorem.

Aside from the positive results listed above, some negative results are also known if we consider the problem of consistency of the axiom schema of comprehension together with some other axioms of set theory. For instance, let V be the sentence which expresses the axiom of extensionality. Then it is known that

$$\mathrm{Mod}(\Sigma_2 \cup \{V\}, \{1\}) = 0$$

and

$$\mathrm{Mod}(\Sigma_3 \cup \{V\}, \{1\}) = 0.$$

For a reference see [7], where some alternatives to the classical 2-valued interpretation of the identity $\equiv$ are discussed.

As a last remark about the results (D)–(G), we point out the following curious fact. It follows from the method of proof of (D)–(G) that the formulas $U(t, x_1, \cdots, x_m)$ in each one of the sets $\Sigma_1 - \Sigma_4$ may be allowed to contain free occurrences of the variable y as well. This is certainly not what one would expect in 2-valued logic.

Problems in syntax. We now discuss some generalizations of the notion of the set of semantical consequences of Σ in 2-valued logic, given by the set

$$\mathrm{Th}(\mathrm{Mod}_2(\Sigma, \{1\}), \{1\}).$$

Let Σ be a set of sentences, and let $Y, Z \subset X$. Consider the set of sentences Δ defined by the equation

$$(**) \qquad \Delta = \mathrm{Th}(\mathrm{Mod}(\Sigma, Y), Z).$$

We would like to know the exact relationship between Σ and Δ when some reasonable assumptions are made on the sets Y and Z. For instance, we know from the Gödel completeness theorem that the set

$$\mathrm{Th}(\mathrm{Mod}_2(\Sigma, \{1\}), \{1\})$$

is recursively enumerable in Σ, and in fact is axiomatizable from Σ. In infinite valued logic problems of this type seem to be very difficult. As an example, even the problem of completeness for infinite valued *propositional* logic has an extremely non-trivial character. On this subject see [14], [18], [5], [12], [3], and the general survey by Rosser [15]. As another example, we have the following negative result of Scarpellini [16]:

(H) Let $Y = X$, $Z = \{1\}$, and let Σ be arbitrary. Then the set Δ given by (**) is not recursively enumerable.

A slight improvement of this result is due to the author and its proof can be found in [1]. The improvement consists in replacing in (H) the set $Z = \{1\}$ by any closed interval $Z = [r, 1]$ where r is a rational in X greater than 0.

As an example of a positive result in problems of this kind we have:

(I) Let r and s be rationals in X, and let $Y = [r, 1]$ and $Z = (s, 1]$. Then the set Δ given by (**) is recursively enumerable in Σ.

Various pieces of this last result are due to Mostowski [13], Hay [10], and Belluce [1]. Actually, (I) can be strengthened by giving a more specific description of Δ in terms of Σ via a certain "axiomatization" of Δ in terms of some logical axioms and sentences depending on Σ. For further references see [1] and [2]; [1] also contains some results when s need not be a rational in X, and [2] contains a more general definition of models which may yet play a role in the kind of problems we are describing.

The two results (H) and (I) are essentially the context of our knowledge in this direction. It is interesting that the methods that are used to prove (H) and (I) do not seem to yield any results either when we have two closed intervals

$$Y = [r, 1] \text{ and } Z = [s, 1],$$

or when we have two open intervals

$$Y = (r, 1] \text{ and } Z = (s, 1].$$

Nothing is known in these two cases even when $r = s$.

The importance of these problems to the study of set theory (or, for that matter, any theory) with infinite valued logic as a basis should be quite evident. For suppose we are able to prove that

$$\mathrm{Mod}(\Sigma_0, Y) \neq 0,$$

where $Y = [r, 1]$ or $Y = (r, 1]$ with r a rational in X. Then we should try to understand the set of sentences

$$\mathrm{Th}(\mathrm{Mod}(\Sigma_0, Y), Y).$$

From the model theoretical results we referred to earlier, it follows that there exists a greatest real number t in X such that

$$\mathrm{Mod}(\Sigma_0, [t, 1]) \neq 0.$$

We can see easily that $\frac{1}{2} \leqq t$. However it is not known if $\frac{1}{2} < t$, nor is it known if

$$\mathrm{Mod}(\Sigma_0, (\tfrac{1}{2}, 1]) \neq 0.$$

REFERENCES

[1] L. P. BELLUCE, Further results on infinite valued predicate logic, *J. Symbolic Logic* **29** (1964), 69–78.

[2] L. P. BELLUCE and C. C. CHANG, A weak completeness theorem for infinite valued predicate logic, *J. Symbolic Logic* **28** (1963), 43–50.

[3] C. C. CHANG, Proof of an axiom of Łukasiewicz, *Trans. Amer. Math. Soc.* **87** (1958), 55–56.

[4] C. C. CHANG, Algebraic analysis of many valued logics, *Trans. Amer. Math. Soc.* **88** (1958), 467–490.

[5] C. C. CHANG, A new proof of the completeness of the Łukasiewicz axioms, *Trans. Amer. Math. Soc.* **93** (1959), 74–80.

[6] C. C. CHANG, Theory of models of infinite valued logics, I-IV, abstracts in the *Notices of Amer. Math. Soc.* **8** (1961), 68 and 141.

[7] C. C. CHANG, The axiom of comprehension in infinite valued logic, *Math. Scand.* **13** (1963), 9–30.

[8] C. C. CHANG and H. J. KEISLER, Continuous model theory, to appear in the *Proc. of the Model Theory Symp.* held at Berkeley, June-July 1963.

[9] J. E. FENSTAD, The consistency of the axiom of comprehension in infinite valued logic, to appear in *Math. Scand.*

[10] L. S. HAY, Axiomatization of the infinite-valued predicate calculus, *J. Symbolic Logic* **28** (1963), 77–86.

[11] J. ŁUKASIEWICZ and A. TARSKI, Untersuchungen über den Aussangenkalkül, *C. R. des Séances de la Société des Sciences et des Lettres de Varsovie*, Classe III, **23** (1930), 30–50.

[12] C. A. MEREDITH, The dependence of an axiom of Łukasiewicz, *Trans. Amer. Math. Soc.* **87** (1958), 54.

[13] A. MOSTOWSKI, Axiomatizability of some many valued predicate calculi, *Fund. Math.* **50** (1961), 165–190.

[14] A. ROSE and J. B. ROSSER, Fragments of many valued statement calculi, *Trans. Amer. Math. Soc.* **87** (1958), 1–53.

[15] J. B. ROSSER, Axiomatization of infinite valued logics, *Logique et Analyse*, n.s. **3** (1960), 137–153.

[16] B. SCARPELLINI, Die Nichtaxiomatisierbarkeit des unendlichwertigen Prädikatenkalküls von Łukasiewicz, *J. Symbolic Logic* **27** (1962), 159–170.

[17] T. SKOLEM, Bemerkungen zum Komprehensionsaxiom, *Z. Math. Logik* **3** (1957) 1–17.

[18] M. WAJSBERG, Beiträge zum Metaaussagenkalkül I, *Monatsh. Math. Physik* **42** (1935), 221–242.

FOUNDATIONS OF HIGHER-ORDER LOGIC[1]

DAVID KAPLAN AND RICHARD MONTAGUE
University of California, Los Angeles, California, U.S.A.

The purpose of this paper is to lay the foundations of the model theory of higher-order logic. This seems not to have been done previously in full generality, in particular, in such a way as to accommodate transfinite types and non-logical predicates of higher-order objects.[2] The investigations lead incidentally to a solution, or perhaps a partial solution, of an old problem of Tarski, to define a general notion of isomorphism appropriate to higher-order structures.

We consider the present investigations as relevant to the foundations of set theory, because from one point of view the various known systems of set theory may be regarded as axiomatic fragments of a single non-axiomatic higher-order logic.[3]

For simplicity we employ a rather rich metatheory. In the first place, we assume the existence of individuals (*Urelemente*, that is, objects other than the empty set which have no members) and a good many of them; the exact assumption is that for every set A there is a set of individuals which can be put into biunique correspondence with A. In the second place, we assume, in addition to the existence of individuals and sets, the existence of *proper classes* (as in the theory of Kelley [3]), as well as of objects, which might be called *superclasses*, of several higher levels. In particular, we wish to be able to construct functions taking proper classes as values. To do this in the natural way we need a set theory with *four levels of classes.*

It would be quite easy to construct an axiomatic metatheory satisfying our two assumptions; clear indications are given in Montague, Scott, Tarski [7], Chapter 6, Sections 22 and 23. Neither assumption, however, is essential; at the expense of complicating certain constructions we could work entirely

---

[1] We are grateful to Professors Alfred Tarski and R. L. Vaught for stimulating suggestions and discussion, and to the U.S. National Science Foundation for its support of this work under Grant No. GP-1603 (Montague).

[2] For higher-order formulas in which no non-logical predicates occur other than those taking individuals as arguments the relevant model-theoretic concepts were introduced in Montague [5] and Montague [6].

[3] This viewpoint is discussed briefly in Section 3 of Montague [5]; to render the viewpoint plausible we must exclude from the realm of set theories the systems of Quine [8] and Quine [9].

within the standard set theory of Bernays-Morse, as presented in the Appendix of Kelley [3]. Individuals could be avoided by a procedure indicated in Montague [5] and [6], or by another procedure indicated in Montague, Scott, Tarski [7], Chapter 6, Section 23; and superclasses could be avoided by using certain proper classes to 'represent' functions whose domain is the universe of sets and whose values are proper classes. The method of representation is rather obvious, but is given in detail in Montague, Scott, Tarski [7], Chapter 6, Section 21.

Nevertheless, the discussion is considerably clearer if conducted in a rich metatheory. We leave to the reader a translation into a more limited framework.

For reasons which will appear later it would amount to a genuine loss of generality if we were to confine attention to monadic higher-order logic. In order, then, to allow for polyadic relations, we characterize the class of *types* as the smallest class containing all ordinals and closed under the formation of non-empty finite sequences. It is assumed that the ordinals are constructed in such a way that no ordinal is a non-empty finite sequence.

As ingredients of formulas we assume the following disjoint classes of symbols to be available: (1) the usual *logical symbols* $\neg$, $\wedge$, $\vee$, $\rightarrow$, $\leftrightarrow$, $\bigwedge$, $\bigvee$, $=$, [,] (respectively 'not', 'and', 'or', 'if...then', 'if and only if', 'for all', 'for some', 'is identical with', left bracket, right bracket), which are assumed to be distinct, (2) for each type τ, a denumerable infinity of *variables* of type τ, and (3) for each type τ, a class of *constants* of type τ. For example, what is usually called an individual constant would be a constant of type 0, and a 2-place predicate of objects of types σ and τ would be a constant of type $\langle\sigma, \tau\rangle$.

Each symbol is assumed to be a 1-place sequence. By an *expression* is understood a finite sequence each of whose 1-place subsequences is a symbol. Concatenation of sequences is indicated by juxtaposition. The class of (*higher-order*) *formulas* is the least class Γ such that (1) Γ contains all expressions $u = v$, Fu, $Pu_0 \ldots u_n$, where n is a natural number, u, v, u_0, ..., u_n are variables or constants of arbitrary type, F is a variable or constant whose type is an ordinal, and P is a variable or constant whose type is an (n + 1)-place sequence, (2) Γ is closed under the application of sentential connectives, and (3) $\bigwedge u\,\varphi$, $\bigvee u\varphi$ are in Γ whenever φ is in Γ and u is a variable of any type. The *atomic formulas* are those comprehended under (1). Thus only very minimal stratification conditions are imposed.

The notion of a *free* variable of a formula is understood in the expected way, and a (*higher-order*) *sentence* is a formula without free variables.

Before considering the models that correspond to higher-order formulas we must introduce a few preliminary set-theoretical notions. By (x, y), or the *ordered pair* of x and y, is understood the set $\{\{x\}, \{x, y\}\}$, whose members are the unit set of x and the unordered pair of x and y. If n is a positive integer, then $(x_i)_{i<n}$, or the *ordered n-tuple* of $x_0, \ldots, x_{n-1}$, and $\Pi_{i<n} A_i$, or the *Cartesian product* of the sets $A_0, \ldots, A_{n-1}$, are defined by the following recursions: $(x_i)_{i<1}$ is x_0; $\Pi_{i<1} A_i$ is A_0; if n is a positive integer, then $(x_i)_{i<n+1}$ is $((x_i)_{i<n}, x_n)$, and $\Pi_{i<n+1} A_i$ is the set of ordered pairs (x, y), where x is a member of $\Pi_{i<n} A_i$ and y is a member of A_n. Clearly, if n is a positive integer and $x_i \in A_i$ for all $i < n$, then $(x_i)_{i<n} \in \Pi_{i<n} A_i$.

With each set A and each type τ we may correlate a set $U(\tau, A)$, called the *τth universe* on A, by means of the following recursion[4]: $U(0, A)$ is A; if α is an ordinal, then $U(\alpha + 1, A)$ is the union of $U(\alpha, A)$ with the power set (set of all subsets) of $U(\alpha, A)$; if α is a limit ordinal, then $U(\alpha, A)$ is the union of all sets $U(\beta, A)$ for $\beta < \alpha$; if n is a positive integer and τ an n-place sequence of types, then $U(\tau, A)$ is the power set of $\Pi_{i<n} U(\tau_i, A)$. Thus the universes corresponding to ordinals are cumulative, whereas those corresponding to other types are in general not. Notice that if α is an ordinal, then $U(\langle\alpha\rangle, A) \subseteq U(\alpha + 1, A)$; here $\langle\alpha\rangle$ is the 1-place sequence of α. If Γ is a class of types, understand by the *closure* of Γ the smallest class including Γ and closed under formation of non-empty finite sequences. Then if Γ is a set of ordinals, α is a limit ordinal greater than all members of Γ, and τ is a member of the closure of Γ, then $U(\tau, A) \subseteq U(\alpha, A)$.

By a *model* (or a *structure*, or a *higher-order structure*) is understood a 2-place sequence $\langle A, I\rangle$ such that A is a set (possibly empty), I is a function whose domain is a set of constants, and whenever c is a constant of type τ in the domain of I, $I(c) \in U(\tau, A)$; we call A the *basic universe*, and the domain of I the *language*, of the model $\langle A, I\rangle$. For example, *first-order relational systems* coincide exactly with those models whose languages consist of constants having types among $\langle 0\rangle$, $\langle 0, 0\rangle$, $\langle 0, 0, 0\rangle$, and so on; every topological space (in the sense of Kelley [3]) may be construed as a model whose language consists of a single constant of type $\langle\langle 0\rangle\rangle$ (or equally well of type 2), regarded as the predicate of being an open set of points; every uniform space (again in the sense of Kelley [3]) may be construed as a model whose language consists of a single constant of type $\langle\langle 0, 0\rangle\rangle$; and every uniform covering system (in the sense of Tukey [10]) may be construed as a model whose language consists of a single constant of type $\langle\langle\langle 0\rangle\rangle\rangle$ (or of type 3). Every Fréchet space may be construed as a model with a language consisting of a single constant of type $\langle\langle\omega, 0\rangle, 0\rangle$, regarded as a 2-place predicate whose two arguments are respectively an infinite sequence of points and a point; by an infinite sequence is understood a many-one

---

[4] This is of course *not* a recursion on the less-than relation among ordinals, but rather a recursion on a somewhat more complicated *well-founded relation*. For a general justification of such recursions see Montague, Scott, Tarski [7], Chapter 6, Section 21.

relation whose domain is the set of finite ordinals. (In this connection observe that if A is any set, then all finite ordinals are members of $U(\omega, A)$, and hence all infinite sequences of members of A are members of $U(\langle\omega, 0\rangle, A)$.) Every system of classical particle mechanics (in the sense of McKinsey, Sugar, and Suppes [4]) may be construed as a model with four constants, of types $\langle\omega + 2\rangle$, $\langle 0, \omega + 2\rangle$, $\langle 0, \omega + 2, \omega + 2\rangle$, and $\langle 0, \omega + 2, \omega, \omega + 2.\rangle$

By a *normal model* we understand one whose basic universe consists of individuals, and by a *finitary non-cumulative model* one whose language consists of constants having types in the closure of $\{0\}$. For models of both these kinds there is no difficulty in defining isomorphism. Indeed, if f is a function with domain A and τ a type in the closure of $\{0\}$, introduce f^+ and f^τ by the following recursions[5]: if x is an individual, then f^+x is $f(x)$; if x is a set, then f^+x is the set of objects f^+y for $y \in x$; $f^0 = f$; if n is a positive integer and τ an n-place sequence of types in the closure of $\{0\}$, then f^τ is the function with domain $U(\tau, A)$ such that, for all $R \in U(\tau, A)$, $f^\tau(R)$ is the set of n-tuples $(f^{\tau_i}(x_i))_{i<n}$ for which $(x_i)_{i<n} \in R$. Then we have:

Definition 1. If $\mathfrak{A}(= \langle A, I\rangle)$ and $\mathfrak{B}(= \langle B, J\rangle)$ are normal models, then f is an *isomorphism* from $\mathfrak{A}$ to $\mathfrak{B}$ if and only if $\mathfrak{A}$ and $\mathfrak{B}$ have the same language, f is a biunique correspondence between A and B, and $f^+I(c) = J(c)$ for each constant c in the language of $\mathfrak{A}$.

Definition 2. If $\mathfrak{A}(= \langle A, I\rangle)$ and $\mathfrak{B}(= \langle B, J\rangle)$ are finitary non-cumulative models, then f is an *isomorphism* from $\mathfrak{A}$ to $\mathfrak{B}$ if and only if $\mathfrak{A}$ and $\mathfrak{B}$ have the same language, f is a biunique correspondence between A and B, and $f^\tau(I(c)) = J(c)$ whenever c is a constant of type τ in the language of $\mathfrak{A}$.

Definition 3. If $\mathfrak{A}$ and $\mathfrak{B}$ both satisfy the hypothesis of one of the two preceding definitions, then $\mathfrak{A}$ and $\mathfrak{B}$ are *isomorphic* if and only if there is an isomorphism from $\mathfrak{A}$ to $\mathfrak{B}$.

It is clear that Definitions 1 and 2 coincide on their common domain of applicability, and coincide with the usual first-order notion of an isomorphism when applied to first-order relational systems.

There is also no difficulty in defining truth and satisfaction for normal models. For such models the following preliminary sense of these notions is adequate.

---

[5] The first recursion is on membership, and the second on another well-founded relation; for justifications see Montague, Scott, Tarski [7].

DEFINITION 4. Let $\mathfrak{A}(=\langle A, I\rangle)$ be a model, φ a formula all of whose constants are in the language of $\mathfrak{A}$, and a a function whose domain contains all free variables of φ and which is such that whenever u is a variable of type τ in the domain of a, $a(u) \in U(\tau, A)$. Then *a satisfies* φ in $\mathfrak{A}$ *in the preliminary sense* if and only if, roughly speaking, φ is true when each of its constants c is understood as denoting $I(c)$, each of its free variables u is understood as denoting $a(u)$, and, for every type τ, each of the bound variables of φ of type τ is understood as ranging over the set $U(\tau, A)$. (An exact recursive definition corresponding to this rough indication, and hence an explicit definition as well, can easily be constructed using the method of Tarski; for an example of such a definition, in a context limited to first-order formulas but admitting, as here, the possibility of an empty universe, see Kalish and Montague [1].) If φ is a sentence, then φ is *true in* $\mathfrak{A}$ *in the preliminary sense* if and only if the empty function satisfies φ in $\mathfrak{A}$ in the preliminary sense.

Thus the basic model-theoretic notions can be defined in a completely natural and obvious way for normal models. As the mathematical examples above indicate, however, we definitely do not wish to restrict attention to such models; for instance, it is important to admit topological spaces and Fréchet spaces whose points are the real numbers or other entities having a complex set-theoretical structure. When we pass beyond normal models, serious difficulties arise in characterizing satisfaction, truth, and isomorphism.

To illustrate the difficulties connected with truth, we need not depart from the domain of first-order relational systems. Indeed, let Λ be as usual the empty set, a an individual, and $\mathfrak{A}, \mathfrak{B}$ isomorphic first-order relational systems with the respective basic universes $\{a\}$, $\{\Lambda\}$; since we are dealing with first-order systems, the notion of isomorphism is here not problematic. Let u, v, w be distinct variables of type 1, and let φ be the sentence

$$\bigvee u \bigvee v \bigvee w[\neg u = v \wedge \neg u = w \wedge \neg v = w] .$$

In the sense given by Definition 4, φ is seen to be true in $\mathfrak{A}$ but not in $\mathfrak{B}$; this is because $U(1, \{a\})$ has three members and $U(1, \{\Lambda\})$ only two. It is of course unacceptable for two isomorphic models to differ in their true sentences.

Such difficulties as the one encountered here could be avoided if we were willing to restrict attention to sentences satisfying a very strong stratification condition:

THEOREM 1. *Let* $\mathfrak{A}, \mathfrak{B}$ *be isomorphic finitary non-cumulative models. Let*

Γ be the class of all limit ordinals together with 0. Let φ be a sentence all of whose constants are in the language of $\mathfrak{A}$, all of whose variables are in the closure of Γ, and each of whose atomic subformulas has the form $x = y$, where x, y are variables or constants of the same type, or else the form $Px_0 \ldots x_n$, where P, $x_0, \ldots, x_n$ are variables or constants of the respective types $\langle \tau_0, \ldots, \tau_n \rangle$, $\tau_0, \ldots, \tau_n$. Then φ is true in $\mathfrak{A}$ in the preliminary sense if and only if φ is true in $\mathfrak{B}$ in the preliminary sense.

(Proofs will be omitted because of limitations of space.)

The stratification condition imposed in this theorem is, however, so strong as to rule out a number of natural constructions. For instance, reference to the finite ordinals could naturally take place by way of variables of type ω. Now the property of being an object of type ω which is a finite ordinal, as well as the basic relations and operations connected with such objects, can be expressed by higher-order formulas not involving (non-logical) constants, but not, it seems, by formulas satisfying the condition in Theorem 1. (On the other hand, if one is willing to regard the finite ordinals as objects of type $\langle\langle\omega\rangle\rangle$, suitable constructions, not violating the stratification condition in Theorem 1, can be performed on the basis of suggestions in Whitehead and Russell [11].)

In any case, Theorem 1 offers no help in defining isomorphism generally, and there are difficulties here. Suppose that $\mathfrak{A} = \langle A, I \rangle$, $\mathfrak{A}$ is a Fréchet space, P is the only member of the language of $\mathfrak{A}$, and A contains all finite ordinals. $\mathfrak{A}$ would seem to be isomorphic to some normal model $\mathfrak{B} = \langle B, J \rangle$ in which $J(P)$ is a non-empty 2-place relation between infinite sequences of members of B and members of B. On the other hand, let f be a biunique correspondence between A and some set C of individuals. Let $\mathfrak{C} = \langle C, K \rangle$, and let $\mathfrak{C}$ be a model such that $K(P) = f^{\langle\langle 0,0\rangle,0\rangle}(I(P))$. (Note that although P has type $\langle\langle\omega, 0\rangle, 0\rangle$, $I(P)$ will in this case be a member of $U(\langle\langle 0, 0\rangle, 0\rangle, A)$.) Then $\mathfrak{A}$ would seem to be isomorphic to $\mathfrak{C}$. But $\mathfrak{B}$ and $\mathfrak{C}$ are normal models which in view of Definition 3 are clearly not isomorphic.

The problem of defining truth can be approached in the following way. When we say that a higher-order sentence φ is true in a model $\mathfrak{A}$, what we mean intuitively is that φ *would be* true in $\mathfrak{A}$ in the preliminary sense if $\mathfrak{A}$ *were* a normal model. The difficulty lies in eliminating the subjunctive conditional in this criterion. Such an elimination, however, is rather simple in the following restricted case.

DEFINITION 5. If $\mathfrak{A}$ is a finitary non-cumulative model and φ a sentence all of whose constants are in the language of $\mathfrak{A}$, then φ is *true in* $\mathfrak{A}$ if and only if there is a normal finitary non-cumulative model $\mathfrak{B}$ such that $\mathfrak{A}$ is isomorphic to $\mathfrak{B}$ and φ is true in $\mathfrak{B}$ in the preliminary sense.

This method of eliminating the subjunctive conditional cannot be extended beyond the case of finitary non-cumulative models, because for other models we do not yet have an adequate notion of isomorphism.

It turns out that the problems of defining satisfaction, truth, and isomorphism can be solved in a uniform way. We shall find a method of associating with each model an 'idealization', which will be a normal model; and we shall reduce the model-theoretic notions as applied to arbitrary models to the same notions as applied to idealizations.

By an *idealization function* for a set A is understood a biunique function f such that the domain of f is A and the range of f is a set of individuals. (One of our set-theoretical assumptions guarantees that every set has an idealization function.) If f is an idealization function for some set and x is an individual or set, then the *extension* of x with respect to f is $g^{+}x$, where g is the converse of f. (We think of x as an 'ideal object' or 'intension', that is, an object in the hierarchy built on the genuine individuals in the range of f, and the extension of x as the corresponding object in the hierarchy built on the 'pseudo-individuals' in the domain of f. There may in general be many intensions with the same extension; this lies at the heart of the difficulties we have considered.) By an *analysis* corresponding to a set A is understood a function F such that the domain of F is the class of all types, F_0 is an idealization function for A, and, for every type τ, we have: (1) F_τ is a function, (2) the domain of F_τ is $U(\tau, A)$, and (3) for each $a \in U(\tau, A)$, $F_\tau(a)$ is a member of $U(\tau, B)$ (where B is the range of F_0) and a is the extension of $F_\tau(a)$ with respect to F_0. A *total analysis* is a function $\mathscr{F}$ whose domain is the class of all sets and which is such that, for each set A, $\mathscr{F}(A)$ is an analysis corresponding to A. (It is to accommodate total analyses that we have acknowledged functions taking proper classes as values.) If $\mathscr{F}$ is a total analysis, $\mathfrak{A} = \langle A, I \rangle$, $\mathfrak{A}$ is a model, and $F = \mathscr{F}(A)$, then the *$\mathscr{F}$-idealization* of $\mathfrak{A}$ is the model $\langle B, J \rangle$, where B is the range of F_0, the domain of J is the language of $\mathfrak{A}$, and J is such that whenever c is a constant of type τ in the language of $\mathfrak{A}$, $J(\mathrm{c})$ is $F_\tau(I(\mathrm{c}))$.

DEFINITION 6. If $\mathfrak{A}$ is a model, $\mathscr{F}$ a total analysis, and φ a sentence all of whose constants are in the language of $\mathfrak{A}$, then φ is *true in $\mathfrak{A}$ with respect to $\mathscr{F}$* if and only if φ is true in the preliminary sense in the $\mathscr{F}$-idealization of $\mathfrak{A}$.

DEFINITION 7. If $\mathfrak{A}\,(= \langle A, I \rangle)$ and $\mathfrak{A}'(= \langle A', I' \rangle)$ are models and $\mathscr{F}$ is a total analysis, then f is an *$\mathscr{F}$-isomorphism* from $\mathfrak{A}$ to $\mathfrak{A}'$ if and only if $\mathfrak{A}$

and $\mathfrak{A}'$ have the same language, f is a biunique correspondence between A and A', and the set of ordered pairs $(F_0(a), F'_0(f(a)))$ for a in A is an isomorphism from $\mathfrak{B}$ to $\mathfrak{B}'$; here F and F' are to be $\mathscr{F}(A)$ and $\mathscr{F}(A')$, respectively, and $\mathfrak{B}$ and $\mathfrak{B}'$ are to be the respective $\mathscr{F}$-idealizations of $\mathfrak{A}$ and $\mathfrak{A}'$. Moreover, $\mathfrak{A}$ and $\mathfrak{A}'$ are *isomorphic with respect to* $\mathscr{F}$ if and only if there is an $\mathscr{F}$-isomorphism from $\mathfrak{A}$ to $\mathfrak{A}'$.

DEFINITION 8. Let $\mathfrak{A}(= \langle A, I \rangle)$ be a model, $\mathscr{F}$ a total analysis, φ a formula all of whose constants are in the language of $\mathfrak{A}$, and a a function whose domain contains all free variables of φ and which is such that whenever u is a variable of type τ in the domain of a, $a(u) \in U(\tau, A)$. Then *a satisfies φ in $\mathfrak{A}$ with respect to* $\mathscr{F}$ if and only if b satisfies φ (in the preliminary sense) in the $\mathscr{F}$-idealization of $\mathfrak{A}$; here b is to be that function whose domain is the set of free variables of φ and which is such that whenever u is a variable of type τ in the domain of b, $b(u)$ is $F_\tau(a(u))$, where $F = \mathscr{F}(A)$.

It is obvious that *logical truth* (truth in all models) does not depend on the choice of a total analysis:

THEOREM 2. *If $\mathscr{F}$ is a total analysis and φ a sentence, then the following conditions are equivalent*: (1) *φ is true with respect to $\mathscr{F}$ in every model whose language contains all constants of* φ; (2) *φ is true in the preliminary sense in every normal model whose language contains all constants of* φ.

A similar remark applies to the model-theoretic relation of *logical consequence* (between a sentence and a set of sentences). Observe that in connection with logical truth non-logical constants are unimportant: they can be replaced in a biunique way by universally quantified variables of the same types, and the resulting sentence will be logically true if and only if the original sentence is logically true. For sentences without constants the property of logical truth has been investigated to a certain extent in Montague [6]. There it is shown that for a large class of higher-order sentences the question of logical truth can be reduced to the corresponding question for sentences of pure dyadic second-order logic (that is, sentences without constants whose variables have types 0, $\langle 0 \rangle$, or $\langle 0, 0 \rangle$), and hence (by way of the usual reduction of relations to sets) to the corresponding question for sentences of pure monadic fourth-order logic (that is, sentences without constants whose variables have types 0, $\langle 0 \rangle$, $\langle\langle 0 \rangle\rangle$, or $\langle\langle\langle 0 \rangle\rangle\rangle$). Kaplan has recently, in [2], accomplished a further reduction: with each sentence of pure dyadic second-order logic can be effectively associated a logically equivalent sentence of pure monadic *third-order* logic.

In considering other notions than logical truth, the relativization to a total analysis which occurs in Definitions 6–8 is undesirable. We shall now see to what extent this relativization can be avoided.

Let α be an ordinal. By $\lim_{\gamma<\alpha} A_\gamma$ we understand the unique object B for which there exists an ordinal $\beta < \alpha$ such that, for every ordinal γ, if $\beta \leqq \gamma < \alpha$, then $A_\gamma = B$ — provided that there is exactly one such object B; otherwise we identify $\lim_{\gamma<\alpha} A_\gamma$ with the empty set. If F is an analysis corresponding to some set, then an object x is said to be *α-stable* with respect to F if $\lim_{\gamma<\alpha} F_\gamma(x)$ exists, that is, if there is exactly one object B satisfying the condition above, with 'A_γ' replaced by '$F_\gamma(x)$'.

A *standard total analysis* is a total analysis $\mathscr{F}$ such that if A is any set and F is $\mathscr{F}(A)$, we have the following (for all α, x, n, $\tau_0, \ldots, \tau_{n-1}, \tau$): (1) if α is an ordinal, x is a subset of $U(\beta, A)$ for some ordinal $\beta < \alpha$, and every member of x is α-stable with respect to F, then $F_\alpha(x)$ is the set, for $y \in x$, of objects $\lim_{\gamma<\alpha} F_\gamma(y)$; (2) if n is a positive integer, $\tau_0, \ldots, \tau_{n-1}$ are types, $\tau = \langle \tau_0, \ldots, \tau_{n-1} \rangle$ and $x \in U(\tau, A)$, then $F_\tau(x)$ is the set of ordered n-tuples $(F_{\tau_i}(a_i))_{i<n}$ for which $(a_i)_{i<n} \in x$. Intuitively speaking, a standard total analysis is one that assigns to each entity with several possible intensions the 'highest' of these.

With each set A we may associate two new hierarchies of universes by means of the following recursions: $V(0, A) = W(0, A) = A$; if α is an ordinal, then $V(\alpha + 1, A)$ is the union of $V(\alpha, A)$ with the power set of $V(\alpha, A)$, and $W(\alpha + 1, A)$ is the power set of $W(\alpha, A)$; if α is a limit ordinal, then $V(\alpha, A)$ is the intersection of $U(\alpha, B)$ (where B is the set of individuals in A) with the union of the sets $V(\beta, A)$ for $\beta < \alpha$, and $W(\alpha, A)$ is the intersection of $U(\alpha, \Lambda)$ with the union of the sets $W(\beta, A)$ for $\beta < \alpha$; if n is a positive integer and τ an n-place sequence of types, then $V(\tau, A)$ is the power set of $\Pi_{i<n} V(\tau_i, A)$ and $W(\tau, A)$ is the power set of $\Pi_{i<n} W(\tau_i, A)$. It is clear that if τ is any type and A any set, then $W(\tau, A) \subseteq V(\tau, A) \subseteq U(\tau, A)$.

By a *quasi-mathematical structure* is understood a model $\langle A, I \rangle$ such that whenever c is a constant of type τ in the domain of I, $I(c) \in V(\tau, A)$, and by a *mathematical structure* a model $\langle A, I \rangle$ satisfying the stronger condition that $I(c) \in W(\tau, A)$ whenever c is a constant of type τ in the domain of I.

THEOREM 3. *Assume that* (1) $\mathscr{F}$, $\mathscr{G}$ *are standard total analyses,* (2) $\mathfrak{A}$ *is a quasi-mathematical structure, and* (3) φ *is a sentence all of whose constants are in the language of* $\mathfrak{A}$. *Then* φ *is true in* $\mathfrak{A}$ *with respect to* $\mathscr{F}$ *if and only if* φ *is true in* $\mathfrak{A}$ *with respect to* $\mathscr{G}$.

THEOREM 4. *Assume* (1), (2) *of Theorem* 3, *that* (3) $\mathfrak{A} = \langle A, I \rangle$, *that* (4)

φ is a formula all of whose constants are in the language of $\mathfrak{A}$, and that (5) *a is a function whose domain contains all free variables of φ and which is such that whenever u is a variable of type τ in the domain of a, $a(u) \in V(\tau, A)$. Then a satisfies φ in $\mathfrak{A}$ with respect to $\mathcal{F}$ if and only if a satisfies φ in $\mathfrak{A}$ with respect to $\mathcal{G}$.*

THEOREM 5. *Assume* (1) *of Theorem* 3, *and that* (2) *$\mathfrak{A}$, $\mathfrak{B}$ are quasi-mathematical structures. Then f is an $\mathcal{F}$-isomorphism from $\mathfrak{A}$ to $\mathfrak{B}$ if and only if f is a $\mathcal{G}$-isomorphism from $\mathfrak{A}$ to $\mathfrak{B}$.*

Theorems 3–5 present the possibility of introducing absolute (or unrelativized) notions of truth, satisfaction, and isomorphism provided that we limit attention to quasi-mathematical structures.

DEFINITION 9. (i) Under assumptions (2) and (3) of Theorem 3, we say that φ is *true in* $\mathfrak{A}$ if and only if φ is true in $\mathfrak{A}$ with respect to some standard total analysis. (ii) Under assumptions (2)–(5) of Theorem 4, we say that a *satisfies* φ *in* $\mathfrak{A}$ if and only if a satisfies φ in $\mathfrak{A}$ with respect to some standard total analysis. (iii) Under assumption (2) of Theorem 5, we say that f is an *isomorphism* from $\mathfrak{A}$ to $\mathfrak{B}$ if and only if f is an $\mathcal{F}$-isomorphism from $\mathfrak{A}$ to $\mathfrak{B}$, for some standard total analysis $\mathcal{F}$.

Observe that a limitation to quasi-mathematical structures (or to mathematical structures) restricts only the possible values of constants and free variables; bound variables are understood as having their usual ranges. The notion of a mathematical structure is required for the following generalization of an important and familiar property of first-order isomorphism.

THEOREM 6. *If $\mathfrak{A}$ is a mathematical structure and f a biunique function whose domain is the basic universe of $\mathfrak{A}$, then there is a mathematical structure $\mathfrak{B}$ such that f is an isomorphism from $\mathfrak{A}$ to $\mathfrak{B}$.*

Clearly all topological spaces, uniform spaces, uniform covering systems, Fréchet spaces, systems of classical particle mechanics, and first-order relational systems, as well, it seems, as all other structures that naturally arise in mathematics, are what we here call mathematical structures. This assertion would not, however, apply to Fréchet spaces and systems of classical particle mechanics if they were treated within a *monadic* higher-order logic.

It is apparent that our final definition of isomorphism, Definition 9 (iii), coincides with each of Definitions 1 and 2 on the common domain of applicability, and that the notion of truth given in Definition 9 (i) coincides, for normal quasi-mathematical structures, with the notion given in Definition 4, and coincides with the notion of truth introduced in Montague [5] and [6] on the domain of applicability considered in the latter papers. It is apparent also that Definition 9(iii) gives the notion of homeomorphism in the case of topological spaces and Fréchet spaces, and the notion of uniform isomorphism in the case of uniform spaces.

We believe that for the case of mathematical structures Definition 9(iii) provides the general notion of isomorphism sought by Tarski, and doubt that a natural generalization beyond this domain is possible.

REFERENCES

[1] Kalish, D., and R. Montague. On Tarski's formalization of predicate logic with identity. To appear in *Archiv für mathematische Logik und Grundlagenforschung*.

[2] Kaplan, D. Relations reduced to classes. In preparation.

[3] Kelley, J. L. *General Topology*. Princeton, 1955.

[4] McKinsey, J. C. C., A. C. Sugar, and P. Suppes. Axiomatic foundations of classical particle mechanics. *Journal of Rational Mechanics and Analysis*, vol. **2** (1953), pp. 253–272.

[5] Montague, R. Set theory and higher-order logic. To appear in the proceedings of the 1963 Logic Colloquium at Oxford.

[6] Montague, R. Reductions of higher-order logic. To appear in the proceedings of the 1963 Symposium on the Theory of Models at Berkeley.

[7] Montague, R., D. Scott, and A. Tarski. *An Axiomatic Approach to Set Theory*. Amsterdam, forthcoming.

[8] Quine, W. V. New foundations for mathematical logic. *American Mathematical Monthly*, vol. **44** (1937), pp. 70–80.

[9] Quine, W. V. *Mathematical Logic*, revised edition. Cambridge, 1951.

[10] Tukey, J. W. Convergence and uniformity in topology. *Annals of Mathematical Studies*, vol. **2** (1940).

[11] Whitehead, A. N., and B. Russell. *Principia Mathematica*, vol. 1. Cambridge, 1910.

A SURVEY OF ULTRAPRODUCTS

H. JEROME KEISLER
University of Wisconsin, Madison, Wisconsin, U.S.A.

This is an expository article. Our aim is to give an overall view of the work done in ultraproducts and of the part which ultraproducts play in the foundations of mathematics. Some of the theorems we shall state are taken from the literature, but we shall also announce some new results in Sections 3 and 4. We shall usually skip the proofs, and instead merely indicate in a few lines what the proof is like. The proofs of the new results announced in this paper will be published in full elsewhere. There are many interesting open questions about ultraproducts, and we shall mention several of them in this paper.

We have divided the paper into four sections. The first section contains the definition of ultraproducts and a fundamental theorem. However, before giving the definition we prepare the reader for it in a way which will be helpful only if he is familiar with a proof of the Gödel completeness theorem. In the second section we discuss very informally some of the known applications of ultraproducts.

In the last two sections we cover more systematically two particular topics concerning ultraproducts: cardinality theorems in Section 3 and saturated models in Section 4. At the present time these are the two areas in which the theory of ultraproducts has been most fully developed.

Ever since model theory began, methods of constructing models have been of basic importance. The Gödel completeness theorem, for example, depends on the construction of a model. Ultraproducts have become one of the most useful methods of constructing models. The idea behind ultraproducts goes back to the construction by Skolem in 1934 of a non-standard model of arithmetic (see Skolem 17—p 699, Scott 27—# 2423).[1] In 1948, Hewitt 10—p 126 studied an operation in field theory which turns out to be just the ultraproduct operation when applied to fields (cf. Kochen 25—# 1992). The present notion of an ultraproduct was essentially

---

[1] Whenever possible, we shall refer to a paper simply by indicating its volume and page or review number in *Mathematical Reviews*; thus 17 — p 699 indicates page 699, Volume 17, and 27—#2423 indicates review number 2423 of Volume 27. The only papers which appear in the reference list at the end are those which are too recent, or too old, to have been reviewed yet.

given in 1955 by Łoś 17—p 700, who also stated the fundamental theorem. However, the intensive development of the subject began in 1958 with a series of abstracts by Frayne, Morel, Scott, and Tarski (see [7] for the references and additional historical information).

Some basic properties of ultraproducts are developed in detail in [7] and also in Kochen 25—#1992. More recent work can be put roughly into three categories: information about the models which are formed by the ultraproduct operation; applications of ultraproducts to problems in other areas; and generalizations of the ultraproduct. In this survey we shall limit our discussion to the first two categories. However, it should be emphasized that a considerable amount of work has been done on various generalizations of the ultraproduct, and some of these generalizations also have applications. One generalization, the reduced product, is also essentially due to Łoś 17—p 700 and is studied in [7], [3], [11], and earlier papers mentioned in [7]. Other kinds of modifications of the ultraproduct are studied or used, for example, in Kochen 25—# 1992, Keisler 26—# 6054, Robinson 27—# 3533, MacDowell-Specker 27—# 2425, and in [3], [4], [5], [6], [9], [12], [25]; see [7] for further references.

§1. Ultraproducts and the compactness theorem.

The most useful theorem in model theory is probably the compactness theorem: *If every finite subset of a set Σ of sentences in first order logic has a model, then Σ has a model.* The compactness theorem is a corollary of the completeness theorem of Gödel [10], due for uncountable languages to Malcev [20]. However, unlike the completeness theorem, the compactness theorem does not involve the notion of a formal deduction, and so it is desirable to prove it directly without using that notion. Perhaps the most elegant direct proof of the compactness theorem is the proof in [7] which uses ultraproducts.

The definition of an ultraproduct is so "non-constructive" that it may appear unnatural at first sight, and its applications may look like magic. As a corrective measure, we shall try to provide the reader with some helpful intuition before stating the definition. What we shall do is to give a direct proof of the compactness theorem. Although this proof is phrased in such a way that it never mentions ultraproducts or even ultrafilters, it is nevertheless substantially the same proof as the one using ultraproducts in [7].

We shall work with a first order predicate logic L with identity symbol $\doteq$ and the usual connectives $\neg$, $\vee$, $\wedge$, $\rightarrow$ and quantifiers $\forall$, $\exists$. To simplify notation we let L have only a single ternary predicate symbol P. However, all of our definitions and results can be carried over in an obvious way to the case where L has an arbitrary (finite, countable, or uncountable) sequence $P_0, P_1, \cdots, P_\alpha, \cdots$ of finitary predicate symbols. We let $\mathfrak{A} = \langle A, R \rangle$, $\mathfrak{B} = \langle B, S \rangle$, sometimes with subscripts, denote structures for L. $\mathfrak{A}$ is said to be a *model of* a set Σ of sentences of L if every member of Σ is true in $\mathfrak{A}$. If μ is an ordinal number, we let $L(\mu)$ be the language obtained by adding to L individual constants c_α, $\alpha < \mu$. If a is a function on μ into A, we denote by $(\mathfrak{A}, a)$ the structure for $L(\mu)$ obtained from $\mathfrak{A}$ by interpreting each constant c_α by the element a_α.

We say that a set Γ of sentences in $L(\mu)$ is a *complete theory in* $L(\mu)$ if: (i) every sentence in Γ has a model; (ii) the conjunction of any two members of Γ belongs to Γ; and (iii) for any sentence ϕ of $L(\mu)$, either $\phi \in \Gamma$ or $\neg \phi \in \Gamma$.

Lindenbaum showed that every set Γ which satisfies (i) and (ii) can be extended to a complete theory (cf. Tarski 17—p 1171).

Suppose now that $\mu > 0$ and let Γ be an arbitrary complete theory in $L(\mu)$. Γ determines a structure $(\mathfrak{B}, b)$ for $L(\mu)$ in the following way. For each constant c_α, let $\bar{c}_\alpha$ be the set of all c_β, $\beta < \mu$, such that $c_\alpha \simeq c_\beta \in \Gamma$. The universe set B of $\mathfrak{B}$ is the set $B = \{\bar{c}_\alpha : \alpha < \mu\}$. The relation S is defined so that $S(\bar{c}_\alpha, \bar{c}_\beta, \bar{c}_\gamma)$ if and only if $P(c_\alpha, c_\beta, c_\gamma) \in \Gamma$. Finally, the function b is such that $b_\alpha = \bar{c}_\alpha$ for each $\alpha < \mu$. We shall call $(\mathfrak{B}, b)$ the *constant structure* for Γ.

The following lemma is essentially due to Henkin 11—p 487, who used it in his proof of the completeness theorem.

LEMMA 1.1. *Suppose that* $\mu > 0$ *and* Γ *is a complete theory in* $L(\mu)$. *Assume that* Γ *has the property*:

(iv) *For any sentence* $(\exists v\ \phi(v)$ *of* $L(\mu)$, *there is a constant* c_α *of* $L(\mu)$ *such that*

$$(\exists v) \phi(v) \rightarrow \phi(c_\alpha) \in \Gamma.$$

Then the constant structure $(\mathfrak{B}, b)$ *for* Γ *is a model of* Γ.

To prove the lemma we show that each sentence ϕ of $L(\mu)$ has the property: ϕ holds in $(\mathfrak{B}, b)$ if and only if $\phi \in \Gamma$. This is done by induction on the length of ϕ. The crucial step is: if $\phi(c_\alpha)$ has the required property for every constant c_α, then $(\exists v) \phi(v)$ has the required property. That step is taken care of by (iv). We note that since "complete theory" is meant in the semantical sense, the proof of the above lemma uses only the semantical notion of truth and not the notion of deducibility.

Consider now a non-empty set I (the "index set"), and structures $\mathfrak{A}_i$, $i \in I$, for L. Let $C = \Pi_{i \in I} A_i$ be the cartesian product of the universe sets. Let c be a function on an ordinal μ onto C, and for each $i \in I$ let a_i be the function on μ into A_i given by $(a_i)_\alpha = c_\alpha(i)$, $\alpha < \mu$. Thus we may form structures $(\mathfrak{A}_i, a_i)$ for $L(\mu)$. We identify the functions $c_\alpha \in C$ with the constants c_α of $L(\mu)$.

The lemma below gives a wide class of complete theories Γ which do have the property (iv).

LEMMA 1.2. *Let Δ be the set of all sentences ϕ of $L(\mu)$ such that ϕ is true in every structure $(\mathfrak{A}_i, a_i)$, $i \in I$. Then any complete theory $\Gamma \supseteq \Delta$ in $L(\mu)$ has the property* (iv) *of Lemma* 1.1.

PROOF. Consider any sentence $(\exists v)\phi(v)$ of $L(\mu)$. For each $i \in I$, we may pick an element $d_i \in A_i$ so that if $(\exists v)\phi(v)$ is true in $(\mathfrak{A}_i, a_i)$ then d_i satisfies $\phi(v)$ in $(\mathfrak{A}_i, a_i)$. The function d so chosen belongs to C, and hence $d = c_\alpha$ for some $\alpha < \mu$. It follows that the sentence $(\exists v)\phi(v) \rightarrow \phi(c_\alpha)$ is true in every $(\mathfrak{A}_i, a_i)$, and hence belongs to Δ and thus to Γ.

We shall now prove the compactness theorem. Let Σ be a set of sentences of L such that every finite subset of Σ has a model. We may assume Σ is non-empty. Let I be the set of all finite conjunctions of sentences in Σ. For each $i \in I$, we may by hypothesis choose a model $\mathfrak{A}_i$ of i. Let us now adopt the notation of Lemma 1.2. Then for each $i \in I$, $(\mathfrak{A}_i, a_i)$ is a model of both the sentence i and the set of sentences Δ. It follows that every finite subset of $\Sigma \cup \Delta$ has a model. We may therefore extend $\Sigma \cup \Delta$ to a complete theory Γ in $L(\mu)$. Now by Lemma 1.2, Γ has the property (iv) of Lemma 1.1, and applying Lemma 1.1 we conclude that the constant structure $(\mathfrak{B}, b)$ for Γ is a model of Γ. Since $\Sigma \subseteq \Gamma$, $\mathfrak{B}$ is a model of Σ, and our proof is complete.

We now make a fresh start and turn to the definition of the ultraproduct. Let I be any non-empty set.

A set D of subsets of I is said to be an *ultrafilter over I* if: (i′) $0 \notin D$; (ii′) the intersection of any two members of D belongs to D; and (iii′) for any subset x of I, either $x \in D$ or $I - x \in D$.

The analogy between the definitions of an ultrafilter and of a complete theory is obvious. Indeed both notions are special cases of the notion of a prime dual ideal in a Boolean algebra (cf. Tarski 17—p 1171). Tarski [22] showed that any set D satisfying (i′) and (ii′) can be extended to an ultrafilter over I.

Now let $\mathfrak{A}_i$, $i \in I$, be structures for L and let $C = \Pi_{i \in I} A_i$. For f, $g \in C$ we write $f =_D g$ if $f(i) = g(i)$ for D-almost all $i \in I$, i.e., if $\{i : f(i) = g(i)\} \in D$. Let f/D be the set of all $h \in C$ such that $f =_D h$. The *ultraproduct* of the sets A_i, $i \in I$ (*modulo D*) is the set

$$D - \text{prod}\, \lambda i A_i = \{f/D : f \in C\}.$$

The *ultraproduct* of the structures $\mathfrak{A}_i$, $i \in I$ (*modulo D*) is the structure

$\mathfrak{B} = D - \text{prod}\, \lambda i\, \mathfrak{A}_i$ described as follows. The universe set B is the ultraproduct $D - \text{prod}\, \lambda i A_i$. The ternary relation S is defined so that for all $f, g, h \in C$, $S(f/D, g/D, h/D)$ if and only if we have $R_i(f(i), g(i), h(i))$ for D – almost $i \in I$. (The "D – almost" terminology is taken from Vaught 26—#4912.)

We must now tie up some loose ends by establishing the connection between the ultraproduct as we have just defined it and the constant structure for a complete theory Γ. Suppose we are given structures $\mathfrak{A}_i$, $i \in I$, and use the notation of Lemma 1.2. For each ultrafilter D over I, there naturally corresponds a complete theory $\Gamma(D)$ of $L(\mu)$; namely, $\Gamma(D)$ is the set of all sentences ϕ of $L(\mu)$ such that ϕ holds in $(\mathfrak{A}_i, a_i)$ for D – almost all $i \in I$. It is easy to check that $\Gamma(D)$ is not only a complete theory but also includes Δ. Conversely, any complete theory $\Gamma \supseteq \Delta$ is equal to $\Gamma(D)$ for some D; we need only take D to contain, for each $\phi \in \Gamma$, the set of all $i \in I$ such that ϕ is true in $(\mathfrak{A}_i, a_i)$. By comparing the definitions involved we see that if $(\mathfrak{B}, b)$ is the constant structure for $\Gamma(D)$, then $\mathfrak{B}$ is exactly the ultraproduct $D - \text{prod}\, \lambda i \mathfrak{A}_i$.

THEOREM 1.3. (*The Fundamental Theorem*) *For any formula* $\phi(v_0, \cdots, v_n)$ *of* L *and any* $f_0, \cdots, f_n \in C$, *the tuple* $f_0/D, \cdots, f_n/D$ *satisfies* ϕ *in* $D - \text{prod}\, \lambda i \mathfrak{A}_i$ *if and only if* $f_0(i), \cdots, f_n(i)$ *satisfies* ϕ *in* $\mathfrak{A}_i$ *for* D – *almost all* $i \in I$. *In particular, a sentence* ψ *holds in* $D - \text{prod}\, \lambda i \mathfrak{A}_i$ *if and only if* ψ *holds in* $\mathfrak{A}_i$ *for* D – *almost all* $i \in I$.

The Fundamental Theorem was stated by Łoś 17—p 700 and proved in [7] by induction on the length of ϕ. However, the Fundamental Theorem now follows at once from our previous discussion. We have $\mathfrak{B} = D - \text{prod}\, \lambda i \mathfrak{A}_i$, where $(\mathfrak{B}, b)$ is the constant structure for $\Gamma(D)$, and we have shown that $(\mathfrak{B}, b)$ is a model of $\Gamma(D)$. This means that a sentence ϕ of $L(\mu)$ is true in $(D - \text{prod}\, \lambda i \mathfrak{A}_i, b)$ if and only if ϕ is true in $(\mathfrak{A}_i, a_i)$ for D–almost all $i \in I$. Theorem 1.3 follows when we replace constants by free variables in ϕ.

If all the $\mathfrak{A}_i$ coincide with one structure $\mathfrak{A}$, the ultraproduct $D - \text{prod}\, \lambda i \mathfrak{A}_i$ is called an *ultrapower* and is written $D - \text{prod}\, \mathfrak{A}$. We define the *natural embedding*, d, on $\mathfrak{A}$ into its ultrapower $D - \text{prod}\, \mathfrak{A}$ so that for each $a \in A$, $d(a) = f/D$ where f is the constant function with value a. We wish to state a corollary of the Fundamental Theorem which gives an important property of the natural embedding. Following Tarski and Vaught 20—#1627, we say that a structure $\mathfrak{B}$ is an *elementary extension* of $\mathfrak{A}$, and $\mathfrak{A}$ an *elementary substructure* of $\mathfrak{B}$, if $A \subseteq B$ and, for each formula $\phi(v_0, \cdots, v_n)$ of L, every $a_0, \cdots, a_n \in A$ which satisfies ϕ in $\mathfrak{A}$ also satisfies ϕ in $\mathfrak{B}$. An *elementary embedding on* $\mathfrak{A}$ *into* $\mathfrak{B}$ is an isomorphism on $\mathfrak{A}$ onto an elementary substructure of $\mathfrak{B}$.

COROLLARY 1.4. *The natural embedding d on* $\mathfrak{A}$ *into* $D-\mathrm{prod}\,\mathfrak{A}$ *is an elementary embedding.*

The beauty of the ultraproduct construction lies in the fact that it is defined using purely set-theoretical considerations which are independent of the particular logic L or structures $\mathfrak{A}_i$, $i \in I$, involved. New possibilities open up when we define ultraproducts by means of ultrafilters instead of by means of complete theories $\Gamma \supseteq \Delta$; for instance, we may consider what happens when we take a single D and vary the stuctures $\mathfrak{A}_i$, $i \in I$. Such possibilities will be exploited in Sections 3 and 4.

In concluding this section we should point out that there are other directions than the one we have chosen from which one can make a gradual approach to the notion of an ultraproduct. From the viewpoint of abstract algebra, one may prefer to regard the ultraproduct as a modification of the direct product (see [7]), or of a quotient field (see Hewitt 10—p 126, Kochen 25—#1992, Gillman-Jerison 22—#6994). Other approaches may be found in Łoś 17—p 700, Robinson 27—#3533.

§2. Applications of ultraproducts.

In order to give an indication of the part which ultraproducts play in the foundations of mathematics, we shall indicate at this time some applications of them. In the last two sections of the paper we shall give some additional applications of ultraproducts, but our emphasis there will be on the study of ultraproducts for their own sake. By an *application* (*of ultraproducts*) we shall mean a result whose statement does not mention ultraproducts but whose proof uses them. We call an application *essential* if, in addition, every "reasonably clear" proof of the result which we know of uses ultraproducts. (Thus an essential application may become inessential when we find another proof of it.) As a rule, essential applications are more convincing than inessential ones in justifying the study of ultraproducts. One of our main objectives in this paper is to show that ultraproducts have a substantial body of essential applications. We shall mention some examples of essential applications from the literature. We shall also mention some inessential applications which, by introducing new methods of proof, have led to the discovery of new results.

After Skolem's original construction of a non-standard model of arithmetic was announced, other constructions were found. However, Skolem's construction suggested to Rabin 21—#5564 the following result, which is probably the earliest essential application. *If* $\mathfrak{m}$ *is a cardinal and* $\aleph_0 \leqq \mathfrak{m} = \mathfrak{m}^{\aleph_0}$, *then every structure* $\mathfrak{A}$ *of power* $\mathfrak{m}$ *has a proper elementary extension of power* $\mathfrak{m}$. The result can be proved in a simple way from the compactness theorem alone for languages with at most $\mathfrak{m}$ predicates, but for languages with more than $\mathfrak{m}$ predicates ultraproducts really seem to be needed. The result is proved as follows. Let I be a countable set and let D be an ultrafilter over I which is *non-principal*, i.e., contains no set of power one. Form the ultrapower $B = D - \text{prod}\,\mathfrak{A}$. It turns out that the natural elementary embedding d maps $\mathfrak{A}$ properly into $\mathfrak{B}$, so $\mathfrak{B}$ is isomorphic to a proper elementary extension of $\mathfrak{A}$. Moreover, the power of $\mathfrak{B}$ is at least $\mathfrak{m}$ and at most $\mathfrak{m}^{\aleph_0}$, hence exactly $\mathfrak{m}$.

In Keisler 26 – #6054 we give the following essential application of ultraproducts. It is a converse of Rabin's theorem above and is an improvement of another result of Rabin 21 – #5564. *If* $\mathfrak{m}$ *is a non-measurable cardinal and* $\mathfrak{m} < \mathfrak{m}^{\aleph_0}$, *then there exists a structure* $\mathfrak{A}$ *of power* $\mathfrak{m}$ *which has no proper elementary extension of power* $\mathfrak{m}$. A cardinal $\mathfrak{m}$ is said to be *measurable* if there is an ultrafilter D over a set of power $\mathfrak{m}$ which is non-principal and is *countably complete* (i.e., is closed under countable intersections). This time the key step of the proof is to show that if $\mathfrak{A}$ has enough relations, then for every proper elementary extension $\mathfrak{A}'$ of $\mathfrak{A}$, some ultrapower $\mathfrak{B}$ of $\mathfrak{A}$ of power at least $\mathfrak{m}^{\aleph_0}$ is elementarily embeddable in $\mathfrak{A}'$. This idea is exploited to obtain other essential applications in 26 – #6054 and [12]. Ultraproducts are also applied to uncountable logics in [24].

The proof in [7] of the compactness theorem is an inessential application. However, by a similar argument Fuhrken [8] obtained a compactness theorem for logic with the extra quantifier "there exist at least $\mathfrak{m}$", and that application is essential.

Using a result of Hanf 28 – #3943 for infinitary logics, Tarski 27 – #1382 proved that the first measurable cardinal, if it exists, is much greater than the first inaccessible cardinal. Keisler [13] gave a later proof of that result using ultraproducts modulo countably complete ultrafilters (an inessential application). Scott 26 – #1263 then proved, by taking an ultrapower of a whole model of set theory modulo a countably complete ultrafilter in that model, that the existence of a measurable cardinal and the axiom of constructibility contradict each other. Then Gaifman [9], iterating Scott's method, obtained some even more remarkable consequences of the existence of measurable cardinals. Rowbottom [21] independently obtained some similar improvements of Scott's result without using ultraproducts, but Gaifman's results remain essential applications. Other applications of ultraproducts to set theory can be found in [2], [17], [26], and Vopenka 25—≠ 13, 26 – #3611.

It is often convenient to use ultraproducts as a substitute for the compactness theorem, for example in non-standard analysis (see Robinson 27 – #3533). In that way, several classical theorems have been proved using ultrafilters instead of the full axiom of choice; e.g. see Luxemburg, 25 – 3837. For interesting applications of non-standard analysis see [23].

Kochen 25 − #1992 used ultraproducts to give a new proof of Tarski's result in 13 − p423 that the theory of real closed fields is decidable, and recently Ax and Kochen [1] used ultraproducts in a similar way to solve some other important problems, for instance to prove that the theory of the field of p-adic numbers is decidable (an inessential application).

The various applications of ultraproducts depend on the Fundamental Theorem 1.3 and on a few other basic theorems which we shall state later on. These basic theorems all point toward a single central question: *Given a structure* $\mathfrak{A}$, *which structures* $\mathfrak{B}$ *are isomorphic to ultrapowers of* $\mathfrak{A}$? Or more generally: *Given structures* $\mathfrak{A}_i$, $i \in I$, *which structures* $\mathfrak{B}$ *are isomorphic to ultraproducts of the* $\mathfrak{A}_i$?

The question is important because in most of the applications, the ultraproduct is used to show the existence of a model with certain properties, especially properties which cannot be expressed in first order logic. Two such properties, cardinality and saturation, are the subjects of the remainder of this paper.

§3. Cardinality theorems.

Let us assume, once and for all, that I is a set of infinite power $\mathfrak{m}$, and D is an ultrafilter over I. The most natural question concerning ultraproducts of sets is the cardinality question: Given sets A_i, $i \in I$, and an ultrafilter D over I, what is the cardinality of the set $D - \text{prod}\, \lambda i A_i$? It is clear that the cardinality of the set $D - \text{prod}\, \lambda i A_i$ depends only on D and on the cardinalities of the sets A_i, $i \in I$. We may therefore write $D - \text{prod}\, \lambda i \mathfrak{n}_i$ for the cardinality of an ultraproduct of sets of cardinalities $\mathfrak{n}_i$, $i \in I$, modulo D.

We begin with a few general observations. One recurrent theme which will soon become apparent in this paper is the interaction between properties of an ultrafilter D and properties of an ultraproduct $D - \text{prod}\, \lambda i \mathfrak{A}_i$ modulo D. Certain kinds of ultrafilters do not yield interesting ultraproducts. For instance, if D is a principal ultrafilter, say $\{i_0\} \in D$, then $D - \text{prod}\, \lambda i \mathfrak{A}_i$ is always isomorphic to $\mathfrak{A}_{i_0}$. All principal ultrafilters are countably complete. If D is countably complete, then an ultrapower $D - \text{prod}\, \mathfrak{A}$ is isomorphic to $\mathfrak{A}$ unless $\mathfrak{A}$ is so large that its power is measurable, and the situation is similar for ultraproducts (see [7]). We shall be concerned mainly with countably incomplete (and hence non-principal) ultrafilters.

We may also limit our attention to ultrafilters which are *uniform*, in the sense that every element $x \in D$ has the power $\mathfrak{m}$ of I. For if D is not uniform, let J be a member of D of minimum cardinality; then the set $E = D \cap S(J)$ is a uniform ultrafilter over J and it is clear that $D - \text{prod}\, \lambda i \mathfrak{A}_i$ is always isomorphic to $E - \text{prod}\, \lambda j \mathfrak{A}_j$ (where $S(J)$ is the power set of J).

Let us now turn to the cardinality question. For simplicity we shall, for the time being, consider only ultrapowers. Let $\mathfrak{n}, \mathfrak{p}, \cdots$ be arbitrary cardinals.

THEOREM 3.1. (i) $\mathfrak{n} \leqq D - \text{prod}\, \mathfrak{n} \leqq \mathfrak{n}^{\mathfrak{m}}$.

(ii) *If D is uniform, then* $\mathfrak{m} < D - \text{prod}\, \mathfrak{m}$.

(iii) $D - \text{prod}\, (\mathfrak{n}^{\mathfrak{p}}) = (D - \text{prod}\, \mathfrak{n})^{\mathfrak{p}}$.

(iv) *If* $\mathfrak{n} \geqq \aleph_0$ *and* D *is countably incomplete, then* $D - \text{prod}\, \mathfrak{n} = (D - \text{prod}\, \mathfrak{n})^{\aleph_0}$.

Parts (i) and (ii) are proved in [7], and parts (iii) and (iv) in [14]. An important open question is the following (from [7]).

Question 3A. *Is it true that* $D - \text{prod}\, \mathfrak{n} = \mathfrak{n}^{\mathfrak{m}}$ *whenever D is uniform and* $\mathfrak{n}$ *is infinite*?

The answer is not known even if we assume the generalized continuum hypothesis. Assuming D is uniform, we may state various weaker questions which are also open: Does $\mathfrak{n} < D - \text{prod}\, \mathfrak{n}$ whenever $\aleph_0 \leqq \mathfrak{n} \leqq \mathfrak{m}$? Does $\mathfrak{n} < D - \text{prod}\, \mathfrak{n}$ imply that $D - \text{prod}\, \mathfrak{n}$ is a regular cardinal (from [7])? Is $\mathfrak{m} < D - \text{prod}\, \aleph_0$?

If we consider only regular ultrafilters, then we can answer question 3A affirmatively. We say that D is *regular* if there exists a subset $X \subseteq D$ of power $\mathfrak{m}$ such that the intersection of any infinite subset of X is empty. It is known that regular ultrafilters exist over I. Every regular ultrafilter, it turns out, is uniform and countably incomplete. For a discussion of regular ultrafilters see [14] or [3]. The following theorem is proved in [7].

THEOREM 3.2. *If D is regular and* $\mathfrak{n}$ *is infinite, then* $D - \text{prod}\, \mathfrak{n} = \mathfrak{n}^{\mathfrak{m}}$.

We state another open question.

Question 3B. *Is every uniform ultrafilter regular*?

An affirmative answer to Question 3B would imply, using Theorem 3.2, an affirmative answer to Question 3A. In just one case, namely $\mathfrak{m} = \aleph_0$, we do know that the answer to Question 3B is yes.

Using regular ultrafilters, Chang and Keisler 26 – # 3606 give another essential application of ultraproducts, a Löwenheim-Skolem type theorem for two cardinals. Let $\mathfrak{A} = \langle A, U, R \rangle$ be a structure where A has power $\mathfrak{n}$ and U has power $\mathfrak{p}$, with $\aleph_0 \leqq \mathfrak{p} \leqq \mathfrak{n}$. Then there exists an elementary extension $\mathfrak{B} = \langle B, V, S \rangle$ of $\mathfrak{A}$ with B of power $\mathfrak{n}^{\mathfrak{m}}$ and V of power $\mathfrak{p}^{\mathfrak{m}}$. The structure $\mathfrak{B}$ is taken to be isomorphic to an ultrapower $D - \text{prod}\, \mathfrak{A}$

where D is any regular ultrafilter, and the cardinalities of the sets B and V are computed using Theorem 3.2.

If n is a finite cardinal, then $D - \operatorname{prod} n = n$. However, if n_i, $i \in I$, are finite cardinals, the ultraproduct $D - \operatorname{prod} \lambda i n_i$ may well be infinite. We now take up the cardinality problem for ultraproducts of finite sets. Let $F(D)$ denote the set of all infinite cardinals $\mathfrak{p}$ such that for some finite cardinals n_i, $i \in I$, we have $\mathfrak{p} = D - \operatorname{prod} \lambda i n_i$. Thus $F(D)$ is the set of all infinite cardinals obtainable as ultraproducts of finite cardinals modulo D. We ask: which sets of cardinals are representable in the form $F(D)$? We summarize below some results known from the literature (see [7]).

THEOREM 3.3. (i) *$F(D)$ is non-empty if and only if D is countably incomplete.*

(ii) *For all $\mathfrak{p} \in F(D)$, $2^{\aleph_0} \leqq \mathfrak{p} \leqq D - \operatorname{prod} \aleph_0$.*

(iii) *If D is regular, then $2^{\mathfrak{m}} \in F(D)$.*

We shall see that the lower bound $2^{\aleph_0}$ of part (ii) need not belong to $F(D)$. Part (iii) implies that, for D regular, the upper bound $D - \operatorname{prod} \aleph_0$ does belong to $F(D)$. More generally we may ask:

Question 3C. *Is $D - \operatorname{prod} \aleph_0 \in F(D)$ for every countably incomplete D?*

Question 3D. *Is it true that $\mathfrak{p} = \mathfrak{p}^{\aleph_0}$ for all D and all $\mathfrak{p} \in F(D)$?* (Compare with Theorem 3.1 (iv) above.)

From now on all of our principal results depend on the generalized continuum hypothesis (G.C.H.). We shall state a new theorem which gives a sufficient condition for a set of cardinals to be of the form $F(D)$. For an arbitrary set C of cardinals, we let $C^+ = \{\mathfrak{n}^+ : \mathfrak{n} \in C\}$ be the set of all cardinal successors $\mathfrak{n}^+$ of elements $\mathfrak{n}$ of C. Our theorem will show, as a special case, that for any finite set C of infinite cardinals there exists an ultrafilter D such that $F(D) = C^+$ (G. C. H.).

THEOREM 3.4 (G. C. H.) *Let C be a non-empty set of infinite cardinals such that:*

(a) *the supremum of any non-empty subset of C belongs to C;*

(b) *for each $\mathfrak{n} \in C$, the set of all cardinals $\mathfrak{p} < \mathfrak{n}$ which belong to C has power less than $\mathfrak{n}$.*

Let $\mathfrak{m}$ be the supremum of C. Then there exists a regular ultrafilter D over a set of power $\mathfrak{m}$ such that $F(D) = C^+$.

Note in particular that any non-empty finite set C of infinite cardinals satisfies (a), (b). The first step in the proof is to show that for each $\mathfrak{n} \in C$ there is an ultrafilter $D_{\mathfrak{n}}$ with $F(D_{\mathfrak{n}}) = \{\mathfrak{n}^+\}$. This is done by using the "good ultrafilters" discussed in the next section. Then we combine these ultrafilters, in a way considered for another purpose in [7], to form the desired ultrafilter D with $F(D) = C^+$.

Additional cardinality results can be found in [21], [7], [14], and Chang-Keisler 26 – #3606.

Question 3E. *Is it true that for every ultrafilter D there exists a set C of cardinals such that* (a), (b) *hold and $F(D) = C^+$?*

Theorem 3.4. leads to another essential application of ultraproducts, where we construct models with surprising properties. A structure $\mathfrak{A}$ is said to be $\mathfrak{n}$ – *universal* if every structure $\mathfrak{B}$ of power at most $\mathfrak{n}$ which satisfies every sentence true in $\mathfrak{A}$ is elementarily embeddable in $\mathfrak{A}$ (cf. Morley-Vaught 27 – #37). To keep in a familiar area, we consider models of arithmetic (i.e. the complete theory of all sentences true in the standard model $\langle \omega, <, +, \cdot \rangle$). If $\mathfrak{A}$ is a model of arithmetic and $a \in A$, let $|a|$ denote the cardinality of the set $\{b \in A : b < a\}$, and let $N(\mathfrak{A})$ denote the set of all infinite cardinals $|c|$, $c \in A$.

THEOREM 3.5. (G. C. H.) *Let C be a non-empty set of infinite cardinals with the properties* (a) *and* (b), *and let $\mathfrak{m}$ be the supremum of C. Then there exists an $\mathfrak{m}^+$ – universal model $\mathfrak{A}$ of arithmetic such that $N(\mathfrak{A}) = C^+$.*

To construct $\mathfrak{A}$, we take an arbitrary $\mathfrak{m}^+$ – universal model $\mathfrak{B}$ of arithmetic and an ultrafilter D with $F(D) = C^+$, and let $\mathfrak{A}$ be the ultrapower $D - \mathrm{prod}\,\mathfrak{B}$. The proof also gives a little more information. We can make the power of $\mathfrak{A}$ exactly $\mathfrak{m}^+$. Moreover, for each infinite cardinal $\mathfrak{n}$, the substructure $\mathfrak{A}_\mathfrak{n}$ whose universe set is $A_\mathfrak{n} = \{a \in A : |a| < \mathfrak{n}\}$ turns out to be an elementary substructure of $\mathfrak{A}$.

Theorem 3.5 should be compared with an unpublished result of Ehrenfeucht, proved by a completely different method and without the G. C. H. Ehrenfeucht showed that for every set C of infinite cardinals there exists a model $\mathfrak{A}$ of arithmetic such that $N(\mathfrak{A}) = C$ and $\mathfrak{A}$ is not $\aleph_0$ – universal.

Ehrenfeucht's result seems to depend strongly on Peano's axioms in arithmetic. Our result, on the other hand, can be proved quite generally. For instance, instead of arithmetic we may take any theory at all which has a predicate $<$ and model $\mathfrak{B}$ in which, for each natural number n, there is an element having exactly n predecessors.

§4. Saturated models.

A structure $\mathfrak{A}$ is said to be $\mathfrak{n}$ – *saturated*, where $\mathfrak{n}$ is an infinite cardinal, if for each ordinal $\mu < \mathfrak{n}$ and function a on μ into A, the structure $(\mathfrak{A}, a)$ is $\mathfrak{n}$ – universal. (See Vaught 26 – # 4912.) Saturated structures are studied in detail in Morley and Vaught 27 – #37, where the following basic result is proved (G. C. H): *Each complete theory has, up to isomorphism, a unique* $\mathfrak{m}^+$ *– saturated model of power* $\mathfrak{m}^+$. They have been an extremely valuable tool in recent research in the theory of models. The theorem below gives a connection between ultraproducts and saturated structures.

THEOREM 4.1. (G. C. H.) *There exists an ultrafilter D over I with the property that for all structures* $\mathfrak{A}_i$, $i \in I$, *the ultraproduct* $D-\text{prod}\ \lambda i \mathfrak{A}_i$ *is* $\mathfrak{m}^+$ *– saturated.*

Theorem 4.1 is stated and proved in [15]; however, the heart of the proof depends on a rather intricate set-theoretic construction by transfinite induction, which is given in [16]. We shall call an ultrafilter D with the property mentioned in Theorem 4.1 a *good ultrafilter*. So Theorem 4.1 says that "good ultrafilters exist." Good ultrafilters can also be defined in a simple set-theoretic way, as is done in [16]. It turns out that every good ultrafilter is regular, and if $\mathfrak{m} = \aleph_0$ then every countably incomplete ultrafilter over I is good. Indeed, for any countably incomplete ultrafilter D over a set I of arbitrary power $\mathfrak{m}$, all ultraproducts $D - \text{prod}\ \lambda i \mathfrak{A}_i$ are $\aleph_1$ – saturated. Nevertheless, if $\mathfrak{m} > \aleph_0$ then there exist regular ultrafilters D over I which are not good (see [15], [16]).

Theorem 4.1 is used in [15] to obtain simple set-theoretic characterizations of some basic model-theoretic concepts. For instance: $\mathfrak{A}$ *and* $\mathfrak{B}$ *are elementarily equivalent if and only if they have some isomorphic ultraproducts. A class K of structures is an elementary class if and only if both K and its complement are closed under isomorphisms and ultraproducts.* Theorem 4.1 shows that every complete theory has an $\mathfrak{m}^+$ – saturated model of power $\leqq \mathfrak{m}^+$; this is an inessential application, for the result is proved in another way by Morley and Vaught in 27– #37.

Theorem 4.1 also has essential applications. It is easily seen that, if D is a good ultrafilter over I and the G. C. H. holds, then $F(D) = \{\mathfrak{m}^+\}$. Curiously we have not been able to find any proof that there exists a D with $F(D) = \{2^{\mathfrak{m}}\}$ except by using good ultrafilters and the G. C. H. and applying Theorem 4.1. Thus Theorem 3.4 above and its application depend on Theorem 4.1.

Question 4A. *Is the condition* $F(D) = \{2^{\mathfrak{m}}\}$ *necessary and sufficient for D to be a good ultrafilter?*

We now prepare to give one more essential application of Theorem 4.1. Let us say that $\mathfrak{A}$ is *maximally* $\mathfrak{n}$ *– saturated* if $\mathfrak{A}$ is $\mathfrak{n}$ – saturated but not $\mathfrak{n}^+$– saturated. We know from Morley and Vaught 27– #37 that every complete theory which has an infinite model has a maximally $\mathfrak{m}^+$ – saturated model of power $\mathfrak{m}^+$ (G. C. H.). Several questions come to mind, for instance:

Question 4B. (G. C. H.) *Which theories have maximally* $\mathfrak{m}^+$ *– saturated models of some power* $\mathfrak{p}^+ > \mathfrak{m}^+$?

Question 4C. (G. C. H.) *If a theory has a maximally* $\mathfrak{m}^+$ *– saturated model of power* $\mathfrak{p}^+ > \mathfrak{m}^+$, *must it also have a maximally* $\mathfrak{n}^+$ *– saturated model of power* q^+ *whenever* $\aleph_0 \leqq \mathfrak{n} < q$?

We shall make a beginning on Question 4B. We remark first that if a theory is categorical in some uncountable power $\mathfrak{n}$ (i.e., all models of power $\mathfrak{n}$ are isomorphic), then by a result of Morley [19] the theory is categorical in all uncountable powers, and consequently it cannot have a maximally $\mathfrak{m}^+$ – saturated model of any power $\mathfrak{p}^+ > \mathfrak{m}^+$. Thus the answer to question 4B must be a condition which implies that the theory is not categorical in uncountable powers. One such condition has been given by Ehrenfeucht 20– #3089.

Consider a structure $\mathfrak{A}$, an infinite subset $X \subseteq A$, and a formula $\phi(v_0, \cdots, v_n)$. ϕ is said to be *connected over* X if for every $x_0, \cdots, x_n \in X$ there is a permutation π of $\{0, \cdots, n\}$ such that the sequence $x_{\pi 0}, \cdots, x_{\pi n}$ satisfies ϕ in $\mathfrak{A}$. Ehrenfeucht proved that a complete theory Γ cannot be categorical in uncountable powers if it has the following property:

(*) There exists a model $\mathfrak{A}$ of Γ, an infinite set $X \subseteq A$, and a formula ϕ such that both ϕ and its negation are connected over X.

We announce the following new result, which apart from the G. C. H. improves Ehrenfeucht's theorem. It is an essential application of ultraproducts.

THEOREM 4.2. (G.C.H.) *Let* Γ *be a complete theory with property*(*). *Then for every cardinal* $\mathfrak{p} \geqq \mathfrak{m}$, Γ *has a maximally* $\mathfrak{m}^+$*–saturated model of power* $\mathfrak{p}^+$.

Ehrenfeucht used Ramsey's theorem on partitions at a crucial point in his proof. Theorem 4.2 follows from Theorem 4.1 and from the theorem below, which also uses Ramsey's theorem.

THEOREM 4.3. (G. C. H.) *Let* Γ *be a complete theory with the property* (*). *Then for any model* $\mathfrak{B}$ *of* Γ *and any countably incomplete* D, *the ultrapower* $D -$ prod $\mathfrak{B}$ *is not* $\mathfrak{m}^{++}$ *– saturated.*

To prove 4.2 from 4.3 we take a model $\mathfrak{B}$ of Γ of power $\mathfrak{p}^+$ and a good ultrafilter D, and the ultrapower $D -$ prod $\mathfrak{B}$ has the required properties.

There are examples of theories Γ which do not have the property (*) but still satisfy the conclusions of Theorems 4.2 and 4.3. There are also theories Γ which satisfy the conclusion of 4.2 but not the conclusion of 4.3. The conclusion of 4.3, of course, implies the conclusion of 4.2. We have two final questions.

Question 4D. (G. C. H.) *Which theories* Γ *satisfy the conclusion of Theorem* 4.3? *Is the answer independent of* $\mathfrak{m}$?

Question 4E. (G. C. H.) *If* $\mathfrak{m} < \mathfrak{p}$, $\mathfrak{B}$ *is* $\mathfrak{p}^+$ *– saturated, D is countably incomplete, and* $D - \text{prod}\,\mathfrak{B}$ *is* $\mathfrak{m}^{++}$ *– saturated, must* $D - \text{prod}\,\mathfrak{B}$ *be* $\mathfrak{p}^+$ *– saturated?*

REFERENCES

[1] AX, J. and KOCHEN, S., Diophantine problems over local fields I, II. Mimeographed, Ithaca, 1964.

[2] CHANG, C. C., Descendingly incomplete ultrafilters (abstract). To appear, *J. Symb. Logic.*

[3] CHANG, C. C. and KEISLER, H. J., *Continuous Model Theory*, Monograph to appear, Princeton Univ. Press.

[4] DAIGNEAULT, A., On automorphisms of polyadic algebras. *Trans. Amer. Math. Soc.*

[5] ENGELER, E., Ultrastructures in first-order model theory (abstract). *Amer. Math. Soc. Notices*, **11** (1964), 465 and 693.

[6] FRAISSÉ, R., Une généralization de l'ultraproduct (abstract). p. 13 of the program of this meeting, Jerusalem, 1964.

[7] FRAYNE, T., MOREL, A. C., and SCOTT, D., Reduced direct products. *Fund. Math.* **51** (1962), 195–228.

[8] FUHRKEN, G., On generalized quantifiers (abstract). *Amer. Math. Soc. Notices* **9** (1962), 132.

[9] GAIFMAN, H., Further consequences of the existence of measurable cardinals (abstract). Presented to this meeting, Jerusalem, 1964.

[10] GÖDEL, K., Die Vollständigkeit der Axiome des logischen Funktionenkalküls. *Monatsh. Math. Phys.* **37** (1930), 349–360.

[11] KEISLER, H. J., Reduced products and Horn classes. To appear, *Trans. Amer. Math. Soc.*

[12] ———, Limit ultraproducts. To appear, *J. Symb. Logic.*

[13] ———, Some applications of the theory of models to set theory. *Logic, Methodology and Philosophy of Science*, Proceedings of the 1960 International Congress (Nagel, Suppes and Tarski eds.), Stanford University Press, 1962, pp. 80–86.

[14] ———, On cardinalities of ultraproducts. *Bull Amer. Math. Soc.* **70** (1964), 644–647.

[15] ———, Ultraproducts and saturated models. *Indag. Math.* **26** (1964), 178–186.

[16] ———, Good ideals in fields of sets. *Ann. of Math.* **79** (1964), 338–359.

[17] ———, Extending models of set theory (abstract). To appear, *J. Symb. Logic.*

[18] KEISLER, H. J. and TARSKI, A., From accessible to inaccessible cardinals. *Fund. Math.* **53** (1964), 225–308.

[19] MORLEY, M., On theories categorical in uncountable powers. *Proc. Nat. Acad. Sci.* **49** (1963), 213–216.

[20] MALCEV, A., Untersuchungen aus den Gebiete der mathematischen Logic. *Rec. Math.* (Math. Sbornik), N. S. **1** (1936), 323–336.

[21] ROWBOTTOM, F., Some strong axioms of infinity incompatible with the axiom of constructibility. Thesis, Univ. of Wisconsin, Madison, (1964). Typewritten.

[22] TARSKI, A., Une contribution à la théorie de la mesure. *Fund. Math.* **15** (1930), 42–50.

[23] BERNSTEIN, A. and ROBINSON, A., Solution of an invariant subspace problem of K. T. Smith and P. R. Halmos. Mimeographed, Univ. of Calif., Los Angeles, 1964.

[24] CHANG, C. C., A lemma on ultraproducts and some applications. *Amer. Math. Soc. Notices* 7 (1960), 635.

[25] CHANG, C. C., Theory of models of infinite valued logic. *Ibid.* **8** (1961), 68 and 141· (Parts I, II, III, IV).

[26] KEISLER, H. J., Extending models of set theory, II. *Ibid.* **12** (1965), to appear.

DEFINABILITY IN AXIOMATIC SET THEORY I*

AZRIEL LÉVY
The Hebrew University, Jerusalem, Israel

§1. Introduction and statement of the results

The problem of effectivity in set theory has already been subject to some discussion (see Sierpiński [16], where further references are given). As an example we shall now mention two particular problems concerning effectivity. By means of the axiom of choice one can prove that the set of all real numbers can be well-ordered; can one give an explicit definition of such a well-ordering? Can one give an explicit definition of a non-denumerable set A of real numbers such that every subset B of A is finite, denumerable or of the same cardinality as A? The answer to these questions is, naturally, going to depend on the axiom system which we use for set theory.

The introduction of the axiom of constructibility $\mathsf{V}=\mathsf{L}$ by Gödel [6] resulted in a dichotomy, which was, until the recent proof of the independence of $\mathsf{V}=\mathsf{L}$ by Cohen [2], rather discouraging. With the aid of the axiom of constructibility the answer to all the questions concerning the existence of definable sets which satisfy given requirements becomes trivially positive (one gets into non-trivial considerations only when one wants to classify those definable sets according to their 'degrees of definability'—see Addison [1]). On the other hand, without the axiom of constructibility (or, in the case of the second problem mentioned above, the continuum hypothesis) one did not see how to get a positive answer to those problems (for good reasons—as we shall see), and in order to get a negative answer one had to prove the independence of the axiom of constructibility. We shall see in later parts of this work that even the addition of the negation of the axiom of constructibility as an axiom is not enough in itself to yield negative results. The prospect of an interesting answer to such questions was so dim that few authors considered it worth their while to mention these questions at all, and those who did so mostly

---

* This work was written with the support of the U.S. National Science Foundation and the U.S. Office of Naval Research, Information Systems Branch, Contract No. 62558–3882, NR 049–130, at the Hebrew University.

The author wishes to thank Professors J. W. Addison, P. J. Cohen, S. Feferman, D. Scott and R. L. Vaught for many helpful conversations.

avoided any rigorous formulation of them. It is rather typical that Sierpiński (in [16, III, §7]) was driven by despair to make the notion of an effective set dependent on the state of human knowledge, thus banishing this notion from the realm of mathematics to the realm of pragmatics.

The situation became completely changed by Paul Cohen's discovery of the forcing method and its application to the construction of models of axiomatic set theory. This method was first applied to questions of definability by Feferman [4], who proved that it is consistent with the axioms of set theory, including the axiom of choice and the generalized continuum hypothesis, to assume that there is no definable well-ordering of the continuum. If one looks at the proofs of Feferman [4] that several consequences of the axiom of choice are independent of the axioms of set theory in the absence of the axiom of choice, it is immediately seen that he has also proved that the following statements are consistent with the axioms of set theory, including the axiom of choice and the generalized continuum hypothesis: There is no definable non-principal prime ideal in the Boolean algebra of all subsets of ω; there is no definable choice function selecting reals from the (additive) cosets of the rationals in the reals; if, for $x, y \subseteq \omega$, we write $x \equiv y$ if x and y differ at only a finite number of places and we denote with $[x]$ the $\equiv$-equivalence class of x, then there is no definable function which selects one member from each pair $\{[x],[\omega - x]\}$ and, a fortiori, there is no definable ordering of the set of all $\equiv$-equivalence classes. In the present paper and in subsequent papers we shall develop the methods of Cohen [2] and Feferman [4] to obtain additional results concerning definability, including a strengthening of Feferman's result on the well-ordering of the continuum.

We shall deal with the set theory formulated in the applied first order predicate calculus (without equality) with $\in$ as the only non-logical symbol; this language will be called *the language of set theory*. The letters ϕ, ψ, χ will denote formulas of this language throughout the paper. $x = y$ will always be an abbreviation of $\forall z(z \in x \leftrightarrow z \in y)$. Whenever we denote a formula with $\phi(x_1, \cdots, x_k)$ we mean, unless otherwise mentioned or implied, that this formula has no free variables other than $x_1, \cdots, x_k$. When we speak of set theory we shall mean the Zermelo-Fraenkel set theory, which consists of the axioms of extensionality, pairing, union, power-set, subsets, infinity, replacement and foundation (axioms Ib, II–V, VII–IX of [5]); we shall denote this set theory with ZF. We shall denote with AC, GCH and V=L the axiom of choice for sets (axiom VI of [5]), the generalized continuum hypothesis ($\forall\alpha(2^{\aleph_\alpha} = \aleph_{\alpha+1})$) and Gödel's axiom of constructibility. We shall denote with + addition of axioms to an axiom system, e.g., ZF + AC is ZF with the axiom AC added. ZF + AC will also be denoted with ZF*. Even though the set theory with which we deal does not admit classes, we shall speak freely of classes and mean by a class the 'collection' of all sets x for which $\phi(x)$ holds, where $\phi(x)$ is a formula which has, possibly, free variables other than x, which are regarded as parameters. This is only a matter of convenience; we could have used the pair $\langle x, \phi(x) \rangle$ (of the metalanguage) rather than the corresponding class. α, β, γ will stand for ordinal variables, introduced by a definition.

Let us say that a set x is *definable in terms of members of A* (in short, *A-definable*), where A is a given class, if for some formula $\phi(z_1, \cdots, z_k, x)$ there are $z_1, \cdots, z_k \in A$ such that x is the unique set for which $\phi(z_1, \cdots, z_k, x)$ holds. A 0-definable set is said to be *definable.* In general there is no single formula $\psi(x)$ such that $\psi(x)$ holds if and only if x is A-definable; nevertheless we can make various statements concerning all A-definable sets e.g., the statement that every A-definable set is constructible is made by the schema

$$(\forall z_1, \cdots, z_k \in A)(\forall x[\forall y(\phi(z_1, \cdots, z_k, y) \leftrightarrow y = x) \rightarrow x \textit{ is constructible}]).$$

On the other hand, in general one cannot just say that there exists an A-definable set such that $\psi(x)$; one has to produce an actual formula $\phi(z_1, \cdots, z_k, x)$ and claim that

$$\exists x(\exists z_1, \cdots, z_k \in A)[\forall y(\phi(z_1, \cdots, z_k, y) \leftrightarrow y = x) \wedge \psi(x)].$$

As mentioned by Gödel in [7] and shown by Myhill and Scott in [13], the case where A is the class On of all ordinal numbers has remarkable properties which make the notion of On-definability (henceforth *ordinal-definability*) especially interesting. One of the remarkable properties of ordinal-definability is that this notion can be expressed by a single formula $\exists \beta \chi(\beta, x)$ such that $\mathsf{ZF} \vdash \forall \beta \exists x \forall y(\chi(\beta, y) \leftrightarrow y = x)$ and, for every formula $\phi(\alpha_1, \cdots, \alpha_k, x)$,

$$\mathsf{ZF} \vdash \forall \alpha_1, \cdots, \alpha_k \forall x[\forall y(\phi(\alpha_1, \cdots, \alpha_k, y) \leftrightarrow y = x) \rightarrow \exists \beta \chi(\beta, x)].$$

Nevertheless, whenever we shall prove the existence of an ordinal-definable set of some kind we shall try to find the simplest formula ϕ which defines it rather than indiscriminately use the formula χ above.

When one wants to compare the notion of ordinal-definability to constructibility, or to use it for the construction of a model of set theory one has to use, rather, the notion of hereditary-ordinal-definability, which is defined as follows. A set is said to be *hereditarily-ordinal-definable* if it is ordinal-definable and its members are ordinal-definable, and the members of its members are ordinal-definable and so on. The formal definition of this notion seems to make an essential use of the formula χ above. In the present paper we shall prove the following result.

THEOREM 1. *If* ZF *is consistent so is* ZF* + GCH *with the following additional axioms*

(2) *There exists a non-constructible subset of* ω.

(3) *Every hereditarily-ordinal-definable set is constructible.*

Theorem 1 will be proved in Section 2 by means of essentially the same model which Cohen used in [2] to prove the relative consistency of ZF* + GCH + (2)

LEMMA 4. (In ZF) *If a set b has an A-definable well-ordering and the class A contains all ordinals, then every member of b is A-definable.*

PROOF. Let $\prec$ be an A-definable well-ordering of b. Let $x \in b$. Then, for some ordinal α,

(5) *x is the* α*-th member (of b) in the well ordering* $\prec$.

Since $\alpha \in A$, (5) defines x in terms of members of A.

The result of Feferman [4] on the well-ordering of the continuum is strengthened as follows.

THEOREM 6. (In ZF + (3)) *An ordinal-definable set b of real numbers has an ordinal-definable well-ordering if and only if every member of b is constructible. Therefore, if there are non-constructible real numbers then the set of all real numbers has no definable well-ordering.*

PROOF. As is well known, every real number can be regarded as a set of natural numbers (and vice versa). If b has an ordinal-definable well-ordering, then, by Lemma 4, every member of b is ordinal-definable. Since the members of b are sets of natural numbers, they are also hereditarily-ordinal-definable and by (3) they are constructible. On the other hand, the set of all constructible real numbers has a definable well-ordering ($\hat{x}\hat{y}\,(Od\text{‘}x < Od\text{‘}y)$ in the notation of Gödel [6]), hence every ordinal-definable set b of real numbers has an ordinal-definable well-ordering.

If (2) holds, then one can easily construct a set b of non-constructible real numbers which is well-ordered by an ordinal-definable relation whose field properly includes b such as, e.g., the natural $<$-ordering of the real numbers.

THEOREM 7. (In ZF + (2) + (3)) *There is a* Π^1_2*-predicate* $P(f)$ *of number theory (i.e., a predicate of the form* $\forall g\, \exists h Q(f, g, h)$, *where* f, g, h

vary over the set ω^ω of all functions on ω into ω, and $Q(f,g,h)$ is an arithmetical predicate) such that there is a function f which satisfies $P(f)$ but no such function f is ordinal-definable.

PROOF. As shown by Addison in [1, §6], the predicate '*f is constructible*' is a Σ_2^1-predicate, hence its negation '*f is not constructible*' is a Π_2^1-predicate. By (2) there are non-constructible members of ω^ω; by (3) no such member of ω^ω is ordinal-definable.

A subset B of ω^ω is said to be a basis for a set C of predicates on ω^ω if

$$\forall P[P \in C \to [\exists f P(f) \to \exists f(f \in B \wedge P(f))]].$$

Addison proved in [1], using V=L, that for $k>0$ the set of all $\Sigma_{k+1}^1 \cap \Pi_{k+1}^1$-functions is a basis for Π_k^1. For $k=1$ one can dispense with the use of V=L by using the result of Shoenfield [15] and thus one can prove in ZF that the set of all $\Sigma_2^1 \cap \Pi_2^1$-functions is a basis for Π_1^1. As we saw in Theorem 7, V= L is essential in Addison's result for $k>1$. Moreover, if (2) and (3) are assumed, then Π_2^1 has no interesting basis at all since even the set of all ordinal-definable members of ω^ω is not a basis for Π_2^1.

We shall now conclude the present section with a brief discussion of some of the results which will be proved in later parts of this work.

We cannot prove in ZF the existence of sets which are not ordinal-definable since ZF + (V = L) is consistent and, obviously, every constructible set is ordinal-definable. On the other hand, by Theorem 1, we cannot prove in ZF* + GCH that every set is ordinal-definable. The next question which one is likely to ask is whether one can prove that every ordinal-definable set is constructible. An easy consequence of the main result of Easton [3] is that if ZF is consistent so is ZF* + (2) with the additional 'axiom'

(8) *There is a definable subset of ω which is not constructible.*

The proof of this result proceeds, roughly, as follows. It follows from the main result of Easton [3] that if f is a non-decreasing function in ω^ω such that $f(n) > n$ for every $n < \omega$ then one can get a model $\mathfrak{N}$ of ZF* in which $2^{\aleph_n} = \aleph_{f(n)}$ for every $n < \omega$. The function f is obviously definable in $\mathfrak{N}$, thus if we take for f a non-constructible function we get, in $\mathfrak{N}$, a non-constructible function which is definable. By extending the function f beyond the natural numbers one can also obtain any cardinal number of ordinal-definable subsets of ω and thus prove a conjecture of Gödel in [7] that it is consistent to assume that ω has more than $\aleph_1$ ordinal-definable subsets. As hinted by Gödel [7] and proved by Myhill-Scott [13], by restricting the universe of all sets to the class of all hereditarily-ordinal-definable sets one gets a model for ZF*. By what was just said one cannot prove in ZF even that $2^{\aleph_0} = \aleph_1$ holds in that model.

The question which now comes up, and which seems to be quite fundamental, is to what extent can we strengthen (8), namely, how simple can the definition of a non-constructible subset of ω be? Shoenfield proved in [15] that every Σ^1_2- or Π^1_2-subset of ω is constructible. Thus the best strengthening of (8) which one could expect to obtain is that there is a non-constructible $\Sigma^1_3 \cap \Pi^1_3$-subset of ω. The author succeeded only in obtaining much less. By using Easton's method in a somewhat different way than mentioned above, we shall be able to get a non-constructible $\Sigma^2_2 \cap \Pi^2_2$-subset of ω (but we do not know how to prove even this modest result without giving up the continuum hypothesis).

In the beginning of the paper we mentioned also a problem concerning the cardinalities of definable sets of real numbers. The results in this direction are mentioned in [9], and they remain valid also if 'definable' is replaced in them by 'ordinal-definable'.

§2. Proof of Theorem 1

The discussion in the present section will be of somewhat wider scope than needed to prove Theorem 1, so as to enable us to use its contents in later parts of this work.

In order to obtain relative consistency results along with whatever we shall prove concerning models of ZF, we shall follow the remarks of Cohen at the end of [2] but operate in a slightly different setup which will make the discussion more natural. We shall consider the set theory ZFM which is formulated in a language which is like the language of set theory except that it has an additional individual constant M. The axioms of ZFM are those of ZF as well as the following (9)–(11).

(9) $\phi \leftrightarrow \mathrm{Rel}(M, \phi)$, *for every sentence* ϕ *of the language of set theory*,

where $\mathrm{Rel}(M, \phi)$ is the sentence obtained from ϕ by relativizing all its quantifiers to M, i.e., by replacing the quantifiers $\exists x(\cdots$ and $\forall x(\cdots$ by $\exists x(x \in M \wedge \cdots$ and $\forall x(x \in M \rightarrow \cdots$, respectively.

(10) M *is denumerable.*

(11) M *is transitive*, i.e., $\forall x, y(x \in y \in M \rightarrow x \in M)$.

Let us denote with DC Tarski's axiom of dependent choices (as formulated, e.g., in [10, Th. 59, (i)]).

LEMMA 12. *All the theorems of* ZFM *which are formulated in the language of set theory are theorems of* ZF + DC. *Therefore, if* Q *is any extension of* ZF *in the language of set theory such that* Q + DC *is consistent, then* Q + (9) + (10) + (11) *is consistent too.*

Proof. Assume ZFM $\vdash \psi$, where ψ is a sentence of set theory, then we have

$$\text{ZF} \vdash M \textit{ is denumerable and transitive} \wedge \bigwedge_{i=1}^{k}(\phi_i \leftrightarrow \text{Rel}(M, \phi_i)) \to \psi,$$

for some sentences $\phi_1, \cdots, \phi_k$ of the language of set theory. Since M does not occur in ψ or in the ϕ_i's, we have, by the rules of logic,

$$\text{ZF} \vdash \exists z\,[z \textit{ is denumerable and transitive} \wedge \bigwedge_{i=1}^{k}(\phi_i \leftrightarrow \text{Rel}(z, \phi_i))] \to \psi.$$

We have to prove that ZF+ DC $\vdash \psi$, and for this purpose it is now enough to show that

$$(13)\ \text{ZF} + \text{DC} \vdash \exists z[z \textit{ is denumerable and transitive} \wedge \bigwedge_{i=1}^{k}(\phi_i \leftrightarrow \text{Rel}(z, \phi_i))].$$

It follows easily from Montague [11, Th. 1] (see [8, Theorems 2 and 6]), which is a version of the Skolem-Löwenheim theorem, that

$$\text{ZF} \vdash \exists z[\omega \subseteq z \wedge \bigwedge_{i=1}^{k}(\phi \leftrightarrow \text{Rel}(z, \phi_i))].$$

From this we get, by the Skolem-Löwenheim theorem (see [10, Th. 59]),

$$\text{ZF} + \text{DC} \vdash \exists z[\omega \subseteq z \wedge z \textit{ is denumerable} \wedge \bigwedge_{i=1}^{k}(\phi_i \leftrightarrow \text{Rel}(z, \phi_i))].$$

Let u be a denumerable set such that $\omega \subseteq u$ and $\bigwedge_{:=1}^{k}(\phi_i \leftrightarrow \text{Rel}(u, \phi_i))$. Following Mostowski [12] and Shepherdson [14] let us define f on u by $f(x) = \{f(y) \mid y \in x \wedge y \in u\}$; by the axiom of foundation there is a unique such f. Put $z = \{f(y) \mid y \in u\}$. It is easily seen that $f(n) = n$, for every $n \in \omega$; thus $\omega \subseteq z$; since u is denumerable, so is z. z is, obviously, transitive. For $x, y \in u$, $x \in y \leftrightarrow f(x) \in f(y)$, hence f can be said to be an $\in$-homomorphism of u on z and therefore we can easily prove that, for every sentence ϕ of set theory, $\text{Rel}(u, \phi) \leftrightarrow \text{Rel}(z, \phi)$. Thus we get (13), which is what was left to be proved.

As was mentioned in Section 1, the construction used here is essentially identical with that of Cohen in [2]. However, rather than use Cohen's procedure which is like the procedure of Gödel in [6] we shall follow the suggestions of Dana Scott carried out by Feferman in [4] and use a ramified hierarchy like the one used by Gödel in his original proof of the consistency of ZF* + GCH. The particular way in which we use here the ramified hierarchy owes also much to Easton and Solovay [3a]; in particular, Definitions 15 and 42 are very similar to their corresponding definitions, and Definitions 16, 17, 20 and 21, as well as Lemmas 32, 36, 38–40, 43, 44 and 53 are taken almost verbatim from Easton [3].

Even though the language of set theory has only one non-logical symbol $\in$, we shall make in our discussions (in ZF or ZFM) full use of the usual terminology introduced in set theory, including symbols for relations and operations, special variables, etc. We shall not even define those symbols which have here their usual meaning. The language which contains all those additional symbols (but not M) will be called *the extended language of set theory*.,

Let K_0 be a set. By the K_0*-language* we shall mean a formal language of the first-order predicate calculus (without equality) with the binary relation symbol ε and an individual constant $\mathbf{x}$ for each $x \in K_0$ such that the single symbols and the sequences of symbols of the K_0-language are objects of our set theory. We shall not give an explicit definition of this language; it can be easily defined along the same lines as the languages $\mathscr{L}$ and $\mathscr{L}_M$ of Definition 15 below. For an arbitrary set K we denote with $\in_K$ the $\in$-relation restricted to K, i.e., the set $\{\langle x, y\rangle \mid x \in y \wedge x, y \in K\}$. The system $\langle K, \in_K\rangle$ will be denoted with $\mathfrak{K}$. If $K_0 \subseteq K$ then the K_0-language can be applied to the system $\mathfrak{K}$, i.e., one can define the notions of satisfaction and truth of formulas of the K_0-language for the system $\mathfrak{K}$. These definitions will not be given here as they are the standard definitions of the semantics of the first-order predicate calculus. For every formula $\phi(x_1, \cdots, x_n)$ of the language of set-theory we shall denote with $\boldsymbol{\phi}(x_1, \cdots, x_n)$ the corresponding formula of the K_0-language. One can easily prove, by induction on the length of ϕ

$$(14)\quad (\forall x_1, \cdots, x_n \in K_0)(\boldsymbol{\phi}(\mathbf{x}_1, \cdots, \mathbf{x}_n) \textit{ is true in } K \leftrightarrow \mathrm{Rel}(K, \phi(x_1, \cdots, x_n))).$$

Let R be the function defined by $R(\alpha) = \bigcup_{\beta<\alpha} P(R(\beta))$, where $P(x)$ denotes the power-set of x. By the axiom of foundation every set u is included in some $R(\alpha)$. By the *rank* $\|y\|$ of a set y we mean the least ordinal α such that $y \subseteq R(\alpha)$.

From now on our discussion will usually be within the framework of ZFM. Throughout this section τ will be a fixed ordinal such that $\tau \in M$, $\tau \neq 0$ and $\tau \in \omega$ or τ is a limit number; σ will be $\max(\omega, \tau)$.

In Definitions 15–18 and 20 we shall define two very similar languages $\mathscr{L}$ and $\mathscr{L}_M$ and develop some syntactical notions for these languages. The additional clauses which apply to the language $\mathscr{L}_M$ will be enclosed in parentheses.

Definition 15. The primitive symbols of the language $\mathscr{L}$ and $\mathscr{L}_M$ are: $\neg = \langle 0,0\rangle$, $\vee = \langle 0,1\rangle$, $\exists = \langle 0,2\rangle$, $\varepsilon = \langle 0,3\rangle$, $A = \langle 0,4\rangle$, $\mathrm{v}_i = \langle 0, 5+i\rangle$ $i < \omega$; for every ordinal α ($\alpha \in M$), $\exists_\alpha = \langle 1,\alpha\rangle$, $\mathsf{Y}_\alpha = \langle 2,\alpha\rangle$; and for every set s ($s \in M$), $\mathbf{s} = \langle 3, s\rangle$. (If s is an ordinal $\alpha, \beta, \cdots$ then $\mathbf{s}$ will be written as $\boldsymbol{\alpha}, \boldsymbol{\beta}, \cdots$) The symbols v_i will be called *variables*. x, y, z, t will denote arbitrary variables, with the understanding that different letters stand for distinct variables. The symbols **s** will be called *set constants.* We shall denote concatenation of finite sequences of the symbols with juxtaposition. For any finite sequences Φ, Ψ of symbols we shall write $\Phi \vee \Psi$ for $\vee \Phi \Psi$, $\Phi \rightarrow \Psi$ for $\neg \Phi \vee \Psi$, $\Phi \wedge \Psi$ for $\neg(\neg \Phi \vee \neg \Psi)$, $\Phi \leftrightarrow \Psi$ for $(\Phi \rightarrow \Psi) \wedge (\Psi \rightarrow \Phi)$. $\forall x \Phi$ for $\neg \exists x \neg \Phi$, $\forall_\alpha x \Phi$ for $\neg \exists_\alpha x \neg \Phi$, $\hat{x}_\alpha \Phi$ for $\lambda_\alpha x \Phi$, $\Phi \varepsilon \Psi$ for $\varepsilon \Phi \Psi$, $A(\Phi, \Psi)$ for $A \Phi \Psi$, where we follow the usual rules of omission of parentheses.

DEFINITION 16. The notions of a *ranked formula* and of an *abstraction term* of $\mathscr{L}$ ($\mathscr{L}_M$) are defined inductively as follows.

(a) If u and v are abstraction terms, set constants or variables, then $u \varepsilon v$ and $A(u,v)$ are ranked formulas.

(b) If Φ and Ψ are ranked formulas, so are $\neg \Phi$, $\Phi \vee \Psi$ and $\exists_\alpha x \Phi$ $(\alpha \in M)$.

(c) If Φ is a ranked formula with no free variables other than x, and α is an ordinal (in M) such that (*i*) Φ contains no occurrence of $\exists_\beta$ with $\beta > \alpha$, (*ii*) Φ contains no occurrence of λ_β with $\beta \geqq \alpha$, (*iii*) Φ contains no set constant **s** for a set s with $\|s\| \geqq \alpha$, and (*iv*) if $\alpha \leqq \sigma$ then Φ contains no occurrence of A, then $\hat{x}_\alpha \Phi$ is an abstraction term.

The notion of a free variable is defined as usual; a ranked formula without free variables is said to be a ranked *sentence.* We shall refer to the set constants and the abstraction terms as *constant terms* (in short: *terms*). u, v and w will vary over the terms, unless otherwise mentioned.

DEFINITION 17. The *rank* $\rho(u)$ of a term u (not to be confused with the rank $\|u\|$ of the term u as a set) is given by (a) $\rho(\hat{x}_\alpha \Phi) = \alpha$, (b) $\rho(\mathbf{s}) = \|s\|$.

DEFINITION 18. The notion of a *formula* of the language $\mathscr{L}$ ($\mathscr{L}_M$) is defined as follows. (a) If u, v are variables or terms, then $u \varepsilon v$ and $A(u,v)$ are formulas. (b) If Φ, Ψ are formulas, so are $\neg \Phi$, $\Phi \vee \Psi$, $\exists x \Phi$, $\exists_\alpha x \Phi$ $(\alpha \in M)$.

ABBREVIATIONS. Let u, v be terms or variables; then $u = v$ will stand for $\forall x(x \varepsilon u \leftrightarrow x \varepsilon v)$, where x is a variable distinct from u, v. For terms u, v, $u \simeq v$ will stand for $\forall_\gamma x(x \varepsilon u \leftrightarrow x \varepsilon v)$, where $\gamma = \max(\rho(u), \rho(v))$. $u \simeq v$ is, obviously, a ranked sentence.

DEFINITION 19. A *condition* is a function on a finite subset of $\tau \times \omega$ $(= \{\langle \alpha, \beta \rangle \mid \alpha < \tau \wedge \beta < \omega\})$ into $2 (= \{0,1\})$. We shall use the letters p, p', etc. as variables for conditions. If $p \subseteq p'$ we shall say that p' is an *extension* of p.

DEFINITION 20. For a ranked formula Φ we set $\mathrm{Ord}(\Phi) = \omega^2 \cdot \alpha + \omega \cdot e + l$ where: (*i*) α is the least ordinal such that Φ contains no $\exists_\beta$ with $\beta > \alpha$ and no term of rank $\leqq \alpha$. (*ii*) $e = 0$ if Φ has no subformula of the form $v \varepsilon u$, where $\rho(v) = \alpha$ and no subformula $A(u, v)$ other than inside an abstraction term; otherwise $e = 1$. (*iii*) l is the length of the formula Φ, where the atomic formulas $u \varepsilon v$ and $A(u, v)$ are said to have length 1.

DEFINITION 21. $p \Vdash \Phi$ (read: *p forces* Φ) is defined for ranked sentences Φ of $\mathscr{L}$ and $\mathscr{L}_M$ by induction on $\mathrm{Ord}(\Phi)$. Notice that p varies over subsets of a set and that, for a given ordinal β, all the ranked sentences Φ with $\mathrm{Ord}(\Phi) < \beta$ constitute a set; therefore definition of $p \Vdash \Phi$ by induction on $\mathrm{Ord}(\Phi)$ is permissible.

(a) $p \Vdash \neg \Phi$ if for no $p' \supseteq p$ does $p' \Vdash \Phi$.

(b) $p \Vdash \Phi \vee \Psi$ if $p \Vdash \Phi$ or $p \Vdash \Psi$.

(c) $p \Vdash \exists_\alpha x \, \Phi(x)$ if $p \Vdash \Phi(u)$ for some u with $\rho(u) < \alpha$.

(d) $p \Vdash u \varepsilon \mathbf{s}$ if $p \Vdash u \simeq \mathbf{t}$ for some $t \in s$.

(e) $p \Vdash u \varepsilon \hat{x}_\alpha \Phi(x)$ if, for some u' with $\rho(u') < \alpha$, $p \Vdash u \simeq u'$ and $p \Vdash \Phi(u')$.

(f) $p \Vdash A(u, v)$ if for some ordinals $\alpha < \tau$ and $\beta < \omega$ such that $\alpha \leqq \rho(u)$ and $\beta \leqq \rho(v)$, $p \Vdash u \simeq \boldsymbol{\alpha}$, $p \Vdash v \simeq \boldsymbol{\beta}$ and $\langle \alpha, \beta, 0 \rangle \in p$.

To see that $p \Vdash \Phi$ is indeed defined by induction on $\mathrm{Ord}(\Phi)$, notice that in the definition of $\mathrm{Ord}(u \simeq v)$ we have $e = 0$.

DEFINITION 22. $p \Vdash \Phi$ is defined for arbitrary sentences Φ of $\mathscr{L}_M$ by induction on the length of Φ. This definition is valid because p varies over a set and all sentences of $\mathscr{L}_M$ shorter than a given sentence constitute a set (since all the terms of $\mathscr{L}_M$ constitute a set). $p \Vdash \Phi$ cannot be defined for arbitrary sentences Φ of $\mathscr{L}$ because in $\mathscr{L}$ all the sentences shorter than a given sentence do not generally constitute a set.

(a) $p \Vdash u \varepsilon v$ and $p \Vdash A(u, v)$ are defined in Definition 21.

(b) $p \Vdash \neg \Phi$ if for no $p' \supseteq p$ does $p' \Vdash \Phi$.

(c) $p \Vdash \Phi \vee \Psi$ if $p \Vdash \Phi$ or $p \Vdash \Psi$.

(d) $p \Vdash \exists_\alpha x \Phi(x)$ if for some u, with $\rho(u) < \alpha$, $p \Vdash \Phi(u)$.

(e) $p \Vdash \exists x \Phi(x)$ if for some u, $p \Vdash \Phi(u)$.

It is obvious that, for ranked sentences Φ of $\mathscr{L}_M$, $p \Vdash \Phi$ according to Definition 22 if and only if $p \Vdash \Phi$ according to Definition 21.

LEMMA 23. *Every condition is in M and every finite sequence of the primitive symbols of $\mathscr{L}_M$ is in M. There is a formula $\pi(x, y, \tau)$ of the language of set theory such that for $a, b \in M$,* $\mathrm{Rel}(M, \pi(a, b, \tau))$ *if and only if a is a condition, b is a ranked sentence of $\mathscr{L}_M$ and $a \Vdash b$.*

PROOF. In this proof we shall use instead of (9) only the following schema (24), which obviously follows from (9) in ZFM. (This will turn out to be helpful later.)

(24) $\quad\text{Rel}(M,\phi)$, where ϕ is any theorem of ZF.

We can use the language of set theory 'inside M', i.e., we can use the language of set theory with the understanding that whatever we say is relativized to M. By (24) we have 'inside M' as much freedom to define new relations, operations, special variables, etc. as we have in ZF. We shall mark with a subscript M the notions thus obtained. For example $\{x,y\}_M$ is the operation corresponding to $\{x,y\}$, 'x is an ordinal $_M$' is the predicate corresponding to 'x is an ordinal'. Much of what will be proved or mentioned here concerning these notions was proved, under slightly different assumptions, by Shepherdson in [14, 2.1–2.3].

'x is an ordinal' is the formula

$$\forall y,z[(y\in z\in x\to y\in x)\wedge(y,z\in x\to y\in z\vee y=z\vee z\in y)].$$

As immediately seen, using (11), for every $x\in M$, x is an ordinal$_M$ if and only if x is an ordinal. It is a matter of simple checking to verify that $0_M=0,\ 1_M=1,\cdots,\omega_M=\omega$, and, for $x,y\in M$, $\{x,y\}_M=\{x,y\},\langle x,y\rangle_M=\langle x,y\rangle$. Since $x\cup\{y\}=x\cup_M\{y\}_M$, for all $x,y\in M$, one can easily prove that every finite subset of M is a member of M (by induction on the number of its members). Since $\omega=\omega_M\subseteq M$, every finite sequence of members of M also is a member of M. Thus we get that every condition and every finite sequence of symbols of $\mathscr{L}_M$ is in M. One proves now the following statements, the proofs of all of which amount to simple checking.

(a) For $\alpha,\beta\in M$, $\alpha+_M\beta=\alpha+\beta$, $\alpha\cdot_M\beta=\alpha\cdot\beta$, $R_M(\alpha)=R(\alpha)\cap M$; for $x\in M$, $\|x\|_M=\|x\|$.

(b) $\neg_M=\neg$, $\vee_M=\vee$, $\exists_M=\exists$, $\varepsilon_M=\varepsilon$, $A_M=A$, $\mathrm{v}_{i,M}=\mathrm{v}_i$, for $i<\omega$; for $\alpha\in M$, $\exists_{\alpha,M}=\exists_\alpha$, $\Game_{\alpha,M}=\Game_\alpha$; for finite sequences $x,y\in M$, $(xy)_M=xy$.

(c) a is a (ranked formula of $\mathscr{L})_M\Leftrightarrow a$ is a ranked formula of $\mathscr{L}_M$; a is an (abstraction term of $\mathscr{L})_M\Leftrightarrow a$ is an abstraction term of $\mathscr{L}_M$.

(d) For every term u of $\mathscr{L}_M$, $\rho_M(u)=\rho(u)$.

(e) a is a condition$_M\Leftrightarrow a$ is a condition.

(f) For every ranked sentence Φ of $\mathscr{L}_M$, $\text{Ord}_M(\Phi)=\text{Ord}(\Phi)$.

(g) Let $a\Vdash b$ denote here the formula defined in Definition 21 (not 22) for ranked sentences of $\mathscr{L}$, and let us assume that it is formulated so that $a\Vdash b\Rightarrow a$ is a condition and b is a ranked sentence of $\mathscr{L}$; then $a\Vdash_M b\Leftrightarrow b$ is a sentence of $\mathscr{L}_M$ and $a\Vdash b$.

We take for $\pi(x,y,\tau)$ the formula $x\Vdash y$.

In Lemma 23 we characterized forcing of ranked sentences of $\mathscr{L}_M$ by a formula $\pi(x,y,\tau)$ relativized to M; this formula was obtained by considering forcing of ranked sentences of $\mathscr{L}$. When we want to characterize in the same way also the forcing of arbitrary sentences of $\mathscr{L}_M$, we need a definition of the notion of forcing for arbitrary sentences of $\mathscr{L}$ that, as was mentioned in Definition 22, does not seem to be available (which is hardly surprising, since the definition of forcing resembles very much the definition of truth, and by the Epimenides-Tarski paradox we cannot define in ZF — or ZFM — the notion of truth for the sentences of $\mathscr{L}$). However, again like in the case of truth, there is no reason why we should not be able to define forcing for a single given sentence or for some particular family of sentences; this is indeed what will be done in Lemma 26.

DEFINITION 24. A *parametric formula* is the term of the extended language of set theory (with finitely many variables taken from u_i, α_i, $i < \omega$) which is given as follows.

(a) $u_i \varepsilon u_j$, $u_i \varepsilon \mathrm{v}_j$, $\mathrm{v}_i \varepsilon u_j$, $\mathrm{v}_i \varepsilon \mathrm{v}_j$, $A(u_i, u_j)$, $A(u_i, \mathrm{v}_j)$ $A(\mathrm{v}_i, u_j)$, $A(\mathrm{v}_i, \mathrm{v}_j)$, $i,j < \omega$, are parametric formulas, where u_i, $i < \omega$, is a variable of the language of set theory and v_i, $i < \omega$, is the term of the extended language of set theory which denotes the set $\langle 0, 5 + i \rangle$.

(b) If $\mathbf{\Phi}$ and $\mathbf{\Psi}$ are parametric formulas so are $\neg \mathbf{\Phi}$, $\mathbf{\Phi} \vee \mathbf{\Psi}$, $\exists \mathrm{v}_i \mathbf{\Phi}$, $\exists_{\alpha_j} \mathrm{v}_i \mathbf{\Phi}$, where α_j, $j < \omega$, is a variable.

The notion of a free occurrence of v_i in a parametric formula is defined as usual; a *parametric sentence* is a parametric formula in which no v_i occurs free.

For every parametric formula $\mathbf{\Phi}(u_1, \cdots, u_k, \alpha_1, \cdots, \alpha_l)$ we can easily prove in ZFM

(25) *If $u_1, \cdots, u_k$ are terms of $\mathscr{L}$ ($\mathscr{L}_M$) and $\alpha_1, \cdots, \alpha_l$ are ordinals (of M) then $\mathbf{\Phi}(u_1, \cdots, u_k, \alpha_1, \cdots, \alpha_l)$ is a formula of $\mathscr{L}$ ($\mathscr{L}_M$).*

LEMMA 26. *Let $\mathbf{\Phi}(u_1, \cdots, u_k, \alpha_1, \cdots, \alpha_l)$ be a parametric sentence. There is a formula $\pi_{\mathbf{\Phi}}(x, y_1, \cdots, y_k, z_1, \cdots, z_l, \tau)$ of the language of set theory such that for all terms $u_1, \cdots, u_k$ of $\mathscr{L}_M$, for all ordinals $\alpha_1, \cdots, \alpha_l \in M$ and for all $a \in M$, $\mathrm{Rel}(M, \pi_{\mathbf{\Phi}}(a, u_1, \cdots, u_k, \alpha_1, \cdots, \alpha_l, \tau))$ if and only if*

$$a \Vdash \mathbf{\Phi}(u_1, \cdots, u_k, \alpha_1, \cdots, \alpha_l).$$

PROOF. We define $\pi_{\mathbf{\Phi}}$ by induction on the length of $\mathbf{\Phi}$ as follows.

(a) If $\mathbf{\Phi}$ is $u_i \varepsilon u_j$ or $A(u_i, u_j)$, then we take for $\pi_{\mathbf{\Phi}}(x, y_i, y_j, \tau)$ the formula $\pi(x, y_i \varepsilon y_j, \tau)$ or $\pi(x, A(y_i, y_j), \tau)$, respectively, where $\pi(x, y, \tau)$ is as in Lemma 23.

(b) If $\mathbf{\Phi}$ is $\neg \mathbf{\Psi}$, then we take for $\pi_{\mathbf{\Phi}}(x, y_1, \cdots, y_k, z_1, \cdots, z_l, \tau)$ the formula x *is a condition* $\wedge \forall x_1(x_1 \supseteq x \rightarrow \neg \pi_{\mathbf{\Psi}}(x_1, y_1, \cdots, y_k, z_1, \cdots, z_l, \tau))$.

(c) If $\mathbf{\Phi}$ is $\mathbf{\Psi} \vee \mathbf{\Gamma}$ then we take for $\pi_{\mathbf{\Phi}}(x, y_1, \cdots, y_k, z_1, \cdots, z_l, \tau)$ the formula

$$\pi_\Psi(x, y_1, \cdots, y_k, z_1, \cdots, z_l, \tau) \vee \pi_\Gamma(x, y_1, \cdots, y_k, z_1, \cdots, z_l, \tau).$$

(d) If $\mathbf{\Phi}$ is $\exists \mathrm{v}_i \mathbf{\Psi}$, then let $\mathbf{\Psi}'$ be the parametric sentence obtained from $\mathbf{\Psi}$ by replacing each free occurrence of the term v_i by an occurrence of the variable u_{k+1}. We take for $\pi_\Phi(x, y_1, \cdots, y_k, z_1, \cdots, z_l, \tau)$ the formula

$$\exists y_{k+1}(y_{k+1} \textit{ is a term of } \mathscr{L} \wedge \pi_{\Psi'}(x, y_1, \cdots, y_k, y_{k+1}, z_1, \cdots, z_l, \tau)).$$

(e) If $\mathbf{\Phi}$ is $\exists_{\alpha_j} \mathrm{v}_i \mathbf{\Psi}$, then let $\mathbf{\Psi}'$ be as in (d). We take for

$$\pi_\Phi(x, y_1, \cdots, y_k, z_1, \cdots, z_l, \tau)$$

the formula

$$\exists y_{k+1}(y_{k+1} \textit{ is a term of } \mathscr{L} \wedge \rho(y_{k+1}) \in z_j \wedge \pi_{\Psi'}(x, y_1, \cdots, y_k, y_{k+1}, z_1, \cdots, z_l, \tau).$$

That $\pi_\Phi(x, y_1, \cdots, y_k, z_1, \cdots, z_l, \tau)$ does indeed satisfy the requirements of the Lemma in Cases (a)–(e) follows easily from Lemma 23, (25) and (c), (d), (e) in the proof of Lemma 23.

Lemma 27. *For every formula $\Phi(x_1, \cdots, x_k)$ of $\mathscr{L}_M$ there is a formula $\Pi_\Phi(x, y_1, \cdots, y_k)$ of the M-language such that, for all $a \in M$ and all terms $u_1, \cdots, u_k$ of $\mathscr{L}_M, a, u_1, \cdots, u_k$ satisfy $\Pi_\Phi(x, _1, \cdots, y_k)$ in $\mathfrak{M}$ if and only if $a \Vdash \Phi(u_1, \cdots, u_k)$.*

Proof. Without loss of generality we can assume that $\Phi(x_1, \cdots, x_k)$ contains no terms, since if $\Phi(x_1, \cdots, x_k)$ is obtained from

$$\Psi(x_1, \cdots, x_k, x_{k+1}, \cdots, x_{k+m})$$

by substitution of the terms $u_1, \cdots, u_m$ of $\mathscr{L}_M$ for $x_{k+1}, \cdots, x_{k+m}$, respectively, then we can set $\Pi_\Phi(x, y_1, \cdots, y_k) = \Pi_\Psi(x, y, \cdots, y_k, \mathbf{u}_1, \cdots, \mathbf{u}_m)$. The proof is by induction on the length of Φ and follows closely the proof of Lemma 26, except for the following differences: (*i*) Here we consider the formulas in the M-language which correspond to the formulas of the language of set theory used there, and use (14) to prove what is claimed in the lemma. (*ii*) We have no variables here corresponding to $z_1, \cdots, z_l$ of Lemma 26 and, accordingly, in the case where Φ is $\exists_\beta x_{k+1} \Psi(x_1, \cdots, x_k, x_{k+1})$ we take for $\Pi_\Phi(x, y_1, \cdots, y_k)$ the formula

$$\exists y_{k+1}(y_{k+1} \text{ is a term of } \mathscr{L} \wedge \rho(y_{k+1}) \,\varepsilon\, \boldsymbol{\beta} \wedge \Pi_\Psi(x, y_1, \cdots, y_k, y_{k+1})).$$

(*iii*) We use here the term $\boldsymbol{\tau}$ instead of the variable τ.

From now on we shall not deal any more with $\mathscr{L}$. By 'term', 'formula' or 'sentence' we shall mean a term, formula or sentence, respectively, of $\mathscr{L}_M$. Φ, Ψ, Γ will denote sentences of $\mathscr{L}_M$ (unless otherwise mentioned).

LEMMA 28. $p \Vdash \Phi \wedge p' \supseteq p \Rightarrow p' \Vdash \Phi$.

PROOF. For ranked Φ this is proved by induction on $\mathrm{Ord}(\Phi)$. If Φ is $\neg\Psi$ then for no $p'' \supseteq p$ does $p'' \Vdash \Psi$ and, a fortiori, for no $p'' \supseteq p'$ does $p'' \Vdash \Psi$, i.e., $p' \Vdash \neg\Psi$. If Φ is other than $\neg\Psi$ then $p' \Vdash \Phi$ follows easily from the definition of forcing and the induction hypothesis. For arbitrary Φ the lemma is now easily proved by induction on the length of Φ.

LEMMA 29. *No condition forces both* Φ *and* $\neg\Phi$.

DEFINITION 30. $p \Vdash^* \Phi$ (*p weakly forces* Φ) if $p \Vdash \neg\neg\Phi$. $p \| \Phi$ (*p decides* Φ) if $p \Vdash \Phi$ or $p \Vdash \neg\Phi$; $p \|^* \Phi$ (*p weakly decides* Φ) if $p \Vdash^* \Phi$ or $p \Vdash^* \neg\Phi$.

LEMMA 31. *For every condition p and every sentence* Φ, *there is an extension* p' *of p such that* $p' \| \Phi$.

LEMMA 32. (a) $p \Vdash^* \Phi \Leftrightarrow$ *no extension of p forces* $\neg\Phi \Leftrightarrow$ *every extension* p' *of p has an extension* p'' *which forces* Φ. (b) $p \Vdash \Phi \Rightarrow p \Vdash^* \Phi$ (c) $p \Vdash^* \neg\Phi \Rightarrow p \Vdash \neg\Phi$. (d) *If* Φ *is* $\Psi \wedge \Gamma$, $\Psi \leftrightarrow \Gamma$, $\forall x \Psi, \forall_\alpha x \Psi, u = v$ *or* $u \simeq v$, *then* $p \Vdash \Phi \Leftrightarrow p \Vdash^* \Phi$.

(e) $p \Vdash^* \forall x \Phi(x) \Leftrightarrow p \Vdash^* \Phi(u)$ *for all terms u of* $\mathscr{L}_M$.

(f) $p \Vdash^* \forall_\alpha x \Phi(x) \Leftrightarrow p \Vdash^* \Phi(u)$ *for all terms u with* $\rho(u) < \alpha$.

(g) $p \Vdash^* \Phi \leftrightarrow \Psi \Rightarrow p \Vdash^* \Phi \Leftrightarrow p \Vdash^* \Psi$.

PROOF. By direct computation. Notice that all the sentences in (d) are of the form $\neg\Gamma$.

LEMMA 33. *If* $p \| \Phi_i$, $i = 1, \cdots, n$, *c is an n-ary sentential connective of the language of set theory* (*i.e., an operation which is an iteration of the primitive sentential connectives* $\neg$ *and* $\vee$) *and* **c** *is the corresponding sentential connective of* $\mathscr{L}_M$ *then* $p \| \mathbf{c}(\Phi_1, \cdots, \Phi_n)$ *and*

$$p \Vdash \mathbf{c}(\Phi_1, \cdots, \Phi_n) \Leftrightarrow c(p \Vdash \Phi_1, \cdots, p \Vdash \Phi_n).$$

PROOF. By induction on the number of times $\neg$ and $\vee$ are used in c. If $c(\phi_1, \cdots, \phi_n)$ is ϕ_i, $1 \leqq i \leqq n$, then the lemma is trivial. If $c(\phi_1, \cdots, \phi_n)$ is $\neg\, c_1(\phi_1, \cdots, \phi_n)$ or $c_1(\phi_1, \cdots, \phi_n) \vee c_2(\phi_1, \cdots, \phi_n)$, the lemma follows easily from the induction hypothesis by means of Lemma 32(b) and straightforward computation.

LEMMA 34. *If* $c_1, \cdots, c_k, c$, $k \geqq 0$, *are n-ary sentential connectives such that* $c_1(\phi_1, \cdots, \phi_n) \wedge \cdots \wedge c_k(\phi_1, \cdots, \phi_n) \rightarrow c(\phi_1, \cdots, \phi_n)$ *is a tautology for all* $\phi_1, \cdots, \phi_n$ *and for given* $\Phi_1, \cdots, \Phi_n$ $p \Vdash^* \mathbf{c}_i(\Phi_1, \cdots, \Phi_n)$, $i = 1, \cdots, k$, *then* $p \Vdash^* \mathbf{c}(\Phi_1, \cdots, \Phi_n)$.

PROOF. Let $p' \supseteq p$. By Lemmas 28 and 31 there is a $p'' \supseteq p'$ such that $p'' \| \Phi_j$, $j = 1, \cdots, n$. By Lemmas 33 and 32(a) and since $p \Vdash^* c_i(\Phi_1, \cdots, \Phi_n)$ we have $p'' \Vdash c_i(\Phi_1, \cdots, \Phi_n)$, $i = 1, \cdots, k$, and hence, by Lemma 33, $c_i(p'' \Vdash \Phi_1, \cdots, p'' \Vdash \Phi_n)$. Since $\bigwedge_{i=1}^{k} c_i(p'' \Vdash \Phi_1, \cdots, p'' \Vdash \Phi_n) \to c(p'' \Vdash \Phi_1, \cdots, p'' \Vdash \Phi_n)$ is a tautology, we have $c(p'' \Vdash \Phi_1, \cdots, p'' \Vdash \Phi_n)$ and, again by Lemma 33, $p'' \Vdash c(\Phi_1, \cdots, \Phi_n)$. By Lemma 32(a), $p \Vdash^* c(\Phi_1, \cdots, \Phi_n)$.

COROLLARY 35. *If $c(\phi_1, \cdots, \phi_n)$ is a tautology for all $\phi_1, \cdots, \phi_n$ then for all $p, \Phi_1, \cdots, \Phi_n$ $p \Vdash^* c(\Phi_1, \cdots, \Phi_n)$.*

LEMMA 36. (a) $p \Vdash u = u$. (b) $p \Vdash u = v \Rightarrow p \Vdash v = u$. (c) $p \Vdash u = v$ *and* $p \Vdash v = w \Rightarrow p \Vdash u = w$.

PROOF. By Lemmas 32(d), (e), 34 and 35.

LEMMA 37. *If $p \Vdash u \varepsilon v$ then, for some term u' with $\rho(u') < \rho(v)$, $\rho(u') \leqq \rho(u)$, we get $p \Vdash u \simeq u'$ and $p \Vdash u' \varepsilon v$.*

PROOF. One proves first $p \Vdash u \simeq u$, like in Lemma 36(a). If $\rho(u) < \rho(v)$ we choose u for u' and the lemma is proved. Assume now $\rho(u) \geqq \rho(v)$. If v is $\mathbf{s}$ then, by the definition of forcing, we have $p \Vdash u \simeq \mathbf{t}$, for some $t \in s$; we choose $\mathbf{t}$ for u' and get $\rho(\mathbf{t}) < \rho(v) \leqq \rho(u)$ and $p \Vdash \mathbf{t} \varepsilon \mathbf{s}$ (since $p \Vdash \mathbf{t} \simeq \mathbf{t}$). If v is $\hat{x}_\alpha \Phi(x)$ then there is a u' such that $\rho(u') < \alpha$ $(= \rho(v) \leqq \rho(u))$, $p \Vdash u \simeq u'$ and $p \Vdash \Phi(u')$. Since $p \Vdash u' \simeq u'$, we have also $p' \Vdash u' \varepsilon v$.

LEMMA 38. $p \Vdash u \simeq v \Leftrightarrow p \Vdash u = v$.

PROOF. That $p \Vdash u = v \Rightarrow p \Vdash u \simeq v$ follows easily from Lemma 32. We shall now prove $p \Vdash u \simeq v \Rightarrow p \Vdash u = v$ by induction on $\gamma = \max(\rho(u), \rho(v))$. Assume $p \Vdash u \simeq v$. To show that $p \Vdash u = v$ it suffices, by Lemmas 31 and 32, to show that there is no extension p' of p and no term w such that $p' \Vdash w \varepsilon u$ and $p' \Vdash \neg w \varepsilon v$, or else $p' \Vdash w \varepsilon v$ and $p' \Vdash \neg w \varepsilon u$. We prove only the former, the latter is proved similarly.

Let $p' \supseteq p$ and $p' \Vdash w \varepsilon u$. If $\rho(w) < \gamma$ then, by $p \Vdash u \simeq v$ and Lemmas 28, 32(b), (f) and 33, we get $p' \Vdash^* w \varepsilon v$, contradicting $p' \Vdash \neg w \varepsilon v$. If, otherwise, $\rho(w) \geqq \gamma$ then, by Lemma 37, there is a term w', $\rho(w') < \rho(u)$, such that $p' \Vdash w \simeq w'$ and $p' \Vdash w' \varepsilon u$. Since $\rho(w') < \rho(u) \leqq \gamma$, we get $p' \Vdash^* w' \varepsilon v$, by $p \Vdash u \simeq v$ and Lemmas 28, 32(b)(f) and 33. Let p'' be an extension of p' such that $p'' \Vdash w' \varepsilon v$. By the definition of forcing there is a term w'', $\rho(w'') < \rho(v)$, such that $p'' \Vdash w' \simeq w''$ and if v is $\mathbf{s}$ then w'' is $\mathbf{t}$, for some $t \in s$, and if v is $\hat{x}_\alpha \Psi(x)$ then $p \Vdash \Psi(w'')$. By the induction hypothesis $p'' \Vdash w' \simeq w''$ implies $p'' \Vdash w' = w''$, from which we get $p'' \Vdash w \simeq w''$ by means of $p' \Vdash w \simeq w'$, Lemma 32 and $\rho(w) \geqq \gamma > \rho(w'), \rho(w'')$, as in the proof of Lemma 36(c). From $p'' \Vdash w \simeq w''$ we get, by what was said above concerning w'' and by the definition of forcing, $p'' \Vdash w \varepsilon v$, contradicting $p' \Vdash \neg w \varepsilon v$.

LEMMA 39. (a) *$p \Vdash u \varepsilon w$ and $p \Vdash u = v \Rightarrow p \Vdash v \varepsilon w$.* (b) *$p \Vdash w \varepsilon u$ and $p \Vdash u = v \Rightarrow p \Vdash^* w \varepsilon v$.*

PROOF. (a) follows from the definition of forcing by Lemmas 36 and 38. (b) follows from Lemma 32.

LEMMA 40. *If $p \Vdash u = \boldsymbol{\alpha}$ then $\rho(u) \geqq \alpha$.*

PROOF. By induction on α. For every $\beta < \alpha$, $p \Vdash \boldsymbol{\beta} \,\varepsilon\, \boldsymbol{\alpha}$; therefore, if $p \Vdash u = \boldsymbol{\alpha}$ then, by Lemma 39(b), $p \Vdash^* \boldsymbol{\beta} \varepsilon u$ and hence there is a condition p' such that $p' \Vdash \boldsymbol{\beta} \varepsilon u$. By Lemmas 37 and 38, there is a term v such that $\rho(v) < \rho(u)$, $p' \Vdash v = \boldsymbol{\beta}$ and $p' \Vdash v \varepsilon u$. By the induction hypothesis $p' \Vdash v = \boldsymbol{\beta}$ implies $\rho(v) \geqq \beta$. Since $\rho(u) > \rho(v)$, we get $\rho(u) > \beta$, for every $\beta < \alpha$, hence $\rho(u) \geqq \alpha$.

LEMMA 41. *If $p \Vdash A(u,v)$ and $p \Vdash u = u'$, $p \Vdash v = v'$ then $p \Vdash A(u',v')$.*

PROOF. This follows easily from the definition of forcing and Lemmas 38, 36 and 40.

DEFINITION 42. A function p on a subset of $\tau \times \omega$ into 2 is said to be *generic* if, for every formula $\Phi(x)$ of the M-language such that every condition p has an extension p' which satifies $\Phi(x)$ in $\mathfrak{M}$, there exists a subset $p'' \subseteq P$ which satisfies $\Phi(x)$ in $\mathfrak{M}$. (Since, for every $\alpha < \tau$ and $\beta < \omega$, every condition p has an extension p' which satisfies $\langle \boldsymbol{\alpha}, \boldsymbol{\beta}, 0 \rangle \varepsilon x \vee \langle \boldsymbol{\alpha}, \boldsymbol{\beta}, 1 \rangle \varepsilon x$, a generic function must be defined on the whole of $\tau \times \omega$.) For any function P on $\tau \times \omega$ into 2 we write $P \Vdash \Phi$ if, for some condition $p \subseteq P$, $p \Vdash \Phi$.

LEMMA 43. *If P is generic then, for every sentence Φ, $P \Vdash \Phi$ or $P \Vdash \neg \Phi$.*

PROOF. For any given Φ consider the formula $\Pi_{\Phi}(x) \vee \Pi_{\neg\Phi}(x)$ of the M-language. By Lemmas 27 and 31, every condition p has an extension which satisfies it. Hence, again by Lemma 27, there is a condition $p'' \subseteq P$ such that $p'' \Vdash \Phi$ or $p'' \Vdash \neg \Phi$.

LEMMA 44. *Every condition p can be extended to a generic function P.*

PROOF. Since M is denumerable, there are only $\aleph_0$ formulas in the M-language. Let us list all the formulas of the M-language with one free variable as $\Phi_0(x), \Phi_1(x), \cdots$. Put $p_0 = p$; let p_{n+1} be an extension p' of p_n such that p' satisfies $\Phi_n(x)$ in $\mathfrak{M}$ if p_n has such an extension p', and $p_{n+1} = p_n$ otherwise. Put $P = \bigcup_{n<\omega} p_n$; P is obviously generic.

From now on P will always stand for a generic function.

DEFINITION 45. val_P is the function defined on the set of all terms, by induction on their rank, as follows.

$\mathrm{val}_P(u) = \{\mathrm{val}_P(v) \mid v \textit{ is a term}, \rho(v) < \rho(u) \text{ and } P \Vdash v \varepsilon u\}$. We put $N_P = \{\mathrm{val}_P(u) \mid u \textit{ is a term}\}$. We shall usually omit the subscript P from val_P and N_P.

LEMMA 46. *For every n-ary sentential connective* c,

$$P \Vdash c(\Phi_1, \cdots, \Phi_n) \iff c(P \Vdash \Phi_1, \cdots, P \Vdash \Phi_n).$$

PROOF. By Lemmas 43 and 28, there is a $p \subseteq P$ such that $p \| \Phi_i$, $1 \leqq i \leqq n$. By Lemma 33, $p \| c(\Phi_1, \cdots, \Phi_n)$ and

$$p \Vdash c(\Phi_1, \cdots, \Phi_n) \iff c(p \Vdash \Phi_1, \cdots, p \Vdash \Phi_n). \tag{47}$$

Since $p \| \Phi_i$, $1 \leqq i \leqq n$, and $p \| c(\Phi_1, \cdots, \Phi_n)$, we get, by Definition 42 and Lemmas 28 and 29, $p \Vdash \Phi_i \iff P \Vdash \Phi_i$, $1 \leqq i \leqq n$, and

$$p \Vdash c(\Phi_1, \cdots, \Phi_n) \iff P \Vdash c(\Phi_1, \cdots, \Phi_n).$$

Thus (47) implies the lemma.

LEMMA 48. $P \Vdash u = v \iff P \Vdash u \simeq v \iff \mathrm{val}(u) = \mathrm{val}(v)$.

PROOF. $P \Vdash u = v \iff P \Vdash u \simeq v$ follows easily from Lemma 38. We shall now prove the rest by induction on $\gamma = \max(\rho(u), \rho(v))$. Assume $P \Vdash u = v$. Let $x \in \mathrm{val}(u)$; then $x = \mathrm{val}(w)$, for some w with $\rho(w) < \rho(u)$ and $P \Vdash w \varepsilon u$. By Lemmas 39, 28, 43 and 29, $P \Vdash w \varepsilon v$. By Lemmas 37 and 38, there is a term w' with $\rho(w') < \rho(v)$ such that $P \Vdash w = w'$ and $P \Vdash w' \varepsilon v$. Since $\max(\rho(w), \rho(w')) < \max(\rho(u), \rho(v)) = \gamma$, we get, by the induction hypothesis, $x = \mathrm{val}(w) = \mathrm{val}(w') \in \mathrm{val}(v)$. Thus we proved $\mathrm{val}(u) \subseteq \mathrm{val}(v)$; similarly one shows $\mathrm{val}(v) \subseteq \mathrm{val}(u)$ and, therefore, $\mathrm{val}(u) = \mathrm{val}(v)$.

Assume $\mathrm{val}(u) = \mathrm{val}(v)$. In order to prove $P \Vdash u = v$ it is enough, by Lemmas 32, 43 and 46, to show that, for every term w, $P \Vdash w \varepsilon u \iff P \Vdash w \varepsilon v$. Assume $P \Vdash w \varepsilon u$; then, by Lemmas 37 and 38, for some term w', $\rho(w') < \rho(u)$, $P \Vdash w = w'$ and $P \Vdash w' \varepsilon u$, hence $\mathrm{val}(w') \in \mathrm{val}(u) = \mathrm{val}(v)$. Therefore, for some term w'', $\rho(w'') < \rho(v)$, $P \Vdash w'' \varepsilon v$ and $\mathrm{val}(w') = \mathrm{val}(w'')$. $\max(\rho(w')$ $\rho(w'')) < \max(\rho(u), \rho(v)) = \gamma$, hence, by the induction hypothesis, $P \Vdash w' = w''$. Since $P \Vdash w'' \varepsilon v$ and $P \Vdash w = w'$, we get, by Lemma 36(c) and 39(a), $P \Vdash w \varepsilon v$. Similarly, if $P \Vdash w \varepsilon v$ then $P \Vdash w \varepsilon u$.

LEMMA 49. *N is a transitive set. For each* $s \in M$, $\mathrm{val}(\mathbf{s}) = s$, *hence* $M \subseteq N$.

PROOF. N is transitive by the definition of N and val. $\mathrm{val}(\mathbf{s}) = s$ is easily shown by induction on $\rho(\mathbf{s})$, using the definition of $\mathrm{val}(\mathbf{s})$, the definition of forcing and Lemma 48.

DEFINITION 50. For every member $x \in N$, we define $\boldsymbol{\rho}_P(x)$ (or $\boldsymbol{\rho}(x)$) to be the least $\rho(w)$ for which $\mathrm{val}(w) = x$.

LEMMA 51. *If $x, y \in N$ and $x \in y$ then $\boldsymbol{\rho}(x) < \boldsymbol{\rho}(y)$. Therefore, if $x \in N$ then $\|x\| \leqq \rho(x)$. For every ordinal α, if $\alpha \in N$ then $\alpha \in M$ (since $\alpha = \|\alpha\| \leqq \rho(\alpha) \in M$).*

THE SEMANTICS OF $\mathscr{L}_M$. We interpret the language $\mathscr{L}_M$ in $\mathfrak{N}_P$ $(= \langle N, \in_N, P \rangle)$ as follows. We assign to the term u the value $\mathrm{val}(u)$. $u \varepsilon v$ is true if $\mathrm{val}(u) \in \mathrm{val}(v)$. The sentential connectives $\neg$ and $\vee$ and the existential quantifier are interpreted in the usual way. $\exists_\alpha x \Phi(x)$ is true if there exists a set y with $\boldsymbol{\rho}(y) < \alpha$ which satisfies $\Phi(x)$. a and b satisfy $A(x, y)$ if $a \in \tau$, $b \in \omega$ and $P(a, b) = 0$.

LEMMA 52. *Φ is true in $\mathfrak{N}$ if and only if $P \Vdash \Phi$.*

PROOF. By induction on the length of Φ (where the length of $u \varepsilon v$ and $A(u, v)$ is 1).

(a) Φ is $u \varepsilon v$. If $P \Vdash u \varepsilon v$ then, by Lemma 37, for some u' with $\rho(u') < \rho(v)$, $P \Vdash u \simeq u'$, $P \Vdash u' \varepsilon v$. By Lemma 48 and the definition of val, $\mathrm{val}(u) = \mathrm{val}(u') \in \mathrm{val}(v)$; thus $u \varepsilon v$ is true in $\mathfrak{N}$. If, on the other hand, $\mathrm{val}(u) \in \mathrm{val}(v)$ then, for some u' with $\rho(u') < \rho(v)$, $P \Vdash u' \varepsilon v$ and $\mathrm{val}(u') = \mathrm{val}(u)$. By Lemma 48, $P \Vdash u' = u$; hence, by Lemma 39(a), $P \Vdash u \varepsilon v$.

(b) Φ is $A(u, v)$. If $\mathrm{val}(u) = \alpha < \tau$, $\mathrm{val}(v) = \beta < \omega$ and $P(\alpha, \beta) = 0$ then, by Lemmas 49 and 48, $P \Vdash u \simeq \boldsymbol{\alpha}$, $P \Vdash v \simeq \boldsymbol{\beta}$ and, by the definition of forcing, $P \Vdash A(u, v)$. On the other hand, if $P \Vdash A(u, v)$ then, for some $\alpha < \tau$ and $\beta < \omega$, $P \Vdash u \simeq \boldsymbol{\alpha}$, $P \Vdash v \simeq \boldsymbol{\beta}$ and $P(\alpha, \beta) = 0$; hence, by Lemmas 49 and 48, $\mathrm{val}(u) = \alpha$, $\mathrm{val}(v) = \beta$.

(c) Φ is $\neg \Psi$ or $\Psi \vee \Gamma$. In these cases the lemma follows easily from the induction hypothesis by means of Lemma 43.

(d) Φ is $\exists_\alpha x \Psi(x)$. If Φ is true in $\mathfrak{N}$ then, for some term u with $\rho(u) < \alpha$, $\mathrm{val}(u)$ satisfies $\Psi(x)$ in $\mathfrak{N}$, i.e., $\Psi(u)$ is true in $\mathfrak{N}$ and, by the induction hypothesis, $P \Vdash \Psi(u)$; hence $P \Vdash \exists_\alpha x \Psi(x)$. On the other hand, if $P \Vdash \exists_\alpha x \Psi(x)$. then, for some term u with $\rho(u) < \alpha$, $P \Vdash \Psi(u)$. By the induction hypothesis, $\Psi(u)$ is true in $\mathfrak{N}$, i.e., $\mathrm{val}(u)$ satisfies $\Psi(x)$ in $\mathfrak{N}$, and therefore $\exists_\alpha x \Psi(x)$ is true in $\mathfrak{N}$.

(e) Φ is $\exists x \Psi(x)$. The proof of the lemma in this case is similar to the proof in case (d).

LEMMA 53. *$p \Vdash^* \Phi$ if and only if Φ is true in all systems $\mathfrak{N}_P$ where P is generic and $P \supseteq p$.*

PROOF. If $p \Vdash^* \Phi$ and $P \supseteq p$ then, by Lemmas 32(a), 28, 29 and 43, $P \Vdash \Phi$ and, by Lemma 52, Φ is true in $\mathfrak{N}_P$. If, on the other hand, Φ is not weakly forced by p then, by Lemma 32 (a), some extension p' of p forces $\neg\Phi$; by Lemma 44, there is a generic $P \supseteq p' \supseteq p$, consequently $P \Vdash \neg\Phi$; by Lemma 52, Φ is not true in $\mathfrak{N}_P$.

LEMMA 54. $\mathrm{val}(\hat{x}_\alpha \Phi(x))$ *is the set of all members* y *of* N *such that* $\rho(y) < \alpha$ *and* y *satisfies* $\Phi(x)$ *in* $\mathfrak{N}$.

PROOF. If $\rho(y) < \alpha$ then $y = \mathrm{val}(v)$ for some term v with $\rho(v) < \alpha$. If y also satisfies $\Phi(x)$ in $\mathfrak{N}$ then $\Phi(v)$ is true in $\mathfrak{N}$. By Lemma 52, $P \Vdash \Phi(v)$; hence, by the definition of forcing and Lemmas 36 and 38, $P \Vdash v \,\varepsilon\, \hat{x}_\alpha \Phi(x)$; hence, by Lemma 52, $y = \mathrm{val}(v) \in \mathrm{val}(\hat{x}_\alpha \Phi(x))$. On the other hand if $y \in \mathrm{val}(\hat{x}_\alpha \Phi(x))$ then $y = \mathrm{val}(v)$, for some v such that $\rho(v) < \alpha$ and $P \Vdash v \,\varepsilon\, \hat{x}_\alpha \Phi(x)$. Thus $\rho(y) < \alpha$ and, for some term v' with $\rho(v') < \alpha$, $P \Vdash v \simeq v'$ and $P \Vdash \Phi(v')$. By Lemma 52, $\Phi(v')$ is true in $\mathfrak{N}$ and, therefore, $\mathrm{val}(v')$ satisfies $\Phi(x)$ in $\mathfrak{N}$. By Lemma 48, $y = \mathrm{val}(v) = \mathrm{val}(v')$; hence y satisfies $\Phi(x)$ in $\mathfrak{N}$.

LEMMA 55. *For every* $\lambda < \tau$, $\mathrm{val}(a_\lambda) \subseteq \omega$ *and* $\mathrm{val}(a_\lambda) \notin M$, *where* a_λ *is* $\hat{x}_{\sigma+1} A(\lambda, x)$.

PROOF. $\mathrm{val}(a_\lambda) \subseteq \omega$ follows easily from Lemma 54. We shall now prove that $0 \Vdash \neg a_\lambda = \mathbf{s}$, for every $s \in M$, which implies $\mathrm{val}(a_\lambda) \notin M$, by Lemmas 49 and 52. Let p be any condition and let $s \in M$. Let β be the least natural number for which $p(\lambda, \beta)$ is undefined, and let p' be $p \cup \{\langle \lambda, \beta, 0 \rangle\}$ if $\beta \notin s$ and $p \cup \{\langle \lambda, \beta, 1 \rangle\}$ if $\beta \in s$. Assume $\beta \notin s$. Then $\beta \notin \mathrm{val}_P(\mathbf{s}) = s$ for every P; also, if $P \supseteq p'$ then $P(\lambda, \beta) = 0$ and, by Lemma 54, $\beta \in \mathrm{val}_P(a_\lambda)$. Thus $\neg a_\lambda = \mathbf{s}$ is true in every $\mathfrak{N}_P$ with $P \supseteq p'$ and, by Lemmas 53 and 32(c), $p' \Vdash \neg a_\lambda = \mathbf{s}$. Now assume that $\beta \in s$; as above, we get again $p' \Vdash \neg a_\lambda = \mathbf{s}$. Thus we proved that every condition p has an extension p' such that $p' \Vdash \neg a_\lambda = s$; by Lemma 32(a, c), this implies $0 \Vdash \neg a_\lambda = \mathbf{s}$.

At this point the development has been carried out far enough to enable us to dwell on the main features of the proof of Theorem 1. When we set $\tau = 1$ it can be shown that, for every axiom ϕ of ZF* + GCH + (2) + (3), one can prove in ZFM + (V=L) that, for every generic function P, $\mathrm{Rel}(N_P, \phi)$. We shall now see that this suffices to establish Theorem 1. Assume that ZF* + GCH + (2) + (3) is inconsistent; then we can prove in this theory $\psi \wedge \neg\psi$, for some sentence ψ. In ZFM it follows from Lemma 49 that, for every generic P, $N_P \neq 0$ and hence we have $\mathrm{Rel}(N_P, \phi)$ also for every axiom ϕ of first-order predicate calculus. The proof of $\psi \wedge \neg\psi$ in the theory ZF* + GCH + (2) + (3) can now be repeated, with all formulas relativised to N_P, in ZFM + (V=L), where we shall get $\mathrm{Rel}(N_P, \psi) \wedge \neg \mathrm{Rel}(N_P, \psi)$. Thus the existence of a generic function, provable in ZFM by Lemma 44, leads to a contradiction in ZFM + (V = L), and ZFM + (V = L) is, therefore, inconsistent. By Lemma 12 also ZF + (V = L) (which is the same as ZF + DC + (V = L) since V = L implies DC) is inconsistent; as Gödel proved, in essence, in [6], if ZF + (V = L) is inconsistent so is ZF.

The full proofs of $\mathrm{Rel}(N_P, \phi)$ for the axioms ϕ of ZF* + GCH will not be given here; for detailed discussion see Cohen [2] and Easton [3]. We shall only mention here briefly some of the proofs. If ϕ is an axiom of ZF then $\mathrm{Rel}(N_P, \phi)$ can be proved in ZFM for arbitrary τ, as follows. If ϕ is one of the axioms of extensionality and foundation, $\mathrm{Rel}(N_P, \phi)$ is an immediate consequence of Lemma 49. Also if ϕ is the axiom of infinity, $\mathrm{Rel}(N_P, \phi)$ is an immediate consequence of Lemma 49 (since $\omega \in M$ by the proof of Lemma 23). If ϕ is one of the axioms of pairing and union, the proof is also easy. If ϕ is an instance of the axiom schema of replacement, the proof is more difficult and it makes use of Lemma 26—see Cohen [2] and Easton [3]. The axiom schema of subsets is a consequence of the axiom schema of replacement. If ϕ is the axiom of power set which, in the presence of the axiom schema of subsets, can be taken to be $\forall x \exists y \forall z(z \subseteq x \rightarrow z \in y)$, we shall present the following proof due to R. Solovay.

We have to prove $(\forall x \in N)(\exists y \in N)(\forall z \in N)(z \subseteq x \rightarrow z \in y)$. For $x \in N$ let $x = \mathrm{val}(u)$, $\rho(u) = \alpha$. Let $b_\alpha = \mathrm{val}(\hat{x}_\alpha(\neg x \,\varepsilon\, x)) = \{x \in N \mid \rho(x) < \alpha\}$, by Lemma 54. By Lemmas 51 and 54, $x \subseteq b_\alpha$. For every term w of $\mathscr{L}$ we put $[w] = \{\langle p, v\rangle \mid \rho(v) < \alpha \text{ and } \pi(p, v \,\varepsilon\, w)\}$. As easily seen, for all terms w_1, w_2 of $\mathscr{L}_M$, $\mathrm{Rel}(M, [w_1] = [w_2]) \Rightarrow \mathrm{val}(w_1) \cap b_\alpha = \mathrm{val}(w_2) \cap b_\alpha$. Since for all terms w the sets $[w]$ are subsets of a fixed member of M we have

$$\mathrm{Rel}(M, \exists\beta\forall w(w \textit{ is a term of } \mathscr{L} \rightarrow \exists w_0(w_0 \textit{ is a term of } \mathscr{L} \wedge \rho(w_0) < \beta \wedge [w] = [w_0]))).$$

If $y \subseteq x$ and $y \in N$ then $y = \mathrm{val}(w)$ for some w; hence there is a w_0 with $\rho(w_0) < \beta$ such that $\mathrm{Rel}(M, [w] = [w_0])$; therefore $\mathrm{val}(w_0) \cap b_\alpha = \mathrm{val}(w) \cap b_\alpha = y \cap b_\alpha = y$. Put $\gamma = \max(\alpha + 1, \beta)$. $\mathrm{val}(\hat{x}_\gamma(x \,\varepsilon\, w_0 \wedge x \,\varepsilon\, \hat{z}_\alpha(\neg z \,\varepsilon\, z))) = y$; thus $\rho(y) \leqq \gamma$, $y \in b_{\gamma+1} \in N$, where γ is independent of y.

In ZFM + AC one can prove $\mathrm{Rel}(N_P, \mathrm{AC})$ for arbitrary τ, and in ZFM + AC + GCH one can prove $\mathrm{Rel}(N_p, \mathrm{GCH})$ for $\tau = 1$—see Cohen [2] and Easton [3] for both proofs. To complete the proof we still have to prove $\mathrm{Rel}(N_P, (2))$ and $\mathrm{Rel}(N_P, (3))$ in ZFM + (V = L).

Shepherdson proved in [14, 2.329] that if M satisfies (24) then x is constructible$_M$ if and only if x is constructible and $\mathrm{Od}`x \in M$ (in the terminology of Gödel [6]). Actually, the assumptions of Shepherdson are somewhat stronger than (24), but his proof does not use anything additional to (24). Since, as we have seen, (24) holds also with M replaced by N, we get that x is constructible$_N$ if x is constructible and Od'$x \in N$. For every constructible set Od'x is an ordinal and by Lemmas 49 and 51, M and N contain the same ordinals; hence

(56) $\qquad x$ *is constructible*$_N \Leftrightarrow x$ *is constructible*$_M \Rightarrow x \in M$.

By Lemma 55, val$(a_0) \subseteq \omega$ and val$(a_0) \notin M$; hence val(a_0) is not constructible$_N$ and since, by the proof of Lemma 23, $\omega_M = \omega$, we have val$(a_0) \subseteq \omega_M$ and thus we get Rel$(N,(2))$. In Lemma 63 we shall prove that if x is (hereditarily-ordinal-definable)$_N$ then $x \in M$. In ZFM + (V = L) we have, by (9), Rel$(M, \mathrm{V} = \mathrm{L})$, hence $x \in M \Rightarrow x$ is constructible$_M$. Combining this with (56) and Lemma 63 we get that if x is (hereditarily-ordinal-definable)$_N$ then x is constructible$_N$; thus we have Rel$(N,(3))$. The rest of the present section will lead to the proof of Lemma 63. This will be done by the method of Feferman [4].

DEFINITION 57. Let $r \in M$ be a subset of $\tau \times \omega$. We define

$$[r, Q] = \{\langle\alpha,\beta,e\rangle \mid \langle\alpha,\beta,e\rangle \in Q \wedge \langle\alpha,\beta\rangle \notin r \vee \langle\alpha,\beta,1-e\rangle \in Q \vee \langle\alpha,\beta\rangle \in r\},$$

where Q is any function on a subset of $\tau \times \omega$ into 2. For any formula or term Φ we write $[r,\Phi]$ for whatever is obtained from Φ by replacing each occurrence of $A(u,v)$, where u and v are terms or variables, by

$$[r,u]\,\varepsilon\,\boldsymbol{\tau} \wedge [r,v]\,\varepsilon\,\boldsymbol{\omega} \wedge (A([r,u],[r,v]) \leftrightarrow \neg\langle[r,u],[r,v]\rangle\,\varepsilon\,\mathbf{r})$$

where $[r,x]$ is x for every variable x and $\langle[r,u],[r,v]\rangle\,\varepsilon\, r$ stands for

$$\exists_\sigma x(x\,\varepsilon\,\mathbf{r} \wedge \forall_\sigma y(y\,\varepsilon\, x \leftrightarrow (\forall_\sigma z(z\,\varepsilon\, y \leftrightarrow z \simeq [r,u]) \vee$$
$$\forall_\sigma z(z\,\varepsilon\, y \leftrightarrow (z \simeq [r,u] \vee z \simeq [r,v])))))),$$

where $z \simeq [r,u]$ and $z \simeq [r,v]$ stand for $\forall_\sigma t(t\,\varepsilon\, z \leftrightarrow t\,\varepsilon\,[r,u])$ and $\forall_\sigma t(t\,\varepsilon\, z \leftrightarrow t\,\varepsilon\,[r,v])$, respectively.

It follows easily from Definitions 16, 18 and 19 that $[r,p]$ is a condition and that if Φ is a term, a ranked formula or a formula, then $[r,\Phi]$ is a term, a ranked formula or a formula, respectively. (This is why we required condition c (*iv*) in Definition 16.)

LEMMA 58. *For all* $r \in M$, $r \subseteq \tau \times \omega$,
(a) $Q \subseteq Q' \Rightarrow [r,Q] \subseteq [r,Q']$; (b) $[r,[r,Q]] = Q$;
(c) *if* $\Phi'(x) = [r,\Phi(x)]$ *then* $[r,\Phi(u)] = \Phi'([r,u])$; (d) $\rho([r,u]) = \rho(u)$.

LEMMA 59. *If P is generic so is* $[r,P]$.

PROOF. Let $\Phi(x)$ be a formula of the M-language such that every condition p has an extension p' which satisfies $\Phi(x)$ in $\mathfrak{M}$; we shall prove that some $p^{\#} \subseteq [r,P]$ satisfies $\Phi(x)$ in $\mathfrak{M}$. Let $\Psi(x)$ be the following formula of the M-language: '*x is a condition and* $\Phi([r,x])$'. We shall first see that $\Psi(x)$ too has the property that every condition p has an extension p' which satisfies $\Psi(x)$. The condition $[r,p]$ has an extension p^* which satisfies $\Phi(x)$. By Lemma 58(a, b), $[r,p^*] \supseteq [r,[r,p]] = p$ and $[r,p^*]$ satisfies $\Psi(x)$; thus $p' = [r,p^*]$ is an extension of p which satisfies $\Psi(x)$. Since P is generic there is a condition $p'' \subseteq P$ which satisfies $\Psi(x)$. Put $p^\# = [r,p'']$. By Lemma 58(a), $p^\# = [r,p''] \subseteq [r,P]$ and since p'' satisfies $\Psi(x)$, $p^\# = [r,p'']$ obviously satisfies $\Phi(x)$.

LEMMA 60. *For every term* u, $\mathrm{val}_P(u) = \mathrm{val}_{[r,P]}([r,u])$. *For every sentence* Φ, Φ *is true in* $\mathfrak{N}_P$ *if and only if* $[r,\Phi]$ *is true in* $\mathfrak{N}_{[r,P]}$.

PROOF. If u is a set constant $\mathbf{s}$, the lemma is trivial since $[r,\mathbf{s}] = \mathbf{s}$ and, by Lemma 49, $\mathrm{val}_P(\mathbf{s}) = s = \mathrm{val}_{[r,P]}(\mathbf{s})$. For an abstraction term u and for a ranked sentence Φ, we shall prove the lemma by a simultaneous induction on $\omega^2 \cdot (\rho(u)+1)$ and $\mathrm{Ord}(\Phi)$, respectively. If a term u occurs in a ranked formula Φ then $\omega^2 \cdot (\rho(u)+1) < \mathrm{Ord}(\Phi)$ since $\rho(u) < \alpha$ in Definition 20 (of $\mathrm{Ord}(\Phi)$); hence $\rho(u)+1 \leqq \alpha$, and $\mathrm{Ord}(\Phi) > \omega^2 \cdot \alpha$ because $l \geqq 1$. On the other hand, $\omega^2 \cdot (\rho(\hat{x}_\alpha \Phi(x)) + 1) = \omega^2 \cdot (\alpha+1) > \mathrm{Ord}(\Phi(v))$, for every term v with $\rho(v) < \alpha$.

If Φ is $\neg \Psi$ or $\Psi \vee \Gamma$ this follows immediately from the induction hypothesis.

If Φ is $\exists_\alpha x \Psi(x)$ then $[r,\Phi]$ is $\exists_\alpha x \Psi'(x)$, where $\Psi'(x) = [r,\Psi(x)]$. If Φ is true in $\mathfrak{N}_P$ then, for some term u with $\rho(u) < \alpha$, $\Psi(u)$ is true in $\mathfrak{N}_P$; therefore, by the induction hypothesis, $[r,\Psi(u)]$ is true in $\mathfrak{N}_{[r,P]}$; but, by Lemma 58(c), $[r,\Psi(u)]$ is $\Psi'([r,u])$. Since, by Lemma 58(d), $\rho([r,u]) = \rho(u)$ we get that $\exists_\alpha x \Psi'(x)$ is true in $\mathfrak{N}_{[r,P]}$; hence $[r,\Phi]$ is true in $\mathfrak{N}_{[r,P]}$. Now assume, on the other hand, that $[r,\Phi]$ is true in $\mathfrak{N}_{[r,P]}$. Since $\exists_\alpha x \Psi'(x)$ is true in $\mathfrak{N}_{[r,P]}$ there is a term u with $\rho(u) < \alpha$ such that $\Psi'(u)$ is true in $\mathfrak{N}_{[r,P]}$.

$\mathrm{val}_{[r,P]}(u) = \mathrm{val}_{[r,[r,P]]}([r,u])$ by the induction hypothesis,
$= \mathrm{val}_P([r,u])$ by Lemma 58(b)
$= \mathrm{val}_{[r,P]}([r,[r,u]])$ by the induction hypothesis, since, by Lemma 58(d), $\rho([r,u]) = \rho(u) < \alpha$. Therefore also $\Psi'([r,[r,u]])$ is true in $\mathfrak{N}_{[r,P]}$. $[r,\Psi([r,u])] = \Psi'([r,[r,u]])$, by Lemma 58(c). Since, by Lemma 58(d), $\mathrm{Ord}(\Psi([r,u])) = \mathrm{Ord}(\Psi(u)) < \mathrm{Ord}(\exists_\alpha x \Psi(x))$, we have, by the induction hypothesis, that $\Psi([r,u])$ is true in $\mathfrak{N}_P$, and since $\rho([r,u]) < \alpha$ also $\exists_\alpha x \Psi(x)$ is true in $\mathfrak{N}_P$.

If Φ is $u \varepsilon v$ then $[r,\Phi]$ is $[r,u] \varepsilon [r,v]$. By the induction hypothesis, $\mathrm{val}_P(u) = \mathrm{val}_{[r,P]}([r,u])$ and $\mathrm{val}_P(v) = \mathrm{val}_{[r,P]}([r,v])$. Therefore $u \varepsilon v$ is true in $\mathfrak{N}_P$ if and only if $\mathrm{val}_{[r,P]}([r,u]) \in \mathrm{val}_{[r,P]}([r,v])$, i.e., if and only if $[r,u] \varepsilon [r,v]$ is true in $\mathfrak{N}_{[r,P]}$.

If Φ is $A(u,v)$ then $[r,\Phi]$ is

$$[r,u] \, \varepsilon \, \boldsymbol{\tau} \wedge [r,v] \, \varepsilon \, \boldsymbol{\omega} \wedge (A([r,u],(r,v]) \leftrightarrow \neg\langle [r,u],[r,v]\rangle \varepsilon \, \mathbf{r}).$$

By the interpretation of $\mathscr{L}_M$, $A([r,u],[r,v])$ is true in $\mathfrak{N}_{[r,P]}$ if and only if for some $\alpha<\tau$ and $\beta<\omega$ $\mathrm{val}_{[r,P]}([r,u])=\alpha$, $\mathrm{val}_{[r,P]}([r,v])=\beta$ and $[r,p](\alpha,\beta)=0$. By the induction hypothesis, $\mathrm{val}_{[r,P]}([r,u])=\mathrm{val}_P(u)$, $\mathrm{val}_{[r,P]}([r,v])=\mathrm{val}_P(v)$, hence $A([r,u],[r,v])$ is true in $\mathfrak{N}_{[r,P]}$ if and only if, for some $\alpha<\tau$ and $\beta<\omega$, $\mathrm{val}_P(u)=\alpha$, $\mathrm{val}_P(v)=\beta$ and $[r,P](\alpha,\beta)=0$. $[r,p](\alpha,\beta)=0$ if and only if $\langle\alpha,\beta\rangle\notin r$ and $P(\alpha,\beta)=0$ or $\langle\alpha,\beta\rangle\in r$ and $P(\alpha,\beta)=1$. $\langle[r,u],[r,v]\rangle\varepsilon\,\mathbf{r}$ is an abbreviation as in Definition 57. By easy checking (using Lemma 51) we see that if $\mathrm{val}_{[r,P]}([r,u])=\alpha<\tau$ and $\mathrm{val}_{[r,P]}([r,v])=\beta<\omega$ then $\langle[r,u],[r,v]\rangle\varepsilon\,\mathbf{r}$ is true in $\mathfrak{N}_{[r,P]}$ if and only if $\langle\alpha,\beta\rangle\in r$. As a result of all that was said in the present paragraph we get that

$$[r,u] \, \varepsilon \, \boldsymbol{\tau} \wedge [r,v] \, \varepsilon \, \boldsymbol{\omega} \wedge (A([r,u],[r,v]) \leftrightarrow \neg\langle [r,u],[r,v]\rangle \varepsilon \, \mathbf{r})$$

is true in $\mathfrak{N}_{[r,P]}$ if and only if, for some $\alpha<\tau$ and $\beta<\omega$, $\mathrm{val}_P(u)=\alpha$, $\mathrm{val}_P(v)=\beta$ and $P(\alpha,\beta)=0$, i.e., if and only if $A(u,v)$ is true in $\mathfrak{N}_P$.

We shall now prove that $\mathrm{val}_P(\hat{x}_\alpha\Phi(x))=\mathrm{val}_{[r,P]}([r,\hat{x}_\alpha\Phi(x)])$. Let $y\in\mathrm{val}_P(\hat{x}_\alpha\Phi(x))$; then, by Lemma 54, $\boldsymbol{\rho}_P(y)<\alpha$ and y satisfies $\Phi(x)$ in $\mathfrak{N}_P$, i.e., $y=\mathrm{val}_P(v)$, for some v such that $\rho(v)<\alpha$ and $\Phi(v)$ is true in $\mathfrak{N}_P$. By the induction hypothesis, $[r,\Phi(v)]$ is true in $\mathfrak{N}_{[r,P]}$. By Lemma 58(c), $[r,\Phi(v)]=\Phi'([r,v])$, where $\Phi'(x)=[r,\Phi(x)]$. By the induction hypothesis, $\mathrm{val}_{[r,P]}([r,v]) = \mathrm{val}_P(v) = y$; therefore y satisfies $\Phi'(x)$ in $\mathfrak{N}_{[r,P]}$. Since $\boldsymbol{\rho}_{[r,P]}(y)\leqq\rho([r,v])=\rho(v)<\alpha$, $y\in\mathrm{val}_{[r,P]}(\hat{x}_\alpha\Phi'(x))=\mathrm{val}_{[r,P]}[r,\hat{x}_\alpha\Phi(x)]$. On the other hand, if $y\in\mathrm{val}_{[r,P]}([r,\hat{x}_\alpha\Phi(x)])$ then, by Lemma 54, y satisfies $\Phi'(x)$ in $N_{[r,P]}$ and, for some term w with $\rho(w)<\alpha$, $y=\mathrm{val}_{[r,P]}(w)$. By the induction hypothesis and Lemma 58(b, c, d), $\mathrm{val}_{[r,P]}([r,[r,w]])=\mathrm{val}_P([r,w])$ $=\mathrm{val}_{[r,[r,P]]}([r,w])=\mathrm{val}_{[r,P]}(w)=y$; therefore $\mathrm{val}_{[r,P]}([r,[r,w]])$ satisfies $\Phi'(x)$ in $\mathfrak{N}_{[r,P]}$, i.e., $\Phi'([r,[r,w]])$ is true in $\mathfrak{N}_{[r,P]}$. $[r,\Phi([r,w])]=\Phi'([r,[r,w]])$; since $\mathrm{Ord}(\Phi([r,w]))=\mathrm{Ord}(\Phi(w))<\omega^2\cdot(\alpha+1)$, we have, by the induction hypothesis, that $\Phi([r,w])$ is true in $\mathfrak{N}_P$. As we saw, $y=\mathrm{val}_P([r,w])$, also $\boldsymbol{\rho}_P(y)\leqq\rho([r,w])=\rho(w)<\alpha$; hence, by Lemma 54, $y\in\mathrm{val}_P(\hat{x}_\alpha\Phi(x))$.

For an arbitrary Φ the lemma is proved similarly by induction on the length of Φ.

LEMMA 61. *If $P \Vdash \Phi$ and the symbol A does not occur in Φ then $0 \Vdash^* \Phi$.*

PROOF. If $P\Vdash\Phi$ then, for some $p\subseteq P$, $p\Vdash\Phi$. Let Q be any generic function. We denote with $D(p)$ the domain of p; $D(p)$ is a finite subset of $\tau\times\omega$. Let $r=\{\langle\alpha,\beta\rangle \mid \langle\alpha,\beta\rangle\in D(p) \textit{ and } Q(\alpha,\beta)\neq P(\alpha,\beta)\}$. r is a finite subset of $\tau\times\omega$; hence, as in the proof of Lemma 23, $r\in M$. $[r,Q]$ coincides with P on $D(p)$; hence $p \subseteq [r, Q]$ and therefore $[r, Q] \Vdash \Phi$. Φ is the same as $[r, \Phi]$, since A does not occur in Φ; hence $[r, Q] \Vdash [r, \Phi$ and, by Lemma 60, $Q \Vdash \Phi$. Since $Q \Vdash \Phi$ for every generic function Q, we have, by Lemmas 52 and 53, $0 \Vdash^* \Phi$.

LEMMA 62. *Let $\phi(x, y_1, \cdots, y_n)$ be a formula of set theory. Let B be a 'class' of N such that, for some $y_1, \cdots, y_m \in M$, we have that, for all x, $x \in B \Leftrightarrow x \in N \wedge \mathrm{Rel}(N, \phi(x, y_1, \cdots, y_n))$; then there is a formula $\psi(x, y_1, \cdots, y_n, \tau)$ of set theory such that, for the same $y_1, \cdots, y_n$ and for the given τ, $x \in B \cap M \Rightarrow x \in M \wedge \mathrm{Rel}(M, \psi(x, y_1, \cdots, y_n, \tau))$. If, in addition, $B \in N$ then also $B \cap M \in M$.*

PROOF.
$$\begin{aligned} B \cap M &= \{x \mid x \in M \wedge \mathrm{Rel}(N, \phi(x, y_1, \cdots, y_n)\} \\ &= \{x \mid x \in M \wedge \boldsymbol{\phi}(\mathbf{x}, \mathbf{y}_1, \cdots, \mathbf{y}_n) \textit{ is true in } \mathfrak{N}\} \text{ (by (14))} \\ &= \{x \mid x \in M \wedge P \Vdash \boldsymbol{\phi}(\mathbf{x}, \mathbf{y}_1, \cdots, \mathbf{y}_n)\} \quad \text{(by Lemma 52)}. \end{aligned}$$

A does not occur in $\boldsymbol{\phi}(\mathbf{x}, \mathbf{y}_1, \cdots, \mathbf{y}_n)$; hence, by Lemma 61, $P \Vdash \boldsymbol{\phi}(\mathbf{x}, \mathbf{y}_1, \cdots, \mathbf{y}_n)$ if and only if $0 \Vdash \neg\neg\boldsymbol{\phi}(\mathbf{x}, \mathbf{y}_1, \cdots, \mathbf{y}_n)$; thus

$$B \cap M = \{x \mid x \in M \wedge 0 \Vdash \neg\neg\boldsymbol{\phi}(\mathbf{x}, \mathbf{y}_1, \cdots, \mathbf{y}_n)\} =$$
$$\{x \mid x \in M \wedge \mathrm{Rel}(M, \pi_{\neg\neg\phi}(0, x, y_1, \cdots, y_n, \tau))\} \text{ (by Lemma 26)}.$$

If we write $\psi(x, y_1, \cdots, y_n, \tau)$ for $\pi_{\neg\neg\phi}(0, x, y_1, \cdots, y_n, \tau)$, then $\psi(x, y_1, \cdots, y_n, \tau)$ satisfies the requirement of the lemma. If $B \in N$ then, by Lemma 51, $\|B\| \leqq \rho(B) = \alpha$. Thus we have $B \subseteq R(\alpha)$, $B \cap M \subseteq R(\alpha) \cap M = R_M(\alpha) \in M$, by the proof of Lemma 23 (under (a)). $B \cap M = \{x \mid x \in R_M(\alpha) \wedge \mathrm{Rel}(M, \psi(x, y_1, \cdots, y_n, \tau))\}$. By (9) and the axiom schema of subsets we have $\mathrm{Rel}(M, (\forall y_1, \cdots, y_n, \tau) \forall z \exists t \forall x (x \in t \leftrightarrow x \in z \wedge \psi(x, y_1, \cdots, y_n, \tau)))$ from which we get, by taking $z = R_M(\alpha)$, $B \cap M \in M$.

LEMMA 63. *For every $x \in N$ if x is (hereditarily-ordinal-definable)$_N$ then $x \in M$.*

PROOF. As mentioned in Section 1, there is a formula $\chi(\alpha, x)$ of set theory such the formula $\exists \alpha \chi(\alpha, x)$ defines the property of being ordinal-definable and such that $\chi(\alpha, x) \wedge \chi(\alpha, y) \rightarrow x = y$. Assume that there is a member s of $N - M$ which is (hereditarily-ordinal-definable)$_N$. By the axiom of foundation, there is such a set $s \subseteq M$. Since, as is easily seen, s is (ordinal-definable)$_N$ there is an ordinal$_N$ α (i.e., an ordinal $\alpha \in N$) such that $\mathrm{Rel}(N, \forall t (t = s \leftrightarrow \chi(\alpha, t)))$; by Lemma 51, $\alpha \in M$. Obviously, for all z, $z \in s \Leftrightarrow z \in N \wedge \mathrm{Rel}(N, \exists y (z \in y \wedge \chi(\alpha, y)))$. Since $\alpha \in M$ and $s \in N$ we get, by Lemma 62, $s = s \cap M \in M$, which contradicts $s \in N - M$.

REFERENCES

[1] J. W. Addison, Some consequences of the axiom of constructibility, *Fundamenta Mathematicae* **46** (1959), 337–357.

[2] P. J. Cohen, The independence of the continuum hypothesis, *Proceedings of the National Academy of Sciences* **50** (1963), 1143–1148; **51** (1964), 105–110.

[3] W. B. Easton, Powers of regular cardinals, Ph.D. thesis, Princeton University, 1964. See abstract — Proper classes of generic sets, *Notices of the Amer. Math. Soc.* **11** (1964), 205.

[3a] W. B. Easton and R. Solovay, Powers of regular cardinals, in preparation.

[4] S. Feferman, Some applications of the notions of forcing and generic sets, *Proceedings of the* 1963 *International Symposium on the Theory of Models in Berkeley*, Amsterdam, 1965.

[5] A. A. Fraenkel and Y. Bar-Hillel, *Foundations of set theory*, Amsterdam, 1958.

[6] K. Gödel, *The consistency of the continuum hypothesis*, Princeton, 1940.

[7] K. Gödel, Remarks before the Princeton Bicentennial Conference, in *The Undecidable* (M. Davis, ed.), Raven Press, 1964.

[8] A. Lévy, Axiom schemata of strong infinity, *Pacific Journal of Mathematics* **10** (1960), 223–238.

[9] A. Lévy, Independence results in set theory by Cohen's method I, III, IV (abstracts), *Notices of the Amer. Math. Soc.* **10** (1963), 592–593.

[10] A. Lévy, A hierarchy of formulas in set theory, *Memoirs of the Amer. Math. Soc.*, 1965.

[11] R. Montague, Fraenkel's addition to the axioms of Zermelo, *Essays on the Foundations of Mathematics*, Jerusalem, 1961, 91–114.

[12] A. Mostowski, An undecidable arithmetical statement, *Fundamenta Mathematicae* **36** (1949), 143–164.

[13] J. Myhill and D. Scott, Ordinal definability, in preparation.

[14] J. C. Shepherdson, Inner models for set theory — Part I, *Journal of Symbolic Logic* **16** (1951), 161–190.

[15] J. R. Shoenfield, The problem of predicativity, *Essays on the Foundations of Mathematics*, Jerusalem, 1961, 132–139.

[16] W. Sierpiński, *Cardinal and ordinal numbers*, Warsaw, 1958.

METAMATHEMATICAL METHODS IN FOUNDATIONS OF GEOMETRY

WOLFRAM SCHWABHÄUSER
*Humboldt-Universität zu Berlin, Berlin, DDR**

The aim of this address is to give a survey of metamathematical problems, methods, and results for several theories of Euclidean and non-Euclidean geometry. Most of the material (sections 1–4) will deal with *elementary* or *first-order* theories, which have been mainly investigated till now in metamathematics. Section 1 contains a discussion of results for Tarski's (elementary and complete) Euclidean geometry, analogous non-Euclidean geometries, and some subtheories. In section 2, the undecidability of some geometrical theories, and in section 3, some definability problems are discussed. In section 4, an arrangement is discussed for a geometrical construction—common for certain kinds of Euclidean, hyperbolic, and elliptic geometry—of the basic field for each model and development of trigonometry using this field. This construction is thought to demonstrate the connections between the various constructions already known for each particular of these geometries. The final section 5 contains some brief remarks on weak second order theories.

1. A significant example of an elementary theory[1] is Tarski's *elementary plane Euclidean* (or *parabolic*) *geometry*—let us denote it by $\mathscr{P}^*$—which was set up in [31] (for more details see the fundamental paper, Tarski [34]). The language of this theory contains one kind of variables only, in fact, variables for points—thus, $\mathscr{P}^*$ is a theory with standard formalization in the sense of [35, pp. 5 ff]—and two non-logical constants D and B for the equidistance and the betweenness relations. $\mathscr{P}^*$ is based on an axiom system consisting of 12 single axioms, which are in essence equivalent to the groups I–IV of Hilbert's axiom system for plane Euclidean geometry (see Hilbert [6]), and the axiom schema of continuity (contained, e.g., also in Tarski and Szceerba [36]), with the axioms falling under this scheme being special cases—expressible in $\mathscr{P}^*$—of a somewhat modified form of the well-known Dedekind continuity axiom for the field of real numbers.

---

*Present address: Universität Münster, Münster, West-Germany

[1] For other theories, see especially p. 155.

It is well-known that all models of the non-elementary Hilbert axiom system for plane Euclidean geometry are isomorphic with each other and, especially, isomorphic with the two-dimensional Cartesian space $\mathfrak{C}(\mathfrak{R})$[2] over the field $\mathfrak{R}$ of real numbers. The same applies to a system containing the groups I–IV and the above-mentioned form of the Dedekind continuity axiom[3] instead of Hilbert's continuity axioms. It is a much deeper result that each elementary sentence (i.e., each sentence of $\mathscr{P}^*$) which is true in $\mathfrak{C}(\mathfrak{R})$ is already a valid sentence of $\mathscr{P}^*$, i.e., a consequence of the axioms of $\mathscr{P}^*$ (containing merely the elementary cases of the Dedekind continuity axiom). By the Gödel completeness theorem for first-order logic, each such sentence is also derivable from the axioms by means of some familiar rules of inference. This *semantical completeness* (with respect to the fixed model $\mathfrak{C}(\mathfrak{R})$) was stated in Tarski [31] and implies the completeness theorem for $\mathscr{P}^*$ which states that for each sentence σ of $\mathscr{P}^*$, σ itself or the negation $\neg\sigma$ is valid in $\mathscr{P}^*$. (Several equivalent versions of this theorem are known; cf., e.g., Tarski [34, p. 22].)

We know two methods for proving the completeness of $\mathscr{P}^*$, both based on the analogous result obtained by Tarski [31] for the elementary theory $\mathscr{R}$ (arithmetic) of real numbers.

The one has been applied by the author (see [18]) for another theory, in fact, for elementary three-dimensional Euclidean geometry, formalized in another language according to Hilbert's axiom system. However, it is not difficult to carry over this method to $\mathscr{P}^*$ itself and it can then be described as follows.

To each sentence σ of $\mathscr{P}^*$ a sentence $\mathrm{Rda}(\sigma)$ of $\mathscr{R}$ (the "arithmetical reductum of σ"), and to each formula φ of $\mathscr{R}$ a formula $\mathrm{Rdg}(\varphi)$ of $\mathscr{P}^*$ (the "geometrical reductum of φ") is constructed such that the following theorems can be established.

THEOREM 1. *For each sentence σ of $\mathscr{P}^*$, σ is true in $\mathfrak{C}(\mathfrak{R})$ iff[4] $Rda(\sigma)$ is true in $\mathfrak{R}$.*

THEOREM 2. *For each formula φ of $\mathscr{R}$, if φ is derivable from the axioms of $\mathscr{R}$ ($\vdash_{\mathscr{R}}\varphi$), then* $\mathrm{Rdg}(\varphi)$ *is derivable from the axioms of $\mathscr{P}^*$ ($\vdash_{\mathscr{P}^*}\mathrm{Rdg}(\varphi)$).*

THEOREM 3. *For each sentence σ of $\mathscr{P}^*$, σ is equivalent with its double reductum* $\mathrm{Rdg}(\mathrm{Rda}(\sigma))$, *i.e.*

$$\vdash_{\mathscr{P}^*} \sigma \leftrightarrow \mathrm{Rdg}(\mathrm{Rda}(\sigma)).$$

---

[2] See, e.g., Tarski [34, p. 21].

[3] See Tarski [34, p. 18].

[4] We use "iff" as an abbreviation for "if and only if".

By means of these theorems (Theorem 2 applied to $\phi = \mathrm{Rda}(\sigma)$), the semantical completeness of $\mathscr{P}^*$ is directly obtained (without use of the Gödel Completeness Theorem) from the semantical completeness of $\mathscr{R}$.

All models of $\mathscr{P}^*$ can be characterized (see [18] or [34]) up to isomorphy as Cartesian spaces over real closed fields, which are exactly the models of Tarski's arithmetic $\mathscr{R}$. It is remarkable that (as has been emphasized in an analogous treatment of hyperbolic geometry in [19]) just the same geometrical propositions are needed for this characterization of models as for the proof of Theorems 2 and 3. The reason for this remarkable connection can be seen from Tarski's completeness proof for $\mathscr{P}^*$, which was published in [34] and shall be the second method to be reported here.

In Tarski [34], the result dealing with characterization of all models has been stated as the "Representation Theorem" and established, essentially, by means of the above-mentioned geometrical constructions. Then, as a direct consequence of the Representation Theorem, the Completeness Theorem for $\mathscr{P}^*$ has been obtained in a version stating that all models of $\mathscr{P}^*$ are *elementarily equivalent*, i.e., that each sentence of $\mathscr{P}^*$ which is true in one model of $\mathscr{P}^*$ is true in any other model. This result is obtained from the analogous result for Tarski's arithmetic $\mathscr{R}$ by the argument that the truth of a sentence σ in a model of $\mathscr{P}^*$—which has the form $\mathfrak{C}(\mathfrak{F})$ ($\mathfrak{F}$ real closed)—means a certain statement for the field $\mathfrak{F}$. This statement—if formalized—is just that the arithmetical reductum $\mathrm{Rda}(\sigma)$ is true in $\mathfrak{F}$, and the argument, if analysed, is equivalent wich our Theorem 1 for an arbitrary real closed field $\mathfrak{F}$ instead of $\mathfrak{R}$.[5] Now, if σ is true in $\mathfrak{C}(\mathfrak{F})$ then $\mathrm{Rda}(\sigma)$ is true in $\mathfrak{F}$, hence, it is true in each model $\mathfrak{F}'$ of $\mathscr{R}$, hence σ is true in each model $\mathfrak{C}(\mathfrak{F}')$ of $\mathscr{P}^*$, as was to be established. From this version, one can easily obtain the other versions of the Completeness Theorem.

Thus, Tarski's method of establishing both metamathematical results—characterization of all models and completeness of $\mathscr{P}^*$—is much more preferable, since the geometrical constructions needed are used but once and no second kind of reducta is needed while the first kind is but little used. I now see only one advantage of the first method, namely the following. If one is interested not only in the result of (one version of) the Completeness Theorem that each sentence σ true in $\mathfrak{C}(\mathfrak{R})$ (or any other fixed model) is provable but also in a method of constructing, for each such sentence, a proof based on the axioms of $\mathscr{P}^*$, then the methods of

---

[5] In fact, Theorem 1 holds for an arbitrary structure $\mathfrak{F} = \langle F, +, \cdot, < \rangle$ (with binary operations $+$, $\cdot$ and a binary relation $<$) if appropriate definitions for $\mathfrak{C}(\mathfrak{F})$ are used.

establishing our Theorems 2 and 3 (cf. [18] or [19]) yield a method of constructing such a proof from a proof of the corresponding sentence Rda(σ) of $\mathscr{R}$, which can be obtained from Tarski's decision method for $\mathscr{R}$. This method, in general, will not lead to the "best proof" of any sentence but it is more direct than other methods based merely on the result that a proof exists. Beyond this, the method of using one or more kinds of reducta may be useful to carry over metamathematical results from one formalized theory to other ones.

From this characterization of models we can see at once that $\mathscr{P}^*$ is not categorical, i.e., that $\mathscr{P}^*$ has different models which are not isomorphic to each other. This result, however, can be already obtained from the Löwenheim-Skolem Theorem, which states that each elementary theory having an infinite model has models in each infinite power. Thus, no elementary theory having an infinite model is categorical.

A further metamathematical result for $\mathscr{P}^*$ is that this theory is *decidable*. This can be obtained, as was pointed out in Tarski [34], from the general metamathematical result that each complete and axiomatizable theory with standard formalization is decidable, which was stated in Tarski, Mostowski, Robinson [35, p. 14]. A more direct decision method, however, was given in Tarski [31], where his decision method for arithmetic $\mathscr{R}$ of real numbers is carried over to $\mathscr{P}^*$ by means of the arithmetical reducta.

The last fundamental result for $\mathscr{P}^*$ established in Tarski [34] is the "Non-Finitizability Theorem", stating that $\mathscr{P}^*$ is not finitely axiomatizable, i.e. no finite set of sentences of $\mathscr{P}^*$ is equivalent to the original (infinite) axiom system. In the proof, essentially, this is reduced to the analogous result for the arithmetic $\mathscr{R}$ of real numbers. By the "Endlichkeitssatz" of model theory, it is sufficient to establish that no finite subset of an appropriate axiom system Σ for $\mathscr{R}$ is equivalent to Σ, which is done by constructing for each such subset a model which is not a model of Σ itself.

The same metamathematical results as for $\mathscr{P}^*$ (with other kinds of spaces in the Representation Theorem) can be established for a theory $\mathscr{H}^*$ of *elementary plane hyperbolic geometry*—having the same language as $\mathscr{P}^*$ and axioms arising from those for $\mathscr{P}^*$ by replacing Euclid's axiom of parallels by its negation—, for a theory $\mathscr{E}^*$ of *elementary plane elliptic geometry*—containing a constant C for a quaternary relation of cyclic order instead of B and based on appropriate axioms—, and for similar theories for other finite dimensions by the same methods based on other geometrical constructions as will be discussed below (p. 159); (cf. Szmielew [26] and Schwabhäuser [19], [22]).

The important role of the Representation Theorem suggests considering subtheories of $\mathscr{P}^*$, $\mathscr{H}^*$, and $\mathscr{E}^*$. We shall consider theories $\mathscr{P}$, $\mathscr{H}$, $\mathscr{E}$, arising from $\mathscr{P}^*$, $\mathscr{H}^*$, $\mathscr{E}^*$, respectively, by omitting the axioms of continuity, and theories $\mathscr{P}'$, $\mathscr{H}'$, $\mathscr{E}'$ arising from $\mathscr{P}$, $\mathscr{H}$, $\mathscr{E}$, respectively, by adding the so-called circle axiom, which expresses (for $\mathscr{P}'$ and $\mathscr{H}'$) that a straight line containing an interior point of a circle has at least one point in common with this circle (for $\mathscr{E}'$ see [22]); finally, we shall consider a theory $\mathscr{H}''$ arising from $\mathscr{H}$ by adding Hilbert's axiom of hyperbolic parallels (see Hilbert [6, p. 162] or [5, p. 139f]).

One can establish that the models of $\mathscr{P}'$, $\mathscr{H}''$, $\mathscr{E}'$ are—up to isomorphy—the Cartesian or certain Klein spaces, respectively, constructed over Euclidean ordered fields (see [34], [28], [22]). The same applies for $\mathscr{P}$ and $\mathscr{E}$ with Pythagorean ordered fields while, for $\mathscr{H}$ and $\mathscr{H}'$, a more complicated description of models can be obtained from Pejas [11] and Bachmann [1][6]. Thus, all these theories are proper subtheories of $\mathscr{P}^*$, $\mathscr{H}^*$, $\mathscr{E}^*$, respectively, and, consequently, they are incomplete. Moreover, they are finitely axiomatizable by their definitions. The decision problems for these theories, as far as I know, are still open. Of course, they can be reduced to the decision problems for the theories of Euclidean or Pythagorean ordered fields, but we do not know if these theories are decidable.

2. However, we know undecidable geometrical (elementary) theories by the following result of Rautenberg [14] (cf. also [13]). Let $\mathfrak{A}(\mathfrak{R}_0)$ be the affine two-dimensional space over the field $\mathfrak{R}_0$ of rational numbers (where points, straight lines, and the incidence relation are defined as usual). If $\mathscr{T}$ is a theory containing at least a constant for the incidence relation such that one of its models is $\mathfrak{A}(\mathfrak{R}_0)$ or a structure arising from $\mathfrak{A}(\mathfrak{R}_0)$ by adding further relations, then $\mathscr{T}$ is undecidable. For the proof, one establishes that a certain theory $\mathscr{R}_0$ of rational numbers is relatively weakly interpretable in $\mathscr{T}$. By a result of Julia Robinson [15], $\mathscr{R}_0$ is completely undecidable, i.e., it has an essentially undecidable and finitely axiomatizable subtheory. Then, the conclusion follows immediately from a general result in Tarski, Mostowski, Robinson [35, section I.5]. Rautenberg

[6] When the paper was presented at the Congress, there was a discussion in which we did not find out whether the theories $\mathscr{H}'$ and $\mathscr{H}''$ were identical as it was claimed—in different words—in [17]. Some weeks later I was directed by Professor F. Bachmann (Kiel) to the papers [11] and [1], and he could demonstrate that $\mathscr{H}'$ and $\mathscr{H}''$ are not identical by constructing the following model $\mathfrak{U}(\mathfrak{F})$ for $\mathscr{H}'$ which is not a model of $\mathscr{H}''$. Let $\mathfrak{F}$ be a non-Archimedean ordered field which is Euclidean, and $\mathfrak{K}(\mathfrak{F})$ the well-known Klein model (for plane hyperbolic geometry) over $\mathfrak{F}$ (see, e.g., Szmielew [26, p. 48]). Then, $\mathfrak{U}(\mathfrak{F})$ is the submodel of $\mathfrak{K}(\mathfrak{F})$ the points of which are the pairs $\langle x, y\rangle$ of infinitely small elements x, y of $\mathfrak{F}$ (x is called infinitely small iff $-1/n < x < 1/n$ for each natural element n of $\mathfrak{F}$).

shows—applying this result—that the following theories, e.g., are undecidable: a) a theory of affine geometry based on Hilbert's axioms of incidence, order, and parallelity, b) a theory of Euclidean geometry indicated in Rautenberg [14][7] which does not contain the axiom of segment construction, the models of which are Cartesian spaces over arbitrary ordered fields, and c) similar theories with standard formalization containing variables for points only. By a somewhat more complicated proof he also shows that a hyperbolic geometry of incidence (based on the axioms of incidence and order and Hilbert's axiom of hyperbolic parallels) is undecidable.

More involved is the result of Tarski (cf. [36]) that his affine geometry is undecidable.

As regards essential undecidability (cf. [35, p. 14]), we see at once that no one of the geometrical theories considered is essentially undecidable since, for one of the Euclidean or non-Euclidean geometries, $\mathscr{P}^*$, $\mathscr{H}^*$, or $\mathscr{E}^*$ is a decidable extension with (in essence) the same language, while, for affine geometry, a similar extension can be constructed (cf., e.g., [14, §3]).

3. The most important of the definability problems is the one which asks which notions can serve as the *only primitive notions* for the underlying theory, that means that these notions can be defined by the original primitive notions occurring in the language of the theory and conversely.

If one wants to state that a certain notion (a constant within a theory) is definable in terms of some other ones, one has to give an appropriate definition and to establish that it is a valid sentence of the theory considered. For the theories considered here, this is a purely geometrical task. On the other hand, if one wants to show that a certain constant R is not definable by given other ones, one has to show that there is no appropriate definition. For this purpose, the method of Padoa (cf. [2], [8], or [30]) is mostly used, which can be put in the following form: One has to construct models $\mathfrak{M}_1$ and $\mathfrak{M}_2$ of the underlying theory and a 1-1 mapping f from $\mathfrak{M}_1$ to $\mathfrak{M}_2$ such that the relations corresponding to the given constants in $\mathfrak{M}_1$ are transformed by f into the relations corresponding to these constants in $\mathfrak{M}_2$ while the same does not apply for the constant R. (One can take the same universe and the same relations corresponding to the given constants in $\mathfrak{M}_1$ and $\mathfrak{M}_2$ such that f becomes an automorphism with respect to these relations.)

The following results can be obtained in this way.

It is well-known that congruence is definable in terms of Pieri's relation

[7] In fact, Rautenberg states that this elementary theory can be obtained through some modifications from his non-elementary theory of Euclidean geometry over Archimedean ordered fields.

P which holds for points x, y, z iff the distance from x to y equals the distance from x to z (cf. Pieri [12]). This holds in all geometries treated in section 1. Moreover, order is definable in terms of congruence if the basic field is Euclidean; thus, P can be used as the only primitive notion in $\mathscr{P}'$, $\mathscr{H}''$, $\mathscr{E}'$.

On the other hand, order is not definable in terms of congruence and collinearity in $\mathscr{P}$ and $\mathscr{E}$, hence, P cannot be used as the only primitive notion in $\mathscr{P}$ and $\mathscr{E}$ (Royden [17, p. 91, Theorem 1[8]]).

The same applies to a relation T such that $T(xyz)$ iff x, y, z form a right angle with vertex x (cf. [17, sect. 3]).

However, in Tarski [32], another relation P' is considered—such that $P'(xyz)$ iff the distance from x to y at most equals the distance from x to z—for which it is proved that it can be used as the only primitive notion for Euclidean geometry (of any finite dimension) over the field of real numbers—as originally stated. As one can see from the proof, this applies also to $\mathscr{P}$. It also applies to $\mathscr{H}$. A fortiori, it applies to the extensions considered.

In $\mathscr{H}''$, one can even use collinearity as the only primitive notion (Menger [9]).

In $\mathscr{E}'$, a binary relation A can be used as the only primitive notion where $A(xy)$ iff x and y are at the polar distance, which is understood to be the greatest distance possible for points of elliptic geometry.[9] This does not hold in $\mathscr{E}$ but, in $\mathscr{E}$, another binary relation A' of two points being closer than half the polar distance can be used as the only primitive notion (Royden [17, sect. 5]).

Thus, in $\mathscr{P}$ and $\mathscr{H}$, we have a single ternary relation between points available as the only primitive notion while we have a binary such relation for $\mathscr{E}$.

It is a remarkable result that no single binary relation can serve as the only primitive notion for Euclidean and hyperbolic geometry, nor even a finite number of such relations. This was established by R. M. Robinson in [16] for the geometries over the field of real numbers, i.e. $\mathscr{P}^*$ and $\mathscr{H}^*$, and for theories arising from $\mathscr{P}^*$ and $\mathscr{H}^*$ by adding further relations of being two points at given distances. Hence, this also applies to the subtheories $\mathscr{P}$, $\mathscr{P}'$, $\mathscr{H}$, $\mathscr{H}'$, $\mathscr{H}''$. For $\mathscr{P}^*$, this was stated already in Lindenbaum and Tarski [8]. The general proof in [16] is too complicated to be

---

[8] Reference is to that part of this theorem which concerns non-definability (Propositions 1–3 contain errors).

[9] In full (non-elementary) elliptic geometry, this is the distance $\pi/2$ if distances are described by real numbers in the usual way.

presented here but it may be pointed out that a method rather different from Padoa's one is used as follows.[10] Assume that there exist binary relations (constants) $A_1, \cdots, A_n$ which can serve as the only primitive notions of the underlying theory. Then there would exist a definition for equidistance of the form

$$D(xyuv) \leftrightarrow \varphi(xyuv)$$

which is valid in our geometry and where φ is a formula containing merely $A_1, \cdots, A_n$ as non-logical constants. By an induction according to the definition of formulas it is proved that for each such formula φ the corresponding relation has a certain property which the equidistance relation has not. Hence, the assumption must be false.

A surprising result was established by Beth and Tarski in [3] with respect to the relation E of three points forming an equilateral triangle. This relation can serve as the only primitive notion for an n-dimensional Euclidean geometry, analogous to $\mathscr{P}^*$—we shall denote it by $\mathscr{P}_n^*$ [11]—, if $n \geqq 3$, while it cannot serve as the only primitive notion for $\mathscr{P}_n^*$ (equidistance is not definable) if $n=1$ or $n=2$. The proof also applies to theories $\mathscr{P}'_n$ similarly related to $\mathscr{P}'$. A more general result in this direction is contained in Tarski [32].

Thus, we have a result which essentially depends on the dimension while most of the before-mentioned results extend to arbitrary dimensions greater than 2, and it was for convenience only that they have been formulated merely for dimension 2.

The following general result on dimension is contained in Scott [23]. If φ is any sentence formulated in $\mathscr{P}^*$, then there exists a number m such that φ is valid in $\mathscr{P}_m^*$ iff φ is valid in any other theory $\mathscr{P}_k{}^*$ where $k \geqq m$. In fact, if φ contains n distinct variables, one can take $m = n-1$. The proof of a more general result concerning arbitrary structures (or relational systems) is by induction according to the definition of formulas.

4. As has been mentioned above, some geometrical constructions are needed to obtain the metamathematical results for our geometrical theories from similar results for the arithmetic of real numbers. These constructions are

1. construction of an ordered field;
2. introduction of coordinates being elements of this field;
3. description of the primitive notions of the underlying theory by means of these coordinates.

For items 1. and 2., one has several possibilities of constructing an ordered field and a coordinate system. It depends upon the choice of these constructions what description has to be given as item 3. However, not

---

[10] For a further non-definability result (and proof method) see p. 160.

[11] This theory can be obtained from $\mathscr{P}^*$ by changing two dimension axioms only, cf. Tarski [34, footnote 5] and Scott [23].

all such constructions well-known from classical geometry can be used for our purpose since some of these apply to non-elementary theories only and, hence, cannot be based on the axioms considered here.

For Euclidean geometry, such constructions fit for our purpose are contained in Hilbert [6]. For hyperbolic geometry, an ordered field was constructed in the well-known end calculus in Hilbert [5] but not all primitive notions were described there by means of appropriate coordinates. Hilbert's end calculus was applied in solving our task by Paul Szász, who used Weierstrass homogeneous coordinates (see [24] or [25]), and by the author, who used Beltrami coordinates (see [19]). Another solution of our task was given by Wanda Szmielew, who constructed another ordered field, which she refers to as segment calculus, and used Beltrami coordinates (see [26] and [28]). This construction can also be subsumed under a construction common for both Euclidean and hyperbolic geometry (see Szmielew [27] and [29]).

One can also consider another ordered field where addition of two points on a given straight line is defined by adding their distances from a given zero point. Then, in full hyperbolic geometry, multiplication of such points is uniquely determined. But this field cannot be used in elementary hyperbolic geometry since its multiplication is not definable there. Moreover, the length of segments—considered as an element of Hilbert's end calculus—is not definable in elementary hyperbolic geometry. These non-definability results were established by the author in [20] and [21] by reduction to Tarski's results on definable sets of real numbers (see Tarski [31]).

For elliptic geometry, a solution of our task can be found, e.g., in [22] where a special case of projective number scales for the ordered field $\mathfrak{F}$ and a special case of projective homogeneous coordinates are used.

We shall now briefly discuss an arrangement for such constructions—common for Euclidean, hyperbolic, and elliptic geometry. The ordered field is a special case of projective number scales in all our geometries. It is well known how such number scales can be introduced in projective geometry—one scale being determined by three points on a straight line—and that each model of elliptic geometry is also a model of projective geometry. In Hjelmslev [7] it was shown how an arbitrary model of—as we can say now—$\mathscr{P}$ or $\mathscr{H}$ can be extended to a model of projective geometry by adding ideal elements. It turns out that elementary sentences involving ideal elements can be expressed by sentences of $\mathscr{P}$ or $\mathscr{H}$, i.e., by elementary sentences involving elements of the original model ("real elements") only. Now, in the projective space thus obtained, we can choose a projective number scale in such a way that the point u at infinity, the zero point o and the unit point e have the same position as have the corresponding points of a coordinate axis in the space usually considered as a model for our geometry.[12] Then, in each case, the number scale is determined by its zero element o and its unit element e. In hyperbolic geometry, however, not all points of our field (and, in general, not e itself) are real points. Next, we can introduce special projective homogeneous coordinate systems as they are used in the usual models. Let this be done. Now, for our third task of describing primitive notions by means of coordinates it is sufficient to develop a certain part of trigonometry, which can be done in the following way.

First, we define a function T—assigning to each segment pq a non-negative element x of our number scale such that $D(pqox)$ (Fig. 1). It turns out

$$T(p,q) = x$$

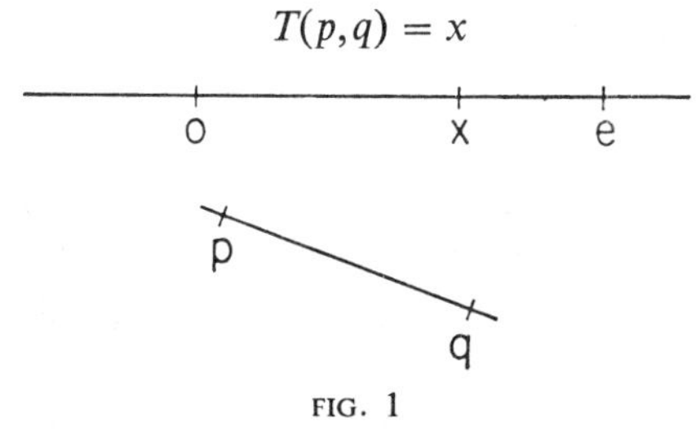

FIG. 1

that $T(p,q)$ represents the tangent of (the length of) pq in elliptic geometry, the hyperbolic tangent in hyperbolic geometry, at least in $\mathscr{H}''$, and the length itself in Euclidean geometry. In fact, $T(p,q)$ is this value, if the basic field is the field $\mathfrak{R}$ of real numbers (up to isomorphism between this field and our number scale). Next, we define the cosine, $\cos(p,q,r)$, of an angle $\sphericalangle\, pqr$ as the quotient $\dfrac{T(q,t)}{T(q,s)}$ where s,t are points on the sides qp, qr, respectively, such that $\sphericalangle\, qts$ is a right angle (Fig. 2). This quotient

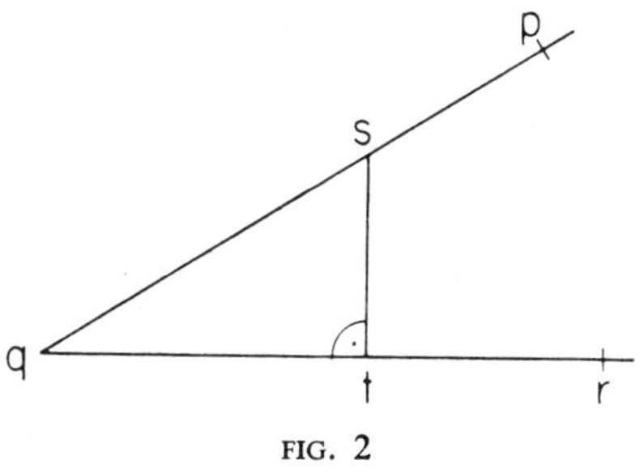

FIG. 2

[12] That means: o is an arbitrary point on the line (number scale) g. u is the (ideal or real) common point of all lines which have a common perpendicular line l, passing through o, with g. e is a) in Euclidean geometry, an arbitrary point different from o and u on g; b) in hyperbolic geometry, the (ideal) common point of all lines which are parallel to a fixed half-line contained in g; c) in elliptic geometry, one of the two points which have the same distance from o and u.

can easily be proved to be independent of the choice of s. Then, we define the sine, $\sin(p, q, r)$, of an angle $\sphericalangle\, pqr$ as the cosine of a complementary angle, and the tangent and the cotangent as the appropriate quotients of sine and cosine. We can also define a function C for segments by putting, e.g., $C(p,q) = \dfrac{T(p,s)}{T(q,r)}$ where p, q, r, s form a Lambert quadrangle with right angles at p, q, r (Fig. 3). This quotient turns out to be independent of the choice of s; it represents the cosine of the segment pq in elliptic geometry, the hyperbolic cosine in hyperbolic geometry $\mathscr{H}''$, and is always

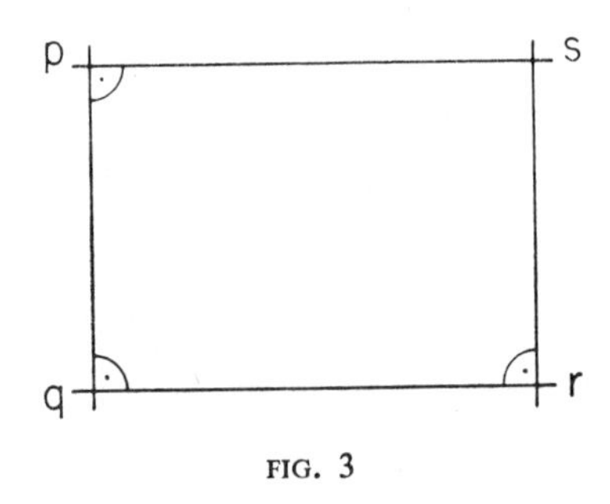

FIG. 3

the unit element in Euclidean geometry. A further function S (sine, hyperbolic sine or length of segments) can be defined by $S(p,q) = T(p,q) \cdot C(p, q)$ and a function Ct, if wanted, as the reciprocal of T.

For these "trigonometric functions," the usual formulas of trigonometry including the addition formulas can be proved. For elliptic geometry this was done in Schwabhäuser [22]. Similar constructions for elliptic geometry, though by means of another calculus, are already contained in Hessenberg [4]. However, this special case seems to be much easier since in elliptic geometry one can use common definitions for the trigonometric functions of segments and angles. For the general case, the author has proofs not all of which are very elegant and, for some propositions, he has no proof common for all the three geometries.

The resulting propositions of trigonometry are the same ones as in the segment calculus in Szmielew [26], [27], [28], [29] for hyperbolic and Euclidean geometry. Thus, W. Szmielew's segment calculus turns out to be our special case of projective number scales which differs from it merely in the choice of definitions. It may turn out that her approach to the trigonometry of all geometries is easier. Even in elliptic geometry, where the constructions in [22] seem to be very simple, one has to consider that the result that the points of a number scale form a field is taken without proof from projective geometry. If one carries out the proof, one needs Pascal's Theorem—the proofs of which from the congruence axioms are somewhat involved in all presentations I know—while, in the development of W. Szmielew, the field property can be proved without using Pascal's theorem. However, it may be an advantage of the development indicated above that one can see the connection between the projective calculus and that of W. Szmielew.

There is also a close connection between Hilbert's end calculus and our projective calculus in hyperbolic geometry $\mathscr{H}''$. Let g be a number scale with zero element o. Then, let Hilbert's ends 0 and ∞ (see Hilbert [5]) lie on the line perpendicular to g in o and Hilbert's end 1 be the (ideal) unit point of g. Then, one obtains an isomorphism between Hilbert's end calculus and the given number scale if one assigns to each end α (different from ∞) the (real or ideal) point in which the line $(\alpha\infty)$ (joining the ends α and ∞) intersects g. This was established, for the Klein space over the field of real numbers, by the author in [19]. The proof can easily be extended to an arbitrary model of $\mathscr{H}''$ by embedding it in a model of plane projective geometry by the method of Hjelmslev [7].

5. At last, we shall briefly discuss a metamathematical result for a non-elementary theory, in fact, Tarski's Euclidean geometry formalized in the weak second order logic (containing also variables for finite sets of points), the axioms being of the same form as for $\mathscr{P}^*$ although the scheme of continuity axioms is more comprehensive now (see Tarski [34, p. 24ff]); let us denote this theory by $\overline{\mathscr{P}^*}$. Mostowski (see [10]) considered another weak second order geometry $\overline{\mathscr{P}^T}$ the valid sentences of which are all the sentences true in the Cartesian space $\mathfrak{C}(\mathfrak{R})$ over the field of real numbers. He established (in the original version for the analogous theory of real numbers) that this geometry is not axiomatizable at all. Thus, it must be different from $\overline{\mathscr{P}^*}$ and, hence, it is a proper extension of $\overline{\mathscr{P}^*}$. From this, we obtain that $\overline{\mathscr{P}^*}$ is incomplete which answers the question raised in Tarski [34] to this effect.

REFERENCES

[1] F. Bachmann, Zur Parallelenfrage, *Abh. Math. Sem. Univ. Hamburg* **27** (1964), 173–192.

[2] E. W. Beth, On Padoa's method in the theory of definition, *Indag. Math.* **15** (1953), 330–339.

[3] E. W. Beth and A. Tarski, Equilaterality as the only primitive notion of Euclidean geometry, *Indag. Math.* **18** (1956), 462–467.

[4] G. Hessenberg, Neue Begründung der Sphärik, *S.-B. Berlin. Math. Ges.* **4** (1905), 69–77.

[5] D. Hilbert, Neue Begründung der Bolyai-Lobatschefskyschen Geometrie, *Math. Ann.* **57** (1903), 137–150 (reprinted as Anhang III in [6]).

[6] D. Hilbert and P. Bernays, *Grundlagen der Geometrie*, 9th ed., Stuttgart, 1962.

[7] J. Hjelmslev, Neue Begründung der ebenen Geometrie, *Math. Ann.* **64** (1907), 449–474.

[8] A. LINDENBAUM and A. TARSKI, Über die Beschränktheit der Ausdrucksmittel deduktiver Theorien, *Ergebnisse Math. Koll.* **7** (1936), 15–22 (English translation contained in [33]).

[9] K. MENGER, A new foundation of non-Euclidean, affine, real projective and Euclidean geometry, *Proc. Nat. Acad. Sci. U.S.A.* **24** (1938), 486–490.

[10] A. MOSTOWSKI, Concerning the problem of axiomatizability of the field of real numbers in the weak second order logic, *Essays on foundations of mathematics*, Jerusalem, 1961, pp. 269–286.

[11] W. PEJAS, Die Modelle des Hilbertschen Axiomensystems der absoluten Geometrie, *Math. Ann.* **143** (1961), 212–235.

[12] M. PIERI, La geometria elementare istituita sulle nozioni di 'punto' e 'sfera'. *Memorie di Matematica e di Fisica della Società Italiana delle Scienze*, ser. 3, **15** (1908), 345–450.

[13] W. RAUTENBERG, Unentscheidbarkeit der euklidischen Inzidenzgeometrie, *Z. Math. Logik Grundlagen Math.* **7** (1961), 12–15.

[14] ———, Über metatheoretische Eigenschaften einiger geometrischer Theorien, *Z. Math. Logik Grundlagen Math.* **8** (1962), 5–41.

[15] J. ROBINSON, Definability and decision problems in arithmetic, *J. Symb. Logic* **14** (1949), 98–114.

[16] R. M. ROBINSON, Binary relations as primitive notions in elementary geometry, *Symposium on the axiomatic method*, Amsterdam, 1959, pp. 68–85.

[17] H. L. ROYDEN, Remarks on primitive notions for elementary Euclidean and non-Euclidean plane geometry, *Symposium on the axiomatic method*, Amsterdam, 1959, pp. 86–96.

[18] W. SCHWABHÄUSER, Über die Vollständigkeit der elementaren euklidischen Geometrie, *Z. Math. Logik Grundlagen Math.* **2** (1956), 137–165.

[19] ———, Entscheidbarkeit und Vollständigkeit der elementaren hyperbolischen Geometrie, *Z. Math. Logik Grundlagen Math.* **5** (1959), 132–205.

[20] ———, On completeness and decidability and some non-definable notions of elementary hyperbolic geometry, *Logic, Methodology and Philosophy of Science*, Stanford, 1962, pp. 159–167.

[21] ———, Über die Nichtdefinierbarkeit einiger Begriffe der hyperbolischen Geometrie mit elementaren Mitteln, *Z. Math. Logik Grundlagen Math.* **8** (1962), 43–55.

[22] ———, On models of elementary elliptic geometry, To appear in *Symposium on the theory of models*, Amsterdam, 1965.

[23] D. SCOTT, Dimension in elementary Euclidean geometry, *Symposium on the axiomatic method*, Amsterdam, 1959, pp. 53–67.

[24] P. SZÁSZ, Unmittelbare Einführung Weierstrassscher homogener Koordinaten in der hyperbolischen Ebene auf Grund der Hilbertschen Endenrechnung, *Acta Math. Acad. Sci. Hungar.* **9** (1958), 1–28.

[25] ———, Direct introduction of Weierstrass homogeneous coordinates in the hyperbolic plane, on the base of the endcalculus of Hilbert, *Symposium on the axiomatic method*, Amsterdam, 1959, pp. 97–113.

[26] W. SZMIELEW, Some metamathematical problems concerning elementary hyperbolic geometry, *Symposium on the axiomatic method*, Amsterdam, 1959, pp. 30–52.

[27] ———, Absolute calculus of segments and its metamathematical implications, *Bull. Acad. Polon. Sci.*, Sér. Sci. Math. Astronom. Phys. **7** (1959), 213–220.

[28] ———, A new analytic approach to hyperbolic geometry, *Fund. Math.* **50** (1961), 129–158.

[29] ———, New foundations of absolute geometry, *Logic, Methodology and Philosophy of Science*, Stanford, 1962, pp. 168–175.

[30] A. TARSKI, Einige methodologische Untersuchungen über die Definierbarkeit der Begriffe, *Erkenntnis* **5** (1935), 80–100 (English translation, in essence, contained in [33]).

[31] ———, *A decision method for elementary algebra and geometry*, 2nd ed., Berkeley and Los Angeles, 1951.

[32] ———, A general theorem concerning primitive notions of Euclidean geometry, *Indag. Math.* **18** (1956), 468–474.

[33] ———, *Logic, semantics, metamathematics*, Oxford, 1956.

[34] ———, What is elementary geometry? *Symposium on the axiomatic method*, Amsterdam, 1959, pp. 16–29.

[35] A. TARSKI, A. MOSTOWSKI, and R. M. ROBINSON, *Undecidable theories*, Amsterdam, 1953.

[36] A. TARSKI and L. W. SZCZERBA, Metamathematical properties of some affine geometries, *this volume*, pp. 166–178.

METAMATHEMATICAL PROPERTIES OF SOME AFFINE GEOMETRIES

L. W. SZCZERBA

University of Warsaw, Warsaw, Poland

and

A. TARSKI

University of California, Berkeley, California, U.S.A.

This paper is closely related to two earlier articles, Szmielew [4] and Tarski [6] (see the references at the end of the paper).[1] In those articles certain geometrical theories have been described and some basic metamathematical properties of these theories have been studied.

The main subject of [6] is elementary n-dimensional metric Euclidean geometry, $\mathscr{E}_n$, where n is an arbitrary integer $\geqq 2$. In [4] two related theories are discussed, the elementary n-dimensional metric hyperbolic geometry, $\mathscr{H}_n$, and the elementary n-dimensional metric absolute geometry, $\mathscr{A}_n$. When referring to these and related theories we shall usually omit the words "elementary" and "metric". The geometries $\mathscr{E}_n$, $\mathscr{H}_n$, and $\mathscr{A}_n$ are treated as theories formalized within first-order predicate logic $\mathscr{L}_{BD}$ which contains just two non-logical constants ("primitive terms"), the ternary predicate $\boldsymbol{B}$ denoting the betweenness relation among three points and the quaternary predicate $\boldsymbol{D}$ denoting the equidistance relation between two pairs of points.[2] The theorems are based upon suitable axiom systems. The axioms for $\mathscr{E}_n$ are obtained by modifying an axiom system which provides an adequate basis for the ordinary (non-elementary) n-dimensional Euclidean geometry $\mathscr{E}'_n$; the modification consists in replacing the continuity axiom, which, as is well known, cannot be expressed in first-order logic, by an infinite collection of axioms falling under the so-called continuity schema — loosely speaking, by all the first-order instances of the continuity axiom. The method of obtaining the axioms for $\mathscr{H}_n$ could be

---

[1] This paper was prepared for publication when both authors were working at the University of California, Berkeley, on a research project in the foundations of mathematics sponsored by the U.S. National Science Foundation (Grant No. GP-1395). Detailed proofs of the results stated in this paper will be published later. The authors wish to express their sincere appreciation to Mr. H. N. Gupta for his help in preparing the paper for publication; due to his critical remarks various defects of the original manuscript have been removed.

[2] We use here $\boldsymbol{B}$ and $\boldsymbol{D}$ instead of the symbols β and δ employed in [4] and [6].

described analogously by referring to the ordinary hyperbolic geometry $\mathscr{H}'_n$. We can say more simply, however, that the axiom system for $\mathscr{H}_n$ has been obtained from that for $\mathscr{E}_n$ by replacing Euclid's axiom by its negation; if instead Euclid's axiom is omitted altogether we arrive at the axiom system for $\mathscr{A}_n$. The main results established in [4] and [6] for these theories are the following: the class of all models of each theory has been characterized in algebraic terms; $\mathscr{E}_n$ and $\mathscr{H}_n$ have been shown to be complete and consistent, while $\mathscr{A}_n$ is incomplete but has $\mathscr{E}_n$ and $\mathscr{H}_n$ as the only complete and consistent extensions (with the same non-logical constants); each of the theories discussed is decidable, and none of them is finitely axiomatizable.

In this paper we discuss analogous problems concerning certain affine geometries closely related to $\mathscr{E}_n$, $\mathscr{H}_n$, and $\mathscr{A}_n$. Just as in [4] and [6] the discussion will be concentrated upon the case $n = 2$; the extension of the results to higher dimension is a routine matter.

In logical and metamathematical notation we shall adhere in general to [7] introducing, however, some changes in logical symbolism. All the expressions written in logical symbols are treated not as expressions of a formalized theory, but as metamathematical designations of such expressions. As the designations of quantifiers we shall use $\forall$ and $\exists$; the symbols representing variables immediately following quantifiers will be written as subscripts, e.g., $\forall_{\boldsymbol{x}}$, $\exists_{\boldsymbol{y}}$. We shall consistently employ bold italics (and other bold types) to denote or represent variables, predicates, and operation symbols occurring in a formalized theory; the corresponding ordinary italics will always represent elements of mathematical structures (e.g., of models of a formalized theory), relations between, and operations on such elements. Given a formula ϕ in which $\boldsymbol{x}_1, \cdots, \boldsymbol{x}_n$ occur as the only free variables the meaning of the expression "the elements $x_1, \cdots, x_n$ satisfy ϕ" is assumed to be understood; to make this expression unambiguous we have to assume that all the variables of the formalism discussed have been arranged in an infinite sequence. We also assume to be known under what condition a relational structure $\mathfrak{A}$ is a model of a sentence σ or of a set Σ of sentences or of a theory $\mathscr{T}$; a theory is identified here with the set of all its valid sentences. Unless explicitly stated to the contrary, all the theories involved in our further discussions are assumed to be formalized within the first-order predicate logic $\mathscr{L}_{\boldsymbol{B}}$, with the ternary predicate $\boldsymbol{B}$ as the only non-logical constant. Thus, e.g., when discussing extensions of a theory we shall refer exclusively to extensions formalized in $\mathscr{L}_{\boldsymbol{B}}$; when speaking of sentences and formulas we shall mean sentences and formulas in the formalism $\mathscr{L}_{\boldsymbol{B}}$. The models of sentences are of course structures

$\mathfrak{A} = \langle A, B \rangle$ where A is a non-empty set and B a ternary relation among elements of A; i.e., a subset of the Cartesian cube $A \times A \times A$. Given such a structure $\mathfrak{A}$, by the *elementary theory of* $\mathfrak{A}$, symbolically $\mathscr{T}(\mathfrak{A})$, we understand the set of all sentences which have $\mathfrak{A}$ as a model; the *elementary theory* $\mathscr{T}(\mathfrak{K})$ *of the class* $\mathfrak{K}$ *of structures* is the intersection of all theories $\mathscr{T}(\mathfrak{A})$ with $\mathfrak{A} \in \mathfrak{K}$.

With any *metric geometry* $\mathscr{G}$ formalized in $\mathscr{L}_{BD}$ we associate the *affine geometry* $\mathscr{AG}$ determined by the following stipulations: $\mathscr{AG}$ is formalized in $\mathscr{L}_B$; a sentence in the formalism of $\mathscr{L}_B$ is valid in $\mathscr{AG}$ if and only if it is valid in $\mathscr{G}$.[3] We shall be interested in particular in affine geometries $\mathscr{AE}_2$, $\mathscr{AH}_2$, and $\mathscr{AA}_2$. The way in which these theories have been defined does not yield directly suitable axiom systems for them. In the case of $\mathscr{AE}_2$, however, we easily obtain an adequate axiomatization by proceeding exactly as in the metric case, i.e., by replacing, in a suitable axiom system for non-elementary affine geometry, the axiom of continuity by the set of all its first-order instances. In this way we arrive, e.g., at the following axiom system for $\mathscr{AE}_2$:[4]

A1 [IDENTITY AXIOM].

$$\forall_{xy}[B(xyx) \rightarrow (x = y)]$$

A2 [TRANSITIVITY AXIOM].

$$\forall_{xyzu}[B(xyz) \wedge B(yzu) \wedge (y \neq z) \rightarrow B(xyu)]$$

A3 [CONNECTIVITY AXIOM].

$$\forall_{xyzu}[B(xyz) \wedge B(xyu) \wedge (x \neq y) \rightarrow B(yzu) \vee B(yuz)]$$

A4 [EXTENSION AXIOM].

$$\forall_{yz} \exists_x[B(xyz) \wedge (x \neq y)]$$

A5 [PASCH'S AXIOM].

$$\forall_{txyzu} \exists_v[B(xtu) \wedge B(yuz) \rightarrow B(xvy) \wedge B(ztv)]$$

A6 [DESARGUES' AXIOM].

$$\forall_{txx'yy'zz'uvw}[B(txx') \wedge B(tyy') \wedge B(tzz') \wedge B(xyu) \wedge B(x'y'u) \wedge B(xzv) \wedge B(x'z'v) \wedge B(yzw) \wedge B(y'z'w) \wedge \neg B(txy) \wedge \neg B(xyt) \wedge \neg B(ytx) \wedge \neg B(tyz) \wedge \neg B(yzt) \wedge \neg B(zty) \wedge \neg B(tzx) \wedge \neg B(zxt) \wedge \neg B(xtz) \wedge (x \neq x') \rightarrow B(uvw)]$$

A7 [LOWER DIMENSION AXIOM].

$$\exists_{xyz}[\neg B(xyz) \wedge \neg B(yzx) \wedge \neg B(zxy)]$$

[3] This approach to affine geometry is known from the literature; see, e.g., [9] (where the term "descriptive" instead of "affine" is used). It is essentially equivalent to the more familiar approach that consists in constructing affine geometry as projective geometry in which a certain line, the line "at infinity", has been singled out.

[4] A related axiom system can be found in [9], pp. 3 ff.

A8 [UPPER DIMENSION AXIOM].

$$\forall_{xyzt} \exists_u \Big[\{B(yuz) \wedge [B(xut) \vee B(xtu) \vee B(txu)]\} \vee \{B(xuy) \wedge [B(zut) \vee B(tzu)]\} \vee \{B(xuz) \wedge [B(yut) \vee B(tyu)]\} \Big]$$

A9 [ELEMENTARY CONTINUITY AXIOMS]. *All sentences of the form* $\forall_{vw\ldots}\{\exists_z \forall_{xy}[\phi \wedge \psi \rightarrow B(zxy)] \rightarrow \exists_u \forall_{xy}[\phi \wedge \psi \rightarrow B(xuy)]\}$ *where* ϕ *stands for any formula in which the variables* $\mathbf{x}, \mathbf{v}, \mathbf{w}, \ldots$, *but neither* $\mathbf{y}$ *nor* $\mathbf{z}$ *nor* $\mathbf{u}$ *may occur free, and similarly for* ψ, *with* $\mathbf{x}$ *and* $\mathbf{y}$ *interchanged.*

E [EUCLID'S AXIOM].

$$\forall_{txyzu} \exists_{vw}[B(xut) \wedge B(yuz) \wedge (x \neq u) \rightarrow B(xyv) \wedge B(xzw) \wedge B(vtw)]$$

It is of course by no means obvious that the set of sentences A1,...,A9, E is an adequate axiom system for the theory $\mathscr{AE}_2$ (defined as affine geometry associated with $\mathscr{E}_2$); however, from Theorem 1 below we can easily conclude that this is actually the case. From the axiom system for $\mathscr{AE}_2$ we obtain an adequate axiom system for $\mathscr{AE}_n$ with $n \geqq 3$ by modifying the dimension axioms A7 and A8; moreover we can then omit A6 since it becomes derivable from the remaining ones.

To obtain an axiomatization for $\mathscr{AH}_2$, or $\mathscr{AA}_2$, we could try to follow the simple procedure applied in the metric case, i.e., to modify respectively the axiom system for $\mathscr{AE}_2$ by negating Euclid's axiom, or by omitting this axiom altogether. It turns out, however, that this method fails, at least if applied to the special axiom system for $\mathscr{AE}_2$ formulated above; the axiomatic theories obtained by this method prove to be proper subtheories of $\mathscr{AH}_2$ and $\mathscr{AA}_2$. Nevertheless these axiomatic theories seem to deserve interest in their own right. This applies especially to the theory obtained by omitting Euclid's axiom from the axiom system for $\mathscr{AE}_2$. The resulting theory, which we denote by $\mathscr{GA}_2$ (*general affine geometry*), distinguishes itself by the simple structure and the clear mathematical content of its axioms, and by a great variety of interesting extensions. While we shall indicate later how to axiomatize $\mathscr{AA}_2$ and $\mathscr{AH}_2$, we wish now to discuss some basic metamathematical problems concerning $\mathscr{GA}_2$.

We begin with the *representation problem* which we interpret here as the problem of providing a simple algebraic characterization for the class of all models of $\mathscr{GA}_2$. We first construct a special class of possible models for $\mathscr{GA}_2$ by means of the following procedure. Let $\mathfrak{F} = \langle F, +, \cdot, \leqq \rangle$ be any ordered field; the elements 0 and 1 of F and the binary operation $-$ on elements of F are defined in the usual way. Consider the set $A_{\mathfrak{F}} = F \times F$ of all ordered couples $x = \langle x_1, x_2 \rangle$ with $x_1, x_2 \in F$. We define the ternary relation $B_{\mathfrak{F}}$ among elements x, y, z of $F \times F$ by stipulating that $B_{\mathfrak{F}}(xyz)$ holds if and only if

$$(x_1 - y_1) \cdot (y_2 - z_2) = (x_2 - y_2) \cdot (y_1 - z_1),$$
$$0 \leqq (x_1 - y_1) \cdot (y_1 - z_1), \text{ and } 0 \leqq (x_2 - y_2) \cdot (y_2 - z_2).$$

The structure $\mathfrak{A}(\mathfrak{F}) = \langle A_{\mathfrak{F}}, B_{\mathfrak{F}} \rangle$ is referred to as the *two-dimensional affine Cartesian space*, or simply *affine plane*, over $\mathfrak{F}$. Sometimes we refer to the set $A_{\mathfrak{F}}$ as the *plane* and to its elements as *points*. In reference to $\mathfrak{A}(\mathfrak{F})$ we shall use ordinary geometrical terminology. Thus, e.g., we assume to be known which subsets of $A_{\mathfrak{F}}$ are called lines, halflines, triangles, interiors of triangles, polygons, convex sets, etc. Using interiors of triangles we can define a natural topology on $A_{\mathfrak{F}}$ and hence we can employ topological notions such as open set and the frontier of a set. Now let S be any non-empty, open, and convex point set. The structure (the ordered couple) formed by the set S and the relation $B_{\mathfrak{F}}$ restricted to the points of S, i.e., the intersection of $B_{\mathfrak{F}}$ and $S \times S \times S$, will be denoted by $\mathfrak{A}(\mathfrak{F}, S)$ and will be called a *restricted* or, more specifically, the *S-restricted (two-dimensional) affine space over* $\mathfrak{F}$. If in particular S consists of all points x such that $x_1^2 + x_2^2 < 1$, $\mathfrak{A}(\mathfrak{F}, S)$ is referred to as *affine Klein space over* $\mathfrak{F}$. When applying these constructions we often take for $\mathfrak{F}$ the ordered field $\mathfrak{R}$ of real numbers.[5]

We can now state our first result:

THEOREM 1. *Every model of the theory $\mathcal{GA}_2$ is isomorphic to a restricted affine space over some real closed ordered field $\mathfrak{F}$.*

The crucial point of the proof is the construction of a real closed ordered field and the introduction of a coordinate system in an arbitrary model of $\mathcal{GA}_2$. This is achieved by means of methods well known from affine and projective geometries; see, e.g., [8].

The converse of this theorem fails unless we restrict ourselves to fields $\mathfrak{F}$ which are isomorphic to the field of real numbers. This is seen from the following

THEOREM 2. (i) *Every restricted affine space over $\mathfrak{R}$ is a model of $\mathcal{GA}_2$.* (ii) *If the real closed ordered field $\mathfrak{F}$ is not isomorphic to $\mathfrak{R}$, then there is a restricted affine space over the field $\mathfrak{F}$ which is not a model of $\mathcal{GA}_2$.*

The proof of Theorem 2(i) is straightforward. To prove 2(ii) notice that in such a real closed ordered field $\mathfrak{F}$ there is a non-empty subset Y

[5] Only affine spaces over a real closed field are involved in our further discussion. The authors know a number of related results on the geometries of affine spaces over arbitrary ordered fields; these results will be reported elsewhere.

of F which is bounded from below but has no greatest lower bound. Now if S is the set of all points $x = \langle x_1, x_2 \rangle \in F \times F$ where x_1 is a lower bound of Y, then S is non-empty, open, and convex, but some of the axioms A9 are not satisfied in $\mathfrak{A}(\mathfrak{F}, S)$. (Axioms A1–A8 are always satisfied if $\mathfrak{F}$ is an arbitrary ordered field.)

Theorems 1 and 2 do not yield a solution of the representation problem for $\mathcal{GA}_2$. A satisfactory solution would be found if we succeeded in providing a purely mathematical (and sufficiently simple) characterization of the class of those restricted affine spaces over an arbitrary real closed ordered field which are models of $\mathcal{GA}_2$.

Let (i) $\mathfrak{K}$ and (ii) $\mathfrak{L}$ be respectively the classes of restricted affine spaces (i) over all real closed ordered fields and (ii) over the field $\mathfrak{R}$. By Theorems 1 and 2(ii), $\mathcal{T}(\mathfrak{K})$ is a proper subtheory of $\mathcal{GA}_2$. By Theorem 2(i), $\mathcal{T}(\mathfrak{L})$ is an extension of $\mathcal{GA}_2$. The problem is open whether $\mathcal{T}(\mathfrak{L})$ and $\mathcal{GA}_2$ coincide; we know, however, that there is a model $\mathfrak{M}$ of $\mathcal{GA}_2$ which is not elementarily equivalent to any space in $\mathfrak{L}$ (i.e., $\mathcal{T}(\mathfrak{M}) \neq \mathcal{T}(\mathfrak{A})$ for every $\mathfrak{A} \in \mathfrak{L}$). Consider the affine geometry $\mathcal{GA}'_2$ which is provided with a set-theoretical basis and the axiom system of which is obtained from that of $\mathcal{GA}_2$ by replacing all the axioms A9 by the non-elementary continuity axiom (for a formulation of this axiom see, e.g., [6], p. 18). From Theorems 1 and 2(i) we easily conclude that the class of all models of $\mathcal{GA}'_2$ consists of all restricted spaces in $\mathfrak{L}$ and their isomorphic images. Thus, in a suitable formalization, the (non-elementary) theory of $\mathfrak{L}$ coincides with $\mathcal{GA}_2$.

We turn to the *problems of completeness, decidability, and finite axiomatizability of* $\mathcal{GA}_2$.

THEOREM 3. *The theory $\mathcal{GA}_2$ is incomplete and has in fact $2^{\aleph_0}$ different, complete, and consistent extensions.*

In fact, for any given real $x > 0$, let S_x be the interior of the pentagon in $\mathfrak{A}(\mathfrak{R})$ with the vertices $a = \langle 2, -1 \rangle$, $b = \langle 0, -1 \rangle$, $c = \langle 0, 1 \rangle$, $d = \langle 2, 1 \rangle$, $e = \langle 4 + x, 0 \rangle$. We show that $\mathcal{T}(\mathfrak{A}(\mathfrak{R}, S_x))$ is always an extension of $\mathcal{GA}_2$ and that $\mathcal{T}(\mathfrak{A}(\mathfrak{R}, S_x)) \neq \mathcal{T}(\mathfrak{A}(\mathfrak{R}, S_y))$ for any real x, y with $0 < x < y$.

THEOREM 4. *The theory $\mathcal{GA}_2$ is undecidable and so are all its subtheories (in other words, $\mathcal{GA}_2$ is hereditarily undecidable).*[6]

This theorem is proved by using the results of [7]. We construct the restricted affine space $\mathfrak{S} = \mathfrak{A}(\mathfrak{R}, S)$ by taking for S the interior of the

[6] This improves a result in [2] where a much weaker affine geometry (without the continuity schema) has been shown to be undecidable.

least convex set which contains the points $a=\langle 1,0\rangle$, $b=\langle(1+\sqrt{3})/2, (1-\sqrt{3})/2\rangle$, $c=\langle 0,-1\rangle$, $d=\langle -1,0\rangle$, $e=\langle(1-\sqrt{3})/2, (1+\sqrt{3})/2\rangle$, $f_n=\langle n/\sqrt{n^2+1}, 1/\sqrt{n^2+1}\rangle$ for $n=0,1,2,\cdots$. We check that $\mathfrak{S}$ is a model of $\mathcal{GA}_2$ and we show that the theory $\mathfrak{Q}$ of [7], p. 51, is relatively interpretable in $\mathcal{T}(\mathfrak{S})$; in showing this we apply (as in the proof of Theorem 1) the familiar method of introducing coordinates in an affine space.

Theorems 1 and 4 when compared with the corresponding results concerning $\mathcal{A}_2$ in [4] (Theorems 3.2 and 3.3 for $n=2$, pp. 50 ff.) exhibit fundamental metamathematical differences between the theories $\mathcal{A}_2$ and $\mathcal{GA}_2$, notwithstanding the fact that the axiom systems of these two theories have been constructed by essentially the same method.

THEOREM 5. *The theory $\mathcal{GA}_2$ is not finitely axiomatizable.*

To prove this theorem, we show that the finite axiomatizability of $\mathcal{GA}_2$ would imply that of $\mathcal{E}_2$. This is a consequence of the following lemma: *If we remove from the axiom system for $\mathcal{E}_2$ all the elementary continuity axioms in which the symbol* **D** *occurs, then the resulting axiom system is equivalent to the original one.*

We wish to discuss some extensions of $\mathcal{GA}_2$. The most interesting examples of such extensions are defined model-theoretically as the elementary theories of some mathematically important classes of restricted affine spaces. We shall state below a fairly general result concerning extensions of this kind. To formulate this result conveniently we refer to the formalism of the elementary theory of ordered fields in which two binary operation symbols, $+$ and $\cdot$, and the binary predicate $\leq$ occur as the only non-logical constants, i.e., to the formalism of the predicate logic $\mathcal{L}_{+\cdot\leq}$. (With practically no change in the formulation of the result we could refer instead to the formalism of $\mathcal{L}_B$; however, the resulting formulation would be somewhat weaker and less convenient in applications.) Let ϕ and ψ be two formulas in this formalism, let $\mathbf{y}_1,\cdots,\mathbf{y}_n$ be all the distinct free variables of ϕ, and $\mathbf{x}_1,\mathbf{x}_2,\mathbf{y}_1,\cdots,\mathbf{y}_n$ all the distinct free variables of ψ. Given an ordered field $\mathfrak{F}=\langle F,+,\cdot,\leq\rangle$, the formulas ϕ and ψ are said to *define uniformly a class K of subsets S of $F\times F$* if the following condition holds: $S\in K$ if and only if there are elements $y_1,\cdots,y_n\in F$ such that $y_1,\cdots,y_n$ satisfy ϕ while S consists of just those points $x=\langle x_1,x_2\rangle\in F\times F$ for which $x_1,x_2,y_1,\cdots,y_n$ satisfy ψ. We could express this condition equivalently in the following way $S\in K$ if and only if the structure $\langle F,+,\cdot,\leq,S\rangle$ is a model of the formula

$$\exists_{\mathbf{y}_1\cdots\mathbf{y}}\{\phi\wedge\forall_{\mathbf{x}_1\mathbf{x}_2}[S(\mathbf{x}_1,\mathbf{x}_2)\leftrightarrow\psi]\},$$

which belongs of course to the formalism of $\mathscr{L}_{+\cdot\leqq}$ enriched by the binary predicate S. If there exist two formulas ϕ and ψ which define K uniformly, we simply say that K is *uniformly definable* (in the elementary theory of ordered fields).

THEOREM 6. *Let $\mathfrak{F}$ be any real closed ordered field, K any non-empty uniformly definable class of non-empty, open, and convex point sets of $\mathfrak{A}(\mathfrak{F})$, and $\mathfrak{K}$ the class of all restricted affine spaces $\mathfrak{A}(\mathfrak{F},S)$ such that $S\in K$. Under these assumptions the theory $\mathscr{T}(\mathfrak{K})$ satisfies the following conditions*:

(i) *$\mathscr{T}(\mathfrak{K})$ is a finite consistent extension of $\mathscr{G}\mathscr{A}_2$;*

(ii) *$\mathscr{T}(\mathfrak{K})$ is not finitely axiomatizable;*

(iii) *$\mathscr{T}(\mathfrak{K})$ is decidable;*

(iv) *if, whenever $S', S''\in K$, the corresponding restricted affine spaces* $\mathfrak{A}(\mathfrak{F},S')$ and $\mathfrak{A}(\mathfrak{F},S'')$ *are isomorphic* (*or at least elementarily equivalent*), *then $\mathscr{T}(\mathfrak{K})$ is complete;*

(v) *if $\mathfrak{F}'$ is any other real closed ordered field, K' is the class of point sets in $\mathfrak{A}(\mathfrak{F}')$ uniformly defined by the same formulas as K, and $\mathfrak{K}'$ is the corresponding class of restricted affine spaces, then $\mathscr{T}(\mathfrak{K})=\mathscr{T}(\mathfrak{K}')$.*

The proof of this theorem is based upon simple ideas, but its rigorous presentation is long and complicated in details. We consider two kinds of structures: restricted affine spaces $\mathfrak{A}\langle\mathfrak{F},S\rangle$ and structures $\langle\mathfrak{F},S\rangle$ or, more precisely, $\langle F,+,\cdot,\leqq,S\rangle$, where $\mathfrak{F}=\langle F,+,\cdot,\leqq\rangle$ is a real closed ordered field and S is a subset of $F\times F$ belonging to K. The elementary theories of those two kinds of structures are formalized in $\mathscr{L}_B$ and $\mathscr{L}_{+\cdot\leqq S}$, respectively. We establish certain formal connections (loosely speaking, a kind of isomorphism) between these theories by translating the sentences of one formalism into the sentences of the other; the translations are based upon the analytic definition of betweenness and upon the construction of a field in affine geometry (as in the proof of Theorem 1). As a consequence, various problems concerning the theory $\mathscr{T}(\mathfrak{K})$ reduce to corresponding problems concerning the elementary theory of real closed ordered fields, and the solution of the latter problems is known from [5].

The theory $\mathscr{T}(\mathfrak{K})$ can be called (*elementary*) *affine geometry of point sets in K*.

It may be noticed that the notion of uniform definability could be referred, not to a class of subsets S of $F\times F$, but to a class of restricted spaces $\mathfrak{A}(\mathfrak{F},S)$) with $\mathfrak{F}$ now ranging over a class of real closed ordered fields (possibly over the class of all such fields). In this way we could obtain a seemingly more general result, which, however, would not essentially improve Theorem 6.

By Theorem 6(i), the theory $\mathscr{T}(\mathfrak{K})$ is axiomatizable, and an axiom system for this theory can be obtained by adjoining finitely many sentences or, what amounts to the same, a single sentence $\chi(K)$ to the axiom system of $\mathscr{GA}_2$. The proof of Theorem 6 gives an effective construction of such a sentence dependent on the defining formulas ϕ and ψ; frequently, however, a simpler sentence serving the same purpose can be obtained by some other method.

We shall give below a few simple and interesting examples of classes K satisfying the hypothesis of Theorem 6 and in each case we shall formulate an appropriate sentence $\chi(K)$. To fix the ideas we may take $\mathfrak{R}$ for $\mathfrak{F}$.

EXAMPLE 1. K_1 contains the whole plane as the only member. The theory $\mathscr{T}(\mathfrak{K}_1)$, the affine geometry of the plane, clearly coincides with $\mathscr{AE}_2$; as we know, for $\chi(K)$ we can take Euclid's axiom. Obviously K_1 satisfies the premiss of Theorem 6(iv) and therefore $\mathscr{T}(\mathfrak{K}_1)$ is complete.

EXAMPLE 2. K_2 consists of all interiors of halfplanes. For $\chi(K_2)$ we can take the conjunction of the sentence

$$\exists_{xy}\forall_{zuvw}\exists_t\Big[(x\neq y)\wedge\{B(xzv)\wedge B(uvw)\wedge(z\neq v)\rightarrow$$
$$[B(uzt)\vee B(wzt)]\wedge[B(xyt)\vee B(ytx)\vee B(txy)]\}\Big]$$

and the negation of Euclid's axiom. Again $\mathscr{T}(\mathfrak{K}_2)$ is complete.

EXAMPLE 3. K_3 consists of all interiors of circles. Clearly $\mathscr{T}(\mathfrak{K}_3)$ coincides with the affine hyperbolic geometry $\mathscr{AH}_2$. It can be shown that for $\chi(K_3)$ we can take the conjunction of the following two sentences:

H1. $$\forall_{xy}\exists_{zuvw}\forall_t\Big[(x\neq y)\rightarrow B(xzv)\wedge B(uvw)\wedge(z\neq v)\wedge$$
$$\{[\neg B(uzt)\wedge\neg B(wzt)]\vee[\neg B(xyt)\wedge\neg B(ytx)\wedge\neg B(txy)]\}\Big]$$

H2. $$\forall_{xyztuvw}[P(txyv)\wedge P(vxyt)\wedge P(uzyw)\wedge P(wzyu)\wedge P(xvzw)\wedge$$
$$P(xtzu)\wedge B(tyw)\wedge B(vyu)\rightarrow B(xyz)]$$

where $P(txyv)$ is an abbreviation of the formula

$$\forall_s[\neg B(txs)\vee\neg B(yvs)]\wedge\forall_{x'}\exists_{s'}[B(xx'v)\wedge(x\neq x')\rightarrow B(tx's')\wedge B(yvs')].$$

Notice that H1 is the negation of the sentence formulated above in Example 2; on the basis of the axioms of $\mathscr{GA}_2$ it is stronger than the negation of Euclid's axiom. On the other hand H2 is a special form of Pascal's theorem involving six pairs of parallel halflines; in fact, as is easily seen, the formula $P(txyv)$ is satisfied by the points t, x, y, v if and only if the halfline through x with the origin t is parallel to the halfline through v with the origin y. Again $\mathscr{T}(\mathfrak{K}_3)$ is complete.

It is important to observe that, from a mathematical viewpoint, $\mathscr{H}_2$ and $\mathscr{A}\mathscr{H}_2$ are essentially the same theories since it is known that $\boldsymbol{D}$ is definable in $\mathscr{H}_2$ in terms of $\boldsymbol{B}$.[7] Hence *the axiom system consisting of* A1–A8, H1, H2, *and the* (*non-elementary*) *continuity axiom can be regarded as an adequate axiom system for the ordinary* (*non-elementary*) *two-dimensional hyperbolic geometry.*

Another consequence of this observation is that the metamathematical properties of $\mathscr{A}\mathscr{H}_2$ resulting from Theorem 6 can also be derived directly from the corresponding theorems on $\mathscr{H}_2$ stated in [4].

EXAMPLE 4. K_4 is the class consisting of the plane and of all interiors of halfplanes and circles. As $\chi(K_4)$ we can take Pascal's theorem in the special form stated above in Example 3. On the other hand let K_4' be the class consisting of the plane and of all interiors of circles; then $\mathscr{T}(\mathfrak{K}_4')$ clearly coincides with $\mathscr{A}\mathscr{A}_2$, and for $\chi(K_4')$ we can take the conjunction of Pascal's theorem and the negation of $\chi(K_2)$. It might be interesting to find a simpler sentence for this purpose. Obviously neither $\mathscr{T}(\mathfrak{K}_4)$ nor $\mathscr{T}(\mathfrak{K}_4')$ is complete; $\mathscr{T}(\mathfrak{K}_4')$ has three complete and consistent extensions, $\mathscr{T}(\mathfrak{K}_1)$, $\mathscr{T}(\mathfrak{K}_2)$, and $\mathscr{T}(\mathfrak{K}_3)$, while $\mathscr{T}(\mathfrak{K}_4')$ has two such extensions, $\mathscr{T}(\mathfrak{K}_1)$ and $\mathscr{T}(\mathfrak{K}_3)$.

EXAMPLE 5. For each $n \geqq 3$ let P_n, or P_n', be respectively the class of all interiors of convex polygons with at most, or exactly, n vertices. For $\chi(P_n)$ we can take

$$\exists_{tu_1\dots u} \forall_{xy} \exists_z \{(t \neq u_1) \wedge \cdots \wedge (t \neq u_n) \wedge$$

$$[B(u_1tz) \vee \cdots \vee B(u_ntz)] \wedge [B(xyz) \vee B(yzx) \vee B(zxy)]\}.$$

For $n=3$ the classes P_n and P_n' coincide; if $n > 3$, we take for $\chi(P_n')$ the conjunction of $\chi(P_n)$ and the negation of $\chi(P_{n-1})$. The classes P_3' and P_4' satisfy the premiss of Theorem 6(iv), and the corresponding theories $\mathscr{T}(\mathfrak{P}_3')$ and $\mathscr{T}(\mathfrak{P}_4')$, affine geometries of triangles and quadrangles, are complete. On the other hand, all the theories $\mathscr{T}(\mathfrak{P}_n')$ with $n \geqq 5$ are incomplete and each of them has $2^{\aleph_0}$ complete and consistent extensions (compare the outline of the proof of Theorem 3).

It may be mentioned that none of the classes discussed in Examples 2–5 satifies the following condition: together with any set S it contains all the sets X such that the spaces $\mathfrak{A}(\mathfrak{R}, S)$ and $\mathfrak{A}(\mathfrak{K}, X)$ are isomorphic. Hence each of these classes K can be enlarged by some subsets of $R \times R$ without affecting the corresponding theory $\mathscr{T}(\mathfrak{K})$.

---

[7] See [1] and [3].

There are many interesting extensions of $\mathcal{GA}_2$ which are not of the form discussed in Theorem 6 but which are obtained by a direct enlargement of the axiom system. Again we wish to give a few examples. It can be shown that all the theories described in these examples ($\mathcal{T}_{(1)}$ - $\mathcal{T}_{(5)}$) are incomplete and each of them has $2^{\aleph_0}$ complete extensions; each of them is hereditarily undecidable and none is finitely axiomatizable.

EXAMPLE 6. In the axiom system of $\mathcal{GA}_2$ we include the negation of Euclid's axiom. Let $\mathcal{T}_{(1)}$ be the resulting theory. Every model of $\mathcal{T}_{(1)}$ is isomorphic to a restricted affine space $\mathfrak{A}(\mathfrak{F}, S)$ where S is different from the whole plane $F \times F$ (and, of course, $\mathfrak{F}$ is a real closed field). The axiom system for $\mathcal{T}_{(1)}$ has been constructed by the same method as the axiom system for $\mathcal{H}_2$ in [4]; nevertheless the two theories differ essentially in some of their metamathematical properties.

In describing the remaining three examples we shall use $\boldsymbol{E}(\boldsymbol{xyz})$ as an abbreviation of the formula

$$\forall_{tu} \exists_{vw} [\boldsymbol{B}(\boldsymbol{xut}) \wedge \boldsymbol{B}(\boldsymbol{yuz}) \wedge (\boldsymbol{x} \neq \boldsymbol{u}) \rightarrow \boldsymbol{B}(\boldsymbol{xyv}) \wedge \boldsymbol{B}(\boldsymbol{xzw}) \wedge \boldsymbol{B}(\boldsymbol{vtw})].$$

As is easily seen, three points x, y, z of an affine space (i.e., a model of $\mathcal{GA}_2$) satisfy this formula if and only if the angle whose vertex is x and whose sides pass respectively through y and z is a "Euclidean angle"—in the sense that every point u inside this angle lies between some points v and w lying on the two sides of the angle. Using the abbreviation $\boldsymbol{E}(\boldsymbol{xyz})$ we can give Euclid's axiom E the following simple form:

$$\forall_{xyz} \boldsymbol{E}(\boldsymbol{xyz}).$$

EXAMPLE 7. We supplement the axiom system of $\mathcal{GA}_2$ by the sentence

$$\forall_{xrs} \exists_{yz} [(\boldsymbol{r} \neq \boldsymbol{s}) \rightarrow \boldsymbol{B}(\boldsymbol{rys}) \wedge \boldsymbol{B}(\boldsymbol{rzs}) \wedge (\boldsymbol{y} \neq \boldsymbol{z}) \wedge \boldsymbol{E}(\boldsymbol{xyz})].$$

Let $\mathcal{T}_{(2)}$ be the theory based upon the extended axiom system. Using Theorem 1 we easily show that every model of $\mathcal{T}_{(2)}$ is isomorphic to a restricted affine space $\mathfrak{A}(\mathfrak{F}, S)$ in which the frontier of S is rectilinear, i.e., is the closure of a union of segments with distinct endpoints; conversely, every restricted affine space (over a real closed ordered field) with this property is a model of $\mathcal{T}_{(2)}$ provided it is a model of $\mathcal{GA}_2$.

EXAMPLE 8. We now enrich the axiom system of $\mathcal{GA}_2$ by including the sentence

$$\forall_{xys} \exists_{z} [(\boldsymbol{y} \neq \boldsymbol{s}) \rightarrow \boldsymbol{B}(\boldsymbol{yzs}) \wedge (\boldsymbol{y} \neq \boldsymbol{z}) \wedge \boldsymbol{E}(\boldsymbol{xyz})].$$

The resulting theory $\mathcal{T}_{(3)}$, which is a subtheory of $\mathcal{T}_{(2)}$, exhibits some interesting model-theoretical properties. A necessary and sufficient condition for a restricted affine space $\mathfrak{A}(\mathfrak{R}, S)$ to be a model of $\mathscr{T}_{(3)}$ is that $\mathfrak{A}(\mathfrak{R}, S)$ be isomorphic to a restricted affine space $\mathfrak{A}(\mathfrak{R}, S')$ where S' is the interior of a convex polygon, i.e., $S' \in P_n$ for some $n > 3$ (see Example 5); if, however, $\mathfrak{F}$ is any real closed field not isomorphic to $\mathfrak{R}$, then there is a restricted affine space $\mathfrak{A}(\mathfrak{F}, S)$ which is a model of $\mathscr{T}_{(3)}$ and is not isomorphic to any $\mathfrak{A}(\mathfrak{F}, S')$ with $S' \in P_n$. If we further enrich the axiom system of $\mathscr{T}_{(3)}$ by including the negations of all the sentences $\chi(P_n)$, $n = 3, 4, \cdots$, formulated in Example 5, then the resulting theory $\mathscr{T}_{(4)}$ is consistent but has no models in the class $\mathfrak{L}$ of all restricted affine spaces over $\mathfrak{R}$. $\mathscr{T}_{(4)}$ is not a finite extension of $\mathscr{GA}_2$; the problem whether every consistent finite extension of $\mathscr{GA}_2$ has models in the class $\mathfrak{L}$ is equivalent to the problem (previously mentioned) whether $\mathscr{GA}_2 = \mathscr{T}(\mathfrak{L})$.

EXAMPLE 9. Consider finally the theory $\mathscr{T}_{(5)}$ obtained by adding to the axioms of $\mathscr{GA}_2$ the sentence

$$\forall_{xyz}[E(xyz) \rightarrow B(xyz) \vee B(yzx) \vee B(zxy)].$$

Every model of $\mathscr{T}_{(5)}$ is isomorphic to a restricted affine space $\mathfrak{A}(\mathfrak{F},S)$ in which (i) the frontier of S is curvilinear, i.e., does not include any segment with distinct endpoints, and (ii) S does not include any two distinct halflines of the space $\mathfrak{A}(\mathfrak{F})$ which have the same origin and do not lie on the same line.[8] Conversely, every restricted affine space (over a real closed ordered field) which has these properties and is a model of $\mathscr{GA}_2$ is also a model of $\mathscr{T}_{(5)}$.

It should be pointed out that we do not know any consistent extension of $\mathscr{GA}_2$ which is finitely axiomatizable; the problem whether such an extension exists is open.

To conclude we state two results of a different character concerning extensions of $\mathscr{GA}_2$.

THEOREM 7. *In any two consistent extensions of $\mathscr{GA}_2$ the classes of valid universal sentences (i.e., sentences of the form $\forall_{x_1 \dots x_n} \phi$ where ϕ is a formula without quantifiers) are identical.*

THEOREM 8. *Let σ be a sentence of the form $\forall_{x_1 \dots x_k} \exists_{y_1 \dots y_l} \phi$ where ϕ is a formula without quantifiers, and let n be any integer $\geqq k$ and $\geqq 3$. The sentence is valid in $\mathscr{GA}_2$ if and only if it is valid in the theory $\mathscr{T}(\mathfrak{P}_n)$ of Example 5.*

---

[8] In the alternative approach to affine geometry mentioned in footnote 3 condition (ii) expresses the fact that the frontier of S does not include any segment "at infinity" (with distinct endpoints) and is thus but a particular case of condition (1).

These two theorems follow easily from Theorems 1 and 6(v). In the proof of Theorem 8 we apply moreover the following lemma: if $\mathfrak{F}$ is any ordered field, S is an open and convex point set in $\mathfrak{A}(\mathfrak{F})$, and $x_1, \ldots, x_k \in S$ $(k \geqq 3)$, then there is a convex polygon P included in S with at most k vertices such that the points $x_1, \ldots, x_k$ lie in the interior of P.

Using Theorem 8 and analyzing the proof of Theorem 6(iii) we obtain

THEOREM 9. *The theory* $\mathcal{GA}_2$ *is decidable with respect to the set of all sentences of the form* $\forall_{x_1 \ldots x_k} \exists_{y_1 \ldots y_l} \phi$ *where* ϕ *is a formula without quantifiers (in other words, the set of all sentences of this form which are valid in* $\mathcal{GA}_2$ *is recursive).*

REFERENCES

[1] KARL MENGER. A new foundation of non-euclidean, affine, real projective and Euclidean geometry. *Proceedings of the National Academy of Sciences* vol. **4** (1938), pp. 486–490.

[2] WOLFGANG RAUTENBERG. Unentscheidbarkeit der Euklidischen Inzidenzgeometrie. *Zeitschrift für mathematische Logik und Grundlagen der Mathematik*, vol. 7 (1961), pp. 12–15.

[3] H. L. ROYDEN. Remarks on primitive notions for elementary Euclidean and non-Euclidean geometry. *The axiomatic method with special reference to geometry and physics*. Edited by Leon Henkin, Patrick Suppes, and Alfred Tarski. Amsterdam 1959, pp. 86–96.

[4] WANDA SZMIELEW. Some metamathematical problems concerning elementary hyperbolic geometry. *The axiomatic method with special reference to geometry and physics*. Edited by Leon Henkin, Patrick Suppes, and Alfred Tarski. Amsterdam 1959, pp. 30–52.

[5] ALFRED TARSKI. *A decision method for elementary algebra and geometry*. Second edition, Berkeley 1951, VI + 63 pp.

[6] ALFRED TARSKI. What is elementary geometry? *The axiomatic method with special reference to geometry and physics*. Edited by Leon Henkin, Patrick Suppes, and Alfred Tarski. Amsterdam 1959, pp. 16–29.

[7] ALFRED TARSKI, ANDRZEJ MOSTOWSKI, and RAPHAEL M. ROBINSON. *Undecidable theories*. Amsterdam 1953, IX + 98 pp.

[8] OSWALD VEBLEN and JOHN WESLEY YOUNG. *Projective geometry*. Vol. 1, Boston 1938, X + 345 pp.

[9] A. N. WHITEHEAD. *The axioms of descriptive geometry*. Cambridge 1907, VIII + 74 pp.

RECURSIVE FUNCTIONS AND ARITHMETICAL FUNCTIONS OF ORDINAL NUMBERS*

GAISI TAKEUTI
University of Illinois, Urbana, Ill., U.S.A.

Machover [7] and the author [13] independently defined the notion of recursive functions of ordinal numbers. Machover's definition of this notion is different from the author's definition. A little later Lévy [6] defined the notion of computable functions of ordinal numbers and proved that Machover's recursive functions, the author's recursive functions and Lévy's computable functions are closely related. In this paper the author will use his terminology of [13] and [16].

Roughly speaking, the concept of recursive functions of ordinal numbers is obtained from Kleene's definition of recursive functions using the least number operator μx by generalizing the recursion schema to a transfinite recursion schema. Since the meaning of the least number operator μx depends on the range of x, a recursive function of ordinal numbers should be called precisely Ω-recursive where Ω is the range of variables (§1).

Recursive function theory of ordinal numbers has some relations with ordinary recursive function theory. For example, let us call an ordinal 'recursively expressible' if it is expressible by means of recursive functions of ordinal numbers. Then, the concept of recursively expressible ordinals is a nice generalization of the Church-Kleene constructive ordinals in the following sense: the former corresponds to both two function quantifier predicates in the same way that the latter correspond to both one function quantifier predicates (§2).

As usual, we shall define arithmetical predicates from the recursive predicate by prefixing quantifiers. In [14], the author presented a hypothesis called transcendency of cardinals (abbreviated by TC). TC means that the next cardinal cannot be reached by any arithmetical function. This hypothesis seems to me plausible and interesting. We shall consider a subhypothesis $(TC)_1$ which means that ω_1 cannot be reached by any arithmetical function from ω. We shall prove that the consistency of $(TC)_1$ with set theory is equivalent to the consistency of Mahlo's axiom with set theory

* Parts of this work were done when the author worked at IBM during the summer of 1963. Parts of this work were supported by the contract NSFGP 1568.

(§3). We shall also consider the extensibility and contractibility of recursive or arithmetical functions. TC and $V=L$ of [2] form a nice contrast in these problems (§4).

§ 1. Definitions of recursive and arithmetical functions.

Let Ω be a cardinal or the class of all the ordinals which will be denoted by On.

We use the concepts of ordinal numbers, $0, \omega, <, =, a'$ (the successor of a), $\max(a, b)$, as usual. 1 stands for $0'$. We also use the following functions: (These functions as well as a' and $\max(a, b)$ are considered to be functions over $\Omega(> \omega)$.)

$$Iq(a,b) = \begin{cases} 0 & \text{if } a < b, \\ 1 & \text{otherwise.} \end{cases}$$

$$j(g^1(a), g^2(a)) = a, \quad g^1(j(a,b)) = a, \quad g^2(j(a,b)) = b.$$

$$j(a,b) < j(c,d) \leftrightarrow \max(a,b) < \max(c,d)$$

$$\vee (\max(a,b) = \max(c,d) \wedge (b < d \vee (b = d \wedge a < c))).$$

$$f(a) = 0 \rightarrow f(\mu x f(x)) = 0 \wedge \mu x f(x) \leqq a.$$

$$f(\mu x f(x)) = 0 \vee \mu x f(x) = 0.$$

(In the following we shall sometimes write $\mu x A(x)$ instead of $\mu x\, f(x)$ if $\forall x(f(x) = 0 \leftrightarrow A(x))$.)

If a function f over Ω is obtained by combining $0, \omega, a', Iq, \max, j, g^1, g^2$ and further function symbols $h_1, \cdots, h_m$ (which may be function variables) and by applications of the operation μ, f is called a *function combination* over Ω containing $h_1, \cdots, h_m$, and is often written as $f(h_1, \cdots, h_m, a_1, \cdots, a_n)$.

We further define the functions:

$$N(a) = \begin{cases} 0 & \text{if } 0 < a, \\ 1 & \text{otherwise,} \end{cases}$$

which turns out to be $Iq(0, a)$;

$$D_j(a,b) = \begin{cases} 0 & \text{if } a = 0 \text{ or } b = 0, \\ 1 & \text{otherwise,} \end{cases}$$

which turns out to be $N(\max(N(a), N(b)))$;

$$Eq(a,b) = \begin{cases} 0 & \text{if } a = b, \\ 1 & \text{otherwise,} \end{cases}$$

which turns out to be $\max(N(Iq(a,b)),\ N(Iq(b,a)))$;

$$\delta(a) = \begin{cases} b & \text{if } a = b', \\ a & \text{otherwise,} \end{cases}$$

which turns out to be $\mu x D_j(Eq(a,x'), Eq(a,x))$;
and

$$f^a(h_1,\cdots,h_m,b,a_1,\cdots,a_n) = \begin{cases} f(h_1,\cdots,h_m,b,a_1,\cdots,a_n) & \text{if } b < a, \\ 0 & \text{otherwise,} \end{cases}$$

which turns out to be

$$\mu x D_j(\max(Eq(x,f(h_1,\cdots,h_m,b,a_1,\cdots,a_n)),Iq(b,a)),\max(Eq(x,0),N(Iq(b,a)))).$$

(Note that in the definition of f^a, a is compared only with the leftmost argument for variables of f.)

DEFINITION. A function f over Ω is called *Ω-recursive in* $h_1,\cdots,h_m$, if it can be defined by a series of applications of the following schemata, provided that $g, g_1,\cdots,g_l$ are given functions Ω-recursive in $h_1,\cdots,h_m$. f is simply called *Ω-recursive,* if $m = 0$ holds in addition.

(I) $f(a) = a'$

(II) $\begin{cases} f(a) = 0 \\ f(a) = \omega \end{cases}$

(III) $f(a) = a$

(IV) $f(a,b) = Iq(a,b)$

(V) $f(a,b) = \max(a,b)$

(VI) $f(a,b) = j(a,b)$

(VII) $\begin{cases} f(a) = g^1(a) \\ f(a) = g^2(a) \end{cases}$

(VIII) $f(h_1,\cdots,h_m,a_1,\cdots,a_n) = h_i(a_1,\cdots,a_n),\ 1 \leqq i \leqq m$

(IX) $f(h_1,\cdots,h_{m^+},a_1,\cdots,a_n) =$
$g(h_1,\cdots,h_m,g_1(h_1,\cdots,h_m,a_1,\cdots,a_n),\cdots,g_l(h_1,\cdots,h_m,a_1,\cdots,a_n)),\ m \leqq m^+$

(X) $\begin{cases} f(h_1,\cdots,h_m,a_1,\cdots,a_n,a) = g(h_1,\cdots,h_m,a_1,\cdots,a_n) \\ f(h_1,\cdots,h_m,a,a_1,\cdots,a_n) = g(h_1,\cdots,h_m,a_1,\cdots,a_n) \end{cases}$

(XI) $f(h_1,\cdots,h_m,a_1,\cdots,a_n,a) = \mu x_{x<a} g(h_1,\cdots,h_m,a_1,\cdots,a_n,x)$

(XII) $f(h_1,\cdots,h_m,a,a_1,\cdots,a_n) = C(f^a,h_1,\cdots,h_m,a,a_1,\cdots,a_n)$,

where $C(h,h_1,\cdots,h_m,a,a_1,\cdots,a_n)$ is a function combination over Ω containing functions already defined to be recursive in $h,h_1,\cdots,h_m$ by applications of the schemata and $h,h_1,\cdots,h_m$.

(XIII) $f(h_1,\cdots,h_m,a_1,\cdots,a_n) = \mu x g(h_1,\cdots,h_m,a_1,\cdots,a_n,x)$, where g must satisfy the condition

$$\forall x_1 \cdots \forall x_n \exists x(g(h_1, \cdots, h_m, x_1, \cdots, x_n, x) = 0)$$

where $x_1, \cdots, x_n, x$ run over Ω.

DEFINITION. A function over Ω is called *Ω-recursive in the narrow sense* if (VIII) is not used in the above definition and $m = 0$.

DEFINITION. A predicate $A(h_1, \cdots, h_m, a_1, \cdots, a_n)$ over Ω is called *Ω-recursive (in $h_1, \cdots, h_m$)* if its representing function is Ω-recursive (in $h_1, \cdots, h_m$).

DEFINITION. A predicate $A(h_1, \cdots, h_m, a_1, \cdots, a_n)$ over Ω is called *Ω-arithmetical (in the narrow sense)* if it can be expressed by a predicate over Ω such that it is obtained from an Ω-recursive predicate (in the narrow sense) by prefixing a sequence of alternating quantifiers.

DEFINITION. A function $f(h_1, \cdots, h_m, a_1, \cdots, a_n)$ over Ω is called *Ω-arithmetical (in the narrow sense)* if there exists an Ω-arithmetical predicate (in the narrow sense) such that

$$\forall x \forall x_1 \cdots \forall x_n(f(h_1, \cdots, h_m, x_1, \cdots, x_n) = x \leftrightarrow A(h_1, \cdots, h_m, x_1, \cdots, x_n, x)).$$

DEFINITION. A function $f(h_1, \cdots, h_m, a_1, \cdots, a_n)$ over Ω is called *Ω-semi-arithmetical in $h_1, \cdots, h_m$*, if it can be defined by a series of applications of the schemata (I)–(XIII), where the condition on g in (XIII) is omitted. If $m = 0$ also holds, then f is simply called *Ω-semi-arithmetical*. A function over Ω is called *Ω-semi-arithmetical in the narrow sense* if (VIII) is not used in the above definition and $m = 0$.

§ 2. Constructibility of arithmetical functions

Assuming set theory we can construct the theory of ordinal numbers within set theory (cf. [12]). In the following we understand that the theory of ordinal numbers is a subtheory of set theory in this sense and the ε-relation in the original set theory is denoted by ε. We have constructed a model of set theory in the theory of recursive functions. This can be seen, e.g., in [10], [11], [15] and [16]. For most notions and notations, cf. [15] and [16].

DEFINITION. An Ω-recursive function n is defined by the following schema.

1) $n(0) = 0$,
2) $n(a) = \mu x(\forall y(y < a \rightarrow n(y) \in x) \wedge \forall y(y < x \wedge y \in x \rightarrow \exists z(z < a \wedge n(z) \equiv y)))$

for $a > 0$, where by $\in$ and $\equiv$ we understand Ω-recursive functions given in [15]. $n(a)$ is the ordinal number which represents a in the model.

DEFINITION. Let $A(a_1,\cdots,a_n)$ be a predicate over Ω. We say that $A(a_1,\cdots,a_n)$ is *L-expressible over* Ω, if there exists a predicate $\phi(a_1,\cdots,a_n)$ on L over Ω such that

1. $A(a_1,\cdots,a_n)\leftrightarrow\phi(a_1,\cdots,a_n)$

2. $\phi(a_1,\cdots,a_n)$ is constructed only by $a_1,\cdots,a_n$, $\neg$, $\wedge$, $\vee$, $\forall\bar{x}$, $\exists\bar{x}$, ε, where $\bar{x}$ ranges over $F''\Omega$ (cf. Gödel [2] for F and $''$).

THEOREM 1. *Every Ω-semi-arithmetical predicate is L-expressible over Ω-*

We shall prove this theorem in the following form:

THEOREM 1′. *Let f be an Ω-semi-arithmetical function of the form $f(h_1.\ldots, h_m, a_1,\ldots, a_n)$. Then there exists a constructible operation A such that, for every series of constructible functions $h_1,\ldots, h_m$,*

$$b=f(h_1,\cdots,h_m,a_1,\cdots,a_n)\leftrightarrow\langle ba_1\cdots a_n\rangle\ \varepsilon\ A(H_1,\cdots,H_m),$$

where H_i is defined by

$$b=h_i(a_1,\cdots,a_{l_i})\leftrightarrow\langle ba_1\cdots a_{l_i}\rangle\ \varepsilon\ H_i\ (1\leqq i\leqq m).$$

We denote one such A by f^* and H_i by h_i^*.

PROOF. We prove this by mathematical induction on the number of stages used to construct f.

1) If f is $0,\omega,a',a,h_i$ $(1\leqq i\leqq m), Iq, \max, j, g^1, g^2$, then the theorem is clear.

2) If $f(h_1,\cdots,h_m,a_1,\cdots,a_n)$ is of the form

$$g(h_1,\cdots,h_m,g_1(h_1,\cdots,h_m,a_1,\cdots,a_n),\cdots,g_l(h_1,\cdots,h_m,a_1,\cdots,a_n)),$$

then $b=f(h_1,\cdots,h_m,a_1,\cdots,a_n)$ is equivalent to $\exists x_1\cdots\exists x_l(\langle x_1a_1\cdots a_n\rangle\ \varepsilon\ g_1^*(h_1^*,\cdots,h_m^*)\wedge\cdots\wedge\langle x_la_1\cdots a_n\rangle\varepsilon\ g_l^*(h_1^*,\cdots,h_m^*)\wedge\langle bx_1\cdots x_l\rangle\ \varepsilon\ g^*(h_1^*,\cdots,h_m^*))$, where $g_1^*,\cdots,g_l^*,g^*$ are previously defined by the induction hypothesis. Hence the theorem is clear by the General Existence Theorem on L in Gödel [2].

3) If $f(h_1,\cdots,h_m,a_1,\cdots,a_n)$ is of the form $g(h_1,\cdots,h_m,a_2,\cdots,a_n)$ or $g(h_1,\cdots,h_m,a_1,\cdots,a_{n-1})$, then the theorem is proved in the same way as in case 2.

4) If $f(h_1,\cdots,h_m,a_1,\cdots,a_n)$ is of the form $\mu x_{x<a_n}g(h_1,\cdots,h_m,a_1,\cdots,a_n,x)$, then $b=f(h_1,\cdots,h_m,a_1,\cdots,a_{n-1},a_n)$ is equivalent to $(b<a_n\wedge\langle 0a_1\cdots a_{n-1}b\rangle\varepsilon\ g^*(h_1^*,\cdots,h_m^*)\wedge\forall x(x<b\rightarrow\neg\langle 0a_1\cdots a_{n-1}x\rangle\ \varepsilon\ g^*(h_1,\cdots,h_m)))\vee(b=0\wedge\forall x(x<a_n\rightarrow\neg\langle 0a_1\cdots a_{n-1}x\rangle\ \varepsilon\ g^*(h_1^*,\cdots,h_m^*)))$.

5) If $f(h_1,\cdots,h_m,a_1,\cdots,a_n)$ is of the form $\mu xg(h_1,\cdots,h_m,a_1,\cdots,a_n)$, then $b=f(h_1,\cdots,h_m,a_1,\cdots,a_n)$ is equivalent to $((\langle 0ba_1\cdots a_n\rangle \,\varepsilon\, g^*(h_1,\cdots,h_m^*) \wedge \forall x(x<b\rightarrow \neg\langle 0xa_1\cdots a_n\rangle \,\varepsilon\, g^*(h_1^*,\cdots,h_m^*))) \vee (b=0\wedge\neg\exists x(\langle 0xa_1\cdots a_n\rangle \,\varepsilon\, g^*(h_1^*,\cdots,h_m^*)))$.

Hence the theorem is proved as before.

6) If f is defined by the schema

$$f(h_1,\cdots,h_m,a_1,\cdots,a_n) = C(f^{a_1},h_1,\cdots,h_m,a_1,\cdots,a_n),$$

then, by the induct on hypothesis, there exists a constructible operation G_1 such that $c \,\varepsilon\, G_1(h_0^*h_1^*,\cdots,h_m^*a_1,\cdots,a_n)\leftrightarrow\langle ca_1\cdots a_n\rangle \varepsilon\, C^*(h_0^*h_1^*\cdots,h_m^*)$. Then there exists a constructible operation G such that $\langle cd\rangle \,\varepsilon\, G(h_1^*,\cdots,h_m^*a_2,\cdots,a_n)\leftrightarrow c \,\varepsilon\, G_1((\{y\}d'y)^*,h_1^*\cdots,h_m^*D)d),a_1,\cdots,a_n)$ where d is constructible. Thus we can find a constructible operation H such that $H(h_1^*,\cdots,h_m^*,a_2,\cdots,a_n)'a_1=G(h_1^*,\cdots,h_m^*,a_2,\cdots,a_n)'(H(h_1^*,\cdots,h_m^*,a_2,\cdots,a_n)\upharpoonright a_1)$. It is seen that $f(h_1,\cdots,h_m,a_1,\cdots,a_n) = H(h_1^*,\cdots,h_m^*,a_2,\cdots,a_n)'a_1$. Our proof is completed.

For the proof of Theorem 5 below, we notice that throughout the proof the notion of 'power set' does not appear in the construction of f^*.

THEOREM 2. *If $\phi(a_1,\ldots,a_n)$ is a predicate on L over Ω, then there exists an Ω-arithmetical predicate $A(a_1,\ldots,a_n)$ in the narrow sense such that $\phi(a_1,\ldots,a_n)\leftrightarrow A(a_1,\ldots,a_n)$.*

To prove the theorem we use the following lemma, the first part of which has already been proved in [15].

LEMMA. *Let F be the operation of Gödel and let α and β be ordinal numbers. Then $F'\alpha\,\varepsilon\, F'\beta\leftrightarrow\alpha\in\beta$, $F'\alpha=F'\beta\leftrightarrow\alpha\equiv\beta$ and (1) $Od'F'\alpha=Od(\alpha)$, where the left side of the equality is the order of $F'\alpha$ in the sense of Gödel, while the right side is an Ω-recursive function of α, (2) If $F'\alpha\,\varepsilon\, On$, $n(F'\alpha)\equiv Od'F'\alpha$, (3) $\beta=F'\alpha\leftrightarrow n(\beta)\equiv\alpha$.*

PROOF. (2) We shall prove it by transfinite induction on α. If $Od'F'\alpha<\alpha$, $n(F'Od'F'\alpha)\equiv Od'F'Od'F'\alpha$ by the induction hypothesis, i.e., $n(F'\alpha)\equiv Od'F'\alpha$. Let $Od'F'\alpha=\alpha$. Then, for every $\delta\in n(F'\alpha)$, there exists an ordinal δ_0 such that $\delta_0<F'\alpha$ and $n(\delta_0)\equiv\delta$ (by the definition of n). By [2,9.52], $Od'\delta_0<Od'F'\alpha=\alpha$. Therefore, $n(F'Od'\delta_0)\equiv Od'F'Od'\delta_0(=Od'\delta_0)$, by the induction hypothesis, which implies $F'\delta=F'Od'\delta_0=\delta_0<F'\alpha$ (by (1)), i.e., $F'\delta\,\varepsilon\, F'\alpha$. Applying (1) again, we have $\delta\in\alpha$. Thus $n(F'\alpha)\subseteq\alpha=Od'F'\alpha$. Conversely, let $\delta\in\alpha=Od'F'\alpha$. Then $F'\delta<F'\alpha$ by (1), which implies $n(F'\delta)\in n(F'\alpha)$ by the definition of n. Since $n(F'Od'F'\delta)$

$\equiv Od'F'Od'F'\delta$ by the induction hypothesis, $n(F'\delta) \equiv Od'F'\delta = Od(\delta)$ (by (1)) $\equiv \delta \in n(F'\alpha)$. Thus we have $Od'F'\alpha = \alpha \subseteq n(F'\alpha)$.

(3) $\beta = F'\alpha \to n(\beta) \equiv \alpha$ by (2). Let $n(\beta) \equiv \alpha$. Since $\alpha \equiv Od(\alpha) = Od'F'\alpha = Od'\beta$, $F'\alpha = F'Od'\beta = \beta$.

PROOF OF THEOREM 2. Let $\phi(a_1, \cdots, a_n)$ be a predicate on L over Ω. We define the *-operation on a predicate on L as follows:

1) $(F'\alpha \,\varepsilon\, F'\beta)^*$ is $\alpha \in \beta$;
2) $(\neg A)^*$ is $\neg(A)^*$;
3) $(A \wedge B)^*$ is $(A)^* \wedge (B)^*$;
4) $(\exists \bar{x}\psi(\bar{x}))^*$ is $\exists \xi \psi^*(\xi)$.

We see that $\phi(F'\alpha_1, \cdots, F'\alpha_n) \leftrightarrow \phi^*(\alpha_1, \cdots, \alpha_n)$, by the lemma. Then we have $\phi(a_1, \cdots, a_n) \leftrightarrow \exists \alpha_1 \cdots \exists \alpha_n (a_1 = F'\alpha_1 \wedge \cdots \wedge a_n = F'\alpha_n \wedge \phi(F'\alpha_1, \cdots, F'\alpha_n)) \leftrightarrow \exists \alpha_1 \cdots \exists \alpha_n (n(a_1) \equiv \alpha_1 \wedge \cdots \wedge n(a_n) \equiv \alpha_n \wedge \phi^*(\alpha_1, \cdots, \alpha_n))$, which is Ω-arithmetical.

As a corollary of Theorems 1 and 2 we have

THEOREM 3. *Every predicate over Ω is Ω-semi-arithmetical if and only if it is Ω-arithmetical in the narrow sense.*

THEOREM 4. *Let f be Ω-recursive and a be an ordinal number $< \Omega$. Then there exists an ordinal number b such that $b < \Omega$ and*

$$\forall x(x < a \to f(x) = (F'b)'x).$$

PROOF. By Theorem 1, we see easily that $f^* \cdot (f^{*\prime\prime} a \times a)$ is a constructible set. Let b_1 be the order of $f^* \cdot (f^{*\prime\prime} a \times a)$. By the proof in Chapter VIII of Gödel [2], we can find b satisfying the theorem from b_1.

THEOREM 5. *Let a be an ordinal number $< \Omega$ and f be an Ω-recursive function. Then there exists an ordinal number b such that $b < \Omega$ and $\forall x(x<a \to f(x) = u(b^+ n(x)))$, where u and $+$ are Ω-recursive. (For the definition of u and $+$, cf., e.g., [16].)*

PROOF. This is proved by interpreting Theorem 4 in our model. Let a be a variable of ordinal numbers $< \Omega$ and $A(a)$ a predicate over Ω. We shall sometimes write $a \in A$ and $cl(\{x\}A(x))$ instead of $A(a)$ and $\forall x \exists y \forall z (z \in A \wedge z \in x \leftrightarrow z \in y)$, respectively. We define *constructive predicate over Ω*, corresponding to '*constructive formula*' in [15], in the same way as in [15]. Let f be an Ω-recursive function of the form $f(h_1, \cdots, h_m, a_1, \cdots, a_n)$. We can find a constructive predicate $A(x_1, \cdots, x_n)$ over Ω such that, for every series of constructible functions $h_1, \cdots, h_m$, $b = f(h_1, \cdots, h_m, a_1, \cdots, a_n) \leftrightarrow \langle n(b), n(a_1), \cdots, n(a_n)\rangle \in A(H_1, \cdots, H_m)$, where H_i is a constructive predicate over Ω satisfying $b = h_i(a_1, \cdots, a_{l_i})$

$\leftrightarrow \langle n(b),\ n(a_1), \cdots, n(a_{l_i})\rangle \in H_i$ $(1 \leqq i \leqq m)$ and $\langle a_1, \cdots, a_n\rangle$ is a recursive function defined by $\langle a_1\rangle = a_1$, $\langle a_1, a_2\rangle = j(1, j(1, a_1, a_1), j(1, a_1, a_2))$, $\langle a_1, a_2, \cdots, a_n\rangle = j(1, a_1, \langle a_2, \cdots, a_n\rangle)$, for $n > 2$, and $cl(\{x\} A(H_1, \cdots, H_m, x))$, by following the proof of Theorem 1′. (Since we have not used the notion of power set in the proof of Theorem 5, this can easily be done.)

For a given f as in the theorem, let A be a constructive predicate over Ω as above. Then by [15, 5.3.13], we can find an ordinal number c such that $c < \Omega$ and $\forall x(x \in c \leftrightarrow \exists y(y \in n(a) \wedge \langle x, y\rangle \in A))$. Since $cl(\{x\}A(x))$, we can find an ordinal number b such that $b < \Omega$ and $\forall x(x \in b \leftrightarrow \exists y \exists z(x \equiv \langle z, y\rangle \wedge y \in n(a) \wedge z \in c \wedge x \in A))$. Thus we see that $\forall x(x < a \rightarrow f(x) = u(b + n(x)))$.

DEFINITION. A function over Ω is called *Ω-quasi-elementary*, if it can be defined by a series of applications of the schemata (I)-(XI) of §1 and the following schemata:

(XIV) $f(a, b) = a + 1$;
(XV) $f(a, b) = b \times a\ (= J(a, b))$;
(XVI) $f(a) = fn(a)$;
(XVII) $f(a) = u(a)$;
(XVIII) $f(a) = n(a)$.

(For the definition of $+, \times, fn$ and u, cf. [15].) A predicate over Ω is called *Ω-quasi-elementary* if it has an Ω-quasi-elementary representing function. A predicate over Ω is called an *Ω-qe-predicate* if it is constructed from Ω-quasi-elementary predicates, propositional connectives and quantifiers over Ω.

Let $P(f, \cdots, g, a, \cdots, b)$ be a predicate obtained from an Ω-quasi-elementary predicate by prefixing a sequence of alternating k quantifiers. Then we call $P(f, \cdots, g, a, \cdots, b)$ a *k-qe-predicate over* Ω. A *k-qe*-predicate over Ω is called *a* $\sum_k^{qe}$- or *a* $\prod_k^{qe}$*-predicate over* Ω according as the outermost quantifier is existential or universal. A predicate over Ω is said to be *expressible in* $\sum_k^{qe}$*-form* or *in* $\prod_k^{qe}$*-form* if it is equivalent to a $\sum_k^{qe}$-predicate over Ω or a $\prod_k^{qe}$-predicate over Ω respectively. A predicate over Ω is said to be *expressible in both k-qe-forms* if it is expressible both in $\sum_k^{qe}$-form and in $\prod_k^{qe}$-form.

COROLLARY OF THEOREM 5. *Proposition* 6 *of* [16] *for gr* (*or qe*)*-predicates* (*over* Ω) *is true without assuming the axiom of constructibility, i.e., for each of these predicates, an unbounded quantifier can be advanced across a bounded quantifier of opposite kind.*

LEMMA. *Every recursive predicate over* Ω *is expressible in both* 1-*qe-forms*. (This corresponds to Theorem 2 of [16] and is proved without assuming the axiom of constructibility.)

PROOF. We follow the proof of Theorem 2 on p. 208 of [16], replacing 'elementary' by 'quasi-elementary'. We have only to change the case of $f(h_1, \cdots, h_m, a_1, \cdots, a_n) = C(f^{a_1}, h_1, \cdots, h_m, a_1, \cdots, a_n)$ as follows:

Let $Fn(b, c)$ be $\forall u \forall v \forall w(u < b \wedge v < b \wedge w < b \wedge \langle v, u \rangle \in b \wedge \langle w, u \rangle \in b \rightarrow v \equiv w) \wedge \forall x(x < c' \rightarrow \exists y(y < b \wedge 0(y) \wedge \langle y, n(x) \rangle \in b))$.

Then $(b = f(h_1, \cdots, h_m, a_1, \cdots, a_n))^*$ is defined to be
$\exists x(Fn(x, a_1) \wedge \forall y(y < a_1' \rightarrow G(\{v\}u(x^+ n(v)), h_1, \cdots, h_m, u(x^+ n(y)), y, a_2, \cdots, a_n)) \wedge b = u(x^+ n(a_1)))$ and (or)
$\forall x(Fn(x, a_1) \wedge \forall y(y < a_1' \rightarrow G(\{v\}u(x^+ n(v)), h_1, \cdots, h_m, u(x^+ n(y)), y, a_2, \cdots, a_n)) \rightarrow b = u(x^+(n(a_1))))$,
where $G(h, h_1, \cdots, h_m, a_1, \cdots, a_n)$ is $(c = C(h, h_1, \cdots, h_m, a_1, \cdots, a_n))^*$. By Theorem 5 we get the lemma.

COROLLARY. *The class of general recursive predicates over Ω coincides with the class of predicates over Ω expressible in both 1-qe-forms. For each $k \geqq 0$, the class of Σ_k^{gr}-predicates over Ω containing no function variables coincides with the class of Σ_k^{grn}-predicates over Ω and the class of $\prod_k^{gr}$-predicates over Ω containing no function variables coincides with the class of $\prod_k^{grn}$-predicates over Ω. (For Σ_k^{gr}, $\prod_k^{gr}$, Σ_k^{grn}, $\prod_k^{grn}$, cf.* [16].)

This is proved by the above lemma and the similar proof of Theorem 1 of [16] without assuming the axiom of constructibility.

THEOREM 6. *Each ω_1-recursive predicate containing no function variables is expressible in both two function-quantifier forms.*

PROOF. By means of the lemma, we have only to show that every quasi-elementary predicate over ω_1 is expressible in both two function-quantifier forms. We shall modify §7 of [16] as follows: In defining 'a system of equations' $[f(a_1, \cdots, a_n) = C(k_1, \cdots, k_j, a_1, \cdots, a_n)]$ and its auxiliary functions, we add

4) Let $C(k_1, \cdots, k_j, a_1, \cdots, a_n)$ be of the form $\mu x C_0(k_1, \cdots, k_j, x, a_1, \cdots, a_n)$. Then $[f(a_1, \cdots, a_n) = C(k_1, \cdots, k_j, a_1, \cdots, a_n)]$ is $h_0(f(a_1, \cdots, a_n), a_1, \cdots, a_n) = 0 \wedge \forall x(x < f(a_1, \cdots, a_n) \rightarrow h_0(x, a_1, \cdots, a_n) \neq 0) \wedge \forall x \forall x_1 \cdots \forall x_n([h_0(x, x_1, \cdots, x_n) = C_0(k_1, \cdots, k_j, x, x_1, \cdots, x_n)])$, where the auxiliary functions of $[f(a_1, \cdots, a_n) = C(k_1, \cdots, k_j, a_1, \cdots, a_n)]$ are h_0 and the auxiliary functions of $[h_0(x, x_1, \cdots, x_n) = C_0(k_1, \cdots, k_j, x, x_1, \cdots, x_n)]$. In defining a system $[f(a_1, \cdots, a_n) = C(k_1, \cdots, k_j, a_1, \cdots, a_n)]^{a_0}$ of equations restricted by a_0, we add $[f(a_1, \cdots, a_n) = \mu x C(k_1, \cdots, k_j, x, a_1, \cdots, a_n)]^{a_0}$ is
$(a_1 < a_0 \wedge \cdots \wedge a_n < a_0 \rightarrow h(f(a_1, \cdots, a_n), a_1, \cdots, a_n) = 0 \wedge$
$\forall x(x < f(a_1, \cdots, a_n) \wedge x < a_0 \rightarrow h(x, a_1, \cdots, a_n) \neq 0)) \wedge$
$\forall x \forall x_1 \cdots \forall x_n([h(x, x_1, \cdots, x_n) = C(k_1, \cdots, k_j', x, x_1, \cdots, x_n)]^{a_0})$.

In defining $[f(a_1, \cdots, a_n) = C(a_1, \cdots, a_n)]^{\hat{\alpha}_0}$, we add: If $C(a_1, \cdots, a_n)$ is of the form $\mu x C_0(x, a_1, \cdots, a_n)$, then $[f(a_1, \cdots, a_n) = C(a_1, \cdots, a_n)]^{\hat{\alpha}_0}$ is
$(D(\alpha_0, \hat{a}_1) \wedge \cdots \wedge D(\alpha_0, \hat{a}_n) \to \hat{h}_0(\hat{f}(\hat{a}_1, \cdots, \hat{a}_n), \hat{a}_1, \cdots, \hat{a}_n) = \psi_1(\hat{a}_1) \wedge \forall \hat{x}(D(\alpha_0, \hat{x}) \wedge \alpha_0(\hat{x}, \hat{f}(\hat{a}_1, \cdots, \hat{a}_n)) = 0 \wedge \hat{x} \neq \hat{f}(\hat{a}_1, \cdots, \hat{a}_n) \wedge \hat{h}_0(\hat{x}, \hat{a}_1, \cdots, \hat{a}_n) \neq \psi_1(\hat{a}_1))) \wedge \forall x \forall x_1 \cdots \forall x_n([h_0(x, x_1, \cdots, x_n) = C_0(x, x_1, \cdots, x_n)]^{\hat{\alpha}_0})$.

By means of Theorem 5, we can prove the existence of an ordinal number closed with respect to any given ω_1-recursive functions. Propositions 10–12 remain valid by omitting 'primitive' there. (They are proved without using the axiom of constructibility from the first in spite of an omission of the mark °.) Thus in the same way as in §7 of [16], we can complete the proof.

As a corollary of this theorem we can prove the following Theorem 8 of [16] without using the axiom of constructibility: Each predicate over ω_1 containing no function variable and expressible in the *k-pr*-quantifier form ($k \geqq 1$) is expressible in the $k+1$-function form keeping the outermost quantifier of the same kind.

THEOREM 7. *A predicate over Ω of natural numbers is Ω-recursive, if and only if it is expressible in both two function-quantifier forms in the Kleene hierarchy.*

PROOF. By means of Theorem 3 of [13], we have only to prove the theorem in case Ω is ω_1. Let $P(a_1, \cdots, a_n)$ be an ω_1-recursive predicate of natural numbers. Then by Theorem 6 it is expressible in both two function-quantifier forms, say $A(\phi_1, \cdots, \phi_n)$. Let $v_a(x, y)$ be the representing function of the recursive predicate $x < y \wedge y < a'$. Then $P(a_1, \cdots, a_n)$ is expressible by $A(v_{a_1}, \cdots, v_{a_n})$, which is a predicate of variables $a_1, \cdots, a_n$ only.

The last half of the theorem is obtained from §8 of [16] and Shoenfield [8].

DEFINITION. An ordinal number is called a Δ_2 *ordinal* if it is the order type of a well-ordering of a set of natural numbers in both two function-quantifier forms.

DEFINITION. An ordinal number is called *Ω-recursively expressible* if it is expressible by means of Ω-recursive functions.

THEOREM 8. *A Δ_2 ordinal is an Ω-recursively expressible ordinal number and vice-versa.*

PROOF. If an ordinal α is a Δ_2 ordinal, then α is the order type of some Δ_2 well-ordering $A(a, b)$ of natural numbers. By Theorem 7, there exists an Ω-recursive function $f(a, b)$ such that $A(a, b)$ is equivalent to $f(a, b) = 0$. Without loss of generality we may assume that $\exists x(f(x, a) = 0 \to 0 < a)$. We define an Ω-recursive function $g(x)$ as follows:

1) $g(0) = \mu x_{x<\omega} \forall y(y < \omega \wedge \neg y = x \rightarrow \neg f(y,x) = 0) \wedge f(x,x) = 0,$

2) $g(a) = \mu x_{x<\omega} \forall y(y < \omega \wedge y = x \wedge$
$\forall z(z < a \rightarrow \neg g(z) = y) \rightarrow \neg f(y,x) = 0) \wedge f(x,x) = 0$ for $a > 0$.

Then α is clearly expressed by $\mu x(g(x) = 0$, namely, it is Ω-recursively expressible.

Conversely, let α be Ω-recursively expressible. Let $D(b;a)$ be $\exists x(x < b \wedge (\langle a,x\rangle \in b \vee \langle x,a\rangle \in b))$ and $R(b)$ be

$\forall x(x < b \wedge x \in b \rightarrow \exists y \exists z(y < \omega \wedge z < \omega \wedge \langle y,z\rangle \equiv x)) \wedge$
$\forall u \forall v \forall w(u < b \wedge v < b \wedge w < b \wedge \langle u,v\rangle \in b \wedge \langle v,w\rangle \in b \rightarrow \langle u, w\rangle \in b) \wedge$
$\forall u \forall v(u < b \wedge v < b \wedge D(b;u) \wedge D(b;v) \rightarrow \langle u,v\rangle \in b \vee \langle v,u\rangle \in b) \wedge$
$\forall u \forall v(u < b \wedge v < b \wedge \langle u,v\rangle \in b \wedge \langle v,u\rangle \in b \rightarrow u \equiv v),$

which is Ω-recursive and means that b gives a linear ordering of natural numbers. Let $M(c;a,b)$ be

$\forall x(x < c \wedge x \in c \rightarrow \exists u \exists v(u < a \wedge v < b \wedge D(b;v) \wedge \langle u,v\rangle \equiv x) \wedge$
$\forall u(u < a \rightarrow \exists v(v < b \wedge D(b;v) \wedge \langle u,v\rangle \in c) \wedge$
$\forall u \forall v(u < a \wedge v < a \rightarrow (u < v \leftrightarrow \exists x \exists y(\langle u,x\rangle \in c \wedge \langle v,y\rangle \in c \wedge$
$\langle x,y\rangle \in b \wedge \neg x \equiv y)))$.

Let $W(a;b)$ be the predicate expressing that b gives a well-ordering of natural numbers whose order-type is a. $W(a;b)$ is expressible in both Σ_1^{ord} and Π_1^{ord} forms as follows, so that it is seen to be Ω-recursive: A Σ_1^{ord}-form of $W(a;b)$ is

$$R(b) \wedge \exists y(M(y;a,b) \wedge \forall u(u < b \wedge D(b;u) \rightarrow \exists v(v < a \wedge \langle v,u\rangle \in y))$$

(which means b is a linear order of natural numbers and there is a one-to-one order preserving onto mapping of a to b). A Π_1^{ord}-form of $W(a;\ b)$ is

$R(b) \wedge \forall y(M(y;a,b) \rightarrow \forall u(u < b \wedge D(b;u) \rightarrow \exists v(v < a \wedge \langle v,u\rangle \in y))) \wedge$
$\forall y(M_1(y;b,a) \rightarrow \forall u(u < a \rightarrow \exists v(v < b \wedge D(b;v) \wedge \langle v,u\rangle \in y))) \wedge$
$\forall y(M(y;\omega,b) \rightarrow \exists x \exists u \exists v(x < \omega \wedge u < b \wedge v < b \wedge D(b;u) \wedge D(b;v) \wedge$
$\langle x,u\rangle \in y \wedge \langle x+1,v\rangle \in y \wedge \neg(\langle v,u\rangle \in b \vee u \equiv v))),$

where $M_1(c;b,a)$ is the abbreviation of

$\forall x(x < c \wedge x \in c \rightarrow \exists u \exists v(u < b \wedge v < a \wedge D(b;u) \wedge \langle u,v\rangle \equiv x) \wedge$
$\forall u(u < b \wedge D(b;u) \rightarrow \exists v(v < a \wedge \langle u,v\rangle \in c)) \wedge$
$\forall u \forall v(u < b \wedge v < b \wedge D(b;u) \wedge D(b;v) \rightarrow$
$(\langle u,v\rangle \in b \wedge \neg u \equiv v \leftrightarrow \exists x \exists y(\langle u,x\rangle \in c \wedge \langle v,y\rangle \in c \wedge x < y)))$.

Now consider $\mu x W(\alpha;x)$ which gives a well-ordering of natural numbers whose order-type is α. Since α is Ω-recursively expressible and $W(a;b)$ is Ω-recursive, this is an Ω-recursively expressible ordinal number. Let $R(a,b)$ be $n(a) < \omega \wedge n(b) < \omega \wedge \langle n(a), n(b)\rangle \in \mu x W(\alpha;x)$. Then $R(a,b)$ is an

Ω-recursive predicate of natural numbers, and hence it is expressible in both two function-quantifier forms by Theorem 7.

THEOREM 9 (Schoenfield [8]). *A set of natural numbers is expressible in both two function-quantifier forms if and only if it is a constructible set of natural numbers whose order is a Δ_2 ordinal.*

PROOF. Let A be a constructible set of natural numbers whose order is a Δ_2 ordinal. Then the order of A, say α, is Ω-recursively expressible by Theorem 8. By the lemma for Theorem 2, we can prove the following: If B is a constructible set of order β, then $c \varepsilon B$ if and only if $\gamma \in \beta$ where γ is the order of c. Then A is expressible by $\{n \mid n < \omega \wedge n \in \alpha\}$ which is Ω-recursive and therefore expressible in both two function-quantifier forms by Theorem 7. Conversely, suppose A is expressible in both two function-quantifier forms. Then A is expressible by an Ω-recursive predicate by Theorem 7, whence it is constructible by Theorem 1. The order of A is expressible by $\mu x(\forall y(y < \omega \to (A(y) \leftrightarrow y \in x)) \wedge \forall y(y < x \wedge y \in x \to y \in \omega))$. Therefore it is a Δ_2 – ordinal by Theorem 8.

Now we shall relativize the results given above with respect to a function from ω into ω. Let e be a function from ω into ω. We define $L(e)$ as follows: Let F_e be defined to be

$$\alpha \varepsilon W(J_0) \to F_e{}'\alpha = W(F_e \restriction \alpha) \text{ if } \alpha \neq \omega+1,$$

$$\alpha \varepsilon W(J_i) \to F_e{}'\alpha = \mathfrak{F}_i(F_e{}'K_1{}'\alpha, F_e{}'K_2{}'\alpha) \text{ if } \alpha \neq \omega + 1 \quad (i = 1,2,\cdots,8).$$

$$F_e{}'\omega + 1 = e.$$

Let $L(e)$ be $W(F_e)$. ($L(e)$ should be considered to be a special case of L_k of [5] and [9].) We can construct the model of $L(e)$ by the primitive recursive function in e (cf. [15]). Then the following relativized theorem of Shoenfield [8] can be proved in the same way as in [8].

THEOREM (Schoenfield). *If $A(\alpha, \beta)$ is a $\prod_1^1$-predicate, β_0 is a function constructible in $L(\gamma)$ and the class of α such that $A(\alpha, \beta_0)$ is non-empty, then this class contains a function constructible in $L(\gamma)$.*

COROLLARY. *If $A(\alpha, \beta, \gamma)$ is arithmetical (in the usual sense) and γ_0 is constructible in $L(\delta)$, then*

(1) $\forall \alpha \exists \beta A(\alpha, \beta, \gamma_0) \leftrightarrow \forall \alpha(\alpha \varepsilon L(\delta) \to \exists \beta(\beta \varepsilon L(\delta) \wedge A(\alpha, \beta, \gamma_0))$,

(2) $\exists \alpha \forall \beta A(\alpha, \beta, \gamma_0) \leftrightarrow \exists \alpha(A \varepsilon L(\delta) \wedge \forall \beta(\beta \varepsilon L(\delta) \to A(\alpha, \beta, \gamma_0))$.

By using these lemmas, if we replace recursive, arithmetical, semi-arithmetical, and L in the above theorems by recursive in e, arithmetical in e, semi-arithmetical in e, and $L(e)$ respectively, then these generalized theorems can be proved in the same way. For example, we have the following theorems:

THEOREM. *A predicate of natural numbers is Ω-recursive in e if and only if it is expressible in both two function-quantifier forms (in Kleene's sense) whose scopes are recursive in e.*

THEOREM. *A Δ_2-ordinal in e is an ordinal number which is Ω-recursively expressible in e and vice-versa.*

THEOREM. *A set of natural numbers is expressible in both two function-quantifier forms relative to e if and only if it is a set of natural numbers constructible in $L(e)$ such that its order is a Δ_2-ordinal in e.*

§3. Mahlo's axiom and the transcendency of ω_1

In this section we shall show that the consistency of $(TC)_1$ is equivalent to the consistency of Mahlo's axiom. The precise forms of TC and $(TC)_1$ are as follows.

TC: If $a \geqq \omega$ and f is arithmetical, then $\overline{\overline{f(a)}} \leqq \overline{\overline{a}}$.

$(TC)_1$: If $\overline{\overline{a}} \leqq \omega$ and f is arithmetical, then $\overline{\overline{f(a)}} \leqq \overline{\overline{\omega}}$.

Note in passing that for recursive f, $\overline{\overline{f(a)}} \leqq \overline{\overline{a}}$ is proved in [13]. As a formulation of Mahlo's axiom, we shall use the following schema of Lévy [4];

$$\mathrm{N}''' \qquad \exists\alpha(\mathrm{In}(\alpha) \wedge \forall x \,\varepsilon\, R(\alpha)\,(\phi \leftrightarrow \mathrm{Rel}(R(\alpha), \phi)))\,,$$

where $\mathrm{In}(\alpha)$ means 'α is inaccessible' and $R(\alpha) = \sum_{\beta<\alpha} P(R(\beta))$ ($P(x)$ is the power set of x) and $\mathrm{Rel}(R(\alpha), \phi)$ is obtained from ϕ by replacing all the quantifiers $\forall y, \exists z, \cdots$ by $\forall y \,\varepsilon\, R(\alpha)$, $\exists z \,\varepsilon\, R(\alpha), \cdots$ respectively and ϕ is any formula which has no free variables except x.

An ordinal α_0 is called 'essentially inaccessible' if $\mathrm{In}(\alpha_0)$ and for any formula ϕ

$$(*) \qquad \forall x \,\varepsilon\, R(\alpha_0)\,(\phi \leftrightarrow \mathrm{Rel}(R(\alpha_0), \phi))\,.$$

This means that $R(\alpha_0)$ is an elementary submodel of V. Clearly the consistency of N''' with set theory is equivalent to the consistency of the existence of an essentially inaccessible number with set theory. (For the formulation of the latter, it is enough to introduce the new constant α_0 and the axioms $\mathrm{In}(\alpha_0)$ and (*).)

From the proof of [14], we easily see that the consistency of $(TC)_1$ implies the consistency of the existence of an essentially inaccessible number. Now, following Cohen [1], we shall assume the existence of a countable complete model of set theory $\mathfrak{M}$ in which $V = L$ and the existence of an essentially inaccessible number holds, and shall prove the existence of a complete model of set theory $\mathfrak{N}$ in which $(TC)_1$ holds.

Let π be an essentially inaccessible number in $\mathfrak{M}$. We know $\pi = \aleph_\pi$ in $\mathfrak{M}$. Let a_α, for $\alpha < \pi$, be a function from ω. Set $U = \{\langle a_\alpha, a_\beta \rangle \mid \alpha < \beta\}$. We define F_β as the sets constructed from the a_α's and U.

DEFINITION.

1) $F_\alpha = a_\beta$ if $\alpha = \aleph_\beta + 1$,

2) $F_\alpha = U$ if $\alpha = \pi + 1$.

3) If 1) and 2) are not the case,

$$F_\alpha = \{F_\beta \mid \beta < \alpha\} \quad \text{if} \quad N(\alpha) = 0,$$
$$F_\alpha = \mathfrak{F}_i(F_{K_1(\alpha)}, F_{K_2(\alpha)}) \quad \text{if} \quad N(\alpha) = i > 0.$$

Set $\mathfrak{N} = \{F_\alpha \mid \alpha \in \mathfrak{M}\}$. As in Cohen's paper, we see that there is a function in $\mathfrak{M}$ independent of a_α, $\alpha' = f(\alpha)$, such that $F_{\alpha'} = \alpha$ and $f(\alpha) < \aleph_\gamma$ if $\alpha < \aleph_\gamma$. We assume that a_α is a function from ω into $f''\aleph_\alpha$. The condition P is a set of finitely many conditions on a_α's. P^β is the subcondition of P about a_β. For example if $P = \{\langle 0\ 0\rangle \varepsilon a_{\alpha_1}, \langle 0\ 0\rangle \varepsilon a_{\alpha_2}\}$, $P^{\alpha_1} = \{\langle 0\ 0\rangle \varepsilon a_{\alpha_1}\}$. Let $\phi_0, \phi_1, \cdots$ be an ω-type well-ordering of all the unlimited statements and $b_0^\alpha, b_1^\alpha, \cdots$ be an ω-type-well ordering of all the elements of $\aleph_\alpha$ and let $\alpha_0, \alpha_1, \alpha_2, \cdots$ be an ω-type-well ordering of all the elements of π.

DEFINITION.

1) $i \varepsilon D(P^\alpha)$ if and only if there exists b such that $\langle bi \rangle \in a_\alpha$ is contained in P^α.

2) $b \in W(P^\alpha)$ if and only if there exists i such that $\langle bi \rangle \in a_\alpha$ is contained in P^α.

Let P_0 be empty. P_{2n+1} is defined by the following: If P_{2n}^α is not empty, then $P_{2n+1}^\alpha = P_{2n}^\alpha \cup \{\langle f(b_K^\alpha), i \rangle \varepsilon a_\alpha\}$ where i is the first element which is not a member of $D(P_{2n}^\alpha)$ and $f(b_K^\alpha)$ is the first element (in the ω-type well-ordering) which is not a member of $W(P_{2n}^\alpha)$. If α is the first element (in the ω-type well-ordering) such that P_{2n}^α is empty, then P_{2n+1}^α is $\{\langle f(b_0^\alpha), 0 \rangle\}$. Otherwise P_{2n+1}^α is always empty. Let P_{2n+2} be the first extension of P_{2n+1} which forces either ϕ_n or $\neg \phi_n$.

We set $a_\alpha = \lim_{n \to \infty} P_n^\alpha$. By Cohen's arguments, we see that $\mathfrak{N}$ is a model of set theory and the axiom of choice. By the construction it is directly seen that if $\alpha < \pi$, then a_α is a one-to-one map from ω onto $f''\aleph_\alpha^m$. Therefore $\pi \leqq \aleph_1^n$. Now we are going to show $\pi = \aleph_1^n$. Repeating Definition 16 in Cohen [1], we define $P_1 \sim P_2$ and W.

Modifying Lemma 16 in Cohen [1], we have

LEMMA. *There exists a set b of P's such that b is definable in* $\mathfrak{M}$ *and* $\overline{\overline{b}}^m < \pi$ *and* $\forall P \in W \exists P' \supseteq P \exists P_1 \in b(P' \sim P_1)$.

PROOF. In $\mathfrak{M}$, we can well-order all the P's: $P^0, P^1, \cdots, P, \cdots$. (This well-ordering should be definable in $\mathfrak{M}$.) Define ${}^0P_0$ to be the first P such that $P \in W$. Assume ${}^0P_0, {}^1P_1, \cdots, {}^2P_{\alpha_2}, {}^2P_{\alpha_2+1}, \cdots, {}^nP_{\alpha_n}, {}^nP_{\alpha_n+1}, \cdots$ have been chosen in W. Define $R_n = \{\langle f(b_j^\alpha), i\rangle \in a_\alpha \mid \exists k \leqq n \exists \beta (i \in D({}^KP_\beta^\alpha))\}$. Let ${}^{n+1}P_{\alpha_n+1}, \cdots, {}^{n+1}P_{\beta_1}, \cdots (\beta < \alpha)$ be the elements of W, such that if P' is any element in W, for some β, $\alpha_{n+1} \leqq \beta < \alpha$, P' has precisely those conditions in R_n which ${}^{n+1}P_\beta$ has. Since the cardinality of R_n is less than π, one can always find such ${}^{n+1}P_{\alpha_{n+1}}, \cdots, {}^{n+1}P_\beta, \cdots (\beta < \alpha)$, where $\alpha < \pi$. Define $b = \bigcup_{n<\omega}\{{}^nP_{\alpha_n}, \cdots\}$. We have $\overline{\overline{b}}^m < \pi$. Now let $P \in W$. We see easily $\exists P' \supseteq P \exists P_1 \in b(P' \sim P_1)$.

By this lemma, we can prove the following in the same way as in Lemma 17 in [1].

LEMMA. $F_{f(\omega)} < \overline{\overline{F_{f(\pi)}}}^n$.

By this lemma we see $\pi = \omega_1^n$. Since π is essentially inaccessible in $\mathfrak{M}$ and arithmetical in $\mathfrak{N}$ means definable in $\mathfrak{M}$ (cf. Theorem 1 in §2) we see easily that $(TC)_1$ holds in $\mathfrak{N}$.

§ 4. Extensibility and contractibility of recursive and arithmetical functions

We shall consider the relation between 'Ω_1-recursive' and 'Ω_2-recursive' for $\Omega_1 \neq \Omega_2$. Let $\Omega_1 < \Omega_2$ and f be a function over Ω_2. $f \restriction \Omega_1$ is defined to be a function over Ω_1 such that $f \restriction \Omega_1(a) = f(a)$ for $a < \Omega_1$. In [13] we have proved that if $\Omega_1 < \Omega_2$ and f is Ω_2-recursive, then $f \restriction \Omega_1$ is Ω_1-recursive.

DEFINITION. Let $\Omega_1 < \Omega_2$ and f be Ω_1-recursive (or Ω_1-arithmetical). f is called *Ω_2-recursively extensible* (or *Ω_2-arithmetically extensible*), if there exists an Ω_2-recursive (or Ω_2-arithmetical) function g such that $g \restriction \Omega_1 = f$.

DEFINITION. An ordinal number is called *Ω-arithmetically definable*, if it is defined by using 0, ω and Ω-arithmetical functions.

THEOREM 1. *If the axiom of constructibility holds (i.e.* $V = L$*), then there exists a recursive function which is not recursively extensible.*

PROOF. We shall assume $V = L$. We have $\forall x \exists y \forall z (C(x, y, z) = 0 \wedge 0 < y)$ (cf. the proof of Theorem 5 in [13]). We have, moreover, $\neg \exists' y \forall' z (C(\omega, y, z) = 0 \wedge 0 < y)$, i.e., $\forall' y \exists' z \neg (C(\omega, y, z) = 0 \wedge 0 < y)$, where $\forall' x$ or $\exists' y$ means that x or y runs over ω_1. Let $P(y,z)$ be the abbreviation of $\neg (C(\omega,y,z)=0 \wedge 0<y)$. The predicate $\forall y \exists z P(y,z)$ is true, only if y and z run over ω_1. Hence $\mu z P(b,z)(=f(b))$ is ω_1-recursive. Suppose there exists an $\omega_\alpha(\alpha>1)$-recursive function f_0 such that $f_0(b)=f(b)$, for $b<\omega_1$. Then $\mu b \neg P(b,f_0(b))=\omega_1$, for $\neg P(b,f_0(b))$ for any $f_0(b)$, if $b=\omega_1$. Since $\mu b \neg P(b,f_0(b))$ is ω_α-recursive, $\mu b \neg P(b,f_0(b))<\omega_1$ (cf. [13]).

THEOREM 2. *TC implies extensibility of recursive functions.*

PROOF. Since the contractibility of recursive function is proved in [13], it is sufficient to prove that every ω_α-recursive function f is *On*-recursive. We shall prove this by mathematical induction on the number of stages to construct f. Since other cases are no problem, we have only to treat the case when f is of the form $\mu y f_1(x,y)$, where

(*) $$\forall x \exists y f_1(x,y)=0,$$

if x and y run over ω_α. Suppose that it were not *On*-recursive, i.e. $\neg \forall x \exists y f_1(x,y)=0$, if x and y run over On. By means of (*), $\mu x \forall y \neg (f_1(x,y)=0) \geqq \omega_\alpha$, which contradicts TC.

Clearly we have

LEMMA. *TC implies that, if f is On-arithmetical and $a<\Omega$, $f(a)<\Omega$.*

THEOREM 3. *TC implies the extensibility of the arithmetical functions.*

PROOF. Let $f(a)$ be Ω_1-arithmetical. Let $g(a)$ be an Ω_2-arithmetical function defined by the same expression as $f(a)$. We shall prove by mathematical induction on the number of stages used to construct f that $a<\Omega_1 \rightarrow f(a)=g(a)$. Since other cases create no problem we have to treat only the case in which f is of the form $\mu y f_1(a,y)$. If $\mu y<\Omega_1 f_1(a,y) \neq \mu y<\Omega_2 f_1(a,y)$ for some $a<\Omega_1$, we see that $\forall y<\Omega_1(f_1(a,y)\neq 0) \wedge \exists y<\Omega_2(f_1(a,y)=0)$. Hence it follows that $\mu x(\exists y<x \ f_1(a,y)=0) \geqq \Omega_1$ for some $a<\Omega_1$, which contradicts TC.

THEOREM 4. *TC implies that, if f is On-arithmetical, $f \upharpoonright \Omega$ is Ω-arithmetical.*

PROOF. We have only to consider the case that f is of the form $\mu x f(x,a)$. By TC, $\exists x<\Omega(f(x,a)=0)$, if $a<\Omega$ and $\exists x(f(x,a)=0)$.

THEOREM 5. *TC implies that if $\Omega_1<\Omega_2$ and f is Ω_2-arithmetical, then $f \upharpoonright \Omega_1$ is Ω_1-arithmetical.*

PROOF. By Theorem 3, f is *On*-arithmetical, which means that $f \upharpoonright \Omega_1$ is Ω_1-arithmetical by Theorem 4.

THEOREM 6. *TC implies that if $\Omega_1 < \Omega_2$ and f is Ω_1-arithmetical, there exists an Ω_2-arithmetical function g such that $g \upharpoonright \Omega_1 = f$.*

PROOF. By Theorem 3, f is extensible to an On-arithmetical function $\tilde{f}$ such that $\forall x(x < \Omega_1 \rightarrow f(x) = \tilde{f}(x))$. By Theorem 4, $\tilde{f} \upharpoonright \Omega_2$ is Ω_2-arithmetical. Clearly $(\tilde{f} \upharpoonright \Omega_2) \upharpoonright \Omega_1 = f$.

THEOREM 7. *If $V = L$, then some arithmetical function is not extensible.*

PROOF. Assume $V = L$. Let α be the first ordinal not arithmetically definable. $\bar{\bar{\alpha}} = \omega \,.\, \omega_\alpha$ is the first cardinal not definable under the assumption of $V = L$. On the other hand α is definable in the set theory constructed on ω_α. Let f be defined as follows:

$$f(\beta) = \begin{cases} 1 & \text{if } \beta = \alpha, \\ 0 & \text{otherwise.} \end{cases}$$

Then f is ω_α-arithmetical. f is not extensible to an On-arithmetical function.

REMARK. Under the assumption of $V = L$, if Ω is On-definable, then every Ω-arithmetical function is extensible to an On-arithmetical function.

THEOREM 8. *If $V = L$, then there exists an On-arithmetical function such that $f \upharpoonright \omega_1$ is not ω_1-arithmetical.*

PROOF. Since $V = L$ implies that 'arithmetical' is equivalent to 'definable in set theory', we have only to show that there exists an ordinal number $\alpha < \omega_1$ such that α is definable but not $F''\omega_1$-definable. The following f satisfies the desired condition:

$$f(\beta) = \begin{cases} 1 & \text{if } \beta = \alpha, \\ 0 & \text{otherwise.} \end{cases}$$

If we construct the model Δ on the ordinal numbers, the notion of satisfaction $F''\omega_1 \models \phi$ on the set theory constructed on ω_1 is definable in set theory. Therefore the proposition 'an ordinal number is definable on $F''\omega_1$' is definable, that is to say, 'the least ordinal number not definable on $F''\omega_1$' is definable.

REFERENCES

[1] P. J. COHEN, *The independence of the axiom of choice*, Mimeographed notes, Mathematics Department, Stanford University (May, 1963), 32 pp. (See also The independence of the continuum hypothesis, *Proceedings of the National Academy of Sciences*, **50** (1963), 1143–1148 and **51** (1964), 105–110).

[2] K. GÖDEL, *The consistency of the continuum hypothesis*, third edition, Princeton, 1953.

[3] S. C. KLEENE, *Introduction to metamathematics*, New York, Toronto and Amsterdam, 1952.

[4] A. LÉVY, Axiom schemata of strong infinity in axiomatic set theory, *Pacific J. Math.*, **10** (1960), 223–238.

[5] A. Lévy, A generalization of Gödel's notion of constructibility, *J. Symbolic Logic*, **25** (1960), 147–155.
[6] A. Lévy and M. Machover, *Recursive functions of ordinal numbers*, Amsterdam, to appear. (See abstract in *Notices Amer. Math. Soc.*, **6** (1959), 826.)
[7] M. Machover, The theory of transfinite recursion, *Bull. Amer. Math. Soc.*, **67** (1961), 575–578.
[8] J. R. Shoenfield, The problem of predicativity. *Essays on the foundations of mathematics*, 132–139, Jerusalem, 1961.
[9] J. R. Shoenfield, On the independence of the axiom of constructibility. *American J. Math.*, **81** (1959), 537–540.
[10] G. Takeuti, Construction of the set theory from the theory of ordinal numbers, *J. Math. Soc. Japan*, **6** (1954), 196–220.
[11] G. Takeuti, On the theory of ordinal numbers, *J. Math. Soc. Japan*, **9** (1957), 93–113.
[12] G. Takeuti, On the theory of ordinal numbers, II, *J. Math. Soc. Japan*, **10** (1958) 106–120.
[13] G. Takeuti, On the recursive functions of ordinal numbers, *J. Math. Soc. Japan*, **12** (1960), 119–128.
[14] G. Takeuti, Transcendency of cardinals, to appear in *J. Symbolic Logic*.
[15] G. Takeuti, A formalization of the theory of ordinal numbers, to appear.
[16] G. Takeuti and A. Kino, On hierarchies of predicates of ordinal numbers, *J. Math. Soc. Japan*, **14** (1962), 199–232.

Section III

Philosophy of Logic and Mathematics

THE JUSTIFICATION OF SET THEORIES

LEO APOSTEL
Ryksuniversiteit, Ghent, Belgium

1. Discussion of set-theoretical axioms.

1.1 This paper analyzes a problem in the philosophy of mathematics. As we understand this discipline, it aims at studying the relationship between mathematics proper, human action in general and the structure of nature as a whole. In order to explain how such a study can be undertaken in a responsible way, we are going to begin by considering one set-theoretical axiom (the comprehension axiom), and comparing different versions of it which appear in various important and influential set-theoretical systems[1].

Our purpose will be to "understand" some of the reasons one might give for or against the proposed axioms of comprehension.

Let us consider the following four statements:

(a) $(Ey)(Ax)\ [(x \varepsilon y) \leftrightarrow F(x)]$
(b) $(Az)(Ey)(Ax)\ [(x \varepsilon y) \leftrightarrow ((x \varepsilon z) \wedge F(x))]$
(c) $(Ey)(Ax)\ [(x \varepsilon y) \leftrightarrow (Ez)((x \varepsilon z) \wedge F(x))]$
(d) $(Ey^i)(Ax^{i-1})\ [(x^{i-1} \varepsilon y^i) \leftrightarrow F(x^{i-1})]$

1.2 We are going to interpret each of these statements as the foundation of specific classification procedures. The exact logical relation between the instructions about classifications and the axioms a-d will be discussed later.

(a′) In classification Kl, it is allowed to form a class whenever some objects have a property (whatever it may be) in common.

(b′) In classification K2, it is allowed to form a class whenever some objects, all of which belong to a class *z* already constructed, have a property in common.

(c′) In classification K3, it is allowed to form a class whenever some objects which are contained in previously constructed classes (not neces-

1 *Foundations of Set Theory*, North-Holland Publ. Co., 1958, by A. Fraenkel and Y. Bar-Hillel, offers both information about and interesting philosophical comments upon the topic.

sarily the same class, in distinction to (b′)) have some property in common.

(d′) In classification K4, it is allowed to form a class whenever some objects have some property in common if and only if (1) these objects have some fundamental property in common (being of the same type) and (2) the class has a fundamental property in common (being of the same type) with all other classes having similar objects as elements.

We claim that, in view of the aims of the action of classification, all four of these instructions are efficient, in specific circumstances.

At the beginning of a classification procedure, when nothing whatever is known in the field, complete freedom is given as to the construction of classes (case a′). When the classification procedure is already a going concern and when the results of earlier grouping are stable, classification continues without rearrangement: classes to be built should keep classes constructed earlier intact (case b′). When a crisis is reached and the procedures hitherto followed must be altered, classification should take into account only the earlier classified objects and free regrouping is allowed (case c′). When classification continues without crisis, but when it is imperative that classification at one level should be strongly different from and yet completely determined by the procedures applied on earlier levels, we have case d′.

All four types of instruction are thus admissible in specific circumstances, and efficient in specific circumstances.

Let us now go back to the relation between a-d, and a′-d′.

Certainly the two series of assertions are not synonymous.

The relation is rather the following one: when the set-theoretical statements of the first sequence are true, it is possible to execute the instructions in the second series and it is useful to do so. The first series is a sequence of necessary conditions for the feasibility and efficiency of the actions prescribed in the second series[2].

Our proposal would then be the following: to select for use in a given science at a given stage of its development those axioms of comprehension that are necessary (strict or probabilistic) conditions for the feasibility and efficiency of classification procedures of that science at that moment.

---

[2] We refer to L. Apostel "Le Problème Formel des Classifications Empiriques," pp. 157–230, in *La Classification dans les Sciences*, 1963 (Ed. Duculot-Gembloux) and to L. Apostel "Logique et Apprentissage," pp. 1–137, in "Logique, Apprentissage et Probabilité," *Etudes d'Epistémologie Génétique*, VIII, for further study of logical and psychological aspects of classification systems.

This proposal is reasonable because we can only discover sets by meeting them in classifications. Those sets exist for a given science at a given moment that, in principle, can be constructed by that science at that moment.

1.3 Let us now leave this first reinterpretation of our four axioms and look at another translation. We are going to speak in the perspective of a very general part of physics. We contrast the following assertions with each other:

A. a1: The basic laws governing the behavior of all types of systems are essentially the same; a2: these basic laws are different for different types of systems.

B. b1: The universe as a whole is a system; b2: the universe as a whole is not a system.

C. c1: The universe as whole can be ordered (c1.1) unidimensionally, or (c1.2) multidimensionally; c2: it cannot be so ordered.

We claim that these statements, though extremely general, can receive empirical confirmation and disconfirmation (though certainly not conclusive proof or disproof).

We must now express some relations between set-theoretical axioms and these very general statements about our universe.

If a1, b2 and c2 are simultaneously true and if thus the universe is not a system, cannot be ordered and is composed out of homogeneous collections, then the unrestricted comprehension axiom (a) should be adopted.

If c1, a2 and b2 are simultaneously true, then the comprehension axiom (d) should be adopted.

If a1 and c1.1 are true, comprehension axiom (b) should be adopted and if a1 and c1.2 are true, comprehension axiom (c) should be adopted.

Why do we say this and how can we defend these relationships between very general physical facts and set-theoretical axioms?

In a completely homogeneous universe, sets do not correspond to species or natural kinds: they can thus be formed with maximal liberty.

In a universe ordered in a series of qualitatively different levels, sets should be constructed according to type theory if one wants the order of the sets to mirror the order of nature in this heterogeneous universe.

In a homogeneous universe that can however be ordered (unidimensionally or multidimensionally), one should again have complete freedom of set formation but this freedom should take into account the existence of the unique or multiple ordering principles.

In general we may say that a set theory conforms to a physics in as far as the restrictions on set formation are completely or partially isomorphic to the restrictions on system formation in the physical universe.

The proposal to construct set theories according to such an isomorphism principle is reasonable because in the end our sets must be used to classify real objects (so the relativization of set theory to physics should if possible be a well confirmed general part of physics)[3].

1.4 Finally, and briefly we want to mention a third type of reinterpretation.

We may inquire, for any given natural language, what type of words are used to designate what we call "sets" (the study of collectives in English and other languages for instance would help in this direction). We would then see that sometimes collectives are used in the plural and sometimes in the singular (so that sometimes they are treated as names of entities, and sometimes rather as simple collection-words) and would conclude that in natural language certain contexts imply the substantification of sets and other contexts reject it. The comparative study of both types of contexts should then yield information about the conditions under which given set-theoretical axioms would be acceptable.

We cannot go here into details however.

1.5 We claim the following: set theory is a linguistic tool, developed in the course of human action, with the aim of explaining and controlling nature. To understand set theory, one should thus try to study its relation to language, to action and to nature. We should not be satisfied to justify set theory by its psychological effect upon us, or by the usefulness of some of its far reaching consequences but should look for a direct relationship between its fundamental notions and those of physics, theory of action and linguistics.

Some suggestions in that direction have been made here and it seems to us that the possibilities of the method cannot be denied. The method should however also be applied to the other fundamental axioms of set theory (such as, for instance, the axiom of infinity, the axiom of predicativity or impredicativity, the axiom of choice, and many others).

It will be impossible to analyse these very intricate axioms sufficiently here. If we mention them in spite of this, it is to point out that very different behavioral linguistic or physical justifications can and should be given for the different axioms. We take as two illustrative examples the axioms of infinity and of predicativity.

---

[3] We refer to *General Systems Yearbooks* I-VIII (edited by L. von Bertalanffy and A. Rapoport) for further comment on systems theory. See also L. Apostel "Théorie des Systèmes et Théorie des Prévisions" in *Colloque Prévisions*, Gauthier-Villars, Paris, 1964.

1.6 The axiom of infinity states that there exists at least one infinite set. We can discuss it supposing either that there is no infinite system in real nature or that such an infinite system really exists. Let us take the first alternative. In the case that no infinite system exists, the acceptability of an axiom of infinity (now, strictly speaking, false) depends upon certain properties of the universe and of human thinking. This is no isolated case. Indeed: there has never been completely empty space, yet it is useful at first to develop mechanics without friction. There has never been a pure free market economy, yet it is useful to develop economy from this standpoint. Many sciences are typological sciences, studying complex reality through studying simplified models of it (schematizations, idealizations, fictions). Saying this however only states the problem: the philosopher has to explain *why it is necessary to study what is not in order to understand what is and how it is possible to understand what is through the study of what is not.*

One answer would be the following: the exact number of objects in the universe is unknown to us, and many properties of our universe are independent of the number of objects in it. For the human mind on the other hand it may be impossible to understand the consequences of given statements in which parameters appear, although the statements may be independent of these parameters, if we do not set these parameters equal to an extreme value (zero or infinite). This statement has a physical part and a psychological part, and both these parts are empirically confirmable statements.

If the universe is infinite (and some empirical arguments seem to point in the direction of this second alternative) then the justification of the axiom of infinity has to be undertaken along entirely different lines.

1.7 Other types of justification procedures can be applied to the choice between predicative and impredicative set theory. It is well known that the F in the statements (a)-(d) can be more or less restricted: if F contains a bound variable of the same type as the y occurring in the left half of the formulae (a)-(d), the defining characteristic of the class y refers to the set of all classes y, and the definition of y is impredicative (y being defined by its relation to a totality to which it itself belongs). The comprehension axiom is predicative if no such y occurs in F.

It has been claimed both that impredicativity is to be completely rejected as meaningless and that predicativity is entirely insufficient to develop mathematics. However we want to come to a conclusion on the basis of a direct interpretation, not on pragmatic grounds or on *a priori* evidence.

Let us call a region of reality holistic if and only if it contains wholes that can only be defined with reference to their parts, parts that again can be defined only with reference to the wholes. The non-holistic regions could be called "elementaristic." We claim that both types of regions occur in reality (inorganic aggregates are elementaristic; organic systems and psycho-social systems are holistic). An impredicative set theory is analogous to the structure of a holistic region. A predicative set theory is analogous to the structure of an elementaristic region.

In human action, it is certain that both features occur (perceptive and learning behaviour being holistic, manipulative behaviour being elementaristic).

If we engage in classificatory behaviour, we can decide simultaneously, in an ongoing process, what we are classifying and the classes we use to classify, climbing as it were continuously up and down in the hierarchy, or we can take the earlier levels as given and build upon them. Both these types of behaviour occur in the construction of classifications. Impredicative set theory justifies the first practice; predicative set theory justifies the second practice (and both are justified by what they justify).

The relation between set-theoretical axiom and practical instruction or general physical property is similar to the ones discussed in 1.2 and 1.3.

Just as we are compelled to accept (without any reference to utility, personal preference or external applications) both predicative and impredicative analysis, so we are also compelled, upon a non-formalist basis, to accept both the classical point of view and the intuitionistic point of view. If we are classifying a region in strong and constant evolution and if our own classification schemes are equally engaged in such development, then the set theory adapted to such a situation is intuitionistic set theory. But not all regions and not all sciences are constantly in such a state and thus intuitionism has its own limited validity, just as the more classical points of view.

If we had more time it would be possible to point out various interesting mixtures of set theories that have never been developed, yet appear natural from the point of view of the interpretation here suggested.

1.8 We want only to conclude that for the axiom of infinity, as for the axiom of comprehension, physical and psychological arguments are relevant; and once more what would be a good model or type in one science at one moment, would be a bad one for another science at another moment. Yet the arguments that influence our attitude to the axiom of comprehension are very different in content from those which influence our attitude to the axiom of infinity, moreover the latter do not very much resemble those discussing predicativity.

2. Pluralistic realism.

2.1 The preceding discussion leads to the following point of view:

1. Reality presents different regions and different set theories are true for these different regions.

2. Man needs various types of acting and constructing and different set theories are true for these types of action and construction.

3. Man uses various ways of designating by means of language and various set theories are true for these ways of designating.

4. For each reality type, language type or action type, various models can be constructed, simplifying the studied object in different directions so as to point mainly towards one or a few features.

The first three remarks can be called the thesis of "regional realism"; the fourth remark can be called the thesis of "schematic realism"; both taken together constitute "pluralistic realism".

2.2 A. Pluralistic realism has something in common with both nominalism and platonism. It interprets set theory in such a way that it is meaningful to say that set theory speaks about certain real entities and is wholly or approximately true of them.

Pluralistic realism also has something in common with conventionalism or formalism. Many systems of set theory that are incompatible can be true in various regions or contexts, and therefore useful. With formalism it shares the rejection of one unique set-theoretical system as adequate for the whole of action and reality; with platonism and nominalism, the rejection of a set theory that is a pure formal system without interpretation.

B. Pluralistic realism has something in common with classical Cantorism. If reality is not my construction and if sets are either real or at least founded in reality, then it is not necessary for all sets to be constructible.

But pluralistic realism also has something in common with constructivism. In as far as statements about sets must be related to instructions for building classifications, to confirmable statements about physical systems, or to observable ways of using language, assertions about sets that cannot be defined or named constitute serious problems.

C. Though these affinities exist, pluralistic realism deviates fundamentally from all stated positions in the following respects: a) it requires a direct interpretation of the set theoretical concepts in terms of physical nature and human action and the interaction of both (claiming that the applicability of set theory is not understandable if such an interpretation is not forthcoming); b) it thus accepts empirical arguments for and against the acceptability of set theory; c) it claims to be able to validate both Cantorism and constructivism in various circumstances and various regions; but

d) it looks for the system of all set theories, the whole of which would be sufficient and necessary for nature and action as far as we know them. (In this system, not all of the set theories already proposed will be justifiable members, while many set theories would be added to those we already known, if we knew the "Mendeleev" natural table for theories of sets.)

3. Possible developments of pluralistic realism.

3.1 In order to complete its development, any theory in the philosophy of mathematics strives towards expression in a well developed formal system. Pluralistic realism cannot be an exception. It will very naturally be expressed by means of mathematical linguistics (linguistic root of the multiplicity of set theories), by means of general systems theory (physical root of the multiplicity of set concepts) and by means of automata theory (furnishing partial models for the human activity of classification). In fact, the foundations of set theory (and mathematics) as developed in pluralistic realism would consist in the building up of these very general sciences on the basis of a few specific so-called applications (linguistics, automata theory and general systems theory). Needless to say, these developments do not exist. We arbitrarily select the theory of classification by means of automata, and ask what type of set theories could be built upon this basis.

We shall try to make three remarks:

1) We define the concept of a classificatory automaton, and relate set theoretical statements to it.

2) We mention and relate to set theory various classificatory strategies that have been used by automata constructors.

3) We look for the type of set theory that could be constructed using the general definition of a finite automaton.

3.2 Uttley[4] has described a classification mechanism of great simplicity. It consists of a) n input elements: $ABC\ldots$, b) r levels of internal units, c) each internal unit having two possible states: active and passive and connected to 2, 3 or n input elements in a disjunctive way (the internal unit being active when one of its connecting wires is active, and only then), d) the mechanism being so organised that all possible n-ads of input elements are connected to at least one internal unit.

For our discussion it is interesting that various modifications of the Uttley mechanism would correspond to various set-theoretical axioms becoming false:

---

[4] A. M. Uttley "Conditional Probability Machines and Conditioned Reflexes," pp. 253–257 in *Automata Studies*, Princeton Univ. Press, 1956, and B.L.M. Chapman "A Self-Organising Classification System," *Cybernetica*, vol. 2, 3, 152–61.

a) if more than one internal unit with the same connections to the external or input elements is introduced, there are two different set forming operators on the same elements: extensionality fails to hold;

b) if the first internal level forming all dyads is incomplete, certain dyads are not combined into one set and the pair set axiom does not hold for the sets introduced by our mechanism;

c) if, for any given level, some later or earlier internal level is incomplete, axioms analogous to the power axiom or to the union set axiom fail;

d) if feedback is introduced into the classification system, the axiom of foundation fails;

e) if the classification mechanism is a growing automaton adding more and more levels, some axiom of infinity (of potential infinity) holds; if it is a fixed automaton, no axiom of infinity holds.

REMARK: We could combine Uttley classification mechanisms in such a way that the input elements of a second mechanism were activated by properties of sequences of operations of a first mechanism. This would perhaps be the best way to mirror the operation by which the law of construction of one set itself becomes an element of another set (the transfinite operation *par excellence*).

If we add counters, the number of times a given input element has been stimulated can be counted and thus the various elements of the input set can be individualised. Elementhood then has its expression in the mechanism.

The various features of the mechanism mirror certain special cases of a comprehension axiom. If we want to introduce a mechanical equivalent of a general comprehension axiom itself (let us remember that in Bernays' set theory a general comprehension axiom is dispensed with) we must again connect various classification mechanisms among themselves. The unrestricted axiom of comprehension corresponds to a set of Uttley automata so connected that each feature of the operation of at least one automaton is connected with an input element in some other automaton of the set.

We are not claiming that this alone gives us a clear approximate model of set theory in the theory of classification automata, but the reader will perhaps see that progress can be made along these lines.

If this is conceded, we can now ask the crucial question: can the justification problem be studied in this way?

We propose the following procedure. Let us start with a randomly connected net, and introduce learning operators (adding or subtracting, weakening or reinforcing connections). The problem of what type of learning operators and what sequences of experiences would cause the randomly connected net to evolve in the direction of an Uttley classification mechanism, possessing or lacking one of the features that we made correspond to the set-theoretical axioms, can then be examined.

We see this as one possible sharpening of some of the ideas of pluralistic realism.

3.3 Sebestyen[5] has studied the specific classification procedures useful in mechanical classification of configurations or patterns.

Objects, to be classified, are seen as points in an n-dimensional space, on which a distance is defined. The selection of the distance in question is very important for the rest of the procedure. However, when this distance is selected, a choice of strategy is still to be made:

a) the space can be partitioned in such a way that the mean distance of points belonging to the same sets is minimal;

b) the space can be partitioned in such a way that the mean distance of points belonging to different sets is maximal;

c) the space can be partitioned so that one or both of these properties is present after addition or subtraction of dimensions in the given space;

d) the space can be partitioned in such a way that there is a minimal number of elements simultaneously in different sets.

Our second evaluation proposal for set theories would be to transcribe the set-theoretical axioms as axioms about distances, and to ask for what classification or partition procedures these axioms would completely or approximately hold.

Let us give only one example (presupposing that the concepts of distance between points, between sets, and between points and sets have been defined).

The extensionality axiom can be rewritten as follows: If, for all objects, the distances from a given couple K and L of sets are either above a threshold n, for both K and L, or below a threshold n, for both K and L, then, for any set M, the distances of M from K and L are also simultaneously above or below threshold $f(n)$.

Luczewska-Romahnowa[6] has shown that a distance can be defined upon any classification tree. Here Sebestyen suggests we base any classification upon a distance in the object space. The study of the properties of these distances correlated with the set-theoretical axioms would be a second way to make pluralistic realism more precise.

3.4 The two earlier approaches have an immediate connection with

---

[5] G. S. Sebestyen *Decision Making Processes in Pattern-Recognition*, Macmillan & Co., New York, 1962.

[6] S. Luczewska-Romahnowa "Classification as a Kind of Distance Function; Natural Classification," *Studia Logica*, vol. 12, 1961.

the classification problem. However, it is desirable to give the most general version the problem is capable of.

We define the concepts of "objective set" and of "subjective set".

An objective set is a collection of automata linked by a relation R (R being a relation between their inputs, outputs, or states).

A subjective set is a collection of parts of a given automaton linked by a relation R (these parts may be inputs, outputs or states of the automaton).

A complete set is an objective set produced by a subjective set or a subjective set produced by an objective set.

The evaluation procedure would be, in this context:

a) the study of various types of automata and linkages,
b) what the properties of subjective, objective and complete sets would be.

The set-theoretical axioms that were rewritten in our evaluation 1 as statements about an Uttley mechanism, in our evaluation procedure 2 as statements about distances and partitioning strategies, would now be rewritten as statements about subjective, objective and complete sets.

The reader will realize that we meet here a very real problem[7]: automata are, in fact, sequences of recursive (or even more special) functions. Set theories are in general non-finitary. If we do not wish the very problem we are concerned with (the selection of adequate set theories and the vindication of even strongly infinitistic ones as adequate in certain contexts) to disappear in consequence of our method, then it must be possible to describe how a finite automaton could and should develop structures isomorphic to strongly infinitistic Cantorian set theory. Such a possibility is by no means excluded and the task may thus be undertaken.

4. Conclusion.

In our last part, we stressed the possibility of only one way of making the point of view of pluralistic realism more precise. This does not in the least imply that we consider the linguistic justification of set theory, or its justification in general system theory to be less important, or to be more difficult. Indeed, it is only by means of the collaboration of the three approaches that a rigorous implementation of pluralistic realism will eventually be reached.

---

[7] As far as we know, Prof. A. Mostowski's views on the foundations of mathematics (even though only known to us from the book mentioned in note 1) seem to agree somewhat with the ones expressed here. Let us also refer to G. Martin "Über die inhaltliche Bedeutung der arithmetischen Axiome", *Gesammelte Abhandlungen, Köln* 1961, pp. 110–124.

MATHEMATICS AND THE CONCEPT OF THEORY*

FRIEDRICH KAMBARTEL
University of Münster, Münster, West Germany

In this talk, mathematics or parts of it (formal theories) will not be treated as a system of scientific theorems, but as some sort of scientific activity. We do not want to discuss whether a certain mathematical statement is provable (or examine given proofs), but rather whether it *should*, if possible, be proved. In other words, we are looking for a justification of the mathematician's professional activity qua theorem prover. Using Kantian categories: in the following argumentation "justification" is a *practical*, not a *theoretical* concept.

Furthermore (in view of the fact that mathematicians get public support for their professional occupation) I propose to confine the justification concept to justifications which can be upheld against an arbitrary reasonable and well-informed opponent and hence involve a public interest. So the fact that some mathematical occupation *pleases* me, or some Intuitionists, or the Bourbaki group, or even all mathematicians, is not acceptable as a justification in our sense. It is only satisfaction of the rational interest of mankind which constitutes justification in the sense intended here. To use another Kantian distinction for the purpose of abbreviation: only *practical*, no *pathological* interests are to be admitted.[1]

In radical formulation: If any formal theory is justifiable in the sense proposed here, then this must be possible, even if it happened that there

---

* I wish to express my gratitude to Professors Y. Bar-Hillel, F. Kaulbach and P. Lorenzen for valuable advice and to Dr. H. Mainusch who checked my English.

1 Kant: *Grundlegung zur Metaphysik der Sitten* B39, note (*Akad. Ausg.* IV, p. 413 f.): "Die Abhängigkeit *eines zufällig bestimmbaren* Willens . . . von Prinzipien der Vernunft heisst ein *Interesse.* Dieses findet also nur bei einem abhängigen Willen statt, der nicht von selbst jederzeit der Vernunft gemäss ist; beim göttlichen Willen kann man sich kein Interesse gedenken. Aber auch der menschliche Wille kann woran ein *Interesse nehmen*, ohne darum *aus Interesse zu handeln.* Das erste bedeutet das *praktische* Interesse an der Handlung, das zweite das *pathologische* Interesse am Gegenstande der Handlung. Das erste zeigt nur Abhängigkeit des Willens von Prinzipien der Vernunft an sich selbst, das zweite von den Prinzipien derselben zum Behuf der Neigung an, da nämlich die Vernunft nur die praktische Regel angibt, wie dem Bedürfnisse der Neigung abgeholfen werde. Im ersten Falle interessiert mich die Handlung, im zweiten der Gegenstand der Handlung (so fern er mir angenehm ist)."

were no formal mathematicians. An example of a justification attempt which does not fulfil our criterion is the following remark of A. Heyting concerning intuitionistic mathematics: "It is easy to mention a score of valuable activities which in no way support science, such as the arts, sports, and light entertainment. We claim for intuitionism a value of this sort, which is difficult to define beforehand, but which is clearly felt in dealing with the matter."[2] Such assurances are of some value for the justification problem only if a concrete practical interest is indicated in "the arts, sports, and light entertainment" and if it is furthermore guaranteed that the similarity of intuitionism to the arts etc., is valid also for the rational basis of this interest. Whoever, nevertheless, insists on admitting "pathological" interests not only contradicts the current meaning of "justification" but is also able to justify nearly everything, so that his concept of justification becomes uninteresting. Furthermore, even dogmatic adherents of justification by pathological interests must admit that the question, what part of formal mathematics may be justified by practical interests only, deserves separate discussion. This is not a profound statement. Our only claim is that we have made a meaningful distinction. It is this distinction and not the words which denote it I want to insist on.

A definite group of actions may be justified by their subordination under some general concept comprising only justified actions or by their value in supporting such actions. For formal mathematics, the concept of *theory* or *theoretical science* is meant to fulfil this purpose. This opinion is caused not only by the term "formal *theories*", but also by the connection of theoretical science with the institution of a university rooted in the Greek Academy, so that the generally assumed theoretical character of mathematics, especially of formal mathematics, has led to public sponsorship of research work and teaching in this field, which in its turn, by feedback, tended to strengthen this character.

The last point suggests the following argument: The institutional acceptance of formal theories by society is a fact, and expresses general need for this form of theoretical activity.—But the social acceptance of formal mathematics presupposes that it falls under the classical theory concept, because classical mathematics fell under this concept and all formal theories are extensions of it. It would now have to be proved that the classical theory concept may be extended along this line *without losing its established sense.* So the argument leads back to the very same problem it was meant to avoid.

Now: A general justification of formal mathematics by characterizing

---

[2] *Intuitionism—An Introduction* (Amsterdam 1956), p. 10.

it as a theoretical science would be successful if we were able to satisfy the following two conditions:

1. There are definitions of the concept of theory which involve a practical interest in theoretical activity.

2. The so-called formal theories can be interpreted in such a way that they fulfil one of these definitions or at least have an application within a justified theoretical activity.

Leaving aside Platonic intentions, we might say that the concept of theory, inasfar as it determined the history of philosophy and science, got its first relatively precise explication by Aristotle. Aristotle uses the terms "θεωρία" and "επιστήμη θεωρητική" to distinguish knowledge of the principles of, or reasons for a certain state of affairs from the mere empirical establishment of this state of affairs. In a short but precise formula we might say: Theories in the Aristotelian sense are (true) answers to questions beginning with "why" (διὰ τί).[3] We shall call theories in the Aristotelian sense "reason-founded theories," "explanation theories" or, in short, "why-theories."

Aristotle stresses that theoretical knowledge, i.e., knowledge of reasons, is desirable even if this knowledge is not useful in a utilitarian sense.[4] This could lead to the view that theoretical science is, by the Aristotelian definition, exempt from any justification postulate at all. But this is apparently not the case. Aristotle himself, in the Nicomachean Ethics, gives a practical, though not utilitarian or pragmatical justification: Knowledge of principles (επιστήμη θεωρητική) is the actualization of human reason. Therefore, theoretical activity, especially metaphysics as research and knowledge of the first principles, realizes the highest aim of human life and leads to happiness (εὐδαιμονία) analogous to the happiness of an artisan in the perfect practice of his specific ability.[5]

It is difficult to distinguish the meaning of the terms *theory* and *philosophy* in the Aristotelian texts.[6] Nowadays, we do not consider theoretical and

---

3 Cf. *Met.* I, 1-3,

4 Cf. e.g. *Nic. Eth.* X 7, 1177b 1 ff. Generally Aristotle includes even this non-utilitarian character in the definition of *theory* or *theoretical life* (*βίος θεωρητικός*), taking up the original meaning of *θεωρητικός* (= spectator at the public games and festivals). This connexion and its relevance for the concept of "academic freedom" is analysed by J. Ritter in: *Die Lehre vom Ursprung und Sinn der Theorie* (Arbeitsgemeinschaft für Forschung des Landes Nordrhein-Westfalen, Heft 1, p. 32 ff). For a short history of the word *θεωρία* cf. also *L'Ethique à Nicomaque–Introduction, Traduction et Commentaire* par R. A. Gauthier et J. Y. Jolif (Louvain, Paris 1959), II 1, p. 848 ff.

5 *Nic. Eth.* X 7, 1177 a 12 ff.

6 *Met.* I 1 ff. tends to call only metaphysics (*σοφία, πρώτη φιλοσοφία*) as the knowledge of the first principles and causes *theoretical*, *Met.* VI 1 also includes mathematics and physics.

philosophical reasoning identical; we rather vaguely call only a most fundamental and general reasoning process a philosophical one. This may well be supported by Aristotelian texts, but it was not common usage before the middle of the nineteenth century. Before this time any why-theory of some importance was called "philosophia," the opposite being "historia," a mere collection of facts. So there is a *historia naturalis*, an enumeration of geographical, biological and physical dates, monstrosities and curiosities, and a *philosophia naturalis* containing metaphysical and physical why-theories. The division of historical and philosophical science in the encyclopaedic systems of Bacon and D'Alembert[7] belongs to this tradition just as it is assumed in Descartes' title *Principia philosophiae* or in Newton's title *Philosophiae naturalis principia mathematica*. Bolzano still uses the term "philosophical presentation of a scientific subject (e.g. of mathematics)" to denote a way of presenting the theorems such that the derivations represent the ground-consequence-relations between the theorems. i.e., inform you of the διότι, not only of the ὅτι of their statements[8]. So the Artistotelian concept of theory continues to live not only where the term "theoria" appears, but also in the sense of "philosophia".

The separation of natural science from natural philosophy in the Aristotelian sense was accompanied by a modification of the concept of theory and reason or principle. I quote Christian Huyghens, who was one of the first to be precise on this point. In the preface to the *Treatise Concerning Light* he says that one whould find in this work proofs that differ from those of mathematics in that "here the principles are verified by the conclusions drawn from them, whereas the mathematicians prove their theorems from certain and irrefutable principles."[9] What Huyghens announces in 1690 is the prognostical foundation of physical explanations. These principles are no longer evident in the Aristotelian sense, i.e., valid independent of their conclusions, but justified by the conclusions they make possible. To distinguish these physical theories from the reason-founded or explanation-theories, let us call them "implication-founded theories" or "application-theories." This modification of the Aristotelian concept of theory changes the justification-context of this concept. Whoever understands application-theories as the immediate actualization of human reason (perhaps as a signal

---

[7] F. Bacon: *De Augmentis scientiarum* (1623). — D'Alembert: *Discours Préliminaire de l'Encyclopédie* (1751); cf. also the Survey of the Encyclopedic System at the beginning of the great French Encyclopedia and the articles "histoire" and "philosophie."

[8] Bolzano: *Was ist Philosophie*? (Wien 1849; 2nd ed., Stuttgart 1960); cf. *Wissenschaftslehre* II, § 198.

[9] "... au lieu que les géomètres prouvent leurs propositions par des principes certains et incontestables, ici les principes se vérifient par les conclusions qu'on en tire. . . " (*Oeuvres Complètes* publ. par la Soc. Hollandaise des Sciences t. XIX, p. 454).

of its finite character, see *Pascal*[10]) may justify them by taking over the Aristotelian argument. Whoever includes only why-theories in this argument may use the following justification of the non-Aristotelian physicist: application-theories, by their prediction value, make man master and governor of nature (to use the Cartesian formula). At least some ways of controlling nature make life (ζῆν) and good life (εὖ ζῆν) possible. Echoing Aristotle, one may add: Application-theories free man from being absorbed by the necessities of life and enable him to reason in the Aristotelian sense. It is clear enough that these arguments justify the occupation with at least some application-theories by attaching a practical interest to them.

Now we shall consider formal mathematical theories in the light of the concepts of theory already discussed. By a formal mathematical theory considered as an activity we understand the derivation of theorems from a formal axiomatic system. *Formal* here implies that this axiomatic system is considered independent of the possibility of interpreting it as a description of facts of some kind, although this interpretation is a major motivation for the formal axioms and theorems. Formal theories in this general sense do not, in a trivial way, fulfil the Aristotelian concept of theory. They could be interpreted as theoretical or as supporting theories only if they are limited to those formal axiomatic systems that possess a concrete model. In this case formal theories could be said to be a technique for easy manipulation of answers to why-questions, i.e., why-theories. Insofar as why-theories are justified, formal theories with why-theoretical applications can pass for justified, too. The same is true for formal theories which are tools for forming justified implication-founded theories. In both cases one might say: Formal theories are justified if, after interpretation, they become justified theories.

This seems to give a *universal* justification argument for formal theories. Even if a certain formal theory is not a theory form at present, it still may become one in the future. Therefore it is good to have it ready for such an eventuality. Yet this argument becomes somewhat absurd if one looks more closely at it. The formal theorist here passes as some sort of intellectual tool maker who manufactures not only tools which answer a certain demand or emerging necessity, but also more or less arbitrarily formed pieces of material which he stores for purposes not yet clearly foreseen. In this last essential point the analogy fails: *Potential tools are no tools*.

There is another argument which explains formal theories as reason- or implication-founded theories. This argument is suggested, often unconsciously, by most of formal mathematical instruction today and is used especially

10 Cf. e.g. *Pensées et Opuscules* (ed. L. Brunschvicg, Paris, Classiques Hachette), p. 350 ff.

to justify unrestricted second order arithmetic. It is linked with the use of Cantor's set-theoretical concepts and realities for the interpretation of formal theories. Most mathematicians are accustomed to taking all those formal theories as justified which may be interpreted as describing systems of set-theoretical facts[11], especially including facts in an unrestricted ontology of real numbers, or sets of natural numbers; e.g., the supposed actual, though perhaps ideal or intellectual existence of the unrestricted power set of natural numbers is thus meant to justify the unrestricted arithmetical comprehension schema with the term A possibly containing quantified set variables (cf. p. 221).

The first well-known difficulty implied by this standpoint is that set theory in the sense of Cantor and Dedekind has since been transformed into a formal axiomatic theory to avoid antinomies. Now: formal set theory does not itself contain the objects and facts which formal mathematics (e.g., unrestricted second order arithmetic) is assumed to apply to. Axiomatic set theory cannot serve as a foundation for the theoretical character, and so for the justification of mathematical theories, because it belongs itself to the sphere of actions the justification of which is called in question. Therefore the justification problem makes it still necessary to take the now so-called naive speculations of Cantor into earnest consideration. Specifically, if the axioms of set theory represent the ontological principles of a region of objects and facts which are a legitimate subject of human knowledge, then we might understand axiomatic set theory as a (formalized) why-theory of this region. Furthermore, this would imply that we have enough models to justify almost arbitrary formal theories (among them unrestricted second order arithmetic) as structural description tools for this why-theoretical activity. There are less trivial statements than this one, by the way, because the syntactical possibilities of formal theories are made to agree rather precisely with the unrestricted existence axioms for sets or vice versa.

If we now follow this line of why-theoretical justification for arbitrary formal theories, we are concerned with the ontological dignity of classical set theory, i.e., with some of the philosophical speculations accompanying its formation. The philosophical arguments by which Cantor supports the "doctrine of the transfinite" (*Lehre vom Transfiniten*) are nowadays almost unknown. It does not seem very likely that formal mathematicians would accept the philosophical anachronism involved in his position. To settle this I shall mention the following items: The present mathematical discussion has limited the concepts of infinity to the actual and the potential infinity.

[11] Cf. e.g. *Bourbaki*: "... on sait aujourd'hui qu'il est possible, logiquement parlant, de faire dériver presque toute la mathématique actuelle d'une source unique, la Théorie des Ensembles." (*Théorie des Ensembles* (1954), *Introduction*, p. 4.)

Cantor's philosophico-mathematical spectrum of infinities contains still another variant, which he calls the "absolute infinity," following the metaphysical tradition. The transfinite in the sense of Cantor is said to lie between the absolute infinity and the finite, which according to Cantor comprises the potential infinity.[12] This intermediate character explains why Cantor glorifies so much the discovery of the transfinite regions. They were in Cantor's view an argument against the Kantian assertion that man is not able to rise above his peculiar finite position of knowledge, i.e., leave the limits of possible experience for a cognitive approach to the infinite. Therefore Cantor characterizes the Kantian "transcendental dialectic" in a letter to G. Eneström, dated Nov. 4, 1885, as follows: "Even taking into consideration the Pyrrhonic and Academic scepticism... hardly anyone discredited human reason and its abilities more than Kant by this part of the critical transcendental philosophy."[13] Man—and here Cantor is in agreement with the philosophico-theological tradition—is unable to possess an adequate knowledge of absolute infinity, i.e., God. But this, according to Cantor, does not mean that man cannot approach the infinite at all. In Cantor's view, the transfinite mathematical concepts prove that the Kantian limitation was unnecessary.[14] To support this, Cantor even revives arguments belonging to rational theology which Kant had completely excluded from scientific reasoning. In a letter to Cardinal Franzelin, dated Jan. 22, 1886, for instance, he hints at proofs for the existence of an "Infinitum creatum": "One proof proceeds from the concept of God and concludes first from the divine perfection that God has the possibility of creating a Transfinitum ordinatum. It then goes on to conclude from the divine all-goodness and glory that a Transfinitum has actually been created."[15]

Were these speculations and the eager communication with the neo-scholastic movement, as far as he found agreement there, only philosophical

---

12 *Über die verschiedenen Standpunkte in bezug auf das aktuelle Unendliche*, *Ges. Abh.* ed. E. Zermelo (Berlin 1932), p. 372 ff.

13 "Es dürfte kaum jemals, selbst bei Mitberücksichtigung der Pyrrhonischen und Akademischen Skepsis, mit welcher *Kant* so viele Berührungspunkte hat, mehr zur Diskreditierung der menschlichen Vernunft und ihrer Fähigkeiten geschehen sein, als mit diesem Abschnitt der 'kritischen Transzendentalphilosophie'." (*Ibid.* p. 375.).

14 Cf. e.g. Cantor's argument against the finite character of the human intellect ("Endlichkeit des menschlichen Verstandes") in: *Grundlagen einer allgemeinen Mannigfaltigkeitslehre* (1883), §1 (*Ges. Abh. p.* 176).

15 "Ein Beweis geht vom Gottesbegriff aus und schliesst zunächst aus der höchsten Vollkommenheit Gottes Wesens auf die Möglichkeit der Schöpfung eines Transfinitum ordinatum, sodann aus seiner Allgüte und Herrlichkeit auf die Notwendigkeit der tatsächlich erfolgten Schöpfung eines Transfinitum. Ein anderer Beweis zeigt a posteriori, dass die Annahme eines Transfinitum in natura naturata eine bessere, weil vollkommenere Erklärung der Phänomene, im besondern der Organismen und der psychischen Erscheinungen ermöglicht als die entgegengesetzte Hypothese." (*Mitteilungen zur Lehre vom Transfiniten* V, *Ges. Abh.* p. 400.)

accessories of an essentially mathematical refutation of Aristotle and Kant? Sometimes Cantor may have thought along that line. This is suggested especially by this extreme lack of understanding of the mathematical "horror infiniti," which was really nothing but a doubting attitude towards the specific kind of philosophy brought into mathematics by Cantor. Actually Cantor needs rational theology or equivalent philosophical arguments to pass from the speculative conception of the transfinite in his own head to its theoretical relevance in the Aristotelian sense. Cantor distinguishes the "intrasubjective," or "immanent," reality of mathematical concepts, especially number-concepts, from their "transsubjective, "or "transient," reality. "Immanent reality of concepts" means for Cantor something like their being well defined: that "they have, by definition, a well defined place in our understanding, may be well distinguished from all other ingredients of our thinking, are (nevertheless) related to them and thus modify the substance of our spirit in a way."[16] "Transient reality of concepts," on the other hand, means, according to Cantor: that "they must be considered as the picture of events and relations of the world outside our intellect."[17] Cantor calls mathematics "free," insofar as it "does not have to test the transient reality of its concepts."[18] This postulate makes transfinite set-theory a possible mathematical subject, and the unrestricted comprehension schema *A* (p. 226) a possible mathematical statement, even if nothing more is guaranteed than its being "a well defined modification of our thinking," as Cantor puts it. We do not know if for Cantor consistency would have been a sufficient condition for this. The epistemological foundations of Cantor's concepts of transient and immanent reality are somewhat foggy and questionable. One implication however is clear enough: Cantor's conception of immanent or free mathematics, especially of immanent transfinite set theory means simply a refusal to justify this sort of intellectual activity at all. Those demanding such a justification must adopt Cantor's philosophical theology or find an equivalent.

---

[16] "Einmal dürfen wir die ganzen Zahlen insofern für wirklich ansehen, als sie auf Grund von Definitionen in unserm Verstande einen ganz bestimmten Platz einnehmen, von allen übrigen Bestandteilen unseres Denkens aufs beste unterschieden werden, zu ihnen in bestimmten Beziehungen stehen und somit die Substanz unseres Geistes in bestimmter Weise modifizieren; es sei mir gestattet, diese Art der Realitat unserer Zahlen ihre *intrasubjektive* oder *immanente Realität* zu nennen." (*Grundlagen einer allgemeinen Mannigfaltigkeitslehre* (1883), §8 — *Ges. Abh.* p. 181).

[17] "Dann kann aber auch den Zahlen insofern Wirklichkeit zugeschrieben werden, als sie für einen Ausdruck oder ein Abbild von Vorgängen und Beziehungen in der dem Intellekt gegenüberstehenden Aussenwelt gehalten werden müssèn. ... Diese zweite Art der Realität nenne ich die *transsubjektive* oder auch *transiente Realität* der ganzen Zahlen." (*Ibid.*).

[18] *Ibid.* p. 182.

There are still other arguments against the why-theoretical character of set theory. The insistence, for instance, on proving the consistency of the critical set-theoretical principles shows that these are not theoretical foundations in the Aristotelian sense. Consistency proofs are of course unable to provide these principles with an Aristotelian character.

There is still another way left, by which one could try to save the theoretical character of unrestricted second order arithmetic or other formal theories built upon set-theoretical concepts of Cantor, namely, by an application-theoretical interpretation. This would mean considering the axioms of set theory as mere assumptions justified by their scientific applicability. Those who do not believe in Cantor's transfinite paradise but appreciate the heuristic value of its concepts argue along this line. Of course, in order to prevent a circular answer to the justification question we would have to suppose that they do not mean a heuristic value referring to the finding of theorems in formal theories.

This position now amounts to denying either any fundamental difference between physical laws and mathematical statements or refusing to give an explanation of this difference. The first of these alternatives contradicts the facts, the second is unworthy of a theoretical scientist. So this stand can only be a transitory emergency solution; all the more as there is a competing foundation of mathematics, which not only avoids this difficulty, but even, at least partially, restores the why-theoretical character of mathematics. This finite, constructive or operative approach defines mathematics as the theory of schematic intuitive actions (Dingler, Lorenzen). So it explains the peculiar validity of mathematical statements on the basis of their being founded not on empirical facts, but on decisions concerning actions, and hence not involving experience, but excercise. Furthermore it tries to give the real reasons for the arithmetical facts, arguing that arithmetic has as its unquestionable concrete objects the simplest of all schematic operation structures, the "do-it and do-it-again-starting-from-the-result" system. Following the current trend in mathematics we would have to refer to such dubious entities as the set which has as its only element the empty set etc., or the set of all one-element sets etc., or we could present a formal axiomatic system (e.g. unrestricted second order arithmetic). The first of these alternatives means adopting questionable philosophical speculations which find little support in philosophy itself, even including Platonic philosophy; following the second we are faced with the difficulty that a formally immanent solution of the justification problem is impossible.

The prognostic and heuristic value of the set-theoretical foundation in comparison with the constructive one remains to be discussed. If set-theoretical Platonism is understood as an implication-founded theory, it should have specific implications of a non-formal character. So far nobody has cared about such implications. At present, then, we have no good argument why science, especially physics, should not be possible without set-theoretical mathematical Platonism. As far as physics is concerned this was indeed Cantor's opinion. He does not mention physics to support the supposition of a "transfinitum in natura naturata" (quoting his own words), but hints at organic and psychological facts.[19] The simple fact that modern physics actually applies the transfinite set-theoretical and not the constructive interpretation of the infinitesimal calculus, for example, does not, as such, prove the existence of a necessary connection. It only means that the task of working out an efficient system of constructive mathematics and introducing it into the mathematical instruction of physicists still remains. After its accomplishment, one would have to put an archangel at the theoretical gate to Cantor's paradise and to the field of those formal theories the only justification of which lies in the application to phenomena in this hypothetical theoretical paradise.

Consequently, if we limit the sphere of justified formal theories to those which apply as structural descriptions to concrete facts, for instance in constructive mathematics, there is still a problem left; namely, to define the justified part of concrete mathematics. In fact, arbitrary formal theories may be understood as well regulated derivation games, i.e., intuitive action systems, and in this form belong to constructive mathematics taken in a wide sense. We have already examined the theoretical character of mathematics. It is now necessary to apply this criterion once more. The invention of an arbitrary intuitive action game and mathematical inquiries attached to it have, by themselves, no theoretical sense, though in a concrete case of instruction they may have a didactical or exercising value. Confining constructive mathematics within the limits of a theoretical science would mean that in general it discusses only questions linked with action systems important outside mathematics. Therefore, considered under the aspects of a *practical* justification postulate, even parts of constructively restricted arith metic are under susp icion: e.g., the question of the existence of a largest pair of twin prime numbers.

Under this limitation the mathematician could justify his activity as detecting the rational foundations for assertions about generally interesting intuitive actions or rules for such actions. This formulation is wide enough not to be misunderstood as crude utilitarianism. Quoting D'Alembert: "La curiosité est un besoin pour qui sait penser." This pure curiosity which is at the bottom of the Aristotelian justification of theoretical "praxis," should by our result not be deprived of its mathematical tools as long as it does not turn its attention to fictitious entities.

[19] Cf. note 15.

DIE RECHTFERTIGUNG EINER BESCHÄFTIGUNG MIT FORMALEN THEORIEN

PAUL LORENZEN
Universität Erlangen-Nürnberg, Erlangen, West Germany

Die Frage nach einer Rechtfertigung der Beschäftigung mit formalen Theorien ist insofern eine philosophische Frage, als sie nach einer Rechtfertigung gewisser menschlicher Handlungen fragt. Sie betrifft das Leben und Treiben der Menschen, z.B.: dass viele Menschen axiomatische Mengenlehre treiben.

Die Frage nach einer Rechtfertigung dieses Treibens hat mehrere Aspekte, über die gegenwärtig—obwohl die Mengenlehre immerhin fast 100 Jahre existiert, die axiomatisierte Mengenlehre über 50 Jahre—noch keinerlei Einigkeit erzielt ist.

Um die Formulierung der philosophischen Fragen nicht durch mathematische Details zu verwirren, möchte ich vorschlagen, die Diskussion auf *eine* formale Theorie zu beschränken, nämlich auf die folgende (uneingeschränkte) Arithmetik der zweiten Stufe.

Diese formale Theorie hat zwei Sorten von Variablen:
$x, y, \ldots$ für natürliche Zahlen (ohne Null);
M, N ... für Mengen von natürlichen Zahlen.

Konstante sind: '1' für natürliche Zahlen;
primitive Terme sind: $x + y$, $x \cdot y$ für natürliche Zahlen;
primitive Formeln sind: $x \in M$.

Zur Term- und Formelbildung sind ausserdem die sprachlichen Mittel der elementaren Logik mit Gleichheit zugelassen (also insbesondere Quantoren $\bigwedge_x$, $\bigvee_x$, $\bigwedge_M$, $\bigvee_M$ und Formeln $x = y$, $M = N$).

Axiome sind:

elementare arithmetische Axiome:
$$\begin{cases} \neg\, x + 1 = 1 \\ x + 1 = y + 1 \rightarrow x = y \\ x + (y + 1) = (x + y) + 1 \\ 1 \cdot y = y \\ (x + 1) \cdot y = x \cdot y + y \end{cases}$$

Extensionalitätsaxiom: $\bigwedge_x (x \in M \leftrightarrow x \in N) \rightarrow M = N$

Komprehensionsschema: $\bigvee_M \bigwedge_x (x \in M \leftrightarrow A(x))$

Im Komprehensionsschema ist $A(x)$ eine beliebige Formel der Theorie. (Wird im Komprehensionsschema der Formel $A(x)$ die Einschränkung auferlegt, keine quantifizierte Mengenvariable zu enthalten, dann heisst die entstehende Theorie: die *eingeschränkte* Arithmetik zweiter Stufe.) Theoreme der formalen Theorie sind alle diejenigen Formeln, die sich aus den Axiomen nach den Regeln der (klassischen) elementaren Logik mit Gleichheit ableiten lassen.

Für das Rechtfertigungsproblem sind nun z.B. folgende Fragen zu klären:

1. Hat es überhaupt Sinn, eine solche Tätigkeit wie das Ableiten von Theoremen zu rechtfertigen?

2. Wenn ja, gibt es eine Rechtfertigung über die Tatsache hinaus, dass das Ableiten von Theoremen in der gegenwärtigen Gesellschaft anerkannt ist—und also honoriert wird?

3. Kann man sich für eine Rechtfertigung auf die "Bewährung" in den aussermathematischen Wissenschaften berufen? Wäre es möglich, dass—nach Umgewöhnung auf z.B. die eingeschränkte Arithmetik—diese sich noch besser "bewähren" würde? Welche Kriterien für "besser" sind möglich?

4. Unter welchen Bedingungen würde ein geglückter Widerspruchsfreiheitsbeweis die Beschäftigung mit der Arithmetik zweiter Stufe rechtfertigen?

5. Kann man sich für eine Rechtfertigung auf die Wahrheit (oder etwa die Wahrscheinlichkeit) der Axiome berufen? (In welche Sprache sind dann die Axiome zu übersetzen? Wie steht es mit der Rechtfertigung der eventuellen Benutzung einer natürlichen Sprache?)

6. Wenn man die Wahrheit gewisser Axiome behaupten wollte, wie würden sich solche Behauptungen rechtfertigen lassen?

7. Wenn man die Wahrheit gewisser Axiome behaupten wollte, müsste man dazu die Axiome als Aussagen über Objekte ('natürliche Zahlen' und 'Mengen von natürlichen Zahlen') interpretieren?

8. Wenn ja, wie liessen sich Behauptungen über die "Existenz" solcher Objekte rechtfertigen?

Falls schon die Frage (1) verneint werden sollte, würden selbstverständlich die Fragen (2)–(8) entfallen. Aber auch dann bliebe das Faktum der Beschäftigung mit formalen Theorien bestehen. Es blieben daher zumindest etwa folgende Fragen:

9. Welchen ontologischen—oder anthropologischen Status hat eine formale Theorie? Ist sie ein Spiel, eine zu leistende Arbeit, ein Werkzeug oder was sonst?

10. Wie ist es zu *verstehen*, dass sich viele Menschen mit formalen Theorien beschäftigen? Ist das ein ästhetisches Vergnügen, eine historisch entstandene Angewohnheit, eine Mode oder was sonst?

11. Muss man sich bei den Aussagen *über* eine formale Theorie auf Ableitbarkeitsaussagen ("aus den Formeln $A_2, \ldots, A_n$ ist die Formel A ableitbar") beschränken oder kann man die Behauptung weitergehender Metaaussagen (etwa: "für jede Ziffer m gibt es eine Ziffer n derart, dass die Formel $A(m, n)$ ableitbar ist") rechtfertigen?

AN EMPIRICIST JUSTIFICATION OF MATHEMATICS

S. KÖRNER
The University, Bristol, England

Some scientific theories involve mathematics. If, as I shall assume, the construction and use of such scientific theories is already justified, then mathematics indirectly partakes of this justification. Its details will depend on the manner in which the scientific theories 'involve' mathematics. In examining this relation, I shall understand by mathematics the system of unrestricted arithmetic of second order and various unspecified extensions of it. The principal example of a scientific theory involving mathematics, will be classical mechanics. By attending to one possible justification of mathematics one does not, of course, deny that there are, or may be, others.

The argument will proceed as follows: First, the orthodox deductivist account of scientific theories and the justification of mathematics as an instrument for the deduction of empirical conclusions from empirical premises will be criticized. These criticisms will, secondly, lead to the proposal of an alternative position. Roughly speaking, it will be argued that scientific theories embedded in mathematics function, and are justified, together with their mathematical framework as syncategorematic constituents of empirical propositions. Thirdly, it will be shown that the proposed alternative is not only incompatible with the deductivist justification of mathematics, but that various other current attempts at justifying mathematics in terms of so-called 'platonist' or 'nominalist' ontologies also have to be seriously qualified. Lastly, by way of summing up, answers will be set down to the questions which the chairman suggested to the symposiasts for consideration (p. 221).

1. The deductivist account of the role of mathematics in science. Within a scientific theory two types of deduction can be distinguished, namely, the deduction of substantive theorems from the substantive postulates of the theory, and the deduction of a state-description, expressed in the vocabulary of the theory, from the conjunction of the postulates and another state-description. The latter deduction, which is mainly relevant here, can be schematically expressed by

(i) $$(b_1 \wedge T_0) \vdash_{L_0} b_2$$

If the scientific theory in question is classical mechanics, then b_1 and b_2 are two state-descriptions of a particle or configuration of particles in terms of momenta and relative positions at different times, and '$\vdash_{L_0}$' indicates deducibility in accordance with the postulates and inference-rules of the logico-mathematical system L_0 underlying classical mechanics.

Now the first point that has to be made about (i) is that b_1 and b_2 are *not* empirical propositions. This becomes clear if one compares the constituent predicates of b_1 and b_2, *i.e.* 'mass', 'velocity', and 'relative position' respectively, with their counterparts, *i.e.* those predicates whose applicability is ascertained by empirical operations. I shall call the former predicates 'theoretical' and the latter 'empirical' and speak accordingly of, *e.g.*, 'theoretical mass' as opposed to 'empirical mass'. The differences between the two kinds of predicate are due to the general restrictions which the logico-mathematical framework L_0 imposes on all predicates incorporated into it and, in particular, on those which are the constituents of T_0, b_1, and b_2.

Thus propositional logic and quantification-theory demand that not only the determinables 'theoretical mass', 'theoretical velocity', and 'theoretical relative position', but also their determinates, expressed as multiples of the relevant units of measurement, be exact (extensionally definite). This requirement is not fulfilled by the corresponding empirical predicates. These and other empirical predicates, which admit of neutral or border-line instances, can be accommodated into a variant of the weak three-valued logic, developed by Kleene. [See S. C. Kleene, *Introduction to Metamathematics*, sec. 64 (Amsterdam 1952).] In order to express the results of empirical measurements in T_0, b_1 and b_2, one has to replace inexact, empirical by exact, theoretical predicates. [For a more detailed discussion of this replacement see: "Deductive Unification and Idealization", in the *British Journal of the Philosophy of Science*, vol. XIV, No. 56, 1964.]

Again, the theory of equality, as used in drawing inferences connecting b_1 and b_2 by means of L_0, employs a transitive relation, whereas the notion of empirical equality or indistinguishability is, as has often been pointed out, not transitive. Thus L_0 requires, if it is to be applied to statements, that two masses, velocities or relative positions are equal, a further idealization of empirical predicates. Similarly, the physical addition of empirically ascertained quantities, their multiplication by a number (understood as repeated physical addition) etc., does not conform to the theoretical principles of measurement which are isomorphic with corresponding theorems of L_0. To achieve this isomorphism further modifications of the empirical predicates are needed, before they, or rather their empirical counterparts, can function as constituents of theoretical state-descriptions. The process of modification must be continued if L_0 is extended so as to allow the employment of the differential and integral calculus in drawing inferences from a state-description and the substantive postulates of a theory to other state-descriptions.

All this shows, even without going into details, that schema (i) represents a deduction, by means of L_0, of non-empirical conclusions from non-empirical premises. The deductivist justification of L_0 as a means of drawing empirical conclusions from empirical premises has to be rejected.

2. An empiricist justification of mathematics. Contrary to the deductivist account of the function of mathematics in scientific reasoning, the theory T_0 and its logico-mathematical framework L_0 do not by themselves connect empirical with empirical propositions. In order to exhibit this connection one must look beyond the theoretical state-descriptions b_1 and b_2 to the empirical propositions of which they are the modified substitutes. For the unmodified empirical counterparts of b_1 and b_2, I shall write 'e_1' and 'e_2' respectively.

Because only the theoretical state-descriptions b_1 and b_2 conform to the restrictive requirements of L_0, neither of them is deducible from e_1 or e_2 in T_0 by L_0. Yet inspite of this logical disconnection, e_1 and b_1, and e_2 and b_2, can be, and are in fact, identified for certain purposes and in certain contexts. The conditions for this identification are twofold: First, the empirical proposition must contain the information which in modified form is expressed in the theoretical state-descriptions. Thus, if b_1 is a theoretical state-description of classical mechanics, e_1 must record the results of empirical procedures for measuring mass, velocity and relative position at a certain time. Second, there must be no known or unknown factors in the empirical situation described by e_1, which would make the identification inadequate. For example, if the bodies on which mechanical experiments are performed are electrically charged, the identification would not in general be borne out by observation. (The disturbing influence of electrical charges on the outcome of mechanical experiments, as predicted by means of classical mechanics and the relevant identifications, was, of course, not always known.)

The conditions for identifications of the first kind might be called 'theory-dependent', since the theory determines what is a state-description in it, and what empirical information is relevant to it. I shall write '$e \underset{T_0}{\approx} b$' or, simply, '$e \approx b$' for the assumption that the theory-dependent conditions are satisfied. The conditions for identifications of the second kind might be called 'context-dependent', since the existence of disturbing factors, which render the identification inadequate, depend on the wider context in which the theory is employed, that is to say on nature. I shall write 'c_0' to indicate the assumption that the context-dependent conditions are satisfied.

On the basis of what has been said about the modification of empirical propositions into theoretical state-descriptions, the logical disconnection between them, and the conditions for their identification, the employment of a scientific theory T_0 and of its logico-mathematical framework in connecting empirical with empirical propositions, in a manner testable by experiment and observation, can be expressed by the following schema:

(ii) $$(e_1 \wedge (e_1 \approx b_1) \wedge (b_1 \wedge T_0 \vdash_{L_0} b_2) \wedge (b_2 \approx e_2) \wedge c_0) \rightarrow e_2$$

In words, if e_1 is true and identified with b_1, if the theoretical derivation is performed and its conclusion b_2 identified with e_2, if, moreover, the context-dependent conditions for the identifications are satisfied, then e_2 is true. (The signs '$\wedge$' and '$\rightarrow$' respectively express conjunction and conditional of the logic of empirical discourse, which is free from the restrictions imposed by L_0 on T_0, b_1 and b_2.) Should experiment or observation reveal e_1 to be true and e_2 to be false, a change must be made either in the theory or in the conditions of identification.

Unlike (i), proposition (ii) is an empirical, and empirically testable, proposition, although it contains the non-empirical T_0, L_0, b_1 and b_2 among its constituents. Any justification of this empirical proposition is thus a justification of its constituents, the relation of which to the whole proposition is exhibited by (ii). Insofar, therefore, as the demand that all justification be based on experiment and observation is characteristic of empiricism, one might say, without thereby being committed to this philosophy, that (ii) provides the possibility of an empiricist justification of mathematics.

3. **Other justifications of mathematics.** I shall say no more about the logicist justification, which assumes that all mathematics is expressible in logical terms and deducible from logical principles, than that the assumption of this deducibility has been shown, and is generally accepted, to be mistaken.

So called 'platonism' accepts as legitimate predicates of logico-mathematical theories, which it acknowledges to be capable of instantiation not in empirical, but only in a non-empirical, reality. The preceding account agrees with platonism in holding that the predicates of L_0 are not empirical, but ideal. But it does not share the ontological commitments of platonism or, at least, of Platonism. The well-known Quine-Church criterion of ontological commitment, i.e., that one is committed to assuming the existence of individuals belonging to an existentially quantified domain, though of great use in combating various confusions, ultimately says no more than that to quantify existentially *is* to quantify existentially. Thus, if I assert that I dreamt of witches, i.e., that witches existed in my dream, I am committed to the assumption that at least one person who dreamt of witches exists, but not also to the assumption that witches exist, although I may be committed to it on other grounds. Similarly the assertion of the empirical proposition (ii), in which T_0 and L_0 occur as syncategorematic constituents, commits me to the intra-theoretical existence of entities the domain of which is existentially quantified within the theory, but not also to their extra-theoretical, 'genuine', existence, although I may be committed to it on other grounds.

The position is similar with regard to intuitionism, which holds that the predicates of mathematics are true not of sense-experience, but of introspectible constructions in pure intuition. Like the platonist, the intuitionist account is compatible with the empiricist justification of mathematics given above. Just as it is incumbent on the platonist to defend the thesis of a non-empirical, mind-independent reality, so it is incumbent on the intuitionist to defend his thesis of a non-empirical, mind-independent intuition.

Whereas the proposed empiricist justification of mathematics required neither the acceptance nor the rejection of platonist realism or intuitionist conceptualism, it does imply the rejection of various nominalistic accounts of mathematics. The dominant nominalism in the philosophy of mathematics is a naive empiricism in that it admits only such mathematical predicates as are instantiated in experience. But it is doubtful whether even the simplest constructive arithmetic lives up to this requirement. The atoms and complexes of atoms, which are formed in accordance with its formation rules, are assumed—or rather postulated—to be well-distinguished from each other, in the sense of being instances of exact predicates, namely 'being a unit', 'being a couple', etc., none of which admits of neutral or border-line cases. However, the empirical predicates of 'being an n-tuple' are, even when n is small, not exact. Again, the assumption —or rather postulate—that one can by the repeated addition of atoms reach complexes of any finite number of atoms, reveals both the mathematical predicates of 'being an atom' and 'being a complex of atoms' to be idealizations of corresponding empirical predicates, for whose instances the assumption is not true. It is thus not merely the assumption of an actual, but already that of a potential, infinity of atoms, which, apart from the elimination of inexactness, enforces the employment of non-empirical predicates in arithmetic. Although neither constructive nor non-constructive mathematics is empirical, one might reasonably argue on the one hand that constructive mathematics is 'nearer' to experience because it rests on less radical idealizations (e.g. weaker comprehension axioms), and on the other hand that some, at least, of the more radical idealizations of non-constructive mathematics are for the purposes of scientific reasoning superfluous.

4. Conclusions. On the basis of the preceding remarks I can now set down my (partial) answers to the chairman's questions (p. 221). Since one can, as I have tried to show, justify mathematics indirectly by appealing to its employment in the extramathematical sciences, the answer to the question whether such justification is possible is 'yes.' *A fortiori* Questions 1 to 3 make sense. A consistency-proof for any mathematical theory, underlying a scientific theory, excludes the possibility that in schema (ii) the constituent '$b_1 \wedge T_0 \vdash_{L_0} b_2$' could be replaced by '$b_1 \wedge T_0 \vdash_{L_0} \neg b_2$' in accordance with the principles of L_0. Since this possibility would make L_0, and therefore T_0 useless or, at least, greatly reduce their usefulness, the exclusion of this possibility is, at least, indirectly justified (Question 4).

The above empiricist justification of mathematics does not require that the axioms and theorems of mathematics be true of anything (Question 5). If one holds them to be true one would have to establish a metaphysical theory, e.g., a Kantian if one is an intuitionist, or Platonic if one is a platonist (Question 6). The objects falling under the predicates of the axioms would then be non-empirical individuals, 'existing' in the sense of the accepted metaphysics (Questions 7–8). Some mathematical theories are—and presumably all may become—constituents of empirical propositions, the search for which is involved in scientific inquiry (Question 9). Although a justification of mathematical research can be given as part of a justification of research in the natural sciences, mathematical inquiries can also be justified as the source of aesthetic pleasure, the satisfaction of personal ambition, etc. (Question 10). An empiricist justification of metamathematics is included in that of mathematics, at least to the extent to which progress in the former enhances progress in the latter (Question 11).

FORMALISM 64

ABRAHAM ROBINSON
University of California, Los Angeles, California, U.S.A.

1. As we look back upon the development of the Philosophy of Mathematics, the fifty years between 1890 and 1940 appear to us as a golden age. We get the impression that the principal opinions which still dominate our thinking in the field were developed during that period, and among them there stand out three well-moulded points of view—Logicism, Intuitionism, Formalism. No doubt, this picture involves a measure of oversimplification. In actual fact, many of the ideas which find expression in one or the other of the above mentioned philosophies have roots which go back beyond, sometimes far beyond, the "golden age." It is also true that these philosophies remained uncommitted on certain important points. For example, the tenets of Logicism seem to be compatible with diametrically opposed views on the problem of existence in Mathematics. This is illustrated by the fact that the early logicists evinced a rather solid belief in the existence of mathematical structures, finite or infinite, while this belief is not shared by the logical positivists (logical empiricists), who represent later Logicism. Again, there are now various constructivist positions related to, but not identical with, Intuitionism. There is also the position of the nominalists which cannot be subsumed under any of the philosophies enumerated above. But in spite of all these qualifications there persists the impression of the period from 1890 to 1940 as an age of renaissance in the Philosophy of Mathematics, an age in which fundamental opinions were vigorously stated—and vigorously attacked.

2. This philosophical activity involved and inspired the development of numerous "technical" methods and theories in Mathematical Logic, and that development continued beyond 1940 and, as we all know, is still going on at an ever increasing pace. At the same time, the general interest in the Philosophy of Mathematics as such has flagged and is only just beginning to regain some of its former vigor. To those who believe that on all matters of principle the correct answers have already been given by one existing school of thought or another, this presents no problem. Personally, I have to admit that I cannot share this optimistic opinion. It seems to me that *all* points of view that have been put forward as a philosophical basis for Mathematics involve serious gaps and difficulties, including the point of view which I now hold and which I propose to expound in this address. As the title indicates my position is, basically, close to that of Hilbert and his school. I have added the year, 64, not only because it is now known (as it was not known in 1925) that Hilbert's program is doomed to failure but also because the present picture in the foundations of Mathematics has been affected by important developments in several other directions. I wish to add that I cannot subscribe to everything that has been stated by Hilbert in this field. In fact, this talk is to be regarded neither as a description of a historical position nor as a manifesto which tries to lay down the law but rather as a confession, as a personal statement of a point of view arrived at over a number of years. For this point of view I claim neither faithful adherence to an existing school of thought nor basic originality. Nevertheless I hope that some of my remarks, particularly those contained in the closing sections, will encourage discussions and developments in new directions. In *choosing* my position, my approach has been empirical. That is to say, I have tried to take into account all the evidence available to me that may have a bearing on the subject, including both basic thinking and "technical" results in Logic and Mathematics. I do not dispose of a general philosophical creed which, among other things, would determine also my Philosophy of Mathematics.

3. Different opinions concerning the foundations of Mathematics may differ in their estimation of the meaning or significance of the established body of Mathematics. Thus, they may differ as to the correct *description* of the contents of a theory, e.g., as regards the basic nature of the number concept, or as to the interpretation of existential statements for which no constructive proof is available. They may also differ with respect to their *deontic* principles, more particularly in their answer to the question what kind of activity is proper and reasonable for a mathematician *as* mathematician. Such principles normally involve value judgements which affirm that certain activities, or results of activities, are more important, or worthwhile, or relevant, than others. In this connection, we should not forget that even mathematicians with the same view (or with no views at all) on the foundations of Mathematics may differ sharply as regards the value of a particular piece of work.

Discussions between different schools of thought in the Philosophy of Mathematics usually involve divergences both on the descriptive and on the deontic aspects of the problem. The respective attitudes concerning these aspects may well be interrelated but nevertheless it will be helpful to distinguish between them.

4. My position concerning the foundations of Mathematics is based on the following two main points or principles.

(i) Infinite totalities do not exist in any sense of the word (i.e., either really or ideally). More precisely, any mention, or purported mention, of infinite totalities is, literally, *meaningless*.

(ii) Nevertheless, we should continue the business of Mathematics "as usual," i.e., we should act *as if* infinite totalities really existed.

Of the two principles just stated, the first is descriptive while the second is deontic or prescriptive. I proceed to discuss the first principle.

The problem of Infinity has been debated since the dawn of Philosophy. It is not my purpose to give a historical survey of this discussion. But I may recall here that much of the criticism that was directed by Aristotle (in the third book of his "Physics") against the notion of actual infinity is still topical.

The problem of Infinity is related to, and is sometimes regarded as part of, the problem of Existence in Mathematics, or of the problem of the existence of abstract notions in general. To a nominalist, the existence of a set of five elements is no less illusory than the existence of the totality of all natural numbers. At the other end of the scale are the so-called platonic realists or platonists who believe in the ideal existence of mathematical entities in general, including the existence of transfinite sets of arbitrarily large cardinal numbers to the extent to which they can be introduced at all by means of suitable axioms. It has been said that the platonists do not believe in the objective existence of mathematical entities but rather in the objective truth of mathematical theorems. However, it seems to me that as a matter of empirical fact the platonists believe in the objective truth of mathematical theorems *because* they believe in the objective existence of mathematical entities. Since in their conception a mathematical structure is rather like a physical object, such as a house or a tree, or like a collection of physical objects, which can be examined at leisure and in detail, they conclude that any meaningful question that can be asked concerning such a structure must by necessity possess an absolute answer.

K. Gödel, who may be regarded as the outstanding platonist of our time, has emphasized the similarity between the investigation of physical objects on one hand and of mathematical objects on the other. He sees no reason why we should affirm the objective existence of the former but deny that of the latter. I am in sympathy with this point of view to a very limited extent. It appears to me that the notion of a particular *class* of five elements, e.g., of five particular chairs, presents itself to my mind as clearly as the notion of a single individual (a particular chair, a particular table). Thus, I do not feel compelled to follow the nominalists, who seem to have little trouble in grasping the notion of an individual but feel incapable of proceeding from there to the notion of a class. At any rate, so far as the theory of finite sets is concerned the position of the nominalists leads only to modifications which are of little depth from the mathematical point of view. (I do not wish to deny that the discussion of the nominalistic thesis is still of interest on a different philosophical plane.)

By contrast, I feel quite unable to grasp the idea of an actual infinite totality. To me there appears to exist an unbridgeable gulf between sets or structures of one, or two, or five elements, on one hand, and infinite structures on the other hand or, more precisely, between terms denoting sets or structures of one, or two, or five elements, and terms purporting to denote sets or structures the number of whose elements is infinite. As stated here, this point of view concerns the notion of infinity in Abstract Set Theory, with particular reference to infinite cardinality. However, other types of infinity may be relevant to our discussion, more particularly ordinal infinity, as in the theory of ordinal numbers, and geometrical infinity, such as infinite length. While these can be reduced to cardinal infinity by mathematical considerations one may well argue that, from a philosophical point of view, ordinal infinity or geometrical infinity are more basic than cardinal infinity and provide access to an understanding of the notion of infinity as a whole. So I must add that I am just as unable to grasp ordinal infinity or geometrical infinity.

It follows that I must regard a theory which refers to an infinite totality as *meaningless* in the sense that its terms and sentences cannot possess the direct interpretation in an actual structure that we should expect them to have by analogy with concrete (e.g., empirical) situations. This is not to say that such a theory is therefore pointless or devoid of significance.

An opponent to my position might put forward the following arguments.

(i) He might say that I am unable to grasp the idea of an actual infinite totality merely because my brain suffers from a peculiar limitation. He might argue that he, on the contrary, has a clear conception of all sorts of infinite totalities or, at the very least, of the totality of natural numbers.

(ii) Alternatively, my opponent may concede that he, also, is unable to grasp the idea of an infinite totality. But he may say that this does not in any way prove that infinite totalities do not exist. In order to show that they do exist he may appeal to the physical world. Or, if he does not wish to, or feels that he cannot, appeal to the physical world, he may affirm the existence of a platonic world which contains infinities of all sorts, or of some sort.

The first argument, which asserts that there exists an immanent appreciation of infinity of which I am incapable does not permit any further direct debate of the issue. The second argument may be discussed on a philosophical, more precisely, epistemological level. Thus, we may question whether it has any meaning to affirm the existence of an entity while admitting that we are incapable of understanding its basic characteristics. This criticism is particularly cogent if such existence is asserted in the platonic sense. However, we shall not pursue this line here. Instead, we shall continue the discussion by examining how the different points of view stand up in the light of present day mathematical knowledge.

We have said that to the platonist the existence of a mathematical structure is primary, and the theorems about the structure are secondary. Nevertheless the degree of (potential or actual) completeness of our picture of a particular mathematical structure, or of the set-theoretic universe as a whole (to use platonic language) may reasonably be said to affect the strength of the case of Platonism against Formalism. That is to say (using platonic terms) if one is faced with a mathematical concept that one believes to be categorical in an absolute sense (i.e., realized by a unique structure, up to isomorphism), or with the entire universe of sets, also absolute, then one will naturally be upset when confronted with a sentence X which is meaningful for the theory under consideration and such that both X and not-X are consistent with all assumptions which seem intuitively correct with regard to the structure or universe in question. As you know, this is precisely the situation in Set Theory today. If X is the continuum hypothesis, $2^{\aleph_0} = \aleph_1$, then we know from the complementary results of K. Gödel and P. Cohen that both X and non-X are compatible with all known "natural" assumptions regarding the universe of sets (to use platonic language). While this suggests to the formalist that the entire notion of the universe of sets is meaningless (in the sense indicated by our first principle) the platonist merely concludes that the basic and commonly accepted properties of the universe of sets which are known to us at present are insufficient to decide the continuum hypothesis one way or the other. He will maintain, in this and similar cases, that at any rate only one of the alternatives that offer themselves is the correct one, i.e., is in agreement with *the truth.* At this point he has to face the question whether in this matter the truth is by necessity *discernible* by the human mind. As far as I know, only a small minority of mathematicians, even of those with platonist views, accept the idea that there may be mathematical facts which are *true* but unknowable. If, on the contrary, a platonist maintains that every mathematical truth, about the universe of sets or about a specific structure, is by necessity discernible then he is still called upon to analyze his own modal manner of speech. In particular, he has to ponder the question whether a truth is discernible only if it will (by necessity) be discerned some day and if so, whether this must involve the discovery of new "natural" assumptions or forms of argument which are acceptable to all or most mathematicians. At any rate, it seems to me that the present situation in Set Theory favors the Formalist.

The situation is different in Number Theory. For although Gödel's theorem shows that any explicitly specified set of axioms for Number Theory is incomplete, the examples which have been produced in order to demonstrate this incompleteness are such as to bias us towards a decision as to their truth or falsehood. Thus, the Gödel sentence which asserts its own improvability at the same time affirms its own truth. To this extent, the present situation in Number Theory favors the Platonist.

The question discussed here may be called the *bifurcation problem* in Set Theory and Number Theory. At this moment in history, the path of Set Theory seems to have bifurcated, but the platonist believes that this bifurcation is illusory since *in reality* only one of the alternatives covering the continuum hypothesis can be true. On the other hand, while the bifurcation of either Set Theory or Number Theory is not an essential part of the point of view of the formalist he regards both as entirely possible. Thus, although the problem of bifurcation is certainly relevant to our problem, its consideration does not lead to a clear-cut decision of the case for or against Platonism.

5. I pass on to the discussion of the second basic principle of my formalist philosophy. This is the prescription, or suggestion, or advice, to continue to do Mathematics in the classical way, i.e., in particular, to use terms which purport to refer to infinite totalities as if they really existed. It is at this point that Formalism is in direct conflict with the various constructivist or operationist schools of thought and it will be illuminating to develop our position by way of contrast with these.

I recall that the starting point of Intuitionism is, like my own, the rejection of the naive notion of an infinite totality. The rejection of the law of the excluded middle which is, perhaps, a more famous peculiarity of the intuitionists is regarded by them explicitly as a consequence of their critique of infinitary mathematical concepts. There are also those philosophers of Mathematics whose opposition to classical Mathematics is directed chiefly against the use of impredicative definitions, i.e., definitions in which the definiens involves totalities that include the definiendum. But here again the problem is a serious one only in relation to infinite totalities. At any rate all these schools of thought reject the development of theories that involve actual infinite totalities as a regrettable aberration which will be superseded in due course by saner methods. Sometimes the adherents of these views look for support to the history of Science and point out, correctly, that the Greeks were well aware of the dangers of infinity and, accordingly, did their best to use constructive methods and terminologies. It seems to them that the formalist who admits frankly that much of existing Mathematics is *meaningless* in the sense explained above but who nevertheless encourages the development of this kind of Mathematics adopts a position which is altogether indefensible.

I believe that this criticism is based on an attitude which, though at first sight quite reasonable, is nevertheless too narrow. Those who adopt this attitude think that a concept, or a sentence, or an entire theory, is acceptable only if it can be *understood* properly and that a concept, or sentence, or a theory, is understood properly only if all terms which occur in it can be interpreted directly, as explained. By contrast, the formalist holds that direct interpretability is not a necessary condition for the acceptability of a mathematical theory. Evidently, this issue can be argued only if the constructivist does not regard the direct interpretability of all terms of a theory as axiomatic. Supposing that he is indeed willing to concede this point, let us consider in turn several other criteria of acceptability which are frequently regarded as basic.

(i) A mathematical theory shall be regarded as acceptable only if it is consistent.

Comment. It is indeed a regrettable fact that no version of classical Mathematics is provably consistent, and that there is no *unique* remedy for the inconsistencies (antinomies) which have arisen. On the other hand, a constructivist may indeed *believe* that his theory is safe from inconsistencies, but so far as I have been able to follow the matter, this cannot be proved by arguments which are conclusive from a finitistic point of view.

(ii) A mathematical theory is acceptable if it can serve as a foundation for the Natural Sciences.

Comment. Again, this criterion does not imply that the terms of the theory should be interpretable directly and in detail. It is sufficient that we should have rules which tell how to apply certain relevant parts of our theory to the empirical world. From this point of view the acceptability of Abstract Set Theory is affected by the question whether the theory of real numbers provided by it can serve as an adequate foundation for physical measurements but is independent of the interpretability of infinitary notions such as the totality of real numbers or the totality of natural numbers.

You will observe that this comment has a positivistic, more particularly an instrumentalist, flavour. However, it does not imply the acceptance of the positivistic point of view as far as the Empirical sciences as such are concerned and remains uncommitted in this respect.

(iii) A mathematical theory is to be judged by aesthetic standards such as its beauty or internal relevance.

Comment. The standards mentioned under (iii) have so far defied any sort of scientific approach. The attempts that have been made by notable mathematicians to give a precise expression to their ideas on the subject are regarded by others merely as monuments to the prejudices of these mathematicians. At the same time, it is a fact that the organized world of Pure Mathematics is regulated to a very large extent by our vague intuitive ideas on mathematical beauty and purely mathematical importance. The situation being what it is I will say only that my own intuitive judgement regarding the beauty of a mathematical theory is not affected by the question of the existence or interpretability of the infinitary notions which occur in it. Moreover, I believe that this is also the attitude of most pure mathematicians. To be sure, there are some whose admiration for Cantor's achievement is due to the impression that he discovered an immense world of totalities of enormous size which had been there waiting since the beginning of time. Personally I must confess that the creation of the theory of transfinite numbers impresses me neither more nor less than other bold mathematical innovations such as the theory of Hilbert Space, or Non-archimedean Geometry, or the Theory of Groups.

To sum up, the direct interpretability of the terms of a mathematical theory is not a necessary condition for its acceptability; a theory which includes infinitary terms is not thereby less acceptable or less rational than a theory which avoids them. To *understand* a theory means to be able to follow its logical development and not, necessarily, to interpret, or give a denotation for, its individual terms. However, we may grant at this point that a constructive theory may well have an interest of its own, irrespective of any criticism of the formalist approach. The direct applicability of the procedures suggested by such a theory may be important in practice and philosophically relevant.

6. In Hilbert's view the formal or uninterpreted part of a theory belonged entirely to Mathematics. At the same time, the Metamathematics of the theory was supposed to be strictly finitistic and directly interpretable, as explained above. However, eventually the use of infinitary modes of expression imposed itself on Metamathematics just as it had imposed itself previously on Mathematics. This is very much in evidence in the contemporary development of theories which include a non-countable number of symbols, or infinitely long well-formed formulae. However, it is important to note that even in a more restrained metamathematical approach the actual infinite may enter the scene. For example, in the classical treatment of recursive functions, the notions of *consistency* and *completeness* of a recursive scheme involve the totality of natural numbers. All such infinitary metamathematical theorems are, from our point of view, subject to the two principles enumerated earlier, i.e., to the extent to which they involve infinite totalities they are *meaningless* (in the sense indicated previously) but there may be good reasons for developing them all the same. These remarks apply, in particular, to Semantics. At first sight, the Theory of Models gives a perfectly satisfactory account of the notion of truth in relation to both finite and infinite structures. However, to the formalist any alleged truth definition relative to an infinite structure merely amounts to the formal application of the rules of interpretation to which we are accustomed from concrete, more particularly, empirical examples, to infinite totalities.

Even within standard syntax there are references to infinite totalities such as the totality of well-formed formulae in a given language, or the totality of provable formulae (theorems). A classical mathematician or logician has no hesitation in referring to these totalities as if such references were meaningful and, among other things, in applying to them the laws of the excluded middle. More particularly, the inductive definitions of these totalities presuppose a system of natural numbers or, alternatively or conjointly, a system of infinitary Set Theory, and these are notions which come under our first basic principle. Here as elsewhere the formalist considers that the logician who indulges in this kind of analysis may well understand what he is doing as long as he does not fall into the trap of believing that his theory is meaningful (directly interpretable). But the question remains how to correlate this abstract logic with the *use* of logical principles in actual fact. We shall return to this problem presently.

7. At the beginning of this address, I stated that all known positions in the Philosophy of Mathematics, including my own, still involve serious gaps and difficulties. Among these, the gap due to the absence of consistency proofs for the major mathematical theories appears to be inevitable and we have learned to live with it. Nevertheless, the fact that the development of certain infinitary branches of Mathematics such as Abstract Set Theory has gone as far as it did with only relatively minor, and remediable difficulties, raises the question whether these theories are not, after all, more significant than the formalist is willing to concede. The view that Abstract Set Theory, for example, is workable because it is obtained by extrapolation from a theory of finite sets, which is interpretable even according to the formalist, is open to criticism. Indeed, in laying down the axioms of Set Theory we accept some properties which hold in the realm of finite sets, such as the axiom of power sets, and reject others, such as the irreflexivity of sets. Thus, the process of extrapolation is actually selective. A possible but perhaps not convincing explanation of the apparently satisfactory nature of so many infinitary theories is that we have arrived at them by trial and error, covering most deductions of moderate length. At any rate there is no doubt that on this score the platonist occupies a more comfortable position.

However, it is the nature of the language and rules actually *used* in the development of (possibly infinitary) mathematical theories that presents us with our greatest problem. Since the formalist cannot rely on the semantic interpretation of his theories, their complete formalization is essential to him. This is part of the original formalist approach due to Hilbert. In this approach the part of Logic and Arithmetic which appear already in the Metamathematics, i.e., which are actually *used* in the development and analysis of a formal theory are supposed to be entirely finitistic (see section 6 above) and such that their validity is self-evident. A formal theory is then obtained by the adjunction of "ideal" (uninterpreted) elements to the formal version of the interpreted finite theory. Thus, the part of a theory that is actually *used* is supposed to be given explicitly and there is no need for its identification a posteriori.

Logical positivists and formalists are in agreement in adopting a pragmatic approach to the development of a formal theory and my sympathy with the positivistic point of view, as far as infinitary theories are concerned, may have been apparent. However, the logical positivists extend this approach to all of metamathematics and maintain that the syntax of a language is a matter of choice at all levels. It is at this point that I am unable to follow them. It seems to me that the rules of logic and of certain parts of Arithmetic which are *used* in the analysis of a formal theory are by no means arbitrary. For example, even when considering some non-standard, e.g., three-valued, logic, the logician ultimately assumes that a given concrete situation either obtains or does not obtain, and that these possibilities are mutually exclusive. In other words, at this point the logician actually *uses* two-valued propositional logic or, in syntactical terms, he assumes the law of contradiction, not-(X and not-X), and the law of the excluded middle, X or not-X.

Thus, it appears to me that there are certain basic forms of thought and argument which are prior to the development of formal Mathematics. I propose to show in the remaining sections that the further discussion of the scope of this basic Metamathematics beyond the explication given by Hilbert is both vital and fruitful.

8. A formal theory may be presented by means of a *primitive frame* (a term due to Curry) which delimits (atomic) symbols, terms, well-formed formulae, and theorems (provable formulae). An example of a primitive frame is provided by the following set of rules (A) for a rudimentary form of recursive arithmetic, which is adapted from [3].

(A) *Symbols.* $0, ', =$.
Terms. 0 is a term. If a is a term then a' is a term.
Formulae. If a and b are terms then $a = b$ is a (well-formed) formula.
Theorems. $0 = 0$ is a theorem. If $a = b$ is a theorem then $a' = b'$ is a theorem.

What is the significance of a primitive frame such as (A)? The answer to this question is ambiguous. In spite of the grammatical form of the sentences of (A), a mathematician will *in practice* interpret them as permissive instructions which enable him to record or communicate terms, formulae, and theorems of the theory defined by the frame if he so desires. Putting it in the most concrete form, (A) tells us that if we take three sheets of paper and label them "terms", "formulae", theorems", then we may enter "0" on the sheet of terms, and we may enter additional inscriptions on the three sheets in accordance with the following rules. If we find that a is an inscription on the sheet of terms then a' may be entered on the sheet of terms. (Here, "a" denotes an inscription, and "a'" denotes the inscription obtained from the former by affixing " ′ " to it.) Similarly, if we find a and b on the sheet of terms then we may enter $a = b$ on the sheet of formulae. As for the sheet of theorems, we may enter $0=0$ on it, and if we read $a=b$ on this sheet we may enter on it also $a' = b'$. We observe that in this interpretation no rules of Logic are involved at all.

Many existing primitive frames for formal theories are less simple. For example, in some versions of the Predicate Calculus overlapping quantifications with respect to the same variable are not permitted, i.e., do not yield well-formed formulae. Thus, in order to know whether or not certain operations are permissible we have to be able to check whether an inscription already available does or does not contain configurations of certain kinds. It may also occur that the natural numbers make their appearance in a primitive frame. Such is the case in Gödel's representation of the language of type theory which includes variables of all types, 0, 1, 2, .. and so on, for all n. This sort of formulation should be ruled out explicitly if the language of the primitive frame is to remain at a strictly finitistic level. It is in fact implicit in the present approach that the primitive frame for a formal theory should be such as to be interpretable immediately in terms of concrete instructions. As already stated this does not commit us to any particular logic, not does it commit us to any a priori assumptions on Arithmetic or Set Theory.

The question what Logic or Arithmetic or Set Theory we actually *use* intuitively, arises when we try to interpret a primitive frame as it stands, i.e., as a *description*, or purported description of a universe of "terms", "well-formed formulae" and "theorems". If we were called up to describe only the terms, well-formed formulae, and theorems *actually* written on a particular sheet of paper, or all terms, etc., *actually* recorded at a given time, then I would not hesitate to say that the full first order Predicate Logic as well as a certain fragment of Arithmetic (which is appropriate to the number of symbols etc. already recorded) are applicable here. However, in actual fact we are required to deal not only with the expressions that *have been* written down but also with those that *might possibly be* written down. To regard those as actual totalities would be contrary to our second principle. If, nevertheless, we decide to act *as if* they constituted actual totalities, we thereby move our metamathematical theory to the realm of uninterpreted formal systems. How far can we go without undertaking this fateful step?

Both formalists and constructionists do not doubt that at least some of the rules of the propositional and predicate calculi are applicable in the interpreted (material, "inhaltlich") kind of metamathematics now under consideration. More precisely these may be taken to include all axioms and rules of the Predicate Calculus which are accepted by those intuitionists to whom a codification of the laws of reasoning is at all acceptable. But there the unanimity ends. Even among the formalists there are differences of opinion as to which parts of Arithmetic may be said to be intuitive and acceptable as finitistic. Personally, I cannot see at this time how a form of reasoning which attempts to escape the consequence of Gödel's second theorem (such as Gentzen's consistency proof for Arithmetic or any other consistency proof for Arithmetic) can remain strictly finitistic and, hence, interpreted. Thus, there is room for further discussion on the precise delimitation of finitistic metamathematical reasoning, with particular reference to the arithmetical and set-theoretical arguments contained in it.

9. On the other hand, we may also take the "fateful step" of removing Syntax itself to the realm of uninterpreted formal theories. If we adopt this policy then the basic syntactical notions such as connectives, variables, quantifiers, well-formed formulae, are not regarded as inscriptions which are created gradually at the whim of a writer but are supposed to constitute rigid totalities or sets, which are connected by certain operations such as concatenation. Similarly, the model-theoretic (semantic) interpretation of sentences takes place in a wider axiomatic framework which includes in addition also sets of individuals and of relations. The connection between these individuals and relations and the symbols (in the abstract sense, as above) which denote them is again given by metamathematical relations, more particularly by the relation or relations of designation. Accordingly, designation appears here as a particular kind of correspondence between abstract sets. In many contexts it is perfectly legitimate to suppose that this correspondence reduces to the identity, in other words, that the notion is autonymous. The somewhat dogmatic approach to the problem of denotation which requires a rigid distinction between name and object is no doubt appropriate to cities and names of cities (e.g., Jerusalem and "Jerusalem") but is not essential when transferred to mathematical entities within the above framework. Consider for example the assertion that there is a one-to-one correspondence between numerals and natural numbers (or, alternatively, a many-one correspondence). Evidently, the notion of a numeral here does not refer to inscriptions (or tokens) since the number of inscriptions that have been written down is finite and can even be estimated. Accordingly, even a numeral must be an abstract entity and may be, for example, the corresponding number. However, we are still faced with the problem of describing the connection between numbers or numerals and the related inscriptions or tokens.

It is important to appreciate that the formal approach to Metamathematics sketched above has been used by the majority of logicians either consciously or unconsciously, for a long time. It is conspicuous in Gödel's famous paper on the incompleteness of Arithmetic and as we have said already, the present high tide in the development of Model Theory is based on the same approach when applied to Semantics. From the point of view of the formalist, this type of Syntax and Semantics of a given formal system T may be accommodated within a single system T' which is an extension of T. The procedure may be repeated so as to yield an extension T'' of T' within which are accommodated the Syntax and Semantics of T'. Further repetitions of this procedure are equally possible.

But although T and T' (and T'') are formal and, in general, uninterpreted theories it is undeniable that in certain important instances they include fragments which are indeed capable of interpretation, more particularly, empirical interpretation. For example, suppose that T is the axiomatic Set Theory of Zermelo and Fraenkel. According to our formalist point of view this theory is uninterpreted. But it is nevertheless the case that the rules for the addition and multiplication of finite cardinals, which form part of that theory, are applicable to concrete, more particularly, empirical situations. Also, if T is Peano's Arithmetic and X is the (formal) theorem of T which states that $x^3 + y^3 = z^3$ has no solution with $xyz \neq 0$ we conclude that no such integers will be found in practice (or else that Peano's Arithmetic is inconsistent). Or again, at the metamathematical level, if a sentence X has been proved undecidable (unprovable and irrefutable) within a certain theory T—this being a result of the formal theory T' as above—then we are confident that indeed we shall never *find* a proof or either X or not-X within T. Thus, both at the mathematical and at the metamathematical level we are inclined to believe that our formal theories have concrete implications. In Hilbert's scheme, the extraction of these concrete implications from a formal theory amounts to the separation of the material (*inhaltlich*) elements of the theory from its ideal elements. However, the identification and interpretation of these material elements may be quite difficult. Suppose for example that X has been proved, within T', to be unprovable within T, *unless T is inconsistent*. Then we are inclined to conclude that we shall either never find a proof for X or if we do find a proof for X then we may also find a proof for not-X. But so long as we have not circumscribed the possibility of finding a proof for not-X more closely this *possibility* of finding a proof for not-X need not in itself be clearly defined in constructive terms so that the material or concrete applicability of our metamathematical result remains in doubt.

The idea of supplementing abstract mathematical or metamathematical results with concrete procedures and algorithms is, of course, quite common. It has both practical and theoretical aspects, as in the development of numerical methods on one hand and in the theory of recursive procedures on the other. But the precise limits of such an activity for a given formal theory and its significance in philosophical terms remain matters for further discussion. With a slight inflexion the idea fits well into the scheme of the logical positivists, to whom a formal theory is *adequate* if it can be linked with the empirical world by means of rules of correspondence, and who dispense with the material Metamathematics discussed previously. By contrast it appears to me that the material (*inhaltlich*) aspects of Logic and Mathematics appear both in the basic Metamathematics discussed in section 7 and in the concrete elements which can be extracted from a formal theory as indicated just now.

10. Thus, it remains a fact that, in ways which we do not fully understand, theorems which to the formalist come under the heading of uninterpreted Mathematics or Metamathematics yield results which can be used in material thought and for empirical purposes. However, since we are under no illusions concerning the concrete meaning of the infinitary assumptions included in such formal theories, our philosophy directs us to investigate whether or not such an assumption should be retained universally. In particular, I will mention here the assumption that there exists a *standard* or *intended model* of Arithmetic or (alternatively, but relatedly) of Set Theory. Clearly, to the formalist, the entire notion of standardness must be meaningless, in accordance with our first basic principle. Accordingly, it is perfectly reasonable to posit that the system of Arithmetic for a mathematical theory T and for the syntax of T are not isomorphic, that is to say, to drop the categoricity of Arithmetic when applied to a mathematical theory and to its syntax simultaneously. On the contrary, it is interesting and fruitful to consider languages in which, for example, the length of a sentence may be n, where n is a natural number in the arithmetic of the syntax whic may be infinite from the point of view of the arithmetic incorporated in the original theory [14].

10. Earlier I stated my view that there are forms of thought which are prior to any formal mathematical theory and I maintained that the logic which is applicable to a given concrete, hence finite, system includes all of the Lower Predicate Calculus. I was less definite in discussing the logic which applies to systems of unbounded extent, i.e., which are potentially infinite. I now wish to suggest that for these a form of Modal Logic may be appropriate. In the Appendix, I show how the semantics of an infinite structure can be defined by means of a concept of potential truth for a set of finite structures. In particular, truth in Arithmetic can thus be defined in terms of potential truth in initial segments of natural numbers. The argument of the Appendix is infinitary but, in keeping with the remarks of section 8 above, it may point the way to a corresponding approach which is syntactical and finitary. In any case, it is still the abiding task of the Philosophy of Mathematics to gain a deeper understanding of (potential) infinity.

REFERENCES

[1] P. Benacerraf and H. Putnam (eds.), *Philosophy of Mathematics*, Selected Readings, Prentice-Hall, 1964.

[2] P. Cohen, The independence of the continuum hypothesis, *Proceedings of the National Academy of Sciences*, vol. 50, 1963, pp. 1143–1148; vol. 51, 1964, pp. 105–110.

[3] H. B. Curry, *Outline of a Formalist Philosophy of Mathematics*, Studies in Logic and the Foundations of Mathematics, North-Holland Publishing Company, 1951.

[4] A. A. Fraenkel and Y. Bar-Hillel, *Foundations of Set Theory*, Studies in Logic and the Foundations of Mathematics, North-Holland Publishing Company, 1958.

[5] K. Gödel, Über formal unentscheidbare Sätze der Principia Mathematica und verwandter Systeme, *Monatshefte für Mathematik und Physik*, vol. 38, 1931, pp. 173–198.

[6] ———, *The Consistency of the Axiom of Choice and of the Generalized Continuum-Hypothesis with* the *Axioms of Set Theory*, Annals of Mathematics Studies No. 3, Princeton 1940.

[7] ———, What is Cantor's continuum problem? *American Mathematical Monthly*, vol. 54, 1947, pp. 515–525. Revised version in [1], pp. 258–273.

[8] A. HEYTING, *Intuitionism, an Introduction*, Studies in Logic and the Foundations of Mathematics, North-Holland Publishing Company, 1956.

[9] D. HILBERT, *Gesammelte Abhandlungen*, vol. 3, Julius Springer, 1935.

[10] ———, Über das Unendliche, *Mathematische Annalen*, vol. 95, 1925, pp. 161–190, Partial translation in [1], pp. 134–151.

[11] G. KREISEL, Hilbert's programme, in [1], pp. 157–180.

[12] P. LORENZEN, *Einführung in die operative Logik und Mathematik*, Grundlehren der mathematischen Wissenschaften, vol. 78, Springer-Verlag, 1955.

[13] A. ROBINSON, *Introduction to Model Theory and to the Metamathematics of Algebra*, Studies in Logic and the Foundations of Mathematics, North-Holland Publishing Company, 1963.

[14] ———, On languages which are based on non-standard arithmetic, *Nagoya Mathematical Journal*, vol. 22, 1963, pp. 83–117.

APPENDIX

A NOTION OF POTENTIAL TRUTH

1. We shall consider sets or *structures* (or, relational systems) and we shall use the ordinary symbol of inclusion, $A \subset B$, in order to indicate that the structure B is an *extension* of the structure A or, which is the same, that A is a substructure of B (see [13], p. 24).

Let Δ be a set of structures which is *directed* under the relation $\subset$. That is to say, if A and B are elements of Δ, then Δ contains a structure C such that $A \subset C$ and $B \subset C$. It is implicit in this definition that the same relations are defined in all structures of Δ. The *union* of Δ is a structure M which is defined in the following way. The set of individuals of M is the set-theoretic union of the sets of individuals of the elements of Δ. Now let $R(x_1, \ldots, x_n)$ be an n-ary relation which is defined for the elements of Δ, and let $a_1, a_2, \ldots, a_n$ be individuals of M. Then there exist elements $A_1, \ldots, A_n$ of Δ such that a_i belongs to A_i, $i = 1, \ldots, n$. Moreover, since Δ is directed under the relation $\subset$ there exists an element A of Δ such that $A_i \subset A$, $i = 1, \ldots, n$, and hence, such that a_i belongs to A, $i = 1, \ldots, n$. We now claim that if A' and A'' are any elements of Δ such that a_i belongs to A and to A'' for $i = 1, \ldots, n$, then either $R(a_1, \ldots, a_n)$ holds in both A' and A'' or $R(a_1, \ldots, a_n)$ holds neither in A' nor in A''. Indeed, since Δ is directed under the relation $\subset$ there exists a structure A'''' in Δ such that $A' \subset A''$ and $A'' \subset A'''$. If $R(a_1, \cdots, a_n)$ holds in A'''' then it must hold in both A' and A''; if $R(a_1, \ldots, a_n)$ does not hold in A''' then $R(a_1, \ldots, a_n)$ holds neither in A' nor in A''. This proves our assertion.

We *define* that $R(a_1, \ldots, a_n)$ holds in M if $R(a_1, \ldots, a_n)$ holds in *some* structure A of Δ which contains $a_1, \ldots, a_n$, and hence in *all* structures of this kind. If there is no such structure A then we define that $R(a_1, \ldots, a_n)$ does not hold in M. This determines the structure M completely.

A sentence X is said to be *defined* in a structure A if the relations and individual constants of X (denote entities that) are contained in A.

2. For given Δ, and for any element A of Δ and any sentence X which is defined in Δ, we now introduce the notion of the *potential truth of X in A* as follows.

If X is an atomic formula, $X = R(a_1, \ldots, a_n)$, where a_i belongs to A_i, $i = 1, \ldots, n$, then X is *potentially true* in A if and only if X is true in A in the usual sense.

If $X = \sim Y$ then X is potentially true in A if and only if Y is not potentially true in A; if $X = Y \vee Z$ then X is potentially true in A if and only if at least one of the sentences Y and Z is potentially true in A; and so on, as by the use of truth tables, for the remaining connectives.

If X is obtained by existential quantification, $X = (\exists y)\, Z(y)$, then X is potentially true in A if and only if there exists an element B of Δ such that $A \subset B$ and such that for some individual b in B, $Z(b)$ is potentially true in all elements of Δ which are extensions of B. If X is obtained by universal quantification, $X = (\forall y)\, Z(y)$, then X is potentially true in A. if and only if $(\exists y)\, [\sim Z(y)]$ is not potentially true in A.

These rules determine the notion of potential truth uniquely for all X and A as described.

THEOREM. *Let A and A' be elements of Δ such that A' is an extension of A and let the sentence X be defined in A and hence in A'. Then X is potentially true in A' if and only if it is potentially true in A.*

PROOF. The assertion is evidently correct if X is atomic, for in this case potential truth coincides with truth. For all other X we shall prove our assertion by induction following the construction of X. If $X = \sim Y$, where the assertion of the theorem has been proved already for Y, and X is potentially true in A, then Y is not potentially true in A. Hence, Y is not potentially true in A' and X is potentially true in A', as required. Similarly, if X is not potentially true in A, Y is potentially true in A and A', hence X is not potentially true in A', as required. If $X = Y \vee Z$, where the assertion of the theorem has been proved already for Y and Z, and X is potentially true in A, then one of Y and Z, e.g., Y, is potentially true in A. It follows that Y is potentially true in A' and hence that X is potentially true in A', as required. If X is not potentially true in A then Y and Z are not potentially true in

A and, hence, are not potentially true in A'. It follows that X is not potentially true in A'. A similar procedure applies to the remaining connectives.

Now suppose that $X = (\exists y)\, Z(y)$, where the assertion of the theorem has been proved for all sentences $Z(a)$. Suppose also that X is potentially true in A. Then for some B in Δ which is an extension of A and for some individual b in B, $Z(b)$ is potentially true in all extensions of B which belong to Δ. But Δ is a directed set, so there exists an element B' of Δ such that $A' \subset B'$ and $B \subset B'$. Then b is contained in B' and $Z(b)$ is potentially true in all extensions of B' which belong to Δ. This shows that X is potentially true in A'.

Suppose on the other hand that X is defined in A and is potentially true in A'. Then there exists a $B' \in \Delta$, $A' \subset B'$ such that, for some individual b' in B', $Z(b')$ is potentially true in all extensions of B' which belong to Δ. But B' is also an extension of A, so X is potentially true in A, as required.

Suppose finally that $X = (\forall y)\, Z(y)$ where the assertion of the theorem has been proved for all sentences $Z(a)$. Then, as shown previously, the assertion of the theorem is true also for all sentences $\sim Z(a)$. It then follows from the preceding argument that $(\exists y)\,[\sim Z(y)]$ is potentially true in A if and only if it is potentially true in A', i.e., that $(\forall y)\, Z(y)$ is *not* potentially true in A if and only if it is not potentially true in A'. This completes the proof of the theorem.

COROLLARY. *A sentence X is either potentially true in all structures of Δ in which it is defined or it is not potentially true in any structure of Δ.*

3. Let M be the union of Δ as defined in section 1 of the appendix. The following result provides the link between the ordinary truth definition in M and the above potential truth definition with respect to Δ.

THEOREM. *Let X be any sentence which is defined in M. Then X is true in M if and only if it is potentially true in all elements of Δ in which it is defined.*

PROOF. The assertion of the theorem is evidently correct if X is atomic. For other X the proof proceeds again by induction following the construction of the formula. Omitting the case of connectives, for which the argument is trivial, we suppose that $X = (\exists y)\, Z(y)$ is a sentence which is defined in M and such that the assertion of the theorem has been proved already for all $Z(b)$ where b is (or denotes) an individual in M.

If X is true in M then $Z(b)$ is true for some b in M. Let A be any element of Δ in which X is defined. Since Δ is directed it includes an element B which is an extension of A and contains b. Then $Z(b)$ is potentially true in all extensions of B which belong to Δ (since the theorem is supposed to have been proved already for each $Z(b)$) and so $(\exists y)\, Z(y)$ is potentially true in A.

Conversely, suppose that X is potentially true in some element A of Δ. (Thus, by the result of the preceding section, X is potentially true in all elements of Δ in which it is defined.) Then $Z(b)$ is potentially true in some extension B of A which belongs to Δ. Hence, $Z(b)$ is true in M and X is true in M, as required.

The case where X is obtained by universal quantification can be reduced immediately to that considered just now. This completes the proof of the theorem.

The conditions imposed on Δ are satisfied in the following two examples.

(i) Let N be the set of natural numbers with the relations of addition, multiplication, and equality, and let Δ be the set of structures which are obtained by restricting N to the sets $N_k = \{0, 1, \ldots, \mathrm{k}\}$, $k = 0, 1, 2, \ldots$. Then $N = M$ is the union of Δ.

(ii) Let M be any structure and let Δ be the set of structures which are obtained by restricting M in all possible ways to its *finite* subsets. Then M is the union of Δ.

In both cases, we arrive at the usual truth definition for M by means of a notion of potential truth for *finite* substructures of M. It will be seen that the most important part of our notion of potential truth concerns existential quantification. It corresponds to the intuitive idea that an existential sentence is potentially true in a given structure if we can find an element that satisfies it by extending the structure far enough in the right direction.

Section IV

Methodology and Philosophy of Science

THE EXPLANATORY FUNCTION OF METAPHOR

MARY HESSE
University of Cambridge, England

The thesis of this paper is that the deductive model of scientific explanation should be modified and supplemented by a view of theoretical explanation as metaphoric redescription of the domain of the explanandum. This raises two large preliminary questions: first, whether the deductive model requires modification, and second, what is the view of metaphor presupposed by the suggested alternative. I shall not discuss the first question explicitly. Much recent literature in the philosophy of science (for example [4, 5, 10, 14]) has answered it affirmatively, and I shall refer briefly at the end to some difficulties tending to show that a new model of explanation is required, and suggest how the conception of theories as metaphors meets these difficulties.

The second question, about the view of metaphor presupposed, requires more extensive discussion. The view I shall present is essentially due to Max Black, who has developed in two papers, entitled respectively 'Metaphor' and 'Models and Archetypes' [3], both a new theory of metaphor, and a parallelism between the use of literary metaphor and the use of models in theoretical science. I shall start with an exposition of Black's *interaction view* of metaphors and models, taking account of modifications suggested by some of the subsequent literature on metaphor [1, 2, 11, 13, 15]. It is still unfortunately necessary to argue that metaphor is more than a decorative literary device, and that it has cognitive implications whose nature is a proper subject of philosophic discussion. But space forces me to mention these arguments as footnotes to Black's view, rather than as an explicit defence *ab initio* of the philosophic importance of metaphor.

The interaction view of metaphor.

1. We start with two systems, situations, or referents, which will be called respectively the primary and secondary systems. Each is described in literal language. A metaphoric use of language in describing the primary system consists of transferring to it a word or words normally used in connection with the secondary system: for example, 'Man is a wolf', 'Hell is a lake of ice'. In a scientific theory the primary system is the domain of the explanan-

dum, describable in observation language; and the secondary is the system, described either in observation language or the language of a familiar theory, from which the model is taken: for example, 'sound (primary system) is propagated by wave motion (taken from a secondary system)'; 'gases are collections of randomly moving massive particles'.

Three terminological remarks should be inserted here. First, 'primary' and 'secondary system', and 'domain of the explanandum' will be used throughout to denote the referents or putative referents of descriptive statements; and 'metaphor', 'model', 'theory', 'explanans' and 'explanandum' will be used to denote linguistic entities. Second, use of the terms 'metaphoric' and 'literal', 'theory' and 'observation', need not be taken at this stage to imply a pair of irreducible dichotomies. All that is intended is that the 'literal' and 'observation' languages are assumed initially to be well understood and unproblematic, whereas the 'metaphoric' and 'theoretical' are in need of analysis. The third remark is that to assume initially that the two systems are 'described' in literal or observation language does not imply that they are exhaustively or accurately described or even that they could in principle be so in terms of these languages.

2. We assume that the primary and secondary system each carry a set of associated ideas and beliefs that come to mind when the system is referred to. These are not private to individual language-users, but are largely common to a given language community and are presupposed by speakers who intend to be understood in that community. In literary contexts the associations may be loosely knit and variable, as in the wolf-like characteristics which come to mind when the metaphor 'Man is a wolf' is used; in scientific contexts the primary and secondary systems may both be highly organized by networks of natural laws.

A remark must be added here about the use of the word 'meaning'. Writers on metaphor appear to intend it as an inclusive term for reference, use, and the relevant set of associated ideas. It is, indeed, part of their thesis that it has to be understood thus widely. To understand the meaning of a descriptive expression is not only to be able to recognize its referent, or even to use the words in the expression correctly, but also to call to mind the ideas, both linguistic and empirical, which are commonly held to be associated with the referent in the given language community. Thus a shift of meaning may result from a change in the set of associated ideas, as well as in change of reference or use.

3. For a conjunction of terms drawn from the primary and secondary systems to constitute a metaphor it is necessary that there should be patent

falsehood or even absurdity in taking the conjunction literally. Man is not, literally, a wolf; gases are not in the usual sense collections of massive particles. In consequence some writers have denied that the referent of the metaphoric expression can be identified with the primary system without falling into absurdity or contradiction. I shall return to this in the next section.

4. There is initially some principle of assimilation between primary and secondary systems, variously described in the literature as 'analogy', 'intimations of similarity', 'a programme for exploration', 'a framework through which the primary is seen.' Here we have to guard against two opposite interpretations, both of which are inadequate for the general understanding of metaphors and scientific models. On the one hand, to describe this ground of assimilation as a *programme* for exploration, or a *framework* through which the primary is seen, is to suggest that the secondary system can be imposed *a priori* upon the primary, as if *any* secondary can be the source of metaphors or models for *any* primary, provided the right metaphor-creating operations are subsequently carried out. Black does indeed suggest that in some cases 'it would be more illuminating... to say that the metaphor creates the similarity than to say it formulates some similarity antecedently existing' (p. 37), and he also points out that some poetry creates new metaphors precisely by itself developing the system of associations in terms of which 'absurd' conjunctions of words are to be metaphorically understood. There is however an important distinction to be brought out between such a use of metaphor and scientific models, for, whatever may be the case for poetic use, the suggestion that *any* scientific model can be imposed *a priori* on *any* explanandum and function fruitfully in its explanation must be resisted. Such a view would imply that theoretical models are irrefutable. That this is not the case is sufficiently illustrated by the history of the concept of a heat fluid, or the classical wave theory of light. Such examples also indicate that no model even gets off the ground unless some antecedent similarity or analogy is discerned between it and the explanandum.

But here there is a danger of falling into what Black calls the *comparison* view of metaphor. According to this view the metaphor can be replaced without remainder by an explicit, literal statement of the similarities between primary and secondary systems, in other words, by a simile. Thus, the metaphor 'Man is a wolf' would be equivalent to 'Man is like a wolf in that...', where follows a list of comparable characteristics; or, in the case of theoretical models, the language derived from the secondary system would be wholly replaced by an explicit statement of the analogy between secondary and primary systems, after which further reference to the secondary system would be dispensible. Any interesting examples of the model-using in science will show, however, that the situation cannot be described in this way. For one thing, as long as the model is under active consideration as an ingredient in an explanation, we do not know how far the comparison extends—it is precisely in its extension that the fruitfulness of the model may lie. And a more fundamental objection to the comparison view emerges in considering the next point.

5. The metaphor works by transferring the associated ideas and implications of the secondary to the primary system. These select, emphasize, or suppress features of the primary; new slants on the primary are illuminated; the primary is 'seen through' the frame of the secondary. In accordance with the doctrine that even literal expressions are understood partly in terms of the set of associated ideas carried by the system they describe, it follows that the associated ideas of the primary are changed to some extent by the use of the metaphor, and that therefore even its original literal description is shifted in meaning. The same applies to the secondary system, for its associations come to be affected by assimilation to the primary; the two systems are seen as more like each other; they seem to interact and adapt to one another, even to the point of invalidating their original literal descriptions if these are understood in the new, post-metaphoric sense. Men are seen to be more like wolves after the wolf-metaphor is used, and wolves seem to be more human. Nature becomes more like a machine in the mechanical philosophy, and actual, concrete machines themselves are seen as if stripped down to their essential qualities of mass in motion.

This point is the kernel of the interaction view, and is Black's major contribution to the analysis of metaphor. It is incompatible with the comparison view, which assumes that the literal descriptions of both systems are and remain independent of the use of the metaphor, and that the metaphor is reducible to them. The consequences of the interaction view for theoretical models are also incompatible with assumptions generally made in the deductive account of explanation, namely that descriptions and descriptive laws in the domain of the explanandum remain empirically acceptable and invariant in meaning to all changes of explanatory theory. I shall return to this point.

6. It should be added as a final point in this preliminary analysis that a metaphoric expression used for the first time, or used to someone who hears it for the first time, is intended to be *understood*. Indeed it may be said that

a metaphor is not metaphor but nonsense if it communicates nothing, and that a genuine metaphor is also capable of communicating something other than was intended and hence of being *mis*understood. If I say (taking two words more or less at random from a dictionary page) 'A truck is a trumpet' it is unlikely that I shall communicate anything; if I say 'He is a shadow on the weary land', you may understand me to mean (roughly) 'He is a wet blanket, a gloom, a menace', whereas I actually meant (again roughly) 'He is a shade from the heat, a comfort, a protection'.

Acceptance of the view that metaphors are meant to be intelligible implies rejection of all views which make metaphor a wholly non-cognitive, subjective, emotive, or stylistic use of language. There are exactly parallel views of scientific models which have been held by many contemporary philosophers of science, namely that models are purely subjective, psychological, and adopted by individuals for private heuristic purposes. But this is wholly to misdescribe their function in science. Models, like metaphors, are intended to communicate. If some theorist develops a theory in terms of a model, he does not regard it as a private language, but presents it as an ingredient of his theory. Neither can he, nor need he, make literally explicit all the associations of the model he is exploiting; other workers in the field 'catch on' to its intended implications, indeed they sometimes find the theory unsatisfactory just because some implications which the model's originator did not investigate, or even think of, turn out to be empirically false. None of this would be possible unless use of the model were inter-subjective, part of the commonly understood theoretical language of science, not a private language of the individual theorist.

An important general consequence of the interaction view is that it is not possible to make a distinction between literal and metaphoric descriptions merely by asserting that literal use consists in the following of linguistic rules. Intelligible metaphor also implies the existence of rules of metaphoric use, and since in the interaction view literal meanings are shifted by their association with metaphors, it follows that the rules of literal usage and of metaphor, though they are not identical, are nevertheless not independent. It is not sufficiently clear in Black's paper that the interaction view commits one to the abandonment of a two-tiered account of language in which some usages are irreducibly literal and others metaphoric. The interaction view sees language as dynamic: an expression initially metaphoric may become literal (a 'dead' metaphor), and what is at one time literal may become metaphoric (for example the Homeric 'he breathed forth his life', originally literal, is now a metaphor for death). What is important is not to try to draw a line between the metaphoric and the literal, but rather to trace out the various mechanisms of meaning-shift and their interactions. The interaction view cannot consistently be made to rest on an initial set of absolutely literal descriptions, but rather on a relative distinction of literal and metaphoric in paticular contexts. I cannot undertake the task of elucidating these conceptions here (an interesting attempt to do so has been made by K. I. B. S. Needham [12]), but I shall later point out a parallel between this general linguistic situation and the relative distinctions and mutual interactions of theory and observation in science.

The problem of metaphoric reference.

One of the main problems for the interaction view in its application to theoretical explanation is the question what is the *referent* of a model or metaphor. At first sight the referent seems to be the primary system, which we choose to describe in metaphoric rather than literal terms. This, I believe, is in the end the right answer, but the process of metaphoric description is such as to cast doubt on any simple identification of the metaphor's reference with the primary system. It is claimed in the interaction view that a metaphor causes us to 'see' the primary system differently, and causes the meanings of terms originally literal in the primary system to shift towards the metaphor. Thus 'Man is a wolf' makes man seem more vulpine, 'Hell is a lake of ice' makes hell seem icy rather than hot, and a wave theory of sound makes sound seem more vibrant. But how can initial similarities between the objective systems justify such changes in the meanings of words and even, apparently, in the things themselves? Man does not in fact change because someone uses the wolf-metaphor. How then can we be justified in identifying what we see through the framework of the metaphor with the primary system itself? It seems that we cannot be entitled to say men *are* wolves, sound *is* wave motion, in any identificatory sense of the copula.

Some recent writers on metaphor [2, 11, 15] have made it the main burden of their argument to deny that any such identification is possible. They argue that if we allow it we are falling into the absurdity of conjoining two literally incompatible systems, and the resulting expression is not metaphoric but meaningless. By thus taking a metaphor literally we turn it into a *myth*. An initial misunderstanding may be removed at once by remarking that 'identification' cannot mean in this context identification of the referent of the metaphoric expression, taken in its *literal* sense, with the primary system. But if the foregoing analysis of metaphor is accepted, then it follows that metaphoric use is use in a different from the literal sense, and furthermore it is use in a sense not replaceable by any literal expression. There remains the question what it is to identify the referent of the metaphoric expression or model with the primary system. As a preliminary to answering this question it is important to point out that there are two ways, which are often confused in accounts of the 'meaning of theoretical concepts', in which such identification may fail. It may fail because it is in principle meaningless to make any such identification, or it may fail because in a particular case the identification happens to be *false*. Instances of false identification, e.g. 'heat is a fluid' or 'the substance emitted by a burning object is phlogiston', provide no arguments to show that other such identifications may not be both meaningful and true.

Two sorts of argument have been brought against the view that metaphoric expressions and models can refer to and truly describe the primary system. The first depends on an assimilation of poetic and scientific metaphor, and points out that it is characteristic of good poetic metaphor that the images introduced are initially striking and unexpected, if not shocking; that they are meant to be entertained and savoured for the moment and not analysed in pedantic detail nor stretched to radically new situations; and that they may immediately give place to other metaphors referring to the same subject matter which are formally contradictory, and in which the contradictions are an essential part of the total metaphoric impact. Any attempt to separate these literal contradictions from the nexus of interactions is destructive of the metaphor, particularly on the interaction view. In the light of these characteristics there is indeed a difficult problem about the correct analysis of the notion of metaphoric 'truth' in poetic contexts. Scientific models, however, are fortunately not so intractable. They do not share any of the characteristics listed above which make poetic metaphors peculiarly subject to formal contradictoriness. They may initially be unexpected, but it is not their chief aim to shock; they are meant to be exploited energetically and often in extreme quantitative detail and in quite novel observational domains; they are meant to be internally tightly knit by logical and causal interrelations; and if two models of the same primary system are found to be mutually inconsistent, this is not taken (*pace* the complementarity interpretation of quantum physics) to enhance their effectiveness, but rather as a challenge to reconcile them by mutual modification or to refute one of them. Thus their truth criteria, although not rigorously formalizable, are at least much clearer than in the case of poetic metaphor. We can perhaps signalize the difference by speaking in the case of scientific models of the (perhaps unattainable) aim to find a 'perfect metaphor', whose referent is the domain of the explanandum, whereas literary metaphors, however adequate and successful in their own terms, are from the point of view of potential logical consistency and extendability often (not always) intentionally imperfect.

Secondly, if the interaction view of scientific metaphor or model is combined with the claim that the referent of the metaphor is the primary system (i.e. the metaphor is true of the primary system), then it follows that the thesis of meaning-invariance of the literal observation-descriptions of the primary system is false. For, the interaction view implies that the meaning of the original literal language of the primary system is changed by adoption of the metaphor. Hence those who wish to adhere to meaning-invariance in the deductive account of explanation will be forced to reject either the interaction view or the realistic view that a scientific model is putatively true of its primary system. Generally they reject both. But abandonment of meaning-invariance, as in many recent criticisms of the deductive model of explanation, leaves room for adoption of both the interaction view, and realism, as I shall now try to spell out in more detail.

Explanation as metaphoric redescription.

The initial contention of this paper was that the deductive model of explanation should be *modified* and *supplemented* by a view of theoretical explanation as metaphoric redescription of the domain of the explanandum. First, the association of the ideas of 'metaphor' and of 'explanation' requires more examination. It is certainly not the case that all explanations are metaphoric. To take only two examples, explanation by covering-law, where an instance of an *A* which is *B* is explained by reference to the law 'All *A*'s are *B*'s', is not metaphoric, neither is the explanation of the working of a mechanical gadget by reference to an actual mechanism of cogs, pulleys, and levers. These, however, are not examples of *theoretical* explanation, for it has been taken for granted that the essence of a theoretical explanation is the introduction into the explanans of a new vocabulary or even of a new language. But introduction of a metaphoric terminology is not in itself explanatory, for in literary metaphor in general there is no hint that what is metaphorically described is also thereby explained. The connection between metaphor and explanation is therefore neither that of necessary nor sufficient condition. Metaphor becomes explanatory only when it satisfies certain further conditions.

The orthodox deductive criteria for a scientific explanans (for example in [6]) require that the explanandum be deducible from it, that it contain at least one general law not redundant to the deduction, that it be not empirically falsified up to date, and that it be predictive. We cannot simply graft these requirements on to the account of theories as metaphors without investigating the consequences of the interaction view of metaphor for the notions of 'deducibility', 'explanandum', and 'falsification' in the orthodox account. In any case, as has been mentioned already, the requirement of deducibility in particular has been subjected to damaging attack, quite apart from any metaphoric interpretation of theories. There are two chief grounds for this attack, both of which can be turned into arguments favourable to the metaphoric view.

In the first place it is pointed out that there is seldom in fact a deductive relation strictly speaking between scientific explanans and explanandum, but only relations of approximate fit. Furthermore, what counts as sufficiently approximate fit cannot be decided deductively, but is a complicated function of coherence with the rest of a theoretical system, general empirical acceptability throughout the domain of the explanandum, and many other factors. I do not propose to try to spell out these relationships in further detail here, but merely to make two points which are relevant to my immediate concern. First, the attack on deducibility drawn from the occurrence of approximations does not imply that there are *no* deductive relations between explanans and explanandum. The situation is rather this. Given a descriptive statement D in the domain of the explanandum, it is usually the case that the statement E of an acceptable explanans does not entail D, but rather D', where D' is a statement in the domain of the explanandum only 'approximately equivalent' to D. For E to be acceptable it is necessary both that there be a deductive relation between E and D', and that D' should come to be recognized as a *more acceptable* description in the domain of the explanandum than D. The reasons why it might be more acceptable—repetition of the experiments with greater accuracy, greater coherence with other acceptable laws, recognition of disturbing factors in arriving at D in the first place, metaphoric shifts in the meanings of terms in D consequent upon the introduction of the new terminology of E, and so on—need not concern us here. What is relevant is that the non-deducibility of D from E does not imply total abandonment of the deductive model unless D is regarded as an invariant description of the explanandum, automatically rendering D' empirically false. That D cannot be so regarded has been amply demonstrated in the literature. The second point of contact between these considerations and the view of theories as metaphors is now obvious. That explanation may modify and correct the explanandum is already built into the relation between metaphors and the primary system in the interaction view. Metaphors, if they are good ones, and *ipso facto* their deductive consequences, do have the primary system as their referents, for they may be seen as correcting and replacing the original literal descriptions of the same system, so that the literal descriptions are discarded as inadequate or even false. The parallel with the deductive relations of explanans and explananda is clear: the metaphoric view does not abandon deduction, but it focusses attention rather on the interaction between metaphor and primary system, and on the criteria of acceptability of metaphoric descriptions of the primary system, and hence not so much upon the deductive relations which appear in this account as comparatively uninteresting pieces of logical machinery.

The second attack upon the orthodox deductive account gives even stronger and more immediate grounds for the introduction of the metaphoric view. It is objected that there are no deductive relations between theoretical explanans and explanandum because of the intervention of correspondence rules. If the deductive account is developed, as it usually is, in terms either of an uninterpreted calculus and an observation language, or of two distinct languages, the theoretical and the observational, it follows that the correspondence rules linking terms in these languages cannot be derived deductively from the explanans alone. Well-known problems then arise about the status of the correspondence rules and about the meaning of the predicates of the theoretical language. In the metaphoric view, however, these problems are evaded, because here there are no correspondence rules, and this view is primarily designed to give its own account of the meaning of the language of the explanans. There is *one* language, the observation language, which like all natural languages is continually being extended by metaphoric uses, and hence yields the terminology of the explanans. There is no problem about connecting explanans and explanandum other than the general problem of understanding how metaphors are introduced and applied and exploited in their primary systems. Admittedly, we are as yet far from understanding this process, but to see the problem of the 'meaning of theoretical concepts' as a special case is one step in its solution.

Finally, a word about the requirement that an explanation be predictive. It has been much debated within the orthodox deductive view whether this is a necessary and sufficient condition for explanation, and it is not appropriate here to enter into that debate. But any account of explanation would be inadequate which did not recognize that, in general, an explanation is required to be predictive, or, what is closely connected with this, to be falsifiable. Elsewhere [8] I have pointed out that, in terms of the deductive view, the requirement of predictivity may mean one of three things:

(i) That general laws already present in the explanans have as yet unobserved instances. This is a trivial fulfilment of the requirement, and would not, I think, generally be regarded as sufficient.

(ii) That further general laws can be derived from the explanans, *without* adding further items to the set of correspondence rules. That is to say, predictions remain within the domain of the set of predicates already present in the explanandum. This is a weak sense of predictivity which covers what would normally be called *applications* rather than extensions of a theory (for example, calculation of the orbit of a satellite from the theory of gravitation, but not extension of the theory to predict the bending of light rays).

(iii) There is also a strong sense of prediction in which new observation predicates are involved, and hence, in terms of the deductive view, additions are required to the set of correspondence rules. I have argued [7, 8] that there is no rational method of adding to the correspondence rules on the pure deductive view, and hence that cases of strong prediction cannot be rationally accounted for on that view. In the metaphoric view, on the other hand, since the domain of the explanandum is redescribed in terminology transferred from the secondary system, it is to be expected that the original observation language will both be shifted in meaning and extended in vocabulary, and hence that predictions in the strong sense will become possible. They may not of course turn out to be *true*, but that is an occupational hazard of any explanation or prediction. They will however be rational, because rationality consists just in the continuous adaptation of our language to our continually expanding world, and metaphor is one of the chief means by which this is accomplished.

REFERENCES

[1] M. C. Beardsley, *Aesthetics*, New York, 1958.
[2] D. Berggren, The Use and Abuse of Metaphor, *Rev. Met.* xvi, 1962, 237 and 450.
[3] M. Black, *Models and Metaphors*, Ithaca, 1962.
[4] P. K. Feyerabend, An Attempt at a Realistic Interpretation of Experience, *Proc. Aris. Soc.* lviii, 1957, 143.
[5] P. K. Feyerabend, Explanation, Reduction and Empiricism, *Minnesota Studies* III, ed. H. Feigl and G. Maxwell, 1962.
[6] C. G. Hempel and P. Oppenheim, The Logic of Explanation, reprinted in *Readings in the Philosophy of Science*, ed. H. Feigl and M. Brodbeck, New York, 1953, 319.
[7] Mary Hesse, Theories, Dictionaries and Observation, *Brit. Journ. Phil. Sci. ix*, 1958, 12 and 128.
[8] Mary Hesse, *Models and Analogies in Science*, Sheed and Ward, London, 1963.
[9] Mary Hesse, A New Look at Scientific Explanation, *Rev. Met.* xvii, 1963, 98.
[10] T. S. Kuhn, *The Structure of Scientific Revolutions*, Chicago, 1962.
[11] Mary A. McCloskey, Metaphors, *Mind*, lxxiii, 1964, 215.
[12] K. I. B. S. Needham, Synonymy and Semantic Classification, Cambridge Ph. D. Thesis, 1964.
[13] D. Schon, *The Displacement of Concepts*, London, 1963.
[14] W. Sellars, The Language of Theories, *Current Issues in the Philosophy of Science*, ed. H. Feigl and G. Maxwell, New York, 1961.
[15] C. Turbayne, *The Myth of Metaphor*, New Haven, 1962.

Section V

Foundations of Probability and Induction

WHY IS IT REASONABLE TO BASE A BETTING RATE UPON AN ESTIMATE OF CHANCE?

R. B. BRAITHWAITE
University of Cambridge, Cambridge, England

§ **1. Introduction.** This paper is an attempt to connect two concepts of measurable probability. The first is that of the objective probability of a member of a class β (e.g. the class of radium atoms now existing) being also a member of a class α (e.g. the class of things existing in 1700 years time). A proposition stating that such a probability has a given numerical value, or a given range of numerical values, is a statistical hypothesis which is an empirical proposition to be explicated by considering the observable circumstances under which it would be rejected, these being given by knowledge of the proportions (frequencies) of α-specimens in observed subclasses (samples) of β. Objective probabilities will be called *chances* in this paper. The second concept is that of subjective (or personal) probability, which is the rate at which a particular man would be prepared to bet on a particular event's happening, i.e. on a particular proposition's being true. Subjective probabilities will be called (as by Carnap) *betting quotients*. No reference will be made in this paper to any concept of probability, whether measurable or not, regarded as a degree of confirmation or as a logical or syntactical relation.

For a philosopher who, like myself, accepts and uses in his daily life both the concepts of chance and of betting quotient the problem as to how they are related is inescapable. It presents itself in the clearest form in the question: Why is it reasonable for a man who believes that the chance of a β-specimen being an α-specimen is m/n to take m/n as his betting quotient when an occasion arises for his betting, for or against, a β-specimen being an α-specimen?

Two answers have been given to this question, neither of them satisfactory. The historic answer makes reference to possible repetitions of the bet. The Law of Great Numbers of probability theory shows, it is said, that, since a bettor's 'mathematical expectation' will be positive if he bets at a rate more favourable to him than the chance, and negative if he bets at a less favourable rate, a betting quotient equal to the chance is the rate at which in the long run he will break even. But the Law of Great Numbers only derives chances from chances: all that a correct application of it shows is that, in a long series of bets, the derived chance is very large that the bettor's average gain per bet (losses being treated as negative gains) will be near zero, if the original chance provides his betting quotient. So a similar question arises as to why it is reasonable to fix a betting quotient by reference to a chance, even if this is a derived chance estimated to be very large.

A second and recent answer to the question is that there is no satisfactory method for measuring values (utilities) except in terms of mathematical expectations; and, if this is done, a betting quotient becomes analytically identical with an estimated chance. But this, by itself, is too facile a solution. The Ramsey-von Neumann-Morgenstern explication of the measurement of values by means of chances (which in other contexts I should follow) requires that it should be reasonable to choose an action (and betting is an action) by making comparisons which involve chances; and this is one of the things which is in dispute. So in attempting to answer our question I shall simplemindedly take bets and lotteries to be in terms of money, and all the values concerned to be monetary values. If a satisfactory answer can be given to the question stated in these terms, the answer can be used to improve upon the monetary way of measuring value. But for us to be able to do this, our question must first be answered.

To avoid circularity, therefore, the notion of mathematical expectation must not be used in our answer. The following sections attempt such an answer, the formal substitute for mathematical expectations being linear functions corresponding to the 'one-way' utilities of R.J. Aumann, whose recent paper [1] has inspired this work.

§ 2. Mathematical apparatus: scores and their degrees. Let $E = [e_1, e_2, \cdots, e_n]$ be a class of n propositions (with $n \geqq 2$) which are such that one and only one of them can be true.

Define a *score* on E, written as e.g. $A_E = (a_1, a_2, \cdots, a_n)_E$, as an n-vector whose components are rational numbers, positive, negative or zero. If, in the ordinary vector way, $A_E + B_E$ is defined as $(A+B)_E = (a_1 + b_1, a_2 + b_2, \cdots, a_n + b_n)_E$ and $k.A_E$, where k is any rational number, is defined as $(kA)_E = (ka_1, ka_2, \cdots, ka_n)_E$, the system of scores on E is closed under addition and multiplication by a rational scalar.

Every score on E is a unique linear function of the n *unit scores* $(U_1)_E = (1, 0, \cdots, 0)_E$, $(U_2)_E = (0, 1, \cdots, 0)_E, \cdots, (U_n)_E = (0, 0, \cdots, 1)_E$, i.e., for every A_E, there is one and only one class of rational numbers $k_1, k_2, \cdots, k_n$ such that $A_E = k_1 \cdot (U_1)_E + k_2 \cdot (U_2)_E + \cdots k_n \cdot (U_n)_E$.

Define $\succcurlyeq$ as the dyadic relation between scores satisfying the condition: $A \succcurlyeq B$ if and only if $a_i \geqq b_i$, for $i = 1, 2, \cdots, n$. This relation yields a partial ordering of the scores on E, since the condition defining it implies that it is transitive, reflexive and antisymmetric.

To apply this system of scores to the problem with which this paper is concerned, we require not only to impose a complete ordering upon this partial ordering of the scores on E, but also to assign a unique rational number to each score A_E, to be called its *degree* and written $D(A_E)$. Satisfaction of the second requirement will enable the first to be satisfied, since the complete ordering of degrees will provide a complete ordering of equivalence classes of scores having the same degree.

In order that the complete ordering of these equivalence classes should be compatible with the basic partial ordering of the scores, it is necessary that the degree function D should satisfy the conditions

$$A_E \succcurlyeq B_E \text{ implies } D(A_E) \geqq D(B_E);$$
$$D(A_E + B_E) = D(A_E) + D(B_E);$$
$$D(k \cdot A_E) = kD(A_E), \text{ for every rational } k.$$

A consequence of this last condition is that $D(0, 0, \cdots, 0)_E = 0$.

Our degree function D corresponds, with unimportant differences, to Aumann's 'one-way' utility function; and it follows immediately from his work [1] that D must be a linear function of the components of the score of the form $D(A_E) = \lambda_1 a_1 + \lambda_2 a_2 + \cdots + \lambda_n a_n$, where $\lambda_1, \lambda_2, \cdots, \lambda_n$ are any positive rational numbers. If D is normalized in the way most convenient for use in this paper, it must be required to satisfy the extra condition $D(1, 1, \cdots, 1)_E = 1$, which implies that $\Sigma \lambda_i = 1$. In what follows a degree function D will be taken to be a degree function thus normalized.

Call an n-vector $\Lambda = (\lambda_1, \lambda_2, \cdots, \lambda_n)$ whose components are positive rationals with $\Sigma \lambda_i = 1$ an *Aumann vector*. Every Aumann vector will yield a distinct degree function D_Λ assigning the Λ-degree $D_\Lambda(A_E)$ to a score A_E. Thus there are an (enumerable) infinity of ways of assigning a degree to a score consistently with the four required conditions.

Each one of these, e.g. D_Λ, will impose a weak ordering upon scores on E and consequently a complete ordering upon equivalence classes of scores on E. Scores on E will be weakly ordered by the relation $\succcurlyeq_\Lambda$ between scores defined by: $A_E \succcurlyeq_\Lambda B_E$ if and only if $D_\Lambda(A_E) \geqq D_\Lambda(B_E)$; for this relation is transitive, reflexive and connected. Then if the Λ-equivalence class comprising A_E, to be written $[A_E]_\Lambda$, is defined as the class of all scores X_E which are such that both $X_E \succcurlyeq_\Lambda A_E$ and $A_E \succcurlyeq_\Lambda X_E$ (which, by the definition of $\succcurlyeq_\Lambda$, is equivalent to $D_\Lambda(X_E) = D_\Lambda(A_E)$), the relation $\geqq_\Lambda$ between Λ-equivalence classes, defined by $[A_E]_\Lambda \geqq_\Lambda [B_E]_\Lambda$ if and only if $A_E \succcurlyeq_\Lambda B_E$, is a complete ordering, since it is transitive, reflexive, antisymmetric and connected. This complete ordering of Λ-equivalence classes of scores, as well as the weak ordering of scores from which it is derived, properly includes the basic partial ordering which is independent of Λ, in the sense that $A_E \succcurlyeq B_E$ implies both $[A_E]_\Lambda \geqq {}_\Lambda[B_E]_\Lambda$ and $A_E \succcurlyeq_\Lambda B_E$, while neither of these implies $A_E \succcurlyeq B_E$.

In order to reduce the infinite plurality of degree functions and of complete orderings to a unique degree function giving a unique complete ordering, it is necessary to settle which Aumann vector Λ is to be used. This can be done directly by choosing n positive rationals whose sum is 1, or indirectly by requiring that certain Λ-equivalence classes of scores on E should be identical. We shall follow this indirect method for scores on a special class of propositions whose existence will be postulated.

A class of scores $[A_E, B_E, \cdots, Z_E]$ will be said to be a class of *linearly independent scores* if no one of them is a linear function of the others, i.e. if there is no class of rational numbers $a, b, \cdots, z$, not all zero, such that $a \cdot A_E + b \cdot B_E + \cdots + z \cdot Z_E = (0, 0, \cdots, 0)_E$.

A_E will be said to be *permutational* to B_E if the components of A_E are a permutation of the components of B_E. The relation of being-permutational-to is symmetric, transitive and reflexive. A class of scores will be called a *permutational class* if each of them is permutational to every one of them. It will be called a *complete permutational class* if it comprises every score permutational to any score of the class.

POSTULATE. *There is a class F of n propositions, one and only one of which can be true, which is such that, for some permutational class Γ of n linearly independent scores, there is an Aumann vector Λ such that $[X_F]_\Lambda = [Y_F]_\Lambda$ for every X_F, Y_F of Γ.*

THEOREM. *If the postulate is satisfied for F, there is only one Aumann vector Λ satisfying the condition in the postulate, namely*

$$\Omega = (1/n, 1/n, \cdots, 1/n).$$

PROOF. Let Λ be an Aumann vector satisfying the condition for the permutational class Γ of n linearly independent scores. Then there is a rational number d such that $D_\Lambda(X_F) = d$ for every X_F of Γ.These n equations, each of the form $\Sigma \lambda_i a_i = d$, together with the normalization requirement $\Sigma \lambda_i = 1$, yield a system of $n+1$ linear equations in the $n+1$ unknowns $\lambda_1, \lambda_2, \cdots, \lambda_n, d$. Algebra of linear equations over the field of rationals shows that these equations are consistent and have the unique solution $\lambda_1 = \lambda_2 = \cdots = \lambda_2 = 1/n$; $d = (\Sigma x_i)/n$, where Σx_i is the sum of the components of any X_F of Γ, and hence of every X_F of Γ, since if A_E is permutational to B_E, $\Sigma a_i = \Sigma b_i$.

For any score A_F, $D_\Omega(A_F) = D_\Omega(X_F)$, and hence $[A_F]_\Omega = [X_F]_\Omega$, for every X_F permutational to A_F. In particular $D_\Omega(U_1)_E = D_\Omega(U_2)_E = \cdots$

$= D_\Omega(U_n)_E = 1/n$. The class $[(U_1)_E, (U_2)_E, \cdots, (U_n)_E]$ is a class of linearly independent scores.

The theorem holds equally if the numbers which are components of scores, and the scalar multipliers of scores, are not restricted to being rationals, so that the scores on E are the vectors in an n-dimensional Euclidean space. [Aumann's work makes no such restriction.] The restriction has been made because of the way scores will be used in this paper and because the theorem holds subject to this restriction.

§ 3. **Lotteries and betting quotients.** Suppose that a man (the lotterer) is certain that one and only one of a class $E = [e_1, e_2, \cdots, e_n]$ of n propositions ($n \geqq 2$) is true, and is not certain that any one of them is false. Then he is uncertain of the truth or falsity of each of the n propositions, and his condition will be called *uncertainty over E.*

For a lotterer in uncertainty over E, a lottery on E, written as $\$A_E = \$(a_1, a_2, \cdots, a_n)_E$, is a right to be paid, conditionally, one of a class of n sums of money, for convenience given in dollars, $\$a_1$, $\$a_2, \cdots$, $\$a_n$, the condition being that $\$a_1$ will be paid him if e_1 is true, $\$a_2$ if e_2 is true, $\cdots$, $\$a_n$ if e_n is true. Any one or more of $a_1, a_2, \cdots, a_n$ may be negative or zero. If a_r is negative, the lotterer will have an obligation to pay $\$(-a_r)$ if e_r is true; if a_r is zero, he will neither have a right to be paid nor an obligation to pay anything if e_r is true. We will suppose that the monetary unit $\$1$ is divisible, for every positive integer m, into m equal parts; so $a_1, a_2, \cdots, a_n$ can be any rational numbers, positive, negative or zero. The problem before us is to discover a method to determine what, for a lotterer in uncertainty over E, is the monetary value of a lottery on E. This value will be the maximum sum which he would be prepared to pay to purchase the right which is the lottery, and the minimum sum for which he would be prepared to part with this right did he possess it.

To attack this problem consider the structure of the system of lotteries and what order can naturally be found in it. Addition of lotteries, and multiplication of a lottery by a rational number, can be defined in a natural way. To purchase two lotteries $\$(a_1, a_2, \cdots, a_n)_E$ and $\$(b_1, b_2, \cdots, b_n)_E$ is the same as to purchase $\$(a_1 + b_1, a_2 + b_2, \cdots, a_n + b_n)_E$. So if this is used to define $\$(A+B)_E$, $\$A_E + \$B_E = \$(A+B)_E$. If $\$\{l/m\}A_E$ represents a lottery identical with $\$A_E$ except that the monetary unit is worth $\$(l/m)$, to purchase m lotteries of $\$\{l/m\}A_E$ is the same as to purchase l lotteries of $\$A_E$. But $\$\{l/m\}A_E = \$((l/m)a_1, (l/m)a_2, \cdots, (l/m)a_n)_E$; so if this is taken to define $\$(l/m)A_E$, $l \cdot \$A_E = m \cdot \$(l/m)A_E$. If, like the monetary unit, any lottery is regarded as divisible into any m equal parts, both sides of this identity may be divided by m, which gives $(l/m) \cdot \$A_E = \$(l/m)A_E$, or

$k \cdot \$A_E = \$kA_E$, where k is any positive rational. If $k = 0$, $\$kA_E = \$(0a_1, 0a_2, \cdots, 0a_n)_E = \$(0, 0, \cdots, 0)_E$, which may be regarded as no lottery at all and thus be identified with $0 \cdot \$A_E$. Finally, to purchase $\$(-a_1, -a_2, \cdots, -a_n)_E$ comes to the same thing as to sell $\$(a_1, a_2, \cdots, a_n)_E$; so if $-\$A_E$ is defined as $\$(-a_1, -a_2, \cdots, -a_n)_E$, $-\$A_E = \$-A_E$. So $k \cdot \$A_E = \$kA_E$ for every rational k, positive, negative or zero. Lotteries on E can thus be treated as the vectors in an n-dimensional 'dollar space', since their system is closed under addition and under multiplication by a rational scalar.

Now consider the ordering of lotteries on E. If in two lotteries $\$A_E$ and $\$B_E$, the sum to be paid the lotterer will never be greater in $\$B_E$ than in $\$A_E$, whichever one of the propositions $e_1, e_2, \cdots, e_n$ is true, whereas, for at least one proposition e_r, the sum to be paid him in $\$A_E$ will be greater than in $\$B_E$ if e_r is true, then $\$A_E$ is *obviously preferable* (for the lotterer) to $\$B_E$. And if the sums to be paid him are exactly the same in all cases in $\$A_E$ as in $\$B_E$ (i.e. if $\$A_E = \$B_E$), $\$A_E$ and $\$B_E$ are *obviously indifferent* (for the lotterer). If neither of these comparisons hold, there is no obvious reason for one lottery to be preferable to another, or for the two lotteries to be indifferent. Using the symbol $\succcurlyeq$ to stand for the relation: is obviously preferable or is obviously indifferent to, all these facts are expressed by: $\$A_E \succcurlyeq \$B_E$ if and only if $a_i \geqq b_i$ for $i = 1, 2, \cdots, n$. This relation yields a partial ordering of the lotteries on E.

The system of lotteries on E partially ordered by this relation differs formally from the partially ordered system of scores on E of §2 only in that the components of scores are rational numbers while the components of lotteries are dollars or fractions of dollars. So the problem of finding the method to determine the *value* of a lottery is exactly the same as that of finding one to determine the *degree* of the corresponding score. For the conditions which a value function V must satisfy in order that the complete ordering of classes of lotteries on E determined by it should be compatible with the basic partial ordering by $\succcurlyeq$, namely

$$\$A_E \succcurlyeq \$B_E \text{ implies } \$V(\$A_E) \geqq \$V(\$B_E);$$
$$\$V(\$A_E + \$B_E) = \$V(\$A_E) + \$V(\$B_E);$$
$$\$V(k.\$A_E) = k\,\$V(\$A_E), \text{ for every rational } k,$$

correspond exactly to the three conditions which degrees on scores must satisfy. And since the value of an 'improper' lottery in which the lotterer will be paid \$1 whichever one of the propositions of E is true is obviously \$1, $\$V(\$(1, 1, \cdots, 1)_E) = \$1$, corresponding to the normalizing condition for the degree function.

The argument of §2 can now be directly applied to the lottery vector-system. For a given E there are an infinite number of value functions V_Λ each associated with a different Aumann vector $\Lambda = (\lambda_1, \lambda_2, \cdots, \lambda_n)_E$, and with $\$V_\Lambda(\$A_E) = \$\sum \lambda_i a_i$. Each V_Λ imposes a weak ordering upon lotteries on E and a complete ordering upon equivalence classes of such lotteries. The lotterer is at liberty to select whichever Aumann vector Λ he likes, and his selection of a particular one fixes the values of all possible lotteries on E.

The theorem of §2 shows that if, instead of directly selecting an Aumann vector, the lotterer judges that the value of any lottery $\$A_E$ is the same as the value of each of $n-1$ other lotteries on E which are permutational to $\$A_E$ and which are such that the n lotteries together form a class of linearly independent lotteries (these conditions imply that $\$A_E$ is not an improper lottery), then logical consistency requires that he should use the Aumann vector $\Omega = (1/n, 1/n, \cdots, 1/n)$ in settling the values of all lotteries on E.

Up to this point the notion of a betting quotient has not been used, since that of a lottery on E is a more general one. If we confine ourselves, as we have been doing, to bets in terms of monetary units (e.g. dollars), a betting quotient of q_1 on e_1 is the proportion of \$1 the bettor is prepared to pay for the right to be paid \$1 if e_1 is true, but nothing if e_1 is false. This is the number of dollars in the value of the lottery $\$(1, 0, \cdots, 0)_E$, which is $\lambda_1 \cdot 1 + \lambda_2 \cdot 0 + \cdots + \lambda_h \cdot 0 = \lambda_1$. So betting quotients of q_1 on e_1, of q_2 on $e_2, \cdots$, of q_n on e_n are $\lambda_1, \lambda_2, \cdots, \lambda_n$, respectively. The condition $\sum \lambda_i = 1$ is the condition $\sum q_i = 1$ which is necessary and sufficient for that *coherence* of the betting quotients which prevents a book being made against the bettor. The choice of Λ may thus be regarded as the choice of a coherent vector of betting quotients. A bettor's judgement that the values of the n lotteries $\$(1, 0, \cdots, 0)_E$, $\$(0, 1, \cdots, 0)_E, \cdots, \$(0, 0, \cdots, 1)_E$ are equal is the same as his taking his n betting quotients as equal to one another, and hence as each equal to $1/n$.

But, whichever way we look at the matter, what is essential for the lotterer's taking the value-determining Aumann vector as being Ω is his judgement that the lotteries in a permutational class of n linearly independent lotteries have all the same value. If he is offered the option of purchasing the lottery $\$(1, 0, 0)_E$ for 30 cents, he may compare it with the lotteries $\$(0, 1, 0)_E$ and $\$(0, 0, 1)_E$ and decide upon the purchase by judging that the values of each of these lotteries, i.e. his betting quotients, multiplied into \$1, on each of e_1, e_2, e_3, are the same. But if he is offered the option of purchasing the lottery $\$(1, 2, 4)_E$ for \$2, he can reasonably decide upon the purchase, without analyzing $\$(1, 2, 4)_E$ into

$$\$(1, 0, 0)_E + 2. \$(0, 1, 0)_E + 4. \$(0, 0, 1)_E,$$

by comparing the values of $\$(1, 2, 4)_E$, $\$(4, 1, 2)_E$, $\$(2, 4, 1)_E$, or by comparing the values of $\$(1,4,2)_E$, $\$(2,1,4)_E$, $\$(4,2,1)_E$, and judging that the values of the three lotteries are equal in either of these classes. Here he is not making any judgement as to the equality of betting quotients: from his judgement that the values of three lotteries are the same, he can deduce, by the reasoning of this paper, that he must use the Aumann vector Ω in calculating the value of the lottery offered him, and that $\$V_\Omega(\$(1,2,4)_E) = \$2\frac{1}{3} > \$2$. In this argument there is no reference to the fact that an appropriate Λ is identical with an appropriate n-vector of betting quotients: the lotterer need not think of betting quotients at all. All that is required is his judgement that certain lotteries which are permutational one to another are equal in value.

§ **4. Chances.** Consider the hypothesis that the chance of a member of a non-empty class β being also a member of a class α is m/n, with m, n non-negative integers, $n \geqq 1$, $0 \leqq m \leqq n$. This hypothesis may be represented by the model of draws from a bag containing n balls m of which are black, where x is a member of β corresponds to y is a draw of one ball from the bag, x is a member of α corresponds to y is a draw of something black (so that x is a member of $\alpha \cap \beta$ corresponds to y is a draw of one black ball from the bag). The logic of chance can be investigated by considering how numbers can be assigned to draws in this model.

These draws can be partially ordered by treating them as forming part of the system of scores of §2. There it was remarked that each score on E is a unique linear function of n unit scores on E, i.e. of the form $k_r \cdot (U_r)_E$, where $(U_r)_E$ is the score whose components are all 0 except for the rth which is 1. Now consider only those scores on E which are linear functions of the n unit scores in which the scalar multipliers $k_1, k_2, \cdots, k_n$ are restricted to being either 0 or 1. The system of these *binary scores* (as they will be called) is not an ordinary vector system, since it is finite and is not closed under addition: although e.g., $(U_1)_E$ can be added to $(U_2)_E$, it cannot be added to itself. But being a part of the system of scores on E, it is partially ordered by the basic relation $\succcurlyeq$. Ordered by this relation it will be a Boolean lattice with 2^n elements, and with $(1,1,\cdots,1)_E \succcurlyeq X_E \succcurlyeq (0,0,\cdots,0)_E$ for every binary score X_E. A degree function D_Λ assigning numbers to scores on E will serve also as a degree function assigning numbers, and imposing a complete ordering, for the binary part of the system.

Draws of a ball from the bag can be represented by binary scores in the following way. Index the n balls in the bag by the numerals $1, 2, \cdots, n$; and let e_r be the proposition that the ball numbered r is the ball drawn from the bag. Then the class of propositions $E = [e_1, e_2, \cdots, e_n]$ is such that one and only one of them can be true. A draw of the ball numbered r is represented by the score $(U_r)_E$, with Λ-degree λ_r; and a draw of a ball numbered either 1 or 2 or $\cdots$ or m by the score $(U_1)_E + (U_2)_E + \cdots + (U_m)_E$, i.e. $(1,1,\cdots,1,0,\cdots,0)_E$ where the first m components are 1 and the remainder 0, with Λ-degree $\lambda_1 + \lambda_2 + \cdots + \lambda_m$.

However the correspondence between the model of draws from a bag with n balls m of which are black and the hypothesis that the chance is m/n makes no reference to any particular numbering of balls in the bag. So the chance that a black ball is drawn will have to be represented not by means of one binary score, but by means of a complete permutational class of binary scores, in order that the chance should be invariant with respect to the numbering. Let $[m_E]$, with $0 \leqq m \leqq n$, be the class of binary scores X_E having any m components 1 and the remainder 0. The number of members of this complete permutational class will be $n!/m!(n-m)!$; this will be 1 if $m = 0$ or $m = n$, and n if $m = 1$ or $m = n - 1$. The Λ-degree of the unique member of $[0_E]$ is 0, and of the unique member of $[n_E]$ is 1; but in other cases there will be no unique Λ-degree assignable to $[m_E]$, but a distinct Λ-degree for every member of $[m_E]$. In order that there should be one and only one Λ-degree assignable to each $[m_E]$, it is necessary that Λ should be such that all the members of some $[m_E]$, with $0 < m < n$, should have the same Λ-degree. This condition is sufficient (and, unless $m = 1$ or $m = n - 1$, more than sufficient) to ensure that E satisfies the condition of the postulate of §2, from which it follows, by the theorem of §2, that

$$\Lambda = (1/n, 1/n, \cdots, 1/n) = \Omega.$$

Then, for any $[m_E]$, the Λ-degree of every member of it will be the same, and so can be written as $D_\Omega[m_E]$. Since $D_\Omega[m_E] = m(1/n) = m/n$, the chance that a black ball is drawn, or more generally that a β-specimen is an α-specimen, may be explicated as being equivalent to $D_\Omega[m_E]$. Since $D_\Omega[0_E] = 0$ and $D_\Omega[n_E] = 1$, chances thus explicated will be rationals between 0 and 1 inclusive.

Parenthetically it may be remarked that the Ω-equivalence class of binary scores comprising A_E is identical with the complete permutational class of scores comprising A_E. So a complete ordering of classes of binary scores is given by a complete ordering of the classes $[m_E]$, which is provided by the complete ordering of $0, 1, 2, \cdots, n$ in numerical order.

In this explication of the notion of chance as the degree of every member of a complete permutational class of binary scores, the equality of the components of the Aumann vector Λ has been derived from the requirement that binary scores which are permutational one to another should have the same degree, if their common degree is to be treated as a chance. The components of Λ have not themselves been treated as chances. If a particular numbering of the balls were to be given, $D_{\Lambda}(U_r)_E = \lambda_r$ could be regarded as the chance that the ball numbered r is the ball drawn from the bag, and equality of the λ's could be interpreted as the equality of the chances that a ball numbered $1, 2, \cdots, n$ is drawn, to be derived perhaps by means of some 'principle of indifference'. Our argument, however, makes no use of any particular numbering of the balls in the bag, and hence makes no use of these chances. The equality of the λ's derives from the permutational invariance required of the degree of a binary score if it is to serve as a chance; it is not derived by first identifying the λ's with chances and then taking these as equal. A principle of indifference only appears in the guise of the essential irrelevance, in explicating chance by means of a balls-in-bag model, of any indexing of the balls in the bag.

§ **5. Connexion between values of lotteries and chances.** Suppose that a man believes that a bag from which one coin is to be drawn contains n coins of which m are silver dollars, each valued at \$1, and $n - m$ counterfeit dollars, each of no value. Suppose that he wishes both to assign a chance to the draw being that of a silver dollar and to assign a value to a lottery in which he will receive whatever coin is drawn. The draw is represented by that complete permutational class of scores on $E = [e_1, e_2, \cdots, e_n]$ each score of which has m components 1 and $n - m$ 0. The lottery is one or other of the $n!/m!(n-m)!$ lotteries on E which have m components \$1 and $n - m$ components \$0; it is some one member of this complete permutational class of lotteries.

In order to assign a chance to the draw it is necessary for the man to believe something equivalent to the proposition: (1) There is a degree-determining Aumann vector yielding the same degree for every score in the complete permutational class. In order to assign a common value to the alternation of the alternative lotteries it is necessary for him to believe something equivalent to the proposition: (2) There is a value-determining Aumann vector yielding the same value for each of the alternative lotteries, i.e. for every lottery in the complete permutational class. And, given the man's beliefs about the contents of the bag and the values of these contents, belief in (1) or something to the same effect, or in (2) or something to the same effect, is in each case sufficient as well as necessary for him to assign a chance of m/n to the draw, in the first case; to assign a unique value of \$$m/n$ to the lottery, in the second case.

For a man to accept (1) while declining to accept (2) would amount to his holding that any numbering of the coins would be irrelevant to his consideration of what chance should be assigned to his being given as a result of the draw something valued at \$1, but would be relevant to a consideration of what value should be assigned to what is to be given him. And *vice versa* for a man who accepted (2) while declining to accept (1). To prove that in either of these cases he would be logically inconsistent would require an analysis of the logical relations between beliefs, value judgements and actions which cannot be attempted here. But, whether or not he would be logically inconsistent, he would surely be irrational to the point of schizophrenia in arguing from a permutational invariance in estimating a chance while rejecting exactly the same argument, applied to values, in judging the value of the corresponding lottery; or *vice versa*.

An estimate of chance presupposes a belief in a permutational invariance. Once having accepted this powerful tool it is irrational not to make use of it again when it is needed for valuing a lottery and thus choosing a betting rate.

Finally, one criticism must be met. The theory of chance is given in terms of classes of things or of events, and applies to an individual thing or event only to the extent that abstraction is made of all the individual's properties which are not implied by its membership of the class of reference. [This is what is frequently, but confusingly, expressed by saying that the individual must be a 'random' member of the class.] However a bet, it is said, is essentially a bet on an individual event: an event which is betted upon is always more than a mere β-specimen; so no belief as to what is the chance of a β-specimen being an α-specimen can be enough to justify choice of a betting quotient on an individual event. This argument neglects the fact that what is in question is the reason for choosing one betting quotient rather than another in circumstances in which betting is unavoidable. A man who was omniscient would have no interest in lotteries nor in their values. But we, who are not omniscient, are interested in choosing between the various lotteries which are offered us. Decision theorists have elaborated various criteria of varying merit for selecting an optimum from among a finite class of lotteries. But their work, and that of Aumann, has shown that there is no general way of assigning numerical values to all lotteries which is not equivalent to making the calculation, in each case, *as if* there were chances to which the betting quotients corresponded. A man who believes he knows that there are these chances, and what these chances are, would be a 'wishful thinker' if he pretended that they were different when calculating betting quotients with which to guide his course of action.

REFERENCE

[1] AUMANN, R. J., Utility theory without the completeness axiom, *Econometrica* **30** (1962), 445–462.

TOWARDS A THEORY OF INDUCTIVE GENERALIZATION

JAAKKO HINTIKKA
University of Helsinki, Helsinki, Finland

In discussing the relation of the notions of probability and rational belief one of the main problems concerns the status of the inductive generalizations of science and of everyday life. A great number of such generalizations are accepted and believed in by scientists and by men of practical affairs. In many cases, their attitude is obviously rational. Can this rationality be interpreted in probabilistic terms? Can the rationality of our acceptance of the generalizations we in fact accept be explained in terms of the high probability of these generalizations, in some interesting sense of probability?

In recent literature a slightly different question often comes to the fore. This literature is characterized by a contrast of two main schools of thought, led by Rudolf Carnap and Karl Popper, respectively[1]. What Popper is primarily concerned with are not our beliefs in various generalizations, but rather our choice of the best possible generalization, best not in the sense of the safest one but rather best in the sense of the most informative and most thoroughly testable (and tested) one[2]. Popper claims that this choice of the best possible generalization cannot be explained in terms of its high probability[3]. On the contrary, Popper maintains that the most informative generalization is typically one with the lowest probability[4]. His answer to our second initial question is thus negative.

---

[1] See Rudolf Carnap, *Logical Foundations of Probability*, The University of Chicago Press, Chicago, 1950, second edition, Chicago, 1963; and Karl R. Popper, *The Logic of Scientific Discovery*, Hutchinson & Co., London, 1959. These two books also supply further references. In discussing Carnap's views in the present paper, I am confining my attention to the views put forward in the first edition of *Logical Foundations of Probability*, disregarding whatever subsequent changes there have been in Carnap's point of view.

[2] Popper, *op. cit.*, p. 399: "Science does not aim, primarily, at high probabilities. It aims at a high informative content, well backed by experience."

[3] According to Popper, *op. cit.*, p. 387, his aim has been "to show that degree of corroboration was not a probability; that is to say, that it was not one of the possible interpretations of the probability calculus." "Degree of corroboration" is Popper's measure "of the severity of tests to which a theory has been subjected, and of the manner in which it has passed these tests, or failed them . . . " (*loc. cit.*).

[4] More accurately, one with what Popper calls the lowest "absolute logical probability." Cf. *op. cit.*, section 35.

Carnap denigrates the importance of (strict, non-statistical) generalizations for our inductive thinking and inductive practice[5]. According to his attempted "rational reconstruction" of our inductive practice we should not strictly speaking believe in the general laws of science and of everyday life, for their probability on evidence (*a posteriori* probability, Carnap's "degree of confirmation") is according to Carnap normally very small. Carnap writes: "Although . . . laws stated by scientists do not have a high degree of confirmation, they have a high qualified-instance confirmation and thus serve as efficient instruments for finding those highly confirmed singular predictions which are needed in practical life."[6] However, this provokes the question whether the acceptance of the general laws which the scientists in fact accept can be justified on Carnap's grounds; more generally, whether accepting generalizations because of their high instance confirmation leads us to choose the generalizations which we in fact choose on sound rational grounds. Carnap does not by any means demonstrate that this is the case, and in this paper I shall argue that in some typical cases he is wrong.

Thus instead of discussing directly one's beliefs in inductive generalizations I am led to discuss the question: Why do we believe that one generalization is better than another? Can our rational preferences in this matter be interpreted in terms of a high probability of the preferred generalizations? The latter question might be called the basic question of this paper. Popper's answer to it is negative. Carnap also gives a negative answer to it in the sense that he does not think that the generalizations themselves are highly probable *a posteriori*. However, he thinks that the preferable generalizations are characterized by a high degree of instance confirmation, and that our choice of a generalization is therefore in a sense guided by probabilistic considerations. This is firmly denied by Popper.

In this paper I shall try to offer some suggestions concerning the basic question just formulated, and thereby to put this conflict of views into a new perspective. In order to fix our ideas, let us start by considering a situation which may look like a rather special case but which will subsequently turn out to be quite representative. Let us assume that we are given k primitive monadic predicates P_j $(j = 1, 2, \cdots, k)$. By means of these predicates and propositional connectives, a partition of our domain of individuals into exactly $K = 2^k$ different kinds of individuals can be defined[7]. Let us

---

[5] See *op. cit.*, pp. 570–577.

[6] *Op. cit.*, p. 575.

[7] They are defined by the expressions $(\pm)P_1(x)$ & $(\pm)P_2(x)$ & . . . &$(\pm)P_k(x)$, where the symbols $(\pm)$ are replaced by negation-signs or by nothing at all in all the different combinations. Some of these kinds of individuals may of course be empty.

assume that these kinds of individuals are defined by the expressions (complex predicates)

(1) $$Ct_1(x), Ct_2(x), \cdots, Ct_K(x).$$

(These are of course simply Carnap's Q-predicates in a new guise[8].) Let us also assume that we have observed exactly n individuals in a domain of individuals (universe of discourse) which contains a totality of N individuals. (In the sequel, it will normally be assumed that $N \gg n \gg K$, i.e. that N is large as compared with n and n large compared with $K = 2^k$.) Let us finally assume that among the n observed individuals exactly c different kinds of individuals are exemplified, say the ones specified by the expressions

(2) $$Ct_{i_1}(x), Ct_{i_2}(x), \cdots, Ct_{i_c}(x)$$

which constitute a subset of all the expressions (1).

Apart from unavoidable oversimplification, this situation does not appear to be entirely unrepresentative of what one is likely to encounter in actual applications.

The question we want to ask is: What generalization concerning the whole universe of discourse ought we to prefer in this situation?

It seems to me that from an intuitive point of view the answer is obvious, provided that $n \gg K$. If n is large in relation to K, there have been plenty of opportunities for the remaining kinds of individuals to prove that they are not empty. Their failure to do so therefore strongly suggests that they are in fact empty. Hence the only rational thing to do in these circumstances is to assume that they are empty, i.e. to prefer the following sentence to all the other comparable generalizations:

(3) $$((Ex)Ct_{i_1}(x) \,\&\, (Ex)Ct_{i_2}(x) \,\&\, \cdots \,\&\, (Ex)Ct_{i_c}(x)) \\ \&\, (x)(Ct_{i_1}(x) \vee Ct_{i_2}(x) \vee \cdots \vee Ct_{i_c}(x)).$$

This is then the desirable hypothesis independently of what $N \geqq n$ is. It also appears to be the preferable generalization on Popper's principles[9].

It is not at all clear, however, how this preference could be justified in terms of inductive logic. The problem is to compare (3) with a competing generalization, say with

---

[8] For the Q-predicates, see *op. cit.*, pp. 124–126.

[9] Some remarks on the connection between this preference and Popper's ideas will be made later. It is also obvious that (3) is the preferable generalization on any reasonable method of maximum likelihood or of "straight rule."

$$(4) \quad ((Ex)Ct_{i_1}(x) \& \cdots \& (Ex)Ct_{i_c}(x) \& \cdots \& (Ex)Ct_{i_w}(x)) \\ \& (x)(Ct_{i_1}(x) \vee \cdots \vee Ct_{i_c}(x) \vee \cdots \vee Ct_{i_w}(x))$$

where $w \geqq c$. (The possibility that $w < c$ is excluded by the evidence.) Carnap's theories offer us two different ways of comparing (3) with (4). We may compare the degrees of confirmation of the two on the evidence in question, or we may compare their respective degrees of instance confirmation[10]. These comparisons are facilitated by the observation that for Carnap the degree of confirmation of the existential part of (4) relative to its universal part (last member of conjunction (4)) is very close to one if n is large in relation to K. Hence we may consider the degree of confirmation of the universal part only, and also apply to it the notion of instance confirmation which does not otherwise apply to (4). According to Carnap's results[11], the degree of confirmation of the universal part of (4) is approximately

$$(5) \quad \left(\frac{n}{N}\right)^{K-w}$$

when n is large in relation to K, and its degree of instance confirmation is

$$(6) \quad \frac{n+w}{n+K} \, .$$

Both (5) and (6) have the greater value the greater w is. Thus the recommendation which Carnap's theory gives us is not to assume that (3) is true, that is to say, not to assume that as few kinds of individuals are exemplified in the world as is compatible with evidence. Instead, we ought to assume that as many of them as possible are exemplified—ideally, all of them. It is obvious that this recommendation is counterintuitive. It is more like a counsel of despair than a rational direction for choosing one's generalization.

It is perhaps worth pointing out that this difficulty of Carnap's theory is largely independent of the possibility that N might be infinite. It appears as soon as $N \gg n \gg K$. It is obvious that this possibility cannot be ruled out in many important types of applications. And it is obvious that this difficulty cannot be eliminated by means of the notion of instance confirmation.

Our example thus seems to add grist to the mill of Carnap's critics. It

---

[10] For the notion of instance confirmation, see Carnap, *op. cit.*, pp. 571–577.

[11] For (5), see Carnap, *op. cit.*, p. 571, formula (11); for (6), see *op. cit.*, p. 573 formula (16), or p. 568, formula (7).

certainly shows that, however much value Carnap's theory has in other respects, there is at least one task it does not perform in a satisfactory manner, to wit, the task of guiding us in our choice of a generalization.

Our example might seem to accomplish even more than this. It might seem to point to the (fallacious) conclusion that there is *no* way of explaining our rational preference of (3) as compared to (4) ($w > c$) in terms of the higher probability of (3). For it might seem that since (4) allows more possibilities for the unexamined individuals of our universe of discourse than (3), (4) must be more probable *a priori* than (3) on any reasonable notion of probability. And since there are usually more unexamined than examined individuals the same would seem to be the case also with *a posteriori* probability. Conversely, since the preferable generalization (3) allows fewer possibilities for the unexamined individuals than any of the competing generalizations (4), it may seem to be the most *im*probable hypothesis in any natural sense of probability. It might thus seem not only that the special form of quantitative inductive logic which Carnap has developed does not account for our reasons for preferring one generalization to another in terms of the higher probability (on evidence) of the former, but also that *no* system of quantitative inductive logic can accomplish this.

It thus seems to me that our example brings out some of the reasons which have led the critics of inductive logic to give a negative answer to our basic question, although their reasons have been derived from somewhat different examples.

Their conclusion is a much too hasty one, however. It is the main purpose of this paper to outline a way of constructing a quantitative inductive logic (for the same kinds of language systems as Carnap) which avoids some of the main disadvantages of Carnap's theory in so far as the problem of dealing with strict (non-statistical) inductive generalizations is concerned.

Such an inductive logic can even be built along lines closely reminiscent of Carnap's. In developing an inductive logic along Carnapian lines the fundamental problem concerns the choice of the underlying measure function. This choice may be thought of as a choice of a method of assigning a weight (or an *a priori* probability, if you prefer) to each of the different state-descriptions that can be formulated in the language we are considering. (These weights are here assumed to be between zero and one and their sum is assumed to equal one.) Now one way in which one comes to choose the particular measure function m^* which Carnap favours may perhaps be described as follows:[12] What we are primarily interested in are not

[12] This is not intended as a reconstruction of Carnap's reasons for preferring m^* (some of these reasons are mentioned briefly in *op. cit.*, pp. 562-566), but rather as an intuitively persuasive line of thought which leads to the same result.

the different "possible worlds" described by state-descriptions but rather the different *kinds of* possible worlds that can be described in our language. These different kinds of worlds are interpreted by Carnap as structurally different worlds. Accordingly, Carnap assigns an equal weight not to all the state-descriptions but rather to what he calls structure-descriptions[13]. Here each structure-description is the disjunction of all the different state-descriptions that can be transformed into each other by permuting free singular terms (names of individuals). The weight which a structure-description receives is then divided evenly among its members.

The main drawback of this procedure seems to me to be its dependence of the domain of individuals which the language in question presupposes. In order to know the weight of a structure-description, one has to know the number of all the structure-descriptions; and this depends on the number of individuals in the domain. In order to apply Carnap's procedure, one therefore has to know the whole universe at least to the extent of knowing its size. In most applications, however, this domain is largely unknown. In general it seems to me perverse to start one's inductive logic from the assumption that one is in some important respect already familiar with the whole of one's world, for this logic is largely designed to reconstruct some of the procedures we use in coming to know it. In view of this perversity, it is small wonder that Carnap's theory in fact leads to great difficulties in connection with generalizations[14].

Hence we cannot use structure-descriptions as an explication of the notion of the description of a possible kind of world. These descriptions must be independent of one's list of individuals. They must be described by general and not by singular sentences[15].

Now the question becomes: Can we find, for each finite set of predicates (each with one or more argument-places) such descriptions of possible kinds of worlds in some natural sense of the word? The answer is easily seen to be affirmative if we limit the number of quantifiers whose scopes may all overlap in our sentences. For a given sentence, we shall call the maximum number of quantifiers whose scopes have a common part in it the *depth* of the sentence in question. More loosely expressed, the depth of a sentence is the number of layers of quantifiers which it contains at its

---

13 For the notion of structure-description, see Carnap, *op. cit.*, pp. 114–117.

14 The worst difficulty is probably the fact that in Carnap's approach a universally quantified sentence (which is not logically true) always has zero as its degree of confirmation when the domain of individuals is infinite.

15 A *general* sentence means in this paper a sentence of applied first-order logic (no predicate variables) without individual constants and without free individual variables. A sentence which is not general is called *singular*.

thickest. The restriction just mentioned is a restriction on this parameter. Let it be restricted to d_0 at most.

With this restriction, how can one arrive at the different descriptions of possible kinds of worlds that we are looking for? An answer is not very difficult to give. Suppose first that we have already arrived at a list of all the possible kinds of individuals that can be described by means of the given set of predicates which we are presupposing and by means of at most $d_0 - 1$ layers of quantifiers, and let these kinds of individuals be described by sentences (1). Then we can arrive at a description of a possible kind of world simply by running through the list and indicating, for each kind of individual, whether individuals of that kind exist or not. Each description of a possible world will in other words be of the form

$$(7) \qquad (\pm)(Ex)\,Ct_1(x)\ \&(\pm)(Ex)Ct_2(x)\ \&\cdots\ \&(\pm)(Ex)Ct_K(x),$$

where in the place of each symbol ($\pm$) there is either a negation-sign or nothing at all. It is obvious that each sentence of form (7) can be rewritten: Instead of specifying, for each possible kind of individual, whether individuals of that kind exist or not, it suffices to list all the kinds of individuals that are exemplified, and then to add that they are *all* the kinds that are not empty. But this means that (7) can be written in form (3) where the sentences $Ct_{i_1}(x)$, $Ct_{i_2}(x)$, $\cdots$, $Ct_{i_c}(x)$ form a subset (proper or improper) of the set of all sentences (1). Each sentence of form (3) will then be called a *constituent*.

But how can we obtain the descriptions (1) of all the possible kinds of individuals? In the monadic case the answer is obvious: The relevant kinds of individuals are specified by Carnap's Q-predicates, i.e. by our sentences (1). Hence in the monadic case our constituents are just the sentences of form (3) considered in our example. Part of the general significance of the example is due to this fact.

Thus it remains to deal with the polyadic (general) case. This case can be handled inductively. Suppose that the following sentences describe all the possible ways in which the value of the variable x_j can be related to the individuals specified by $x_1\ x_2\ \ldots\ x_{j-1}$:

$$(8) \qquad Ct_1^d(x_1, x_2, \cdots, x_{j-1}, x_j),\ Ct_2^d(x_1, x_2, \cdots, x_{j-1}, x_j), \cdots .$$

By "all the possible ways" we here mean all the ways which can be specified by using only the given predicates, propositional connectives and at most d layers of quantifiers. We might also say that (8) is a list of all kinds of individuals x_j that can be specified by these means plus the "reference point" individuals specified by $x_1, \cdots, x_{j-1}$. Then we may obtain a similar list for the preceding variable x_{j-1} and for $d+1$ layers of quantifiers in the same way in which (7) (or (3)) was obtained from (1). This list is given by the sentences

$$(9)\quad (\pm)(Ex_j)Ct_1^d(x_1,x_2,\cdots,x_{j-1},x_j)\,\&\,(\pm)(Ex_j)Ct_2^d(x_1,x_2,\cdots,x_{j-1},x_j)$$
$$\&\cdots\&(\pm)A_1(x_1,x_2,\cdots,x_{j-1})\,\&\,(\pm)A_2(x_1,x_2,\cdots,x_{j-1})\,\&\cdots$$

where at the place of each $(\pm)$ there again may or may not be a negation-sign and where $A_1(x_1, x_2, \cdots, x_{j-1})$, $A_2(x_1, x_2, \cdots, x_{j-1})$, $\cdots$ are all the atomic sentences which can be formed from the given predicates and from the variables $x_1, \cdots, x_{j-1}$, and which contain at least one occurrence of x_{j-1}.

What happens in (9) can be given an intuitive interpretation. Again we run through the list (8) and indicate, for each possible kind of an individual x_j, whether individuals of that kind exist or not. This adds a new layer of quantifiers, but it also transforms the result to something we can assert of the referent of x_{j-1} in relation to the referents of $x_1, x_2, \ldots, x_{j-2}$. Finally, we specify how the referent of x_{j-1} is related to the referents of x_1, x_2, $\cdots$, x_{j-2}.

If $d = 0$, sentences (8) are simply conjunctions of the form

$$(10)\quad (\pm)A_1(x_1,x_2,\cdots,x_j)\,\&(\pm)A_2(x_1,x_2,\cdots,x_j)\,\&\cdots.$$

This gives us a basis for induction.

Obviously we can rewrite each (9) in the same way in which (7) was rewritten as (3). In the sequel, we shall assume that this has always been done. The resulting sentences will be called attributive constituents with depth $d+1$ and with the free variables $x_1, x_2, \ldots, x_{j-1}$. From attributive constituents of depth d_0-1 with one free variable we may obtain constituents of depth d_0 in the way described above.

Every consistent general sentence has a normal form (its *distributive normal form*) which is a disjunction of constituents with the same predicates and with the same depth (or any fixed greater depth). For instance, every consistent constituent of depth d is equivalent to a disjunction of constituents of depth $d+e$, for each $e = 1, 2, 3, \ldots$. These are called its *subordinate* constituents. Similarly, attributive constituents may be split into disjunctions of deeper attributive constituents (their subordinates) with the same free individual variables and constants.

The theory of constituents, attributive constituents, and of distributive normal forms has been developed in greater detail elsewhere[16]. Here I shall only explain briefly how they might be used for the purposes of inductive logic.

Given a finite set of predicates, it may be stipulated that an equal weight be given to all the consistent constituents of some fixed depth d_0 which must be larger than the greatest number of argument-places among our predicates[17]. If constituents of different depths are considered, the weight of each constituent of depth d_0 or more (say of depth d) is then divided evenly among its consistent subordinate constituents of depth $d+1$. There remains an important open problem as to how the different systems obtained by choosing d_0 differently are to be compared with each other.

If we are given, in addition to a finite number of predicates, also a domain of individuals, then weights will also be assigned to the different state-descriptions as equally as is compatible with the previous assignments of weights to constituents. In the monadic case, the weight of each constituent is divided evenly among all the state-descriptions which make it true[18].

It is easily seen that the resulting assignment of *a priori* probabilities to all our general sentences satisfies all the usual axioms of probability calculus (including Kolmogorov's axiom of continuity)[19]. In order to make the requisite set-theoretical concepts applicable, each sentence may be considered as standing for the set of all its models[20].

---

16 See my papers, "Distributive Normal Forms in First-Order Logic," in *Formal Systems and Recursive Functions, Proceedings of the* 1963 *Logic Colloquium in Oxford*, edited by John N. Crossley, Studies in Logic and the Foundations of Mathematics, North-Holland Publishing Company, Amsterdam, 1965, and "Distributive Normal Forms and Deductive Interpolation," *Zeitschrift für mathematische Logik und Grundlagen der Mathematik* vol. 10 (1964), pp. 185–191, as well as the older monograph "Distributive Normal Forms in the Calculus of Predicates," *Acta Philosophica Fennica* vol. 6 (1953).

17 It would also be possible, and for many purposes more natural, to assign to them *different* constant weights, the larger the more attributive constituents they contain. This would make little difference to my arguments in the sequel.

18 In the polyadic case, it may turn out to be desirable to modify this procedure somewhat. It would take us too far, however, to discuss the matter here.

19 These *a priori* probabilities are independent of the domain of individuals. Similar *a priori* probabilities can be assigned to singular sentences only in some given fixed universe of discourse.

20 Suppose that we are given a sequence of general sentences $S_1, S_2, \ldots$ the models of each of which are all also models of its predecessors and which do not possess any single model in common. There will then exist (by the compactness theorem for first-order logic) a member of the sequence which is contradictory and whose probability is therefore zero. This proves the continuity axiom. (For the axiom, see A. N. Kolmogorov, *Grundbegriffe der Wahrscheinlichkeitsrechnung*, Berlin, 1933, ch. 2, §1.)

On this basis, a quantitative inductive logic may be built.[21] Here I can only indicate some of its main features, restricting most of my remarks to the monadic case. I shall try to make the conceptual situation clear rather than to develop the theory itself very far in any particular direction.[22]

For most purposes it suffices to consider only such generalizations as are formulated by a constituent. These generalizations will be called *strong* ones. Since all consistent general sentences are disjunctions of constituents, their probability can be obtained as sums of the probabilities of the constituents occurring in their respective distributive normal forms. In our initial example we already restricted ourselves tacitly to strong generalizations.

What can be said of this example from the point of view of my inductive logic? It is easily seen that in this example our evidence admits of exactly

$$(11) \quad m(w) = w^{N-n} - (w-c)(w-1)^{N-n} + \frac{(w-c)(w-c-1)}{2!}(w-2)^{N-n} - \cdots + (-1)^i \binom{w-c}{i}(w-i)^{N-n} + \cdots$$

state-descriptions which would make (4) true. In the absence of all evidence there would be $M(w)$ such state-descriptions, where $M(w)$ is obtained from (11) by putting $n = 0$, $c = 0$. In particular we obtain $m(c) = c^{N-n}$.

According to Bayes' formula, the probability of (4) on the evidence we have is

---

[21] This quantitative inductive logic is in the case of each finite universe based on the use of a regular and symmetrical measure function. Hence all the results Carnap establishes (*op. cit.*, ch. *viii*) for regular and symmetrical *c*-functions (concerning direct inference, binomial theorem, and Bernoulli's laws) are valid in our inductive logic for each finite system. However, it is easily seen that Carnap's requirement of fitting together (*op. cit.*, pp. 290–292) is not satisfied. Hence the infinite case may have to be dealt with in a way different from Carnap's.

[22] One interesting feature of this conceptual situation is the following: In order to calculate the *a priori* and therefore also the *a posteriori* probability of a general sentence S we have to know which constituents (with the same predicates and the same depth as S) are consistent and which ones are inconsistent. However, the general problem of deciding which constituents are consistent is easily seen to be equivalent to the decision problem for first-order logic (predicate calculus), and hence recursively unsolvable. Hence the general problem of calculating the degree of confirmation of an arbitrary sentence on given evidence is recursively unsolvable. This gives in a sense an answer to the much-debated question: How difficult might induction be? It turns out to be exactly as difficult as the decision problem for first-order logic. The unsolvability of the problem of determining the degree of confirmation of an arbitrary generalization is no argument against my approach, however, but rather for it. Hilary Putnam has shown (roughly speaking) that an optimal inductive strategy, if such a strategy should exist, cannot be computable. (See "Degree of Confirmation and Inductive Logic", *The Philosophy of Rudolf Carnap*, ed. by P. A. Schilpp, Open Court, La Salle, Illinois, 1963.). In other words, any optimal inductive strategy must exhibit recursive undecidability similar to ours.

$$(12)\qquad \frac{\dfrac{m(w)}{M(w)}}{\dfrac{m(c)}{M(c)}+(K-c)\dfrac{m(c+1)}{M(c+1)}+\dfrac{(K-c)(K-c-1)}{2!}\dfrac{m(c+2)}{M(c+2)}+\cdots}$$

A good idea of the behaviour of this expression is obtained by letting N become infinite (infinite domain). Then $\dfrac{m(w)}{M(w)}$ approaches $\dfrac{1}{w^n}$ as a limit and (12) becomes

$$(13)\qquad \frac{\dfrac{1}{w^n}}{\dfrac{1}{c^n}+(K-c)\dfrac{1}{(c+1)^n}+\dfrac{(K-c)(K-c-1)}{2!}\dfrac{1}{(c+2)^n}+\cdots}$$

It is seen that (13) has the greater value the smaller w is, and that (13) assumes its greatest value for $w = c$. Thus the recommendation our inductive logic yields is the intuitively acceptable one: generalization (3) is to be preferred to (4) whenever c is smaller than w.

There is a stronger reason for preferring (3) to (4), however. If we let $n \to \infty$ in (13), the value of (13) is seen to approach zero, except in the case $w = c$ where it approaches one. The more evidence compatible with (3) we thus have, the more probable (3) therefore becomes and the less probable the competing hypotheses (4) become.

In the case $w = c$, (13) assumes the form

$$(14)\qquad \frac{1}{1+(K-c)\left(\dfrac{c}{c+1}\right)^n+\dfrac{(K-c)(K-c-1)}{2!}\left(\dfrac{c}{c+2}\right)^n+\cdots}$$

This expression shows how the inductive probability of (3) grows when more and more confirmation is obtained (in an infinite domain). It approaches one when $n \to \infty$. In the case of an ordinary general implication $(x)(P_1(x) \supset P_2(x))$, with two primitive monadic predicates P_1 and P_2 ($k = 2$, $K = 4$, $c = 3$, presupposing that all the combinations of predicates compatible with the implication have been observed), (14) becomes

$$(15)\qquad \frac{1}{1+\left(\dfrac{3}{4}\right)^n}$$

The "inductive behaviour" suggested by this function is not completely unreasonable qualitatively, although it is clearly far too overoptimistic for small values of n[23].

Preferring (3) to (4) in the situation envisaged in our example may thus be explained in terms of the higher degree of confirmation (*a posteriori* probability) of the preferred generalization. The basic question of this paper thus receives an affirmative answer in the monadic case, which may be expected to be representative of most of the cases for which an inductive logic has so far been developed. In fact, the situation appears essentially similar in the polyadic case.

It is worth pointing out that no difficulties are caused by the possibility that our domain may be infinite; on the contrary, this is usually the most clear-cut case of all. Another disadvantage of Carnap's theory is thus avoided. In general, we may say that the power of a quantitative inductive logic to deal with strict generalizations has been vindicated.

I do not want to suggest, however, that Carnap's critics have been altogether mistaken. On the contrary, the system of inductive logic which I have sketched may be thought of as a partial formalization of ideas closely related to some of their ideas. For instance, it has been maintained by Popper that the best generalizations are typically the *simplest* ones. A closely related idea can now be seen to be incorporated in my system; the fact that our methods prefer (3) to (4) can likewise be related to an interesting notion of simplicity.

Distributive normal forms yield in fact more than one way of measuring the simplicity of a sentence. Among sentences with the same predicates and the same depth, a sentence is naturally said to be the simpler the fewer constituents there occur in its distributive normal form. This criterion of simplicity is closely related to Popper's use of the greater content of a sentence as an indication of its greater simplicity[24].

This does not yet enable us to compare two constituents (with the same predicates and the same depth) for simplicity. A basis for such comparisons is nevertheless seen from (4). The smaller the number w is, the fewer kinds of individuals are allowed to exist by (4), and the simpler the possible world described by (4) therefore is. This number w may therefore serve as a measure of the complexity of (4) in the monadic case.

The polyadic case is in need of further study. It is obvious, in any case, that similar criteria of simplicity are given to us there by the distributive

---

[23] The reasons for this overoptimism are not difficult to diagnose although it would be out of place to discuss them here. It can also be corrected very easily along the lines of footnote 17, e.g. by making the weight of (4) proportional to K^w or perhaps K^{2w}. Then we would have instead of (15) an expression which appears much less overoptimistic.

[24] Popper, *op. cit.*, sections 43, 33–35.

normal forms. For instance, one natural procedure would be to give a weight to each attributive constituent of depth $d-1$ occurring in a constituent of depth d, say in (3). The complexity of (3) is then the sum of these weights. The weight of each of these attributive constituents is in turn obtained by adding the weights of all the attributive constituents of depth $d-2$ occurring in it; and so on, until we reach attributive constituents of depth zero, which all have the same weight.

Although these criteria do not assign a unique measure of simplicity to each closed sentence, they enable us to make several interesting observations. For one thing, our second criterion of simplicity is related rather closely to (although it is certainly different from) Popper's use of what he calls the dimension of a theory as an indication of its complexity[25]. The dimension of a theory is defined by him essentially as the least number of atomic sentences needed to refute it. In the monadic case, the least number of (negated or unnegated) atomic sentences needed to falsify a strong generalization like (4) does not depend on w only. It usually depends also on the order of the predicates which may enter into our atomic sentences[26]. It is easily seen, however, that this number is in no circumstances larger than $1+\log_2 w$, which therefore seems to be the best approximation to Popper's notion of dimension that we can define for the whole of the monadic case. Since $1+\log_2 w$ grows together with w, we obtain by means of it the same simplicity ordering of constituents as was already proposed above.

More important than this is the connection which there exists between our notion of simplicity and the comparison between different generalizations which we have made in the monadic case. Since the complexity of (4) is the greater the larger w is, in preferring (3) to (4) our inductive logic recommends to us the simplest generalization compatible with evidence[27]. Contrary to a frequent suggestion, it is therefore not inevitable that simplicity should always go together with low probability. We have just seen that on a suitable assignment of weights (*a priori* probabilities) to constituents it goes together with high probability on evidence. A simpler constituent does not have a higher *a priori* probability (it would in fact be more natural to give it a lower one)[28], but when more and more evidence comes in, the simplest constituent consistent with it will become more probable than

25 Popper, *op. cit.*, section 38, appendices i, *viii.

26 In other words, the answer to the question: How many of the sentences $(\pm)P_1(a)$, $(\pm)P_2(a), \ldots$, *taken in this order*, are needed to refute (4)? often changes when these sentences are permuted.

27 This recommendation may be thought of as a direct application of Occam's Razor: one should not assume the existence of more kinds of individuals than is made necessary by evidence.

28 Cf. notes 17 and 23.

others. In the long run, this simplest constituent is the only one which is confirmed by the evidence[29].

Interesting further results obtainable in our inductive logic could be mentioned. From a theoretical point of view it is almost as interesting, however, to see what it does *not* enable us to do.

Although it seems to offer a reasonable account of our rational preferences concerning *strict* (inductive) generalizations, it is powerless to cope with inductive generalizations concerning relative numbers of individuals of different kinds. For the sake of a short name, let us call the latter *statistical* generalizations. A simple way of seeing this is to calculate the probability that the next individual observed in our example (in a situation where (3) is the simplest strong generalization compatible with the evidence we already have) should exemplify one fixed kind of individuals, say the one specified by $Ct_{i_1}(x)$. It is easily seen[30] that for an infinite N and for an n large in comparison with K this probability is approximately $(1/c)$. This means that the statistical generalizations which our inductive logic yields take into account the fact which possible kinds of individuals are exemplified in our experience and which ones are not. However, they do not take into account the observed relative frequencies of individuals of the different kinds; as far as the distribution of the individuals among the observed kinds are concerned, our inductive logic adheres simply to the *a priori* recommendation of even distribution.

---

29 When $w > c$, (3) is also easier to falsify than (4) in the sense that fewer (negated or unnegated) atomic sentences are needed to contradict it. This is reminiscent of what Popper says of the connection between simplicity and falsifiability. Perhaps one should not overemphasize the role of falsification here, however, for a certain symmetry obtains here between verification and falsification. Let us write (4) as a conjunction $C_j \& D_j$ of its existential and universal parts. Then in an infinite domain C_j can be finitely verified, but not finitely falsified, whereas D_j can be finitely falsified but not finitely verified. Moreover, C_j is the easier to verify the smaller w is in exactly the same sense in which D_j is the easier to falsify the smaller w is. Hence in a sense a simpler constituent is not only easier to falsify but also easier to verify, namely, easier to verify to the extent it can be finitely verified at all.

30 Let our evidence be e, and let the event that the next individual exemplifies $Ct_{i_1}(x)$ be s. Then we have $p(s/e)$ = the probability of s on the evidence $e = p(s \& e)/p(e)$. Now $p(e) =$

$$(*) \quad \frac{1}{2^K}\left(\frac{m(c)}{M(c)} + (K-c)\frac{m(c+1)}{M(c+1)} + \frac{(K-c)(K-c-1)}{2!}\frac{m(c+2)}{M(c+2)} + \cdots\right).$$

Here the function m is defined as in (11). Clearly, $p(s \& e)$ is obtained from (*) by replacing n by $n+1$. Since (11) is independent of the distribution of the observed n individuals among the c exemplified kinds, the same holds of $p(s/e)$. In fact, an easy computation shows that in the case of an infinite domain it has approximately the value $1/c$ when n is large.

On this basis, it is easily seen that our inductive method does not fall within Carnap's "continuum of inductive methods." The reason is that our method does not satisfy Carnap's condition C9 (see Rudolf Carnap, *The Continuum of Inductive Methods*, University of Chicago Press, Chicago, 1952, p. 13).

This suggests that there are two essentially different problems we have to solve independently of each other in developing a quantitative inductive logic, viz. to formulate the principles that underlie our strict inductive generalizations, and to formulate the principles that underlie our statistical generalizations.[31] Carnap's methods yield undesirable results when applied to the former problem, while our approach does not give much better results in the case of the latter than a strictly *a priori* method. To what extent and in what way the solutions that can be given to these two problems may be combined, remains to be investigated.

If these two problems really turn out to be distinct, we will have to say that some confusion has ensued from the current practice of discussing them as if they were one and the same problem. For instance, we will then have to say that Carnap and Popper have unwittingly addressed themselves to somewhat different problems. Most of the work the former has in fact done pertains to the different methods of statistical generalizations, while the latter appears to be more interested in strict generalizations than in statistical ones.

31 One question that is likely to be affected by a distinction between these two problems concerns Carnap's requirement of "fitting together" (cf. note 21). This requirement may be formulated by saying that the degree of confirmation of a sentence without quantifiers must not change when a new individual is introduced into our language. If the distinction is made, it perhaps suffices to require only that no such change takes place when the new individual does not disturb any of the general laws that hold in our universe.

NEW FOUNDATIONS FOR BAYESIAN DECISION THEORY

RICHARD C. JEFFREY
The City College of the City University of New York, N.Y., U.S.A.

Aristotle makes some use of the ordinal notion of *utility* or *desirability* in his account of deliberation, e.g. in *De Anima* III, 11 (434a):

> whether this or that shall be enacted is . . . a task requiring calculation; and there must be a single standard to measure by, for that is pursued which is *greater*.

But the role of probabilities in deliberation was not clearly seen until much later, e.g. by the authors of the *Port-Royal Logic* (1662; IV, 16):

> to decide what one should do to obtain a good or avoid an evil, it is necessary to consider not only the good and the evil in themselves, but also the probability that they happen or not happen; and to view geometrically the proportion that all these things have together . . .

This is essentially the "Bayesian" account, in which the agent performs an act of maximum ("expected")[1] utility; where the utility of an act is a weighted sum of the utilities of the consequences that it would have in the various possible contingencies, and the weights are the subjective probabilities of (degrees of belief in) those contingencies conditionally upon the act's being performed.

But to use the Bayesian account we need a way of discovering what probabilities the agent does or should attribute to the possible contingencies, and what desirabilities he does or should attribute to the possible consequences of his acts. The values that ought to be attributed often seem easier to obtain or discuss than those that are in fact attributed. Certainly as regards probabilities, the objectivistic relative frequency view has been the one most commonly held, and it has most often been held together with a reluctance to believe that much sense or use can be made of *de facto* subjective probabilities[2].

You can determine someone's degrees of belief if you know the desira-

---

[1] As will appear below, the customary distinction between utility and expected utility plays no role in the present account of deliberation.

[2] The situation has changed considerably in the past decade, largely through the work of Savage [8], which has also had the effect of calling the earlier work of Ramsey [7] and deFinetti [2] to the attention of statisticians. For further references, see the selected bibliography in [4].

bilities he attributes to various prospects; for, roughly speaking, the probability of E is the desirability of what the agent would be just willing to give up in order to be sure of getting something of desirability 1 in case E happens. But how do you measure desirabilities? According to von Neumann and Morgenstern [6] desirabilities are to be measured in terms of probabilities: to say that the desirabilities of A, B, and C are x, y, and z is to say that the agent is indifferent between a guarantee of B and a gamble between A and C with probability $(y - z)/(x - z)$ of getting A and probability $(x - y)/(x - z)$ of getting C.

It was Frank Ramsey [7] who, in 1926, first solved the problem of determining both subjective probability and desirability in terms of the agent's preferences between gambles. For present purposes his procedure may be described as follows. Consider two sets of propositions: the *consequences*, and the *events*, relative to the agent in question. Define a *conditioned event* as the conjunction of a consequence with an event. A *gamble* is a proposition of form

$$[C_1, E, C_2] \tag{1}$$

where C_1 and C_2 are consequences and E is an event. (1) may be read,

$$C_1 \text{ if } E,\ C_2 \text{ if not,} \tag{2}$$

where the two occurrences of 'if' have more than truth-functional significance.

The desirability of such a gamble is a weighted sum of the desirabilities of the possible outcomes, in which the weights are the probabilities of winning and of losing.[3]

$$\mathrm{U}([C_1, E, C_2]) = \mathrm{U}(C_1E)P(E) + U(C_2\bar{E})P(\bar{E}). \tag{3}$$

The fact that the 'if's have more than truth-functional significance in (2) is shown by formula (3), for if we read (1) as

$$(E \supset C_1)\,(\bar{E} \supset C_2) \tag{4}$$

or, equivalently, as

$$EC_1 \vee \bar{E}C_2,$$

[3] I use juxtaposition, the wedge, the horseshoe, and the bar as signs of conjunction, disjunction, material implication, and denial; 'T' for the necessary proposition; and 'F' for the impossible proposition.

the utility of the gamble would be

$$\frac{U(EC_1)P(E)P(C_1/E) + U(\bar{E}C_2)P(\bar{E})P(C_2/\bar{E})}{P(E)P(C_1/E) + P(\bar{E})P(C_2/\bar{E})}.$$

The difference between this expression and the right-hand side of (3) is just the difference between the case in which E and $\bar{E}$ are thought to guarantee C_1 and C_2, and the case in which the probabilities of C_1 and C_2, conditionally on E and $\bar{E}$, need not be 1.

Ramsey showed (*existence theorem*) that if the preference ranking of gambles satisfies certain conditions, there will be functions U and P, defined on the conditioned consequences and on the events, respectively, such that the preference ordering of gambles is identical with the numerical ordering of their expected utilities as computed by (3). He also showed (*uniqueness theorem*) that if two pairs of functions, (U_1, P_1) and (U_2, P_2), both meet the existence conditions, there will be a positive number a and a number b for which we have

$$(5) \quad (a) \qquad P_2 = P_1,$$

$$(b) \qquad U_2 = aU_1 + b.$$

Finally (*equivalence theorem*) if the pair (U_1, P_1) meets the existence conditions and if a is positive, the pair (U_2, P_2) which is defined by (5) will also meet the existence conditions.

My object here is to outline a new theory of subjective probability and utility which is in certain respects simpler and, I think, more satisfactory than Ramsey's[4]. The new theory has certain peculiarities which I shall now list.

1. The theory is *unified* in the sense that probabilities and utilities are attributed to precisely the same objects, viz., to the members of a class of propositions which is closed under the finite truth functional operations (conjunction, disjunction, denial), but from which the impossible proposition has been deleted. In contrast, Ramsey attributes desirabilities to consequences, conditioned consequences, and gambles, but not to events. By Stone's representation theorem, there is no loss of generality if we construe propositions as sets of *possible states of the world*, or *states*, for short[5].

[4] And Savage's [8]. For proofs and fuller expositions, see Bolker [1] and Jeffrey [5].

[5] The states will be the maximal sum ideals in the Boolean algebra of propositions. if propositions are initially construed in some other way than as sets. See Halmos [3], §18.

2. The theory is *non-causal* in the sense that neither [, ,] nor any other such causal notion is taken as primitive[6]. In the theory we can discuss the agent's preference ranking of truth-functional analogues of gambles—of propositions of form (4)—but not of the gambles (1) themselves. In the present theory, the elementary truth-functional operations of conjunction, disjunction, and denial must do the work of Ramsey's operation [, ,], to the extent to which that work is done at all.

3. The role of the linear transformation (5)(b) in Ramsey's theory is played by the *fractional linear transformation*

$$U_2 = \frac{aU_1 + b}{cU_1 + d} \tag{6}$$

in the present theory[7]. The requirement that a be positive in (5)(b) is here replaced by the requirement that the determinant of the transformation be positive:

$$ad - bc > 0. \tag{7}$$

For the equivalence theorem here we also have the requirements

(8) (a) $$cU_1(A) + d > 0,$$

(b) $$cU_1(T) + d = 1,$$

where condition (a) is asserted for all A for which $U_1(A)$ is defined (thus excluding $A = F =$ the empty set of states), and in (b), T is the set of all possible states, i.e., the necessary proposition.

4. Transformation (6) may *change probabilities*: when U_1 is transformed into U_2 by (6), P_1 is transformed into a probability measure P_2 which need not be identical with P_1. The overall situation is most simply described by defining

(9) (a) $$I_1 = U_1 P_1,$$

(b) $$I_2 = U_2 P_2,$$

---

6 The argument following formula (3) above is designed to support the claim that the brackets represent a non-truth-functional operation which, in fact, I take to be a causal operation. This point is discussed further in [5].

7 This was first pointed out to me by Professor Gödel, who also sketched a statement and proof of the uniqueness theorem for the present theory. A somewhat different form of these results had been arrived at earlier (unknown to Gödel and to me) by Ethan Bolker [1].

so that I_n is a measure function, like P_n. In particular, I_n will vanish where P_n does, so that by the Radon-Nykodym theorem there will be an f_n such that

$$(10) \qquad I_n(A) = \int_A f_n \, dP_n .$$

Then by (9) we have

$$(11) \qquad U_n(A) = \frac{1}{P_n(A)} \int_A f_n \, dP_n:$$

the function f_n assigns a "utility" $f_n(s)$ to each state, s, and $U_n(A)$ is then the weighted average of the utilities of the states in which the proposition A is true, the weights being determined by the corresponding probability function, P_n. The transformation (6) can then be represented by a pair of transformations:

$$(12) \quad (a) \quad I_2 = aI_1 + bP_1,$$

$$(b) \quad P_2 = cI_1 + dP_1.$$

Then by (9)(a) we have

$$(13) \qquad P_2 = P_1(cU_1 + d),$$

so that $P_1 = P_2$ if and only if $c = 0$, in which case d must be 1.

I take it that the first two of these peculiarities of the present theory are clear improvements over Ramsey's theory, but that the remaining two require fuller examination and defense.

The unity of the theory is as it should be. Propositions are commonly taken to be the objects of belief, but other sorts of concrete and abstract objects are normally spoken of as the objects of desire. But to desire a certain job, or the love of a good woman, or a ham sandwich is to desire that one or another proposition hold: that the desirer *have* the job, or the woman, or the sandwich, in an appropriate sense of "have."

As to the non-causal character of the present theory, notice first that the relationship between the agent's desirability function, U, and the corresponding probability function, P, is given by

$$(14) \qquad U(A \vee B) = \frac{U(A)P(A) + U(B)P(B)}{P(A) + P(B)} \quad \text{if } P(AB) = 0 \neq P(A \vee B),$$

which is derivable from (9) and the fact that P is a finitely additive probability measure. Putting $\bar{A}$ for B in (14) we have

$$(15) \qquad U(T) = U(A)P(A) + U(\bar{A})P(\bar{A})$$

or setting $P(\bar{A}) = 1 - P(A)$

$$(16) \qquad P(A) = \frac{U(T) - U(\bar{A})}{U(A) - U(\bar{A})_2} \quad \text{if } U(A) \neq U(\bar{A})$$

Equation (16) may seem to reveal a confusion of judgements of fact with judgements of value in the present theory. To see that this appearance is deceptive, notice that by (11), the utility point function, f, can be chosen quite independently of the probability measure P, but that the values of the utility set function, U, will be determined by P as well as by f, for the utility of a proposition is the probability-weighted average of the utilities of the ways in which it might come true. The relationship (16) is no more objectionable than the corresponding relationship in Ramsey's theory,

$$P(E) = \frac{U[C_1, E, C_2] - U(\bar{E}C_2)}{U(EC_1) \;\; - U(\bar{E}C)} \quad \text{if } U(EC_1) \neq U(\bar{E}C_2),$$

which expresses the probability of an event in terms of the utilities of a gamble on that event and of the two possible outcomes of the gamble.

In a sense, any proposition is a gamble, and in particular, given any proposition A which is prefeıred to its denial, the necessary proposition, $A \vee \bar{A}$, is a "natural" gamble on A in which the gain if the agent wins (if A is true) is the truth of A, and the loss if he losses (if A is false) is the truth of $\bar{A}$. It is only such natural gambles that are posited in the present theory. We do not assume, as Ramsey does, that given any pair of consequences C_1, C_2, and any event E, there is a gamble $[C_1, E, C_2]$ on E in which the agent gets C_1 if he wins and C_2 if he loses. Thus, consider the propositions that (C_1) there will be fine weather next week, (C_2) there will be a thermonuclear war next week, and (E) the next card dealt in the Principality of Monaco will be red. In Ramsey's system, the agent is supposed to be in a definite state of preference or indifference between the gamble,

> fine weather next week if the next card dealt in the Principality of Monaco is red, thermonuclear war next week if not,

and (say) the consequence of breaking his ankle tomorrow while playing tennis. But for the agent to seriously entertain the possibility of this gamble would require him to alter his notions of the causes of weather and war so radically as to deprive his state of preference or indifference between it and the consequence in question of all relevance to the world as he actually takes it to be. In confining ourselves to "natural" gambles we work with the world as the agent sees it, including only those causal connections that he actually believes obtain among the propositions in his preference ranking.

It is of some interest to explain the notion of preference with which we are dealing in terms of an ideally self-knowing agent who is being offered options by a person whom I shall call "the operator." To determine the agent's state of preference as to the propositions A, B, the operator offers the agent his choice of (i) having A come true, or (ii) having B come true. If the agent chooses (i), B may come true, or it may not, but A will certainly come true. Similarly, if the agent chooses (ii), A may or may not come true, but B certainly will. The agent must choose (i) or (ii)—one, but not both. Now if the agent chooses (i), he thereby shows that A is at least as high as B in his preference ranking, and if he chooses (ii) he shows that B is at least as high as A in his preference ranking. If B is the necessary proposition, his choice of (ii) is in effect a request that the operator let what will be, be—that the operator not interfere in the course of events. And the choice of (ii) rather than (i) then shows that A is no higher than the necessary proposition in the agent's preference ranking.

It is plausible to define A as *good*, *bad*, or *indifferent* accordingly as A is *above*, *below*, or *with* T in the agent's preference ranking. By (15) it is easy to see that if A is good while $\bar{A}$ is indifferent, then A must have probability 0, and that if two propositions are ranked together but not with T, then the more probable of them is the one whose denial is the further from T. Thus, if the preference ranking is

$$\begin{array}{c} A, B \\ T \\ \bar{A} \\ \bar{B} \end{array}$$

it must be that B is more probable than A. This will give some sense of how the probability ranking is deducible from the preference ranking. The deduction makes heavy use of the fact that the probabilities of propositions play a role in determining the preference ranking of their denials and of other compounds. Our notion of preference already involves the notion of probability; but of course it must, if the notion of probability is to be gotten from it.

To see the exact extent of this "circularity", notice that in explaining preference in terms of the agent's responses to choices offered by an operator, we must suppose that the operator is *neutral* in the sense that his object is neither to gratify nor to frustrate the agent. Suppose the preference ordering of possible states has a bottom rank. If the operator's object were to frustrate the agent as much as possible, he could be relied upon to make whatever happens happen in the worst possible way, and the agent's strategy in choosing between (i) and (ii) above would be to choose (i) if the worst state in A is better than the worst state in B, and to choose (ii) if the worst state in B is better than the worst state in A. The necessary proposition would then be at the bottom of the preference ranking, for it certainly contains the worst states. And since for any proposition A, either A or its denial contains a worst state, either A or its denial will always be ranked at the bottom, with T. Our previous considerations about probability and utility will then go by the board. Thus, by (14), if A and B are incompatible propositions which are not ranked together and which do not have probability 0, then $A \vee B$ will be ranked strictly between A and B; but with a malevolent operator, $A \vee B$ should always be ranked with the worse of A, B.

Then the sort of operator we have in mind when we explain preference in terms of the agent's responses to choices offered by an operator must be described by saying that when the agent chooses (i), say, he believes that the operator will make A happen in a way that is compatible with the agent's existing beliefs: if $A_1, \ldots, A_n$ are n incompatible propositions whose disjunction is A, then the agent believes that when he chooses (i), the probability that the operator will make A_i happen is $P(A_i/A)$, where P is the agent's conditional probability measure . This would vitiate the talk of an operator if it were intended as *the interpretation* of the notion of preference, rather than as a partial explanation. In fact, I do not regard the notion of preference as epistemologically prior to the notions of probability and utility. In many cases we or the agent may be fairly clear about the probabilities he ascribes to certain propositions without having much idea of their preference ranking, which we thereupon deduce indirectly, in part by using probability considerations. The notions of preference, probability, and utility are intimately related; and the object of the present theory is to reveal their interconnections, not to "reduce" two of them to one of the others.

One of the principal uses of the theory of preference is to help the agent see what his state of preference or indifference between two propositions must be, given that his preferences between other related propositions are such-and-such, and that he wishes to be consistent in the Bayesian sense. Thus, consider the thesis that if A is preferred to B then $\bar{B}$ must be preferred to $\bar{A}$. This thesis is false, but its falsity is not obvious. To see it, consider an example in which A is the proposition that the agent dies tomorrow, B is the proposition that the agent is dishonored tomorrow, and the agent is in fact a Roman matron who, subscribing to the slogan, "Death before dishonor", will straightway kill herself if dishonored. For consistency with what we may plausibly take her other preferences to be, she must also prefer $\bar{A}$ to $\bar{B}$: she must prefer a guarantee of living through tomorrow to a guarantee of not being dishonored all day tomorrow, because she takes the probability of living through tomorrow without honor to be zero. In particular, she believes that there are just three real possibilities as to the joint truth and falsity of A and B, which she ranks as follows:

$$\bar{A}\bar{B}$$
$$A\bar{B}$$
$$AB$$

Then $\bar{A}$ is practically the same proposition as $\bar{A}\bar{B}$ and is therefore at the top of this fragment of her preference ranking, while $\bar{B}$, the disjunction of the top and middle-ranking of the three propositions, must be ranked below the top and above the middle, since all three propositions have positive probability.

Let me now consider the third idiosyncracy of the present system—the fact that any fractional linear transformation $U_1 \rightarrow U_2$ as in (6) preserves the preference ranking. To focus attention on essentials, let us arbitrarily set $U_1(T) = U_2(T) = 0$ and $U_1(G) = U_2(G) = 1$, where G is some good proposition of which the denial is bad. Then if (as in Ramsey's theory) the only preference-preserving transformations are linear, all utility values are now determined. But by (6), (7), and (8) we have

$$U_2 = \frac{(c+1)U_1}{cU_1 + 1}: \tag{17}$$

there remains a certain freedom in choosing utility values, corresponding to the parameter c, which can be no smaller than -1.

The situation is shown graphically in Figure 1, where we suppose that c is negative. (The corresponding situation for positive values of c is that in which the point P is below inf U_1.)

In Figure 1, h, the height of the perspective point P, is $-1/c$, and the correspondence between the U_1 scale (vertical axis) and U_2 scale (horizontal axis) is simply this: that for each A, $U_2(A)$ is the perspective image of $U_1(A)$ from P. Since P lies on the line through the units of the two axes, we have $U_1(A) = 1$ if and only if $U_2(A) = 1$; and since the zeros of the two scales coincide we have $U_1(A) = 0$ if and only if $U_2(A) = 0$. But there are no other fixed points unless the perspective point is infinitely high, in which case the rays from P are parallel, at $-45°$, and we have $U_2(A) = U_1(A)$ for all A. This is the case in which h is infinite and therefore c is 0.

The startling fact is that by choosing h equal to the supremum of the U_1 scale we can transform a utility scale which has a finite upper bound into

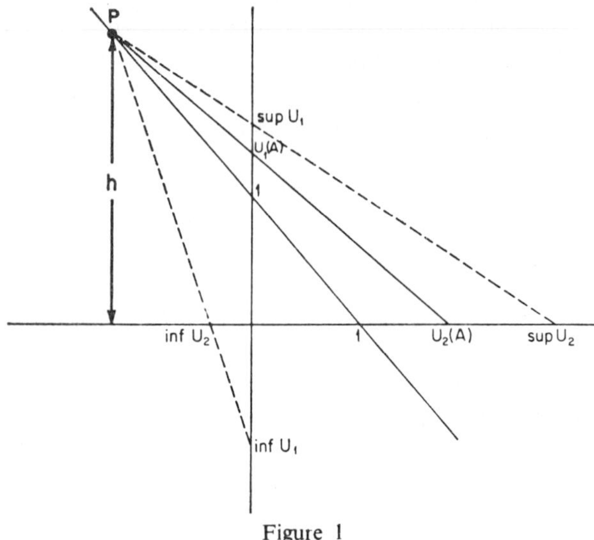

Figure 1

an equivalent utility scale that has no finite upper bound; and similarly (with P below the horizontal axis, on a level with inf U_1) we can transform a scale which has a finite lower bound into a scale that has no finite lower bound. Boundedness above and boundedness below are not preferentially significant properties of utility functions in the present system. But we cannot transform a scale which is finitely bounded above or below or both into an equivalent scale which has neither a finite upper bound nor a finite lower bound; and therefore boundedness (above-or-below-or-both) is a preferentially significant property, even though the three separate species of boundedness are not.

This seems to put us into conflict with the renowned St. Petersburg paradox, as follows. Let $U(C_1)$, $U(C_2)$, ... be an increasing, unbounded sequence, and for each n, let E_n be the proposition that the tail turns up first on the n'th toss of a certain well-balanced coin. Then it is plausible that for each n, $P(E_n) = 2^{-n}$; and we can find a sequence $m_1, m_2, \ldots$ of positive integers such that for each i, $U(C_{m_i})$ is at least 2^i. The *St. Petersburg game* is an arrangement in which the agent tosses the coin in question until the tail turns up on, say, the t'th toss, whereupon the consequence C_{m_t} is actualized. (Suppose that the consequences are payments of money to the agent.) Now the prospect of playing the St. Petersburg game has infinite utility; and the *St. Petersburg paradox* consists in the fact that in systems of the usual sort, a utility function which is unbounded above (below) must actually assume the value ∞ ($-\infty$). And the paradox is disturbing because it is unreasonable for the agent to be a Bayesian if there is a prospect of positive probability and infinite utility. Thus, let A, B, and C be propositions of positive probability; let AC and BC both have probability 0; and let the utilities of A, B, and C be u, v, w, where u and v are finite, w is infinite, and $u < v$. Then the agent, if reasonable, will prefer $B \vee C$ to $A \vee C$; but on Bayesian principles we have

$$U(B \vee C) = U(A \vee C) = w.$$

The usual way of solving the difficulty is to assume that the utility function is bounded above and below; but this move is not open to us, for boundedness is not a preferentially significant property of the utility function in the present theory. Instead, we deny that the prospect of playing the St. Petersburg game occurs in the agent's preference ranking, even if all the relevant E's and C's do. The basic remark is that anyone who offers to let the agent play the St. Petersburg game is a liar, pretending to have an infinite bank. We need not (and cannot) assume that the utility function is bounded above and below; but we do assume that there are finite upper and lower bounds on the utilities of the prospects that it is within anyone's power to make happen at his pleasure. I.e., for each person, we assume the existence of prospects E and F which are ranked above and below all prospects that are within that person's power to make happen at his pleasure. This is not open to Ramsey, who must assume that any two prospects can be the possible outcomes of a gamble[8].

Finally, let us consider the fourth idiosyncracy: that probabilities are not completely determined by the preference ranking. For any proposition A there is a definite probability quantization, corresponding to the difference between the maximum and minimum values that $P_2(A)$ can be forced to assume by suitably adjusting c and d in (13). The difference between this maximum and minimum value for a given A is the preferentially significant quantity

$$P(A)U(A)\left(\frac{1}{\sup U} - \frac{1}{\inf U}\right) \tag{18}$$

—preferentially significant in the sense that it is invariant under all transformations (6), (13) of U and P. Furthermore, the least upper bound of this quantity, for all A in the preference ranking, will be

$$(\sup I)\left(\frac{1}{\sup U} - \frac{1}{\inf U}\right). \tag{19}$$

---

[8] In fact, the paradox need not arise in Ramsey's theory either, even if the utility function is unbounded, for no finite number of applications of the operation [, ,] will yield the St. Petersburg gamble. On the other hand, in Savage's system there are acts of infinite utility if the utility function is unbounded: see [8], p. 81.

Now whether or not U is bounded above or below, I will be bounded both above and below, and the coarseness of the quantization tends toward zero as $\sup U$ and $\inf U$ tend toward $+\infty$ and $-\infty$ with $U(T)$ fixed at 0 and $U(G)$ fixed at 1. Then in practice, the phenomenon of probability quantization need not be noticeable. Certainly it will make no difference to decision-making, since the quantization is precisely the degree of indeterminacy that is left over after all practical questions have been answered. The quantization will be zero if $\sup U = \infty$ and $\inf U = -\infty$, but I know of no reason to suppose that this must generally be the case for a rational agent, unless indeed it be that only then are his probabilities uniquely determined by his preferences.

ACKNOWLEDGMENTS

This work was begun at Stanford University and completed at the Institute for Advanced Study and Princeton University under United States Air Force Grant AFOSR–529–64. I wish to thank Rudolf Carnap and L. J. Savage for valuable criticism at the early stages. As noted above, the unicity problem was first solved by Ethan Bolker (in another context, independently of the present investigation); and he has since solved the mathematically deeper problem of characterizing the conditions under which a preference ranking is Bayesian. These results will appear in Bolker [1]. I am grateful to Bolker for his generosity in allowing me to read his unpublished work and for his patience and skill in explaining it to me. I am especially indebted to Professor Gödel for his generous help and encouragement during my term at the Institute. My first understanding of the mathematical situation described in the equivalence and uniqueness theorems came from him. Equally important has been his guidance concerning the relative philosophical interest and mathematical feasibility of different lines of investigation.

REFERENCES

[1] BOLKER, ETHAN, *Functions resembling quotients of measures*. Ph. D. Dissertation, Harvard University, forthcoming.

[2] DEFINETTI, BRUNO, *La prévision: ses lois logiques, ses sources subjectives. Annales de l'Institut Henri Poincaré*, vol. 7 (1937). English translation in [4].

[3] HALMOS, PAUL R., *Lectures on Boolean algebras*. New York and London, 1963.

[4] KYBURG, HENRY E., JR., and HOWARD E. SMOKLER, editors, *Studies in subjective probability*. New York and London, 1964.

[5] JEFFREY, RICHARD C., *The logic of decision*, forthcoming.

[6] VON NEUMANN, JOHN, and OSKAR MORGENSTERN, *Theory of games and economic behavior*. Princeton, second edition, 1947.

[7] RAMSEY, FRANK P., "Truth and probability", in *The foundations of mathematics and other logical essays*. London and New York, 1931. Reprinted in [4].

[8] SAVAGE, LEONARD J., *The foundations of statistics*. New York and London, 1954.

PROBABILITY, RATIONALITY, AND A RULE OF DETACHMENT

HENRY E. KYBURG
Wayne State University, Detroit, Michigan, U.S.A.

Carnap and many other writers on induction and probability deny that there is a rule of detachment in inductive logic. According to their view the relation among evidence, hypothesis, belief, and action is the following: A scientific hypothesis H is rendered probable to such and such a degree by evidence E. E is understood to include all of the relevant information we have concerning H. We *ought* to have a degree of belief in H corresponding to its probability. We should never *accept* H—not even provisionally. We may *act* in accordance with H, but the basis for our action is a straightforward calculation of the mathematical expectation of our various alternatives.

Another behavioristic school of thought that has been rapidly growing in influence is that of the subjectivistic statisticians. For these writers, probability is no longer even a logical relation between hypothesis and evidence, but a measure of subjective confidence that satisfies certain rules (the axioms of the probability calculus). It would be possible for the subjectivists to talk about the *acceptance* of a hypothesis, but not in any interesting way. To accept a hypothesis H would be to assign to it a personal or subjectivistic probability of unity. But the only way that we can assign a conditional probability of unity to H on evidence E is to assign an *a priori* probability of unity to H. (The probability of E must be the same as the probability of the conjunction of H and E.) This is clearly of no interest to us.

The majority of writers on statistical inference fall into an even more explicitly behavioristic camp. Let us call Carnap and the subjectivists *Bayesians*, and these other writers non-Bayesians. Many non-Bayesians formulate the problem of statistical inference, from the outset, as a problem of choosing between alternative courses of action. Neyman, for example, talks only of inductive *behavior*, not of inductive inference. We have here also the work that stems from Wald's fundamental research in decision theory and the theory of testing hypotheses. Sometimes the behavioristic character of these approaches is thinly veiled by confusing terminology—there are writers who use the phrase "Accept hypothesis H_i," to mean "Follow the course of action A_i which would be most appropriate were

H_i true." "Most appropriate," of course, means most appropriate *under the conditions of the problem*. But I don't think this terminological confusion need bother us; *we* know that when a decision theorist or a hypothesis tester tells us to *accept* hypothesis H_i, he means for us only to adopt a certain course of action in the present circumstances, and nothing more.

One objection to this behavioristic view of science is that it doesn't seem to be true: scientists (and other people) simply do *accept* hypotheses, when the evidence in their favor becomes overwhelming. When there is a good deal of evidence in favor of a hypothesis, so that we surely ought to believe it to some extent, there may still be circumstances in which it would be imprudent to act on it. But on any behavioristic analysis it is absolute nonsense to say "believe *H*, but don't act on it." I am sure that an overpowering ordinary language argument could be constructed in favor of the thesis that scientists accept hypotheses; but in these technical regions I find ordinary language arguments rather less than persuasive and rather more than a little dull, and so I shall spare you one.

There is one argument against the behavioristic approach to scientific inference that does strike me as interesting enough to mention. According to this approach, there are exactly three things we are concerned with in the practice of science: (1) we have a certain body of evidence, consisting of statements asserting the occurrence of particular events; (2) we have a certain number of courses of action open to us; and (3) we have a certain utility function which assigns values to the possible outcomes of the actions that are open to us. There is much that might be said concerning the nature of evidence-statements. (I would claim, for example, that many of the statements that we *accept* as evidence statements are acceptable, or indeed, intelligible, only in virtue of an *accepted* body of theoretical scientific knowledge.) It may also be questioned whether values can be plausibly assigned to the outcomes of actions except within the framework of some general, accepted, theory. But leaving these two points aside, the strikingly paradoxical consequence of the behavioristic approach is that scientific hypotheses—those grand, elegant creations of the human intellect, that have been ranked by some above the works of Bach and Michelangelo—turn out to be altogether redundant and useless. They are, to invert a metaphor of Bergson, the pretty but useless colored sparks that fall away from the tail of the soaring rocket. *Having* hypotheses may be useful—they may help to simplify the computations of the utilities to be associated with various actions—but they add nothing whatever to our view of the world, or to the factual basis of our actions.

In Carnap's *Logical Foundations of Probability*, theories not only turned out to be redundant, but also to have zero probability. They were not only useless, but incredible. Carnap was not satisfied with that, and has since worked on metrics for degree of confirmation which will allow theories and universal hypotheses to have degrees of confirmation greater than zero. But observe that *even then* a universal statement or a theory is redundant. It is a useless intermediary (no matter what its degree of confirmation) between the evidence statements and the statements of outcomes of particular courses of action in a particular concrete situation. We need only be concerned with *evidence*, *utility*, and *action*.

There are those for whom the creation and testing of hypotheses is the very soul of science. In America for many years the prevalent view of scientific practice had it this way: formulate hypotheses; test them; accept them provisionally so long as they continue to pass tests. To accept a hypothesis provisionally is of course to accept a hypothesis. In England Karl Popper and his followers have argued for a view of science in which one of the key elements is the provisional *acceptance* of hypotheses. As a guide to the acceptance of hypotheses, Popper has proposed a concept of 'degree of corroboration'. This is not a probability concept; it is based on a notion of logical measure, but it does not satisfy the axioms of a probability calculus. (Other writers, e.g., Kemeny and Oppenheim, Rescher, Finch, have proposed other measures of factual support; these proposals are systematically compared in my "Recent Work in Inductive Logic," *American Philosophical Quarterly*, I (1964), pp. 1–39.)

The question naturally arises as to the possibility of finding some bridge between the two approaches. Carl Hempel began the attempt to construct such a bridge with the help of the concept of *epistemic utility*. That is, given a body of statements K, including evidence statements, we can consider three courses of epistemic action with respect to a hypothesis H: We may accept H, adding it to the set K; we may reject it, adding its negation to K; or we may suspend judgement and add neither it nor its negation to K.

Isaac Levi has applied this notion of epistemic utility to the analysis of Popper's proposed 'degree of corroboration', and argued that "All utility functions that can be generated from possible analyses of $C(h, e)$ into a measure of expected gain [i.e., into a sum of the utilities of alternatives, each multiplied by its logical probability] require that the utilities assigned to the outcome of accepting h when it is true and to the outcome of accepting h when it is false vary with $P(h, e)$ — the probability of h given e. This amounts to saying that scientists change their preferences with changes in evidence."[1] This violates our intuitions, as well as the

[1] Levi, Isaac. "Decision Theory and Confirmation," *Journal of Philosophy* 60, 1963, p. 623.

conventions of every decision-theoretic approach, Bayesian or otherwise. Indeed Levi's conclusion is quite general: Any Bayesian decision rule which attempts to take such factors as simplicity and explanatory power into account in the calculation of utilities is committed to saying that scientists change their preferences with changes in evidence.

Non-Bayesian strategies may offer some way out of the problem: e.g., perhaps one can plausibly apply a minimax strategy to $C(h, e)$. But this is another problem, and one that I cannot go into here.

At this point it will be helpful to introduce Hempel's criteria of adequacy for an inductive rule of detachment.[2] These criteria, he says, state "*certain necessary conditions of rationality in the formation of beliefs.*" The first three of these conditions seem innocuous enough. They are:

"CR 1. Any logical consequence of a set of accepted statements is likewise an accepted statement; or K contains all logical consequences of any of its subclasses."

"CR 2. The set K of accepted statements is logically consistent."

"CR 3. The inferential acceptance of any statement h into K is decided on by reference to the total system K."

The third condition is merely a requirement of total evidence. The first two conditions are ambiguous as they stand, as I shall show shortly, and except on the weakest interpretation, one of them must be rejected.

The specific rule that Hempel suggests makes use of a content measure function m defined over statements in an appropriate formalized language for science, and a confirmation function c satisfying the usual probability calculus. When h is the hypothesis and k the conjunction of statements in K, he takes the epistemic utility of accepting h when it is true to be $m(h \vee {\sim} k)$ and of accepting h when it is false to be $-m(h \vee {\sim} k)$. The utility of leaving h in suspense is 0. It is shown, after some computation, that this leads to the following rule:

"(12.4) *Tentative rule for inductive acceptance*: Accept or reject h, given K, according as $c(h, k) > 1/2$ or $c(h, k) < 1/2$; *when* $c(h, k) = 1/2$, h may be accepted, rejected, or left in suspense."[3]

It is interesting that the measure of information drops out of the picture, and that we are led to this rule (as Hempel observes) regardless of what confirmation function we start with.

---

[2] Hempel, Carl, "Deductive Nomological vs. Statistical Explanation" in Feigl, H. and Maxwell, G., *Scientific Explanation, Space and Time*, Minnesota Studies in the Philosophy of Science, Vol. III, University of Minnesota Press, Minneapolis, 1962, pp. 150–151.

[3] Hempel, *op. cit.*, p. 155.

Hempel realizes that rule 12.4 is far too liberal. In another context he presents a version of the lottery paradox, but he does not explore the implications of this paradox for his acceptance rules. We can do so easily. Consider this example:

"There are three balls in this urn, a blue one, a red one, and a green one."
"A ball is about to be drawn from the urn by a blindfolded man."

This is surely a plausible basis for a body of knowledge K_1, isn't it? Yet by 12.4 (taking any ordinary confirmation function) we find we must add the following statements to K_1:

"The ball will not be blue".
"The ball will not be red".
"The ball will not be green."

By CR 1, then, we have: "A ball is about to be drawn from an urn containing only a red ball, a blue ball, and a green ball and this ball will be neither red nor blue nor green." But this is in clear violation of CR 2.

Can we fix matters by making the criterion 12.4 mention a higher degree of confirmation? For example, 0.999 instead of 1/2, so that we accept h only if $c(h, k) > 0.999$ and $\sim h$ only if $c(h, k) < .001$, and otherwise suspend judgement? Clearly this won't work. Take any ε you wish; put a number N of counters greater than $1/\varepsilon$ in an urn; the counters to be numbered successively from 1 to N. Take K_2 to include the two sentences:

"There are N consecutively numbered counters in this urn."
"One counter is to be drawn by a blindfolded choirboy."

Again it is easy to show that this K_2 does not satisfy Hempel's conditions.

This already seems to me to be a serious problem; why on earth shouldn't a set of statements such as K_2 be a perfectly plausible set of statements to be believed or accepted?

Wesley Salmon has made one proposal (in correspondence) which would take care of the difficulty in which we find ourselves now; his idea is to limit inductive rules of acceptance to rules for accepting *general* statements; particularly, statistical hypotheses. Hempel suggests (as a cure for the overliberality of his 12.4) that we search for a more plausible measure of epistemic utility. I am sure that many of you are thinking that such sets of statements as K_1 and K_2 are so outrageously silly and arbitrary that it is a waste of time to worry about the fact that they fail to satisfy Hempel's very natural conditions. CR 1 and CR 2 are not only natural, but have been explicitly defended as epistemic principles by Chisholm and by Martin. They hold for some of the concepts defined by Hintikka.

But I shall now show that Salmon's proposal will not correct the situation;

that Hempel's hope that a better definition of epistemic utility would help is unwarranted; that the problem is a real and serious one even in the *practice* of statistics; and that the difficulty stems from CR 1, CR 2, and *any* probabilistic rule of detachment like Hempel's 12.4.

All this is revealed by the following example. Let there be a population of 100,000 balls in a very large urn. Those who are of a very practical turn of mind may think of this as representing the products of an automatic screw machine; telephone receivers; eggs; live births at a hospital; micro-organisms in a bottle; etc. Some of the balls are black. (Some of the parts are defective; some of the births are births of males; etc.) No statistician will deny, I think, that there *are* occasions when Bayesian inference is appropriate. Let this be one of those occasions. For frequency theorists this amounts to stipulating that we have a prior distribution for the parameter p that represents the proportion of black balls in the urn. For subjectivistic and logical theorists this amounts to no stipulation at all. We now draw a sample of 10,000 balls; we stipulate that the sample is *random* in whatever sense is regarded as appropriate. Again this is a condition that can be met in some cases, one way or another, whatever your views on probability. In this sample, 4,867 balls are black. With the help of Bayes' theorem, we calculate a posterior distribution for the parameter p.

Now statements of the form "p lies between p_1 and p_2" are certainly general statistical statements—they meet Salmon's criterion for detachment. They *might* be said to vary in content of information, but surely two such statements in which the difference between the upper and lower bounds was the same would have to be granted the same information content. (I.e., when $|p_1 - p_2| = |p_1' - p_2'|$.) Thus if one of these statements meets a modified Hempelian criterion for detachment, so will the other if its probability is the same. Suppose now that the level of probability chosen as the criterion for detachment of general statistical statements is $1 - \varepsilon$. Then there is no trouble at all, obviously, in finding numbers p_1, p_2, and p_3, where $p_2 < p_3$, such that the statements,

"$.4867 - p_1 < p < .4867 + p_2$"
"$.4867 - p_2 < p < .4867 + p_1$"

have the same probability and are *acceptable*, and also such that the statement,

"It is not the case that $.4867 - p_3 < p < .4867 + p_3$"

is also overwhelmingly probable and must be accepted. (The information content of this statement may not be the same as the information content of the first two; but the fact that we may adjust the parameter p_3 to make this statement as probable as we please allows us to compensate for the effects of any difference in information content. Remember that we are merely constructing a possibility that could occur in a practical situation; we do not have to suppose that this sort of thing occurs frequently, or at all.)

The triple of sentences that I have just described clearly leads to a violation of Hempel's first two conditions. Changing the level of acceptability, changing the definition of information content, restricting ourselves to realistic situations—none of these things have allowed us to avoid this form of the lottery paradox.

If we are to hold on to a probabilistic rule of detachment like Hempel's, there is clearly only one thing to do in this situation, and that is to reject one or both of CR 1 and CR 2. I mentioned before that these conditions are ambiguous. So far I have been interpreting them as Hempel intended; but there is another interpretation in which they hold for my own system of *Probability and the Logic of Rational Belief.* Let us take CR 1 to mean *not* that every logical consequence of the *conjunction* of the statements of K belongs to K, but only that every logical consequence of each *single* element of K belongs to K. Thus if P and Q belong to K, their conjunction may not belong to K, unless it is included in K on independent grounds. In conjunction with a probabilistic rule of detachment this interpretation of CR 1 is very natural—it is clear that P can be overwhelmingly probable, and Q can be overwhelmingly probable, without their conjunction being overwhelmingly probable.

The same interpretation may be given to CR 2. Rather than demand that K be consistent in the sense that no contradiction be deducible from the *conjunction* of all the elements of K, we may rest content with the stipulation that there be no statement S of K such that K contain both S and the negation of S. (An even weaker stipulation would be that K contain no *self-*contradictory statement.)

The rule of detachment in *Probability and the Logic of Rational Belief* is a purely probabilistic one. I take probability to be a logical relation defined for classes of equivalent sentences of a formalized language. I take the probability of a sentence S, relative to a rational corpus K, to be the *interval* $(p; q)$, when there are terms a, b, and c such that in K, S is known to be equivalent to "$a \in c$"; relative to K, a is a random member of b with respect to membership in c; and it is known in K that the proportion of b's that are c's lies in the interval $(p; q)$. I take rational corpora to be based on a certain fund F of 'observation statements', but like Hempel I leave to one side the question of what sort of statements these might be. I define a sequence of rational corpora, based on the set of statements F; the sequence of levels corresponds to levels of 'practical certainty' suitable to different practical situations. The highest level of rational corpus consists of F and the logical consequences of F—including the truths of logic and mathematics. The contents of lower level rational corpora is determined entirely by a general rule of detachment:

D S belongs to the rational corpus of level i and basis F, if and only if the probability of S, relative to the rational corpus of level $i + 1$ and basis F, is $(p; q)$, where $p > r_i$ (r_i being the critical probability level of acceptance for the rational corpus of level i).

It is possible to show that the rational corpora of my system satisfy the modified Hempel conditions CR 1 and CR 2. The lottery paradox fails to arise initially because, even if S_1 and S_2 and ... and S_n are so probable that they are to be included in the rational corpus of level i, it will *not* generally be the case that their conjunction is that probable. We detach any statements that are probable enough—but we do not necessarily accept the conjunctions of the statements we accept.

This sounds quite plausible, but Fred Schick has shown that the lottery paradox still arises, though in a slightly different way. I showed in my book that if S_1 is a statement in the rational corpus of level i, and a conditional whose antecedent is S_1 and whose consequent is S_2 is in the rational corpus of level $i + 1$, then S_2 will also belong to the rational corpus of level i. Since $S_1 \supset (S_2 \supset (S_1 \,.\, S_2))$ is a logical truth that appears in every rational corpus, this means that the conjunction of S_1 and S_2 will appear in the rational corpus of next lower level to that in which S_1 and S_2 appear separately. In other words, if S_1 belongs to the rational corpus of level i and S_2 belongs to the rational corpus of level i, then their conjunction belongs to the rational corpus of level $i - 1$.

So far, so good. This sounds reasonable. But what Fred Schick pointed out was that the process could be continued. Let $S_1, S_2, \ldots, S_n$ belong to the rational corpus of level i. Since $S_1 \supset (S_2 \supset (S_1 \,.\, S_2))$ is logically true, $S_2 \supset (S_1 \,.\, S_2)$ also belongs to the rational corpus of level i. The same argument supports $(S_1 \,.\, S_2) \supset (S_1 \,.\, S_2 \,.\, S_3)$ in virtue of the logical truth of $S_3 \supset ((S_1 \,.\, S_2) \supset (S_1 \,.\, S_2 \,.\, S_3))$; $(S_1 \,.\, S_2 \,.\, S_3) \supset (S_1 \,.\, S_2 \,.\, S_3 \,.\, S_4)$ in virtue of the logical truth of $S_4 \supset ((S_1 \,.\, S_2 \,.\, S_3) \supset (S_1 \,.\, S_2 \,.\, S_3 \,.\, S_4))$; etc. By the principle mentioned above, this leads to the conclusion that the conjunction of all of the statements in the rational corpus of level i will appear in the rational corpus of level $i - 1$.

The lottery paradox now arises in the following way. Let F contain the statements of K_2 above concerning chips in an urn. Let S_i be the statement "Chip number i will not be drawn." Clearly these statements may be so probable as to be detached in a certain rational corpus. But then the rational corpus of the next lower level will contain the explicit contradiction: "One chip will be drawn and it will not be chip number 1, and it will not be chip number 2, and ··· and these are all the chips there are."

I have a solution for this paradox, but it is not one that I am particularly happy with. I should like to think of something better. Nevertheless, I offer it to you for what it is worth.

We can avoid the lottery paradox in my system by dealing with only two levels of rational corpora at a time. The highest one we regard as containing only genuine, unquestioned 'certainties'—it might be regarded as a body of philosophical knowledge. It would, in any given context, correspond to the highest level rational corpus of my original system. The other rational corpus would be of a level corresponding to *practical* (as opposed to philosophical) certainty in the given situation. This level would vary from situation to situation—it might be characterized by a detachment parameter of 0.7 in one situation, by one of 0.95 in another, and by a detachment parameter of 0.99995 in yet another. But we would consider only *one* level of practical rational corpus at a time. The practical rational corpus would contain statements that were appropriately probable relative to the highest level rational corpus. The practical rational corpus would not be logically closed, of course. It, like the rational corpus of level i above, might contain S_1 and S_2, but not their conjunction. Now the only place where we allow an inductive rule of detachment is in getting statements into the practical rational corpus on the basis of their probability relative to the highest level rational corpus. We keep the definition of probability unchanged; we can make probability statements relative to the lower level rational corpus; but we do not allow a statement to be detached from its evidence in virtue of its probability relative to the lower level, practical, rational corpus.

The lottery paradox is now avoided simply by virtue of the fact that the rule of detachment is allowed to operate at only one level in a given context. While the philosophical or highest level rational corpus may contain such statements that for $1 \leqq i \leqq N$, the statement "Chip number i will not be drawn" is an ingredient of the practical rational corpus, there *is* no *next* lower rational corpus in which the conjunction of these statements will appear.

One may argue about whether or not it is reasonable to look for an inductive rule of detachment; I have presented some arguments in favor of looking for one. (But I have not refuted Bar-Hillel's argument, presented at the Wesleyan Conference on Induction. His argument was: "Who *needs* a rule of detachment?") But given that one *is* going to look for a rule of detachment, there is a great deal to be said about the characteristics of the bodies of knowledge to which it leads. Nearly all writers on epistemology, and nearly all proposers of inductive rules of detachment, have accepted conditions of complete logical closure and consistency. Levi has shown that the non-probabilistic proposals involving degrees of corroboration or factual support break down if they are made the basis of a Bayesian approach to inference. I have attempted to show that probabilistic rules—as generally outlined by Hempel—lead to the lottery paradox so long as we hang onto the ideal of consistency and the ideal of logical closure. If this is so, Hempel's conditions *cannot* be satisfied by a probabilistic rule of detachment. In my own system these ideals are abandoned in favor of much weaker conditions of consistency and logical closure. And in this system—as modified above—we have a rule of detachment which can lead to the acceptance of scientific theories and generalizations, as well as of individual predictions, and which yet does not generate the lottery paradox. Defective as the modified system may turn out to be, I still hope that *some* of my beliefs are rational. Hope springs eternal.

Section VI

Methodology and Philosophy of Physical Sciences

IRREVERSIBILITY PROBLEMS

O. COSTA DE BEAUREGARD
Institut Henri Poincaré, Paris, France

In recent years there has been much thinking on the ultimate formulation of the principles of physical irreversibility appearing in the various (though more or less related) fields of general probability theory, phenomenological thermodynamics, statistical mechanics, wave theory, quantum mechanics, and cybernetics. Moreover, a definite connection between the physical irreversibility principles and the general principle of causality has been stressed independently by various authors.

Although I believe that this survey will touch on the major aspects of the so-called statistical theory of time that have been thoroughly discussed in recent years, I am aware that other aspects of the time problem, which have also been considered by valuable authors, will be left out of my scope, for which I apologize.

1. Is physical time directionless?

Let us recall first the early objections of Loschmidt[1] and Zermelo[2] to Boltzmann's[3] deduction of statistical irreversibility.

According to the reversibility of classical mechanics, if, at any instant, all velocities of a system are exactly reversed, the system will go back through the states which it had previously traversed in opposite order; thus, there must be just as many probability decreasing evolutions as there are probability increasing ones.

Also, according to Poincaré's recurrence theorem[4], any mechanical system with a finite number of degrees of freedom approaches any state through which it goes an infinite number of times at a given approximation: in particular, any state of low probability which the system has gone through

---

1 *Wien Ber.* 73, 128 and 366, 1876.

2 *Ann. der Phys.* 57, 485, 1896.

3 *Vorlesungen über Gastheorie*, Leipzig, Barth, 1896.

4 *Acta Math.* 13, 67–72, 1890.

will be very nearly reproduced an infinite number of times in the future (as it has been in the past)[5].

On strictly technical grounds these Loschmidt and Zermelo arguments are irrefutable, as the famous writings of the Ehrenfests[6] and of Smoluchowski[7] have made obvious: *per se*, the evolution in time of a perfectly isolated system with a large, but finite, number of degrees of freedom would by symmetric, with numerous small undulations and rare abrupt canyons in a general plateau, as shown in Figure 1.

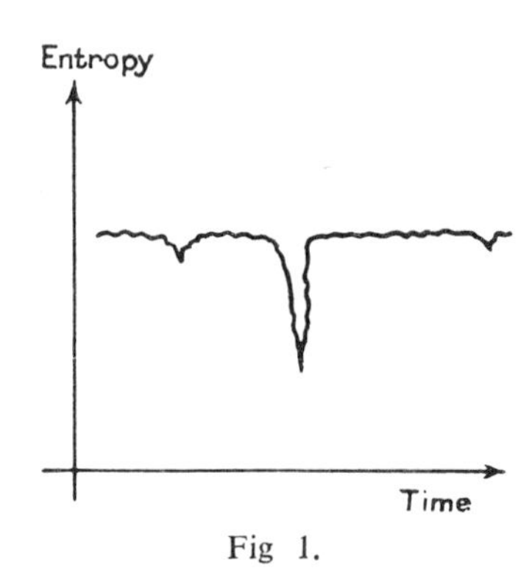

Fig 1.

The problem is thus an epistemological one, raising such questions as: can a finite system be considered as perfectly isolated when long time intervals are considered? Since the days of Loschmidt and Zermelo, there have been many essays which aimed at justifying the physical statistical irreversibility, the older ones of which have, in fact, obscured much more than they have enlightened the critical points involved.

So let us turn to modern writings on the subject.

There have been physicists and physico-chemists among the innovators of the new *information* concept as defined in relation to *entropy*. The first of these was Boltzmann himself, who wrote somewhere that entropy measures "missing information" on a system.

In well known papers of 1929 and 1930 respectively, Szilard[8] and Lewis[9] explicitly introduced the *information* concept into the Maxwell

---

[5] Recently J. A. MacLennan has completed the theorem by showing what was intuitively felt: that the time intervals separating the Poincaré recurrences grow longer as the considered states are further from the equilibrium (or maximal probability) state (*Physics of fluids* 2, 92, 1959).

[6] *Encycl. Math. Wiss.* IV, 4, D (Leipzig 1911) 332 p. 43.

[7] *Oeuvres* t. II, 1927, pp. 361–398.

[8] *Zeits. f. Phys.* 53, 840, 1929.

[9] *Science* 71, 569, 1930.

demon problem. In order to show how the irreversibility *versus* reversibility question is connected, in Lewis' mind, with the information question, I quote directly the following sentences: "In going from the very simple science of mechanics to the very complex science of psychology, we must change from two-way to one-way time. It is important to inquire where this transition comes. The thesis that I wish to elaborate is that throughout the sciences of physics and chemistry, symmetrical or two-way time everywhere suffices." After considering problems of card-shuffling and of scattering of particles, Lewis concludes: "There is no question that these processes involve an element of *loss* which typifies an irreversible process, but this loss implies in no way a dissymmetry in time, nor indeed any temporal implication whatever. *Gain in entropy always means loss of information and nothing else. It is a subjective concept.*" (My italics.)

These assertions convey important elements for understanding our problem. I feel that, on the whole, Lewis nevertheless misses the point he raises by a small margin. He writes for instance that, in the cases he discusses, "there is nothing more formidable than in the proverbial case of a needle dropped into a haystack." I personally believe that it is precisely *there* that the central problem of irreversibility lies, as I will explain in Section 10.

Lewis' statement that the evolution of matter is *per se* time symmetrical, and that temporal asymmetry is merely a projection of mental activity has been independently formulated by Watanabe[10]. His essential argument is that, given a certain state of a system (either classical or quantal), statistical inference towards either the future or the past *using nothing more than the intrinsic data* (that is, prediction and "blind" retrodiction) are formally identical procedures: so, argues Watanabe, entropy is always probably higher at an instant which is *thought of* later than the one where the given state was considered. Although Watanabe does not explicitly use the word, his idea corresponds very exactly to Lewis' *loss of information.*

My personal feeling is that Watanabe's position falls under the same objection as Lewis'. Nobody will seriously doubt that there is *something objective* in the fact that sandheaps get swept away, or that nebular gases gather to build up stars:—that is, that, in a given physical context, more probable situations are generated from less probable ones; this being the case independent of whether there are or are not human observers to detect, for example, the photons a star radiates and then build up theories of the phenomenon! True, such theories may help in devising procedures to recapture some of the energy thus dissipated; this brings the cybernetical

[10] *Revue de Métaphysique et de Morale* 128, 1951; Contribution to *Louis de Broglie, physicien et penseur*, Paris, Albin Michel, 1952, p. 385.

concept of "information" into the picture. But, if one insists (very rightly) on restoring time symmetry to every possible aspect of physical phenomena, so as to ultimately understand where time dissymmetry is located or comes from, the same must certainly be true when the connection between physical facts and subjective representations is considered. In this perspective it is useful to quote from Watanabe himself[10].

"Bergson wrote that human mind recognizes order in both the time directions along which life and matter proceed; the first order might be called order of will, or of finality, while the second one would be the order of passivity, or of causality."

Another significant paper developing the idea of a directionless time is Professor Mehlberg's[11]. He stresses, of course, the gist of the well known Ehrenfest[5] and Smoluchowski[6] argument, but my personal feeling is that he goes too far when flatly dismissing Van der Waals'[12] idea of an essential connection between Carnot's principle and Bayes' principle in general probability theory, Schrödinger's[13] idea of relating the entropy increase in isolated systems with the generation of such systems from the subdivision of a primitive larger one, and also the theory of cybernetics, on the grounds that "information" is merely defined as "negative entropy." It is of course perfectly true, as Mehlberg points out, that neither of these considerations establishes the existence of an intrinsic and "pervasive" time anisotropy; as Mehlberg puts it, the anisotropy of time is definitely not of a "law-like." but merely of a "fact-like" character. It exists, nevertheless, and I will argue, in later sections, that the Van der Waals, the Schrödinger, and the cybernetic considerations are all significant in the discussion of our problem.

A subsequent section of Mehlberg's paper is entitled *Some recent attempts at re-establishing time's arrow*, with three subsections: (a) *Regional or local anisotropy of time*, (b) *Cosmological anisotropy of time*, (c) *Non-entropic (probabilistic) anisotropy of time*.

No long commentary on the first subsection will be necessary: a regional time anisotropy may be understood as either spatial, as in a well known section of Boltzmann's treatise[3], or as temporal as, say, in cyclically oscillating cosmological models. It is obvious, as Mehlberg says, that regional time anisotropy is just the contrary of universal time anisotropy.

I shall return to cosmological anisotropy in Section 8. As for Mehlberg's

---

[11] Contribution to *Current Issues in the Philosophy of Science*, New York, Holt, Rinehart and Winston, 1961, p. 105.

[12] *Physikalische Zeitschrift* 12, 60, 1911. See also P. Hertz, *Ergebnisse der exacten Naturwissenschaften* 9, 60, 1922.

[13] *Proc. Roy. Irish Soc.* 53, 189, 1950.

paper, he is of course entirely right when he says that cosmological irreversibility does definitely not consist of some intrinsic property of the differential equations involved, but rather of an extrinsic choice of the way people decide to follow or go against the usual arrow of physical phenomena (or, may I say, of the choice of the way observers go along the fourth dimension). For instance, the time evolution of the Bondi-Hoyle-Gold models would be reversed if an annihilation instead of a creation of matter were postulated.

An important subsection in Mehlberg's paper concerns Popper's nonentropic but (says Mehlberg, as I do also) probabilistic type of irreversibility. Everybody agrees of course that this kind of irreversibility is "fact-like" rather than "law-like," that is, that its technical formulation is of the boundary condition type, and not an intrinsic evolution law.

After this Mehlberg goes on arguing that phenomenological laws such as Ohm's law, Fock's law for mixing processes, and the Onsager relations, are intrinsically time symmetrical, and that temporal asymmetry arises in them solely through "factual" or boundary type conditions—a conclusion with which everybody should agree, and which has also been excellently stressed in technical papers by Watanabe, E. N. Adams, J. A. MacLennan, Wu and Rivier, and which I will analyse in Section 4.

Thus Mehlberg's overall philosophy is in complete agreement with those of Lewis[9] and Watanabe[10] as the following quotations[11] will make clear: "All those major physical theories which constitute the bulk of our knowledge of the universe in general, and of time in particular, fail to offer any clue to time's arrow. This conspiracy on the part of basic theories which aims at concealing time's arrow becomes even more miraculous when we realize that they have been constructed with no desire to establish or refute time's arrow. Actually, the only plausible way of accounting for that fact is to admit that there is nothing to conceal: time has no arrow. I think that, on the assumption of an isotropic temporal medium pervading all human activities, temporal words like "past," "future," etc., have no independent meaning of their own but become significant according to the circumstances under which they are used. What I remember belongs to my past, what I desire or am planning for belongs to my future, *by definition.* This need not prevent somebody else from desiring what I remember and, thus, having his future overlapping with my past."

So speaks Professor Mehlberg. I daresay that, taken in this literal, provocative fashion, his final statement cannot be reconciled with what we learn from cybernetics, as Wiener[14], among others, has stressed.

---

[14] *Cybernetics,* New York, Wiley, 1948, p. 45.

Finally, Lewis, Watanabe and Mehlberg are quite right when, commenting on the philosophy of the old Loschmidt and Zermelo paradoxes, and on the celebrated Ehrenfest and Smoluchowski papers, they stress that temporal asymmetry in statistical theories is of an "extrinsic" or "fact-like", not of an "intrinsic" or "law-like" character.

What I feel is that the epistemological implications of this important conclusion are not sufficiently discussed in their papers; neither are those of the *subjectivistic* aspects of the time anisotropy, that is, those of the connection between "information" and "entropy."

2. Time's anisotropy as postulated in phenomenological thermodynamics

Classical thermodynamics coordinates a wonderfully large set of irreversible phenomena in physics and chemistry, after explicitly postulating irreversibility in Carnot's principle, the two aspects of which I recall:

1. If two thermostats at different temperatures are put into contact, heat flows from the warmer to the cooler (not the opposite way).

2. In a monothermic transformation, it is possible to produce heat at the expense of physical work (but not the converse).

If the time asymmetry stated in Carnot's principle were reversed, a paradoxical anti-Carnot thermodynamics would be deduced. This has frequently been done in epistemological studies.

It should be noted that the deduction of the anti-Carnot thermodynamics requires great caution, because retrodictability rather than predictability is the rule in it. For instance, as Poincaré[15] and Grünbaum remark, it would be dangerous, in the anti-Carnot world, to get into a lukewarm bath tub, because one could not foretell which end is going to boil and which to freeze. It would be dangerous also to bowl, if friction were an accelerating instead of a damping process. Poincaré says that the anti-Carnot world would be a lawless world; the exact point, however, is that, while the Carnot world is governed by causality (that is, by "blind predictability") the anti-Carnot world would be governed by teleology (that is, by "blind retrodictability"). It will turn out in later sections that the very profound problems thus raised are connected with the information concept.

As is well known, the best axiomatic formulation of classical thermodynamics is due to Carathéodory[16] with interesting *grana salis* of Landé[17] and Born[18]. Irreversibility is introduced there by remarking that

[15] *Science et Méthode*, Paris, Flammarion, Ch. IV, Le hasard.

[16] *Math. Annalen* 67, 365, 1909.

[17] *Handbuch der Physik* 9, 281, 1926.

[18] *Natural Philosophy of Cause and Chance*, Oxford Univ. Press, 1949.

adiabatic expansion of a gas implies an increase of entropy, and by letting it be understood that this is the natural trend of things. But, of course, one must ask why an adiabatic expansion rather than contraction is the way things go; this is definitely a Popper-like question[19].

3. Time's anisotropy as postulated in abstract probability theory

A preliminary question is of course: is the time concept essentially

---

[19] *Nature* 177, 538, 1956; 178, 381, 1956; 179, 1296, 1957; 181, 402, 1958. See also *ibid* commentaries of Schlegel, Hill-Grünbaum and Bosworth.

I personally do not believe that there is something *sui generis* in Popper's examples of irreversibility. To be more specific I assert (1) that Popper's irreversibility postulate is the well known "retarded action" or "retarded causality" postulate and (2) that in fact Popper's argument has recourse to statistics, so that an entropy can be defined (though *not*, of course, a *Boltzmannian* entropy).

(1) Popper writes in his first paper: "It is widely believed that "classical" (non-statistical) mechanics can describe physical processes only in so far as they are reversible. This a myth, however, as a trivial counterexample will show. Suppose a film is taken of a surface of water initially at rest into which a stone is dropped. The reversed film will show contracting waves of increasing amplitudes (and behind) a region of undisturbed water will close in. This cannot be regarded as a classical process. It would demand a vast number of distant coherent generators, the coordination of which, to be explicable, would have to be shown, in the film, as originating from one centre."

What Popper is saying in fact is that a *physical explanation* essentially has recourse to *retarded actions* or, equivalently, to the concept of *active sources*. The symmetrical system of concepts (which is rejected by all physicists as unphysical) would be *teleological explanation*, *advanced actions*, *active sinks*.

(2) Speaking of the coherence of generators is implying their possible incoherence, and thus probability comes into the picture. In his answer to Hill and Grünbaum Popper writes: "*Causes* [my italics] that are not centrally correlated can cooperate only by accident (or by miracle). The probability of such an accident will be zero." The question Popper does not raise, and which would clearly show the fact-like rather than law-like irreversibility postulate in probability theory also, is the question of retrodictive probabilities. In his answer to Schlegel Popper introduced another example (in fact, the Carathéodory one): expansion of a thin gas. I dare say that if the word *expansion* is to have an operational signification, some kind of a target-like marking of space must be conceived (for example, equal cubes if no field is present) so that one can say *where* the molecules are at any time. Then "expansion" means that the occupation numbers of the cells tend to level down at values 0 and 1, which is explicable in terms of probability (see the formula p. 326 below). If the objection were that we are dealing with an indefinite Euclidean space, so that there would be infinitely many cells and no probability is definable, the rejoinder would be that physicists believe the objection is unphysical; for example, quantum physicists, when calculating occupation numbers of a system of plane waves, will imagine the system to be enclosed in a reflecting box, so that they have to deal only with a finite number of plane waves (i.e., phase cells); eventually, they will let the volume of the box go to infinity. So finally, *physically speaking*, I believe that the *expansion* of a thin gas implies an *increase of disorder*, that is of a (non-Boltzmannian) *entropy*.

Grünbaum, in later papers[42], has sharpened his original commentary on Popper. He thus writes (1964): "No "implosions" qualify as the temporal inverses of eternally progressing "explosions." Thus one can assert the *de facto* irreversibility of an *eternal* "explosion" unconditionally, without Popper's restriction [of coherence]. For in an infinite space such a realization would involve a [self-contradiction] akin to that in Kant's First Antinomy: that a process going on for all infinite past time must have had a finite beginning [production by past *initial* conditions]."

A physicist is not very impressed by such metaphysical subtleties. Why should a "creation" be *per se* more necessary than an "annihilation"? An eternally progressing explosion would be self-contradictory as soon as one postulates its necessary "registration" (as Reichenbach[40] would say) by final future conditions.

implied in probability theory or not? In Kolmogorov's[20] celebrated axiomatic presentation, the probability concept is introduced in a completely atemporal way.

However, the physicist is interested in the physical applications of the concept, together with the way his knowledge and action get inserted into the course of phenomena. In this perspective the *probability* concept implies the idea of a *test*, and the very idea of a *test* is of a *temporal* nature. So, for the physicist, it would be completely devoid of meaning to pretend to think of a probability in an atemporal way.

Bayes'[21] principle, in abstract probability theory, is merely a combination of the two principles of addition of partial probabilities and multiplication of independent probabilities. What the physicist is interested in is the temporal aspect of the application of the principle, which has been strongly stressed by Borel[22]. The state of affairs with which the physicist has to deal, and where he has recourse to Bayes' principle for solving his problems, has been stated by Gibbs[23] in an often quoted text: "It should not be forgotten, when our ensembles are chosen to illustrate the probabilities of events in the real world, that while the probabilities of subsequent events may often be determined from those of prior events, it is rarely the case that probabilities of prior events can be determined from those of subsequent events, for we are rarely justified in excluding the consideration of the antecedent probability of the prior events."

While a "blind calculation" starting from a given present state and using only the intrinsic evolution laws of a system gives satisfactory results in prediction (that is, a calculated probability which can be experimentally tested as a frequency), it will fail completely in retrodiction. So, to deal reasonably with retrodictive statistical problems, one must apply Bayes' principle, that is, define a set of *a priori* weights of the possible cases. The important fact is of course that these weights are estimated independently from the intrinsic properties of the statistical system; they may thus be called "extrinsic probabilities."

As to their physical interpretation, the discussion of examples shows that, in the cases we are considering, Bayes' coefficients are used to describe at best the *primitive interaction* out of which the system under study has emerged. But realising this does not tell the whole story. The question then raised is: why do physical interactions produce their statistical effects

20 *Foundations of the Theory of Probability*, Chelsea, New York, 1950.

21 *Essay towards solving a problem in the doctrine of chances*, 1763.

22 *Le Hasard*, Paris, Alcan, 1914, Chap. IV.

23 *Elementary Principles in Statistical Mechanics*, Yale University Press, New Haven, 1914, p. 150.

after they have ended and not before they have begun? Why does shuffling cards destroy, and not select, a significant sequence of the cards (a significant sequence being, of course, one belonging to any low populated sub-ensemble, whatever the definition of such a sub-ensemble).

This problem, to which I shall come back, has been discussed extensively in recent years, because it has become implicitly or explicitly recognized by many authors that the irreversibility problem involved in the statistical interpretation of Carnot's principle is precisely the same as the one involved in Bayes' principle.

4. Time's anisotropy as postulated in classical or quantum statistical mechanics

Though not always explicitly stated, the complete similarity between Bayes' principle in general probability theory and the principle underlying any irreversibility statement in statistical mechanics (either classical or quantum) is the central gist of Watanabe's discussion of prediction and retrodiction in quantum physics together with the *H*-theorem or Onsager's principle[24]; also, in E. N. Adams'[25] and J. A. MacLennan's[26] discussions of transport theories, and in Wu and Rivier's[27] discussion of the so-called "master equations."

Watanabe's philosophy remains of course the same as in his properly philosophical studies[10]. The following quotation is from his 1955 paper: "By assuming a uniform *a priori* initial probability, one can obtain an interesting formulation which exhibits on one hand a complete symmetry with respect to the two directions of "time," but which on the other manifests a definite one-way-ness of the direction of human "inference." In short, the present paper may be said to be an elaboration in the light of quantum physics of the pregnant words due to W. Gibbs." In this same paper, and in the 1951 one, Watanabe shows very clearly that, in quantum theory, retarded and advanced waves are symmetrically related to prediction and retrodiction problems—an important idea to which I shall return in Section 6. In his 1959 paper, Watanabe answers the question "why doesn't Onsager's law apply to the past as it does to the future?" by stating that blind prediction is physical while blind retrodiction is not, and by introducing Bayes' formula explicitly.

In Adams'[25] paper the central idea is (quoting from him) that: "There

---

24 *Phys. Rev.* 84, 1008, 1951, *Rev. Mod. Phys.* 27, 179, 1955; *Transport Processes in Statistical Mechanics*, Interscience Publishers, New York & London, 1958, p. 285. It is in his 1955 paper that Watanabe introduces the expression (if not the concept) of "blind retrodiction."

25 *Phys. Rev.* 120, 675, 1960.

26 *The Physics of Fluids* 3, 493, 1960.

27 *Helv. Phys. Acta* 34, 661, 1961.

are *two* valid transport equations, one causal (or Boltzmannian), the other anticausal (or anti-Boltzmannian), each consistent with the fundamental equations of mechanics. The proper application of the anti-Boltzmann equation is for retrodiction during the progression of a fluctuation, just as that of the Boltzmannian equation is for prediction during the regression of a fluctuation."

Adams also remarks that "the conditions under which most transport processes are observed do not permit the idealization that the observed system is isolated. However, it is possible in certain cases to approach (a certain) idealization of isolation [which he describes as]: (1) the system to be observed is isolated for a long time, after which (2) it is suddenly altered so that the experiment is initiated, (3) the altered system is again isolated and remains so, while the course of the experiment is observed." The question Adams thus raises and does not answer is why the interaction produces a fluctuation just after it has ceased, and does not pick up a fluctuation just before it begins. For instance if, between times t_1 and t_2, a physicist moves a piston in the container of a gas in thermal equilibrium, everybody believes that the Maxwellian distribution is disturbed after time t_2 and not before time t_1. I shall return to this in Section 7.

In Wu and Rivier's paper the stress is laid on an idea very similar to Mehlberg's[11] way of understanding Popper's[19] considerations. They write: "We shall show that a theory containing a time arrow and describing irreversible processes can be founded on a 'probability Ansatz' connecting two sets of probabilities at two instants of time. The probability Ansatz leads to the Master equation." Then Wu and Rivier establish "the existence of the probability Ansatz 'symmetric' to the preceding one and, of course, incompatible with it." Their conclusions are: "This symmetry in the *possibility* of making either choice in the direction of time is inherent in the symmetry in time of the basic theories, namely, classical and quantum mechanics. The two choices of the time arrow are mutually exclusive in the sense that one is valid only for prediction and the other for 'postdiction' on the basis of the information at the present instant. The two master equations are mutually exclusive of each other and do not imply equilibrium for the system."

Fig. 2, inserted in Wu and Rivier's text, clearly pictures the argument. Their final words are "While the two directions of time are on equal footing according to the basic theory, the 'postdiction' about an increase in entropy towards the *past* cannot be verified by comparison with observation, in the same way as a prediction about the future can be. It is difficult to give any operational meaning to the 'postdicted' probabilities. Thus, at least on the basis of our built-in biological time arrow, only the master equation

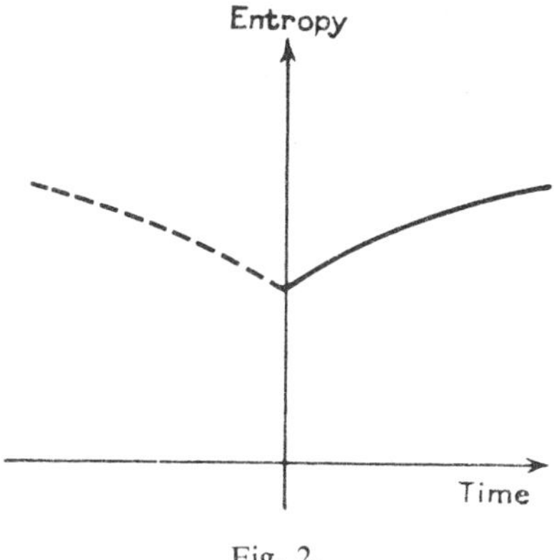

Fig. 2.

describing evolutions towards equilibrium in the future is of practical significance."

On the strictly technical side, the compatibility of the ideas of Watanabe, E. N. Adams, J. A. MacLennan, Wu and Rivier, is complete. This reviewer[28] had independently come to similar conclusions, using as an example Poincaré's[29] formula for the little planets problem.

5. One-to-one correspondence between entropy or probability increase and retarded waves of various kinds

In quite a few of the previously quoted papers, two recurring general ideas are: (1) that the time dissymmetry in statistical theories (either classical or quantum) can be traced back to an application of Bayes' general principle of dissymmetry between prediction and retrodiction, and (2) that, in wave mechanics, retarded and advanced waves are respectively used in prediction and in retrodiction problems. Putting the two ideas together will immediately yield a one-to-one correspondence between retarded solutions and entropy increasing solutions on the one hand, and advanced solutions and entropy decreasing solutions on the other hand—the exclusion of the latter coming from Bayes' principle.

Among the recent papers devoted to transport relations in fluids, MacLennan's[23] is of particular interest in our perspective. The following is directly quoted from it: "We wish to give a discussion of transport in fluids, for states near equilibrium, which is applicable to rapidly varying

[28] *Rev. Quest. Sci.* 8, 171, 1952; Contribution to *Louis de Broglie—physicien et penseur*, Paris, Albin Michel, 1953, p. 400; *Rev. de Synthèse* No. 5–6, 7, 1957; *Cahiers de Phys.* No. 96, 317, 1958; *Ann. Fac. Sci. Clermont*, 12,145, 1962; *Rev. Méta. Mor.* No. 2, 214, 1962. *Le Second Principe de la Science du Temps*, Paris, Ed. du Seuil, 1963, Chap. II.

[29] *La Science et l'Hypothèse*, Chap. XI, §3.

phenomena. The method uses the external non-conservative forces which arise from the interaction between a system and its surroundings, and constrain it to a non-equilibrium state. We will make use of only the "retarded" solution in agreement with causality. However it should be noted that there are other solutions, one being the "advanced" one. In order to obtain local, and approximate, transport relations, we will adopt an expansion in powers of the wave number. The conclusion is that the retarded (resp. advanced) solution leads to a positive (resp. negative) entropy production. Recently, Costa de Beauregard has discussed the connection between retarded interactions and the law of increasing entropy, from a general point of view."

Thus MacLennan's technique provides an appropriate formalism for processes akin to the one previously considered: moving a piston in the wall of a vessel containing a gas near equilibrium. The perturbation in Maxwell's velocity distribution law is propagated as a retarded wave after time t_2, not as an advanced wave before time t_1, so MacLennan's technique demonstrates in that case what has just been stated: a one-to-one correspondence between the Carnot principle and the principle of retarded actions, propagated as waves in the fluid. The demonstration is given by MacLennan either in the classical (or Boltzmannian) and the quantized case.

In 1961 O. Penrose and I. C. Percival[30] proposed a demonstration of a one-to-one correspondence between an appropriate irreversibility postulate in general probability theory and the principle of retarded non-quantized electromagnetic waves. These are quotations from their paper: "The purpose of this paper is to formulate a basic property of probability distributions which is asymmetric under time reversal and from which the time direction of irreversible processes can be deduced. This we call the law of conditional independence. Although the theory is classical we believe that a quantum theory of the direction of time must contain a feature resembling our basic law.

"A starting point in our search for a fundamental asymmetric law is provided by the following 'principle of causality' or 'principle of retarded action' as formulated by Costa de Beauregard: If an otherwise isolated system interacts with its surroundings or with another system, at time t_0, the effect of the interaction is felt after time t_0, not before. This principle will yield information about the direction of time if we can give a symmetrical definition of the word 'effect.' The principles outlined in the beginning, and discussed in more detail by Reichenbach, show that *the definition must*

---

[30] *Proc. Phys. Soc.* 79, 509, 1962.

be statistical. Before time t_0, it is natural to assume that the system is uncorrelated with the rest of the universe, just as it would be if it were to remain isolated for all time. After t_0 the system will in general be correlated with its surroundings because it has interacted with them. Thus *the effect is correlation.* This interpretation enables us to restate the principle of causality in statistical terms: *a system which has been isolated throughout the past is uncorrelated with the rest of the universe*" (my italics). It is clear from these quotations that Penrose and Percival use their "time asymmetric statistical principle of causality" just as we have been using the temporal application of Bayes' principle.

Penrose and Percival give three aplications of their theory. One is classical radiation from a dipole belonging to a statistical ensemble. They show that advanced waves are excluded by their asymmetry principle. Another is collision of two monokinetic beams of particles. The third is entropy increase by coupling and decoupling of two previously uncorrelated systems.

Now I come back to Wu and Rivier's[28] paper already quoted. In Section IV, devoted to *quantum theory of irreversible processes*, they show how Pauli's master equation, which is to be used in predictive problems, has a temporal inverse corresponding to blind retrodictive problems, entailing of course (as Watanabe had pointed out) an entropy increase towards the past. The point I wish to stress is that Wu and Rivier show that the time-reversed Schrödinger equation corresponding to this case is the one defined by Wigner (which of course is no surprise). In other words, if we define retarded and advanced waves as the ones solving the Cauchy problem in the future and the past, respectively, of the time t_0 carrying the boundary conditions, then we must say that Wu and Rivier have shown a one-to-one correspondence between the quantum principle of (fine grained) entropy increase and the principle of retarded waves (this demonstration being given with Klein's[31] definition of fine grained entropy).

Finally I wish to discuss the connection I believe quantum theory introduces between the principle of increasing entropy (or probability) and the principle of retarded waves.

The first indication of such a connection is found in Planck's[32] definition of the entropy of a monochromatic beam—a definition essentially implying the constant h and, thus, the existence of photons. It follows from Planck's definition that the total entropy increases in the scattering of a monochromatic beam, even in the case where no frequency change occurs.

---

31 *Zeits. f. Phys.* 72, 767, 1931.

32 *The Theory of Heat*, London, Macmillan, 1932, Parts III & IV.

The point is that the very word "diffusion" implies that waves are retarded; the paradoxical reversed phenomenon would be a phase-coherent "confusion" of waves, which of course is never observed. So one feels that this theory of Planck's implies a one-to-one correspondence between the two principles of entropy increase and of wave retardation (in the special case of photons).

In this context one must also recall what has been qualified as "an inconclusive but illuminating discussion carried on by Ritz and Einstein[33] in 1909, in which Ritz treats the limitation to retarded potentials as one of the foundations of the second law of thermodynamics, while Einstein believes that the irreversibility of radiation depends exclusively on considerations of probability."

My personal feeling, which I intend to justify, is that Ritz and Einstein were in fact equally right, and that the only thing which prevented them from recognizing that they were looking at the same thing from two opposite directions was simply that, at the time they wrote, the undulatory aspect of mechanics was not yet elucidated (if the corpuscular aspect of light was). If Ritz and Einstein had known, in 1909, that every scattering process in the sense of statistical mechanics *also* implies a scattering of waves, and *vice versa*, then certainly both of them would have recognized that their opposite positions were in fact *reciprocal*, that is, mutually exchangeable.

Now I shall give a striking example illustrating what I have in mind. Suppose we have a linear grating and a monochromatic plane wave falling upon it; it may be any kind of quantized wave, with the quanta, or corpuscles, arriving at long time intervals, so that any two of them do not belong to the same wave train, and are thus *discernible* from each other as macroscopic objects are. Now, classical optics defines a finite number of possible scattered plane monochromatic waves (no generality will be lost if, for simplicity, we assume the intensities of these outgoing waves to be equal). In the inversed problem, to each of the g outgoing waves there corresponds one and the same set of g ingoing waves (to which of course the one first considered belongs).

In quantum theory, classical intensities are reinterpreted as presence probabilities of the particles. So what we are considering is in fact a transition, induced by the grating, between two sets of g states or phase cells. Call n_i the occupation numbers of these cells. The number of complexes corresponding to a given set of occupation numbers n_i obeys the formula

$$N = n!/P(n_i!) \ ;$$

[33] *Ann. de Chimie et de Physique* 14, 145, 1908. *Phys. Zeits.* 8, 903, 1908; 10, 185, 323, 1909; 10, 817, 1910.

Incidentally, the fact-like character of the principle of retarded waves on the one hand, and of the principle of blind retrodiction forbidden on the other, were not explicitly stressed in the discussion.

it decreases if one takes a *ball* from a cell and puts it in a more occupied cell, so that the largest N is the one where all the occupation numbers are equal (with eventual differences of $\pm$ 1).

So what we find is exactly a probabilistic duplication of the principle of retarded waves: here also, phase-coherent scattering implies the presence of retarded waves and is expressed as an increase of probabilities in the distribution of corpuscles. I dare say that this example shows that Bayes' principle is, in quantized waves theory, just another name for the principle of retarded waves—both of them being stated as an appropriate boundary condition.

It is possible[28] (1958) to enunciate this statement in a general abstract form by rewording von Neumann's[34] famous irreversibility theorem pertaining to the quantum mechanical measuring process.

6. Interaction and common cause (Theory of branch systems)

So, as all modern writers in the field have recognized, irreversibility is introduced in every significant physical theory as a *boundary condition* pertaining to one definite side (the *future* side, by definition) of a so-called *initial* instant of time.

A preliminary and much argued question is the following: physically speaking, (1) is the causality principle, or retarded actions principle, a more general principle than the statistical Bayes' or Penrose-Percival postulates? or, (2) is the causality principle an essentially statistical one, ultimately resting on Bayes' or Penrose-Percival's postulates?

My personal feeling is that the second statement is definitely the proper one. The main arguments, all drawn from the epistemology of quantum theory, are the following: a) time symmetry of quantal transitions, even enforced in a Lewis-like[9] direction by the time symmetric role of occupation numbers in superquantized formulas; b) positive and negative frequencies in Fourier representations of spinning particles, and Feynman's[35] interpretation of antiparticles; and c) Einstein's[36] and Schrödinger's[37] so-called scattering correlation paradoxes, showing that knowledge acquired at any time about one of the scattered particles immediately entails a corresponding knowledge about the *other*, though the interaction has taken

---

34 *Mathematische Grundlagen der Quantenmechanik*, Berlin, Springer, 1932, Chap. 5.

35 *Phys. Rev.* 76, 749 and 769, 1949.

36 *Phys. Rev.* 47, 777, 1935.

37 *Naturwiss.* 23, 787, 823, 844, 1935.

place in the past. I[38] cannot imagine any solution of the Einstein and Schrödinger so-called paradoxes other than stating that all physical events are in some sense written once and for all in Minkowski's space time, that is, stating that (as far as individual quantal processes are concerned) there is no question of a law of retarded causality, the interaction law being instead a time symmetric half-retarded and half-advanced one[39]. So, on the grounds of the epistemology of quantum theory, it seems quite reasonable that the law of retarded actions should be a statistical macroscopic emergence rooted in such statistical time asymmetric postulates as Bayes', or Penrose-Percival's ones. This is also Reichenbach's[40] thesis in his famous posthumous book; and Penrose and Percival's[31] as is evident from the following quotation: "The principles outlined in §1, and discussed in more detail by Reichenbach, show that *the definition* (of the word "effect") *must be statistical*" (my italics). It is obvious from all that precedes that the statistical version of the principle of retarded action or causality is just another name for the Carnot-Clausius principle of entropy increase. Many of the authors I have quoted have more or less stated the point, either occasionally (Adams[25], MacLennan[26], Wu-Rivier[29],) or with explicit emphasis (this reviewer[30], Terletsky[41]). The more complex position of such prominent writers as Reichenbach[40] and Grünbaum[42] will need special consideration.

In order to stress the one-to-one connection I believe to exist between the Carnot-Clausius and the (statistical) retarded actions principle I produce a paradoxical statement due to von Weiszäcker[43] (in Grünbaum's translation[42], 1963, p. 243): "*In the absence of other grounds* (my italics) the statistical entropy law provides every reason for regarding a present ordered state as a randomly achieved low entropy state rather than as a veridical trace of an actual past interaction, (for) it is far more probable that the present low entropy states are mere fluctuations rather than the continuous successors of actual earlier states of still lower entropy." To this, Boltzmann[3] (Bd II, p. 257–258) had answered in advance "The fact

[38] *Revue Phil.* 145, 385, 1955; Contribution to *XXème Semaine de Synthèse*, Paris, Albin Michel, 1956, p. 161; *Sciences* No. 6, 31, 1960.

[39] True, there is the accepted time dissymmetric use of Feynman's propagator for describing virtual particles. but it may be argued that this choice is imposed by implicit macroscopic statistical considerations, that is, the necessity of obtaining an exponential decay (and not build-up) of higher energy bound states.

[40] *The Direction of Time*, University of California Press, 1956.

[41] *Journal de Physique* 21, 681, 1960.

[42] *Archiv für Phil.* 7, 165, 1957; *Phil. of Science* 29. 146, 1962; *The Monist* 48, 219, 1964. Contributions to: *Frontiers of Science and Philosophy*, Univers. Pittsburgh Press, 1962, p. 147; *The Philosophy of Rudolf Carnap*. ed. P. A. Schilpp, 1963, p. 599. *Philosophical Problems of Space and Time*, New York, Alfred A. Knopf. 1963, Part II.

[43] *Ann. der Physik* 36, 281, 1939.

that actually a transition from a probable to an improbable state does not occur as often as the opposite one may be sufficiently explained by the *assumption* (my italics) of a very improbable state of the universe around us. As a consequence, a system of bodies entering into interaction is found, in general, in an improbable state" (Reichenbach's[40] translation p. 132). Or, as Grünbaum[42] (1964) puts it. "Among the quasi-closed systems whose entropy is relatively low and which behave as if they might remain isolated, the vast majority have not been and will not remain permanently closed, being branch systems instead." In a somewhat similar vein Watanabe[24] (1955) had written that if somebody finds a pile of playing cards "in order," he will *in fact* not believe that this occurred through shuffling.

All this brings us back precisely to Bayes' principle, showing in addition that Bayes' coefficients are a means of expressing the interaction out of which the observed system has been segregated. In other words, to say that an interaction coupling two systems during a finite time, $t_2 - t_1$, produces its *essentially statistical* "effect" after time t_2 and not before time t_1, or enunciating the statistical version of Carnot's principle, is one and the same thing.

This is exactly Reichenbach's[40] (pp. 125–143) position, as is clear from the following quotations, the first of which is the explanation of a picture I reproduce (Fig. 3). Reichenbach summarizes his theory of branch systems

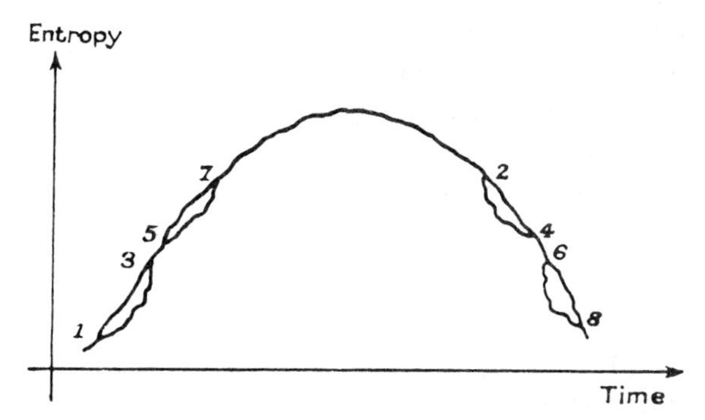

Fig. 3.
An upgrade and a downgrade of the entropy curve of the universe. Some isolated systems branch off and return to the main system.

as follows (p. 136): "1) The entropy of the universe is at present situated on a slope of the entropy curve; 2) there are many branch systems, which are isolated from the main system for a certain period, [and] are connected with [it] at their two ends; 3) . . . ; 4) in the vast majority of branch systems, one end is a low, the other a high [entropy] point; 5) in the vast majority of branch systems, the directions towards higher entropy are parallel to one another and to that of the main system."

In a less formalistic and more philosophical style, this amounts to saying (p.131): "It is through its reiteration in branch systems that the entropy growth of the universe dictates to us the direction of time. The universal increase of entropy is reflected in the behavior of branch systems."[44]

The following quotations from Reichenbach (pp. 151 and 154—155) are perhaps even stronger: "Explanation presents order in the present as the consequence of interaction in the past. The cause is the interaction at the lower end of the branch run through by an isolated system which displays order; and the state of order is the effect. *The convention of defining positive time through growing entropy is inseparable from accepting causality as the general method of explanation.* Once the time direction is assumed in the usual sense, it is not a matter of personal preference, it is a physical law that causality, and not finality, governs the universe. *The word "produces" is a statistical concept* [my italics]. The distinction between cause and effect is revealed to be a matter of entropy and to coincide with the distinction between past and future."

One may incidentally remark that the thesis so adequately formulated by Reichenbach is implicit in the name often given to Bayes' principle[22]: "Principle of probability of causes." One may also wonder how it happened that this one-to-one connection between retarded causality and growing entropy was discovered so late [25, 26, 27, 28, 30, 40, 41], while the symmetrical connection between teleology and decreasing entropy[45] or advanced actions[46] has been obvious for a long time. Grünbaum[42] and I[22] have independently stressed these two opposite parallelisms; Grünbaum's title (1962) "Temporally asymmetric principles, parity between explanation and prediction, and mechanism versus teleology" is especially strong.

Grünbaum[42] (1963 pp. 261–262), in his own presentation of the theory

[44] As I[28] have written extremely similar things I may state that there is one precise point where I disagree with Reichenbach. On pp. 141–142 he writes:

"One might attempt to explain the parallelism of the branch systems as forced upon them by their being not completely insulated from the universe; the small remaining coupling would then (supply) the transmission of a causal influence which makes all (of them) adjust themselves to the main system. But such an interpretation would not be tenable, because . . . ."

It is clear to me that the "branching" "in" and "off" *is* the interaction between the partial and the total system, idealized in a discrete time fashion. Another idealization would be that of a partial system weakly, but continuously, coupled with the main one. In both cases, the gist is that the direction of increasing entropy in both the partial and the total system is also the direction where the influence of the total system on the partial one is *retarded*, i.e., propagated in it in a retarded fashion.

[45] See for instance Bergson, *Creative Evolution*, Chap. III.

[46] See for instance Fantappie, *Teoria Unitaria del Mondo fisico e biologico*, Roma, 1944.

of branch systems, departs from Reichenbach's ideas for two reasons. First, "In view of the reservations Reichenbach himself expressed concerning the reliability of assumptions regarding the universe as a whole, one wonders why he invoked the entropy of the universe at all instead of confining himself to the much weaker assumption of the existence of states of disequilibrium in the universe."

Grünbaum's second objection to Reichenbach is: "More fundamentally, it is unclear how Reichenbach thought he could reconcile the assumption that the branch systems satisfy *initial* [my italics] conditions of randomness during whatever cosmic epochs with the claim of alternation (he had made)." On this very point, I am definitely with Reichenbach against Grünbaum. It may be seen a little further on in Grünbaum's text that he himself feels there is something unclear in his argument. One reads: "If the universe were finite and such that an entropy is defined for it as a whole, then my contention of a *cosmically perversive* statistical anisotropy of time could no longer be upheld. And if one may further assume that the entropy of a finite universe depends additively on the entropies of its (components), then the assumed temporal asymmetry of the branch system would contradict the *time-symmetry* of the one system entropy behavior of the universe. This conclusion, if correct, poses the question—which I merely wish to ask—whether in a closed universe the randomness of the *initial* [my italics] conditions would not hold."

I feel that the weak point in Grünbaum's argument is that it is restricted to the case of predictive probabilities. Had he symmetrically considered (even for rejecting them in a further stage of reasoning) "blind" retrodictive probabilities, he would have obtained the theory of proceeding (and not only receding) fluctuations, as in the writings of Adams[25], MacLennan[26], Wu-Rivier[27], and myself[28].

So my conclusion here is of a definitely Reichenbach style: *There is a one-to-one correspondence between the statistical law of entropy increase (resp. decrease) of branch systems, and the (also statistical) law of retarded (resp. advanced) causality.*

There is nevertheless one important point where, together with Grünbaum, I would amend Reichenbach's assertion that "causality, and not finality, governs the universe."

Grünbaum[42], under the title "The controversy between mechanism and teleology," writes (1964, pp. 311–313) "In view of the demonstrated restricted validity of mechanism, we must therefore deem the statement by Reichenbach as too strong." I shall return to this in Section 10.

7. Irreversibility and cosmology

The theory of branch systems is in itself some kind of a cosmological theory. Since it explains improbable complexions starting Carnot processes as the result of the branching of partial systems from a larger one, and since it shows that the retarded effect of interactions limited in time is explicable in terms of a Carnot process occurring in the total system, thus raising the problem of the generation of the low entropy state existing in the past of the total system, the theory of branch systems essentially refers back the cause of irreversibility in local systems to irreversibility in larger and still larger ones. Thus it may well be called a cosmological theory of irreversibility. But, since the days of relativity the word *cosmology* has a definite sense in the realm of gravitation theory. So the question of a possible connection between large scale features of the observable universe and statistical irreversibility arises naturally.

This problem has been considered by Tolman[47] and also Zanstra[48]. Tolman (1934) discusses the relation between his general-relativistic formulation of phenomenological thermodynamics and various types of cosmological solutions of Einstein's gravitational equation. As he *a priori* assumes independence between the Carnot-Clausius principle and the world's expansion, it is no surprise that, in the particular case of oscillating models, he finds that thermodynamic irreversibility entails a loss in overall gravitational energy and, thus, a progressive alteration in the extrema values of the world's radius; in other words, there would then be a damping of the oscillating universe.

But, after all, phenomenological thermodynamics is a very unsophisticated theory as compared with the problems that are under consideration. It is certainly much more attractive to postulate a one-to-one correspondence between the statistical irreversibility principle and the observed expansion of the universe. The natural way to probe this idea is of course to postulate that all processes of interest such as the emission and absorption of radiation, transitions between elementary particles, etc., are macroscopically reversible in terms of the general-relativistic thermodynamics, and that it is essentially the world's expansion that makes them look as if they were irreversible to an "ordinary observer unfamiliar with relativistic thermodynamics and uninformed as to the general expansion taking place," as Tolman puts it. In this line he discusses the case of a universe filled with black body radiation (and eventually incoherent matter); he shows that the "ordinary observer" will interpret the changes in density, pressure and temperature as radiation leaving his neighborhood towards what would be, in classical conditions, distant regions of low temperature—though

[47] *Phys. Rev.* 38, 797, 1931. *Relativity, Thermodynamics and Cosmology*, Oxford, Clarendon Press, 1934, Chap. 10, Part III, pp. 420–445.

[48] *Proc. Kon. Ned. Akad. Wetens.* 60, 285, 298, 1957.

in fact there is only an expanding homogeneous universe. Previously (1931) he had discussed the case of interacting black body radiation and a perfect monoatomic gas, the interaction consisting of creation or annihilation of matter versus radiation, with the satisfactory result that expansion would be associated with matter annihilation.

It may be recalled that this reviewer[28] (1958), in a somewhat analogous perspective, has briefly discussed, in flat space only, the cases of radioactive disintegration and electron-positron annihilation into two photons, with the result that irreversibility is then understandable in terms of the larger number of phase cells associated with the many products of the radioactive decay, or the higher values of the photon's momenta compared with those of the electron-positron pair:—that is, finally in terms of Bayes' principle.

All these remarks point very definitely in the direction of a one-to-one correspondence between the general irreversibility principle and the world's expansion, the question then being of course: *why* are we observing an expansion and not a contraction?

More recently there have been interesting hints for establishing such a one-to-one correspondence in a more intrinsic fashion. First there has been de Donder's[49] idea that the world's expansion might well be intrinsically associated with the emission of gravitational waves, that is of not subsequently reabsorbed gravitons.

In connection with the remark that the world's expansion provides a natural answer to the old Olbers[50] paradox, there has been a proposal by Hogarth[51] to explain the elimination of advanced potentials in the Wheeler-Feynman[52] radiation theory by the world's expansion, and interesting comments by Gold[53] and Sciama[54]. Hogarth's conclusion (1962) which is given within a class of conformally flat cosmological models including among others the Einstein–de Sitter, and the Bondi–Gold–MacCrea ones, is that the former gives the wrong and the latter the right association between retarded electromagnetic waves and the world's expansion. Narlikar[55] arrives at a somewhat similar conclusion with neutrino theory replacing photon theory.

---

49 Contribution to *Théories nouvelles de Relativité*, Paris, Hermann, 1949, p. 76.

50 *Bode's Jahrbuch*, 1826, p. 110.

51 *Ph.D. Thesis*, London, 1953. *Proc. Roy. Soc.* A 267, 365, 1962.

52 *Rev. Mod. Phys.* 17, 157, 1945.

53 *Proc. 11th Solvay Congress*, Brussels, R. Stoop, 1958, p. 86 and *Amer. Journ. Phys.* 30, 403, 1962.

54 *Ann. Inst. H. Poincaré* 17, 13, 1961.

55 *Proc. Roy. Soc.* A 270, 553, 1962.

These lines of thought, although yet only exploratory, are certainly highly promising.

8. Necessity for introducing both the relativistic space-time and the cybernetical information concepts

Since relativistic kinematics, that is, space-time geometry, has become the universal frame of physical theories (also on the quantal level), every discussion should be formulated in terms appropriate to space-time geometry.

First of all, since at any point-instant, the intrinsic separation of space-time domains is no longer a two-fold one as in Newtonian kinematics—"past" and "future"—but a three-fold one—"past", "future" and "elsewhere"—one is no longer allowed to think of matter as being spatially extended and temporally unextended, for this could not be thought of in a covariant way. Matter *must* be thought of as extended both in space (a traditional idea) and in time (an entirely new idea). Thus, the greatest names in relativity theory have written extremely strong assertions, such as Minkowski's[56] well known one, Weyl's[57] "The objective world simply *is*, it does not *happen*," and Einstein's[58] "The four-dimensional space of special relativity is just as rigid and objective as the space of Newton." This last sentence would be an excellent exergue to Feynman's electrodynamics, or also to what I said before (Section 7) on the Einstein and Schrödinger correlation argument in scattering.

Thus the question in irreversibility is no longer: "Why are physical states more probably generated in the order of increasing than of decreasing probabilities," but: "Probabilities of physical states being generally ordered with reference to the values of the time coordinate, why are living beings bound to explore the time coordinate in the direction where most probabilities are increasing, and not in the opposite one?"

This question implies use of the information concept.

Boltzmann[3] (Bd. II, pp. 257–258) wrote that, on statistical grounds, there ought to be (in a static universe) just as many practically isolated regions of increasing as of decreasing entropies; he then found it reasonable to assume that the time arrows of living beings in these two classes of regions were opposite, so that in every case biological time would follow the arrow of increasing entropy. Incidentally, this conception of Boltzmann certainly implies the idea that the time coordinate and its matter content are displayed

---

56 *The Principle of Relativity*, London, 1923, p. 75.

57 *Philosophy of Mathematics and Natural Science*, Princeton, 1949, p. 173.

58 *Forum Philosophicum* t. 1, 1930, p. 173.

all at once. Then Wiener[14] remarks that exchange of information between the two kinds of regions must be strictly prohibited, as it would raise unsolvable problems.

Brillouin[59], Gabor[60] and Rothstein[61] who have looked carefully at the implications in physics of the cybernetical concepts, have formulated the "generalized Carnot principle" as follows: In an isolated system comprising observers, the sum *negentropy* + *information* cannot go up. Then, if one postulates (which seems reasonable) that life essentially implies an incoming information flux (for instance there is no question of anti-reading a book from the last to the first page, thus erasing one's previous knowledge in the field) then this principle, together with the generalized Carnot principle, will yield a deduction of Boltzmann's postulate[62]. It will also yield, as a corollary, a very sophisticated way for saying that one learns from experience....

Essentially, what spoils Reichenbach's[40] extremely interesting posthumous book is his not recognizing the relativistic necessity of conceiving of matter as extended in both space *and* time and, correlatively, his being unaware of the necessity of introducing man's consciousness explicitly in order to explain the flux of time, that is, becoming. He has written for instance (pp. 269–270): "Man is a part of nature and his memory a registering instrument subject to the laws of information theory. The increase of information defines the direction of subjective time. It is not a human prerogative to define a flow of time, every registering instrument does the same." What is of course wrong in Reichenbach's statements is saying that memory is a recording *instrument*. The recording instrument is the brain; memory is awareness of records. As Grünbaum puts it (1963, pp. 269–270): "The flux of time has a meaning only in the context of the egocentric perspectives of *sentient* organisms and does not also have relevance to the relations between inanimate recording instruments and the events they register. For what can be said of every state of the universe can also be said, *mutatis mutandis*, of every state of an inanimate recorder."

For establishing his own philosophy Grünbaum had incidentally to explain, as this reviewer did independently[38], and Hugo Bergmann[63] had done before, that, contrary to a feeling that prevailed before the explicit recon-

---

59 *Science and Information Theory*, New York, Academic Press, 1956.

60 *M. I. T. Lectures*, 1951.

61 *Communication, Organization and Science*, The Falcon's Wing Press, 1958.

62 This same idea may be found (though in a less formalistic style) in Reichenbach's[40] (pp. 156 and 269–270), Grünbaum's[42] (Chap. 9 § B), and Whitrow's[63] (p. 282) writings.

63 *Der Kampf um das Kausalgesetz in der jüngsten Physik*, Braunschweig, Vieweg, 1929, pp. 27–28.

ciliation of relativistic and quantum formalisms, there is no contradiction between an essentially probabilistic physics and a history of events written in space-time once for all. H. Bergmann has thus written: "Even those who regard the supplanting of determinism by indeterminism as admissible, as we do, will not be willing to admit that the concept of 'now' can be assigned a legitimate place within physics. *'Now' is the temporal mode of the experiencing ego*" (my italics). And Grünbaum (1963, Chap. 10): "The anisotropy of time resulting from irreversible processes consists in the mere structural difference between the two opposite senses but provides no basis for singling out *one* of the two as "the direction" of time. Hence, the assertion that irreversible processes render time anisotropic is not at all equivalent to such statements as "time flows one way." *The theory of relativity does not make any allowance for the transient Now of common sense time*" (my italics).

Like Eddington[64] before him and Whitrow[65] after him, Reichenbach supposed that the exclusion of progress of time by a physical theory is attributable to (its) deterministic character, and believed that indeterministic physics can provide a physical basis for the transient Now. In the same vein, the astronomer Bondi[66] contends that "In a theory with indeterminacy the passage of time transforms statistical expectation into real events." If Reichenbach, Eddington, Bondi, Whitrow and others had merely maintained that indeterminacy makes up for our human inabililty to know in advance what particular kinds of events will in fact materialize, then, of course, there would be no objection. I believe that the issue of determinism *vs.* indeterminism is totally irrelevant to whether becoming is a significant attribute of time independent of human consciousness.

I strongly agree with this discourse of Grünbaum, except on two precise points. I believe the last assertion is too strong, because in fact the determinism *vs.* indeterminism quarrel is *not* irrelevant to the ultimate interpretation of probability and information in physics, and that information is the key to understanding the subjectivistic flux of time. Nor would I have written that it is our conscious *organism* that goes through events in space-time, but merely our *consciousness*, as H. Bergmann[63] and Whitrow[65] have rightly said. It is my feeling that Professor Grünbaum's philosophy, on the precise point we are discussing, is not defined unequivocally, because, in addition to the sentences I have quoted where a strong emphasis is laid

---

64 *Space, Time and Gravitation;* Cambridge University Press, 1953, p. 51.

65 *The Natural Philosophy of Time*, London, Nelson, 1961.

66 *Nature* 159, 660, 1952.

on the essentially subjectivistic basis of the flux of time, some others may be found which go in a much more Reichenbachian direction (1963, pp. 263–264 and 289–290). Here is one of them:

"(The theory of branch systems) explains why the subjective (psychological) and objective (physical) directions of positive time are parallel to one another by noting that man's own body participates in the entropic lawfulness of branch systems in the following sense: man's memory, just as much as physical recording devices, accumulates traces, records or information in a direction dictated by statistics. Contrary to Watanabe's conception of man's psychological time sense as *sui generis*, it will turn out that the future direction of psychological time is parallel to that of the accumulation of traces (increasing information) and hence parallel to the direction (of) entropy increase." My personal feeling is that, if Watanabe had gone too far in the idealistic direction, Grünbaum, in such Reichenbach-like sections, goes too far in the mechanistic one; for, stressing that the arrows of entropy and information increase are parallel to each other is *not* proving that the flow of subjectivistic time has to follow the arrows!

Whitrow[65], who is no more than Popper a member of the statistical theory of time club, wrote very excellent things on this point: "We should concentrate on the fact that, because brain is a material entity, it exists both in three-dimensional physical space and in time, whereas mind, as manifested in consciousness, exists only in time: it is purely a 'process' and not a 'thing.' Consequently the two can interact only in time, and therefore this interaction must occur mentally. Mind is essentially temporal in nature."

Thus Hugo Bergmann, Whitrow, this reviewer and, in some places, Grünbaum put a very strong psychological emphasis upon the well known relativistic statement that the flow of time is not in things, but in the way our consciousness goes through things; somehow like the water flux encountered by a swimmer is not in the pool, but in the way the swimmer goes through the pool.

9. Law-like time symmetry and fact-like time asymmetry in the information and negentropy context

First of all, what is "information?"

In elementary probability problems, such as card shuffling and guessing, information is a fairly obvious concept, the definition of which in terms of the logarithm of a probability is straightforward.

What has been a surprise to physicists is the demonstration by Brillouin[59] that even the gain of information at first sight as factual or conventional as the measure of a length or the decoding of a coded type must be paid for by an equivalent expense of the world's negentropy. This immediately places every experimental physicist, and even every layman, very much in the situation of a gambler; this is a new reason for believing that there is something extremely fundamental concerning the relation between the material world and the perceiving and acting presence of psychic awareness in the probability concept.

Among the new concepts introduced by cybernetics is the idea that the possession of a certain piece of information allows its possessor to decrease the entropy of the physical system to which this information pertains; this is, for instance, the problem of Maxwell's demon, discussed in this context by Smoluchowski[67], Szilard[8], Lewis[9], Demers[68], Gabor[60] and Brillouin[59]. In these texts it is clear that the word "information" is taken in the sense of a finer grained entropy than the one used in classical physics. For example, in order to separate in two communicating vessels the slower and faster molecules of a gas the demon must be able to evaluate with some accuracy the positions and velocities of each of them.

So an amount of information is both a *better knowledge* of a physical situation and a *possibility of intervening* in it in order to *reorganize* it on the macroscopic level. It is my[69] opinion that cybernetics has thus (without having searched for it) rediscovered the old Aristotelian association between two reciprocal aspects of the information concept: *knowledge* and *organizing power* (the second of which had been almost forgotten). The first of the two aspects of the word "information" is of course the one appearing in the *learning transition*

$$\text{negentropy} \longrightarrow \text{information}$$

that is, for example, in any physical observation. The second is the one displayed in the *acting transition*

$$\text{information} \longrightarrow \text{negentropy.}$$

It should be noted that, on the side of psychic awareness, we are perfectly conscious of these two facets of our representation of things: we know very well what is *observational awareness*, in which our representation follows the physical situation in time, and simply *registers* it, and what is *willing awareness*, in which our representation precedes the physical situation in

---

67 *Phys. Zeits.* 13, 1069, 1912.

68 *Canad. J. Res.* 22, 24, 1944; 23, 47, 1945.

69 *Sciences* No. 11, 51, 1961; *Rev. Quest. Sci.* 22, 5, 1961; *Rev. Intern. Philos.* No. 61, 62, 1, 1962; *Le Second Principe de la Science du Temps*, Paris, Ed. du Seuil, 1963, Chap. III.

time, and which our action will contribute to *producing*. Descartes[70], Maine de Biran[71], and Schopenhauer[72] have, among others, insisted that both aspects of our awareness are equally genuine. It should be noted that *observational awareness* is the ultimate solid ground upon which all the building of science rests; if it were fundamentally questioned, the whole of science would be shaken. Observational awareness has been taken by most scientists (until the days of cybernetics) as something evident *per se* and needing no further analysis. This, of course, has not been the feeling of most philosophers, and it is striking that cybernetics has recently supported the view that the observation process is not so simple; as Gabor puts it[60], "We cannot get anything for nothing, not even an observation." But, reciprocally, cybernetics has re-emphasized willing awareness, which tended to be forgotten, and had occasionally been radically questioned, for example by Spinoza[73] or by the theorists of so-called "epiphenomenal consciousness".

We must ask how operational significance of will, or free-will, can be accepted together with the relativistic view that everything is written in space-time once and for all. A few remarks should be made about this. First, H. Bergmann[63], Grünbaum[42], this reviewer[38], and others have pointed out that the essentially probabilistic character of quantum theory is in effect reconciled with Minkowskian geometry. Second, the well-known trait of quantum theory that an *observation* made about a system implies a *perturbation* of that system, is very much in accord with the cybernetical two-fold "information" concept. Third, the apparent antinomy between free-will and a once-and-for-all written history is an old one; it has been much discussed by theologians. The proposed solutions have generally been of the type we would call "complementaristic" today. In this context one may quote a very adequate sentence from Professor Poirier[74]: "The space-time continuum of special relativity, given all at once, brings down from heaven upon earth the old metaphysical theme of a world's history written once for all, *sub specie aeternitatis*, in the divine thought."

The "generalized Carnot principle" of cybernetics has established a previously unsuspected connection between the somewhat passive psychical

---

70 *Letters* published by Adam and Tannery, I, No. 525 p. 222 and III, No. 302, p. 663.

71 *Fondements de la psychologie* (*Oeuvres*, 1959, t. 3, p. 49).

72 *Die Welt als Wille und Vorstellung*.

73 *Ethics*, 2.35 and scolion; 2.48 and scolion.

74 *Rev. Acad. Sci. morales et politiques* 108, 86, 1955.

process of observation and the typical Carnot process of entropy increase in closed systems, that is also, according to previous analyses, with the principle of retarded macroscopic actions. On the other hand, the operation of human will, of animal activity, and of biological organization consists, as Schrödinger[75] has pointed out, of long lasting growing up fluctuations, that is, according to Adam's[25], MacLennan's[26] and Wu-Rivier's[27] analyses, of something that is formally equivalent to advanced actions. Thus we are led to the idea that, just as cognizance awareness has been demonstrated to be associated with the regression of fluctuations, willing awareness ought to be associated with the growing up of fluctuations; and this, of course, is nothing else but a cybernetical formulation of the old idea that life has an anti-Carnot tendency[45], or often looks like an advanced action process[46]. In other words, I am proposing to associate the generation of subjective information by physical negentropy with the idea of causality and the generation of physical negentropy as proceeding from subjective information with the idea of finality.

It should be strongly stressed that, in view of the analyses carried out by so many authors along Loschmidt's[1] lines, all of which concluded with law-like reversibility in statistical theories, it would be quite a sharp turn indeed to pretend to say now that, in the two way transition *negentropy* $\rightleftharpoons$ *information* the upper way is allowed and the lower one forbidden. Moreover, in light of such physical discoveries as antiparticles, that is experimentation precisely finding the thing that fills an open compartment of theory, I believe it would be quite unwise to *a priori* erase the lower arrow.

Things being so, it remains of course true that there is, here as everywhere, a fact-like irreversibility; it is expressed in the chain of inequalities deducible from the generalized Carnot principle

$$\Delta N_1 \geqq \Delta I \geqq \Delta N_2,$$

that is, information gained in a measurement is smaller than the negentropy from which it is borrowed and larger than the negentropy which it would be able to restore. This amounts to saying that observation is easier than action: the upper arrow is easier to follow than the lower one.

This amounts also to saying that causality is much more obvious and general than is finality—a statement which I believe (and Grünbaum[42] 1963, Chap. 9, § D seems to believe also) is more satisfactory than crudely wiping out finality.

Could one make some more significant comments on this fact-like rather than law-like situation? I believe one can, in two different forms.

---

[75] *What is Life*? Cambridge University Press, 1944, chap. VII.

One may comment first upon the smallness of the Boltzmann constant k. This constant relates, among other things, the expression of an entropy in thermodynamical units with that of an information in cybernetical binary units according to the formula

$$\Delta S = k \, Ln2 \, \Delta I.$$

The point is that, when evaluated in practical units, the constant k is very small—a characteristic which it has in common with the relativistic $1/c$ and the quantal h.

To say that a physical constant is very small or very large in practical units is to say that the relations it governs are much outside the realm of usual human experience; that is, the smallness or greatness of such constants expresses an aspect of biological adaptation in the universe or, as existentialism puts it, an aspect of our "situation when being in the world." For example the fact that the relativistic c is so large (as expressed in anthropocentric units) allows people corresponding with each other to feel they are "at the same time." I wonder very much what social life would look like if we were built so as to practically perceive c as a number close to one (that is, if the nervous influx had a near c velocity).

The practical smallness of the constant k certainly has quite analogous implications. Due to this smallness, acquiring information (cognizance information) is very cheap in negentropy terms (so cheap, as Brillouin[59] says, that the corresponding cost had been overlooked before the days of cybernetics). On the other hand, due to the smallness of k, producing negentropy costs very much in information terms. That is, the smallness of k directly expresses a "situation of being in the world" where observation is easy and action painful. In this light, one may say that the theory of epiphenomenal consciousness is obtained by setting $k = 0$, which renders observation gratuitous and action impossible.

So, the fact-like irreversibility according to which we (and all living beings) are exploring the entropy curve upwards and not downwards (as Botzmann[3] puts it) can be traced back to the fact that observation is easier than action. The same argument shows, of course, why causality prevails over finality—a prevalence that is of a statistical rather than absolute nature, because of course there are fluctuations, some of them large and long-lasting.

Finally one may ask why, as a means of explanation, causality is so much more obvious than finality. The answer is, I believe, that, owing to its very nature, that is, *by definition*, causality is the process which is evident to cognitive awareness, just as finality is the process which is obvious to willing awareness.

Concluding remarks

My overall impression is that the statistical theory of physical time has in its favor a tremendous number of operational contacts with physical facts, plus a few quite pleasing insights into some aspects of psychological time. I believe that when the extensions allowed in it by the new theory of cybernetics are taken into account, it has the merit of tracing a reasonable and perhaps far reaching *via media* between the opposite exaggerations of purely materialistic and purely idealistic theories of the anisotropy of physical time, and of the biological experience of "becoming."

PHYSICS AND GEOMETRY*

PETER G. BERGMANN
Syracuse University, Syracuse, N.Y., U.S.A.

Our Symposium consists of three talks, all of them concerned with various aspects of relativity and its foundations. Of the three, my talk is primarily concerned with general relativity, and with the relationship between physics and geometry suggested by that theory.

The general theory of relativity is Einstein's theory of gravitation, right now the only theory of gravitation that, after Newton's, has achieved a measure of universal recognition. It is based on the so-called principle of equivalence, a principle not very easy to formulate, but which I like to state in a relatively weak form, to the effect that inertial and gravitational mass are equal in all physical systems. This principle implies that under the influence of gravitational forces bodies behave alike, pretty much as they do in a so-called inertial field; except for inhomogeneities of the field, different bodies in the same state of motion (i.e. possessing equal velocities) at the same location and time undergo equal accelerations. This principle of equivalence implies that by local techniques alone we cannot establish an inertial frame of reference if a gravitational field is present. The closest approximation possible to an inertial frame of reference is, locally, a freely falling frame. Such a frame cannot be extended over any extended domain without inconsistency; the notion of the freely-falling frame is non-holonomic.

The principle of equivalence is a physical principle, that is to say, it purports to summarize a mass of *observations*, and to predict the outcome of *experiments* to be performed. Its formal counterpart is the principle of covariance, a structural requisite of *theories* to be constructed. The principle of covariance demands that among all the coordinate systems in four dimensions that can be constructed there is to be no privileged class, such as the Cartesian coordinate systems form in a Euclidean geometry. Rather, the laws of nature are to assume the same form in all curvilinear coordinate systems that satisfy certain minimal requirements

* This paper was prepared while the author was at Yeshiva University, where his work was supported by the Air Force Office of Scientific Research and by the Aerospace Laboratories, both under the Office of Aerospace Research.

of differentiability. Einstein "invented" Riemannian geometry a second time (because he was at first unaware of Riemann's work) as the appropriate geometry in a universe in which there were to be no linear coordinate systems. In the presence of a gravitational field, space-time is curved, and in no coordinate system will such a manifold have the appearance of a linear metric space, that is a space with a constant metric form.

In the first blush of success of the new theory, it was claimed that in general relativity physics had been "geometrized". It is difficult to sense accurately just what contemporaries meant by this claim. Presumably it was that the new theory employed as building blocks only elements that possessed an immediate geometric significance, or, perhaps, elements that geometric intuition required to be present regardless of any demands by the physicist. Today we know that, whatever "geometric intuition" might suggest, there is no logical need for a manifold to have any specific structure. If we choose to represent space-time as a differentiable manifold with the attributes of a (pseudo-)Riemannian geometry, then that is not a logical necessity but a choice suggested by empirical data, by observations and experiments. Accordingly, it has occasionally been suggested that general relativity represents the "physicalization of geometry", not vice versa. Leaving aside this play on words, I believe that the geometric structure of space-time has become a legitimate area of inquiry for the natural scientist, precisely because we have come to realize that there is no limit to the number of logical possibilities; our choices must be made on extra-logical grounds.

Very early in the history of general relativity, Kretschmar subjected the principle of covariance to an analysis and attack, which runs approximately as follows. If we state the principle of covariance simply as the requirement that all relationships occurring in the theory are to assume the same form in all coordinate systems, then this requirement is trivial; it is merely a matter of technique to cast all statements of any theory in a covariant (i.e. coordinate-insensitive) form. For instance, if a Riemannian space is flat and, hence, admits rectilinear coordinate systems, then we can add to the field equations in any (curvilinear) coordinate system the additional equations stating that the curvature tensor vanishes; the new, augmented system of equations will assume the same form in all coordinate systems. Accordingly, we must strengthen our statement of the principle of covariance by requiring not only that the equations take the same form in all coordinate systems, but further that no coordinate systems exist in which the equations are simplified. But this requirement is so strong as to exclude the Riemannian geometries as well. We may introduce special coordinate systems in which the geometric objects of Riemannian geometry take a special form; I mention the harmonic coordinate systems as one example, and the so-called intrinsic coordinate systems as another. J. L. Anderson has done a good deal of work to refute Kretschmar's attack on the principle of covariance. Whether or not one accepts his analysis, in any case special coordinate systems cannot be singled out through the algebraic specialization of the gravitational potentials; if field equations can be "simplified" by other special choices, then the underlying concept of simplification is considerably less straight-forward than the one in the minds of physicists early in the twentieth century.

If we grant the transition from linear manifolds to more general structures implied by general relativity, then one principal result is that the identification of world points (the localization of events in space-time) by their coordinate values loses a great deal of its former substance. In Newtonian physics we may identify a frame of reference by indicating at one time the location of a three-dimensional Cartesian coordinate system in space and the state of motion of its origin, combined with the specification that our frame is to be an inertial frame of reference. From then on an event is completely localized if we furnish the numerical values of its four space and time coordinates. Special relativity does not change this arrangement in any way. But in general relativity there are no inertial frames; and there are no other simple means by which we can identify a world point. To be sure, we may locate it by means of a geodesic placed in relation to conditions at a different time, along with a statement of distance along the geodesic. But into this specification enters the geometric-physical structure between our base of operations (the "now") and the world point to be identified. In contrast to the situation in Newtonian or special-relativistic physics, we have no a priori key that permits us to translate localizations by means of different bases into each other. Alternatively, we may identify a world point in terms of properties of the physical fields at that point; this procedure requires consideration of the particular events taking place there. In either case, simple questions such as "How will the field change at a fixed world point as a result of a given change in the Cauchy data?" are no longer simple, or even meaningful in the absence of further specifications. The only clear-cut identifications that are still possible in general relativity are those of whole fields, in extended four-dimensional domains, not of world points.

It seems to me that the interpretation of geometric and of physical statements in general relativity differs fundamentally from that of statements in earlier theories based on rigid rectilinear frames of reference, even though the statements may be composed of the same words. Perhaps, more precisely, many ordinary statements lack meaning, and we must construct a new vocabulary, a subset of the conventional vocabulary, from which alone meaningful statements may be constructed. There exists a subset of physical variables, the "observables", whose values are independent of the choice of coordinate system employed. Thus, any relationship between observables is "meaningful", and conversely, these are the only relationships that are legitimate. A program aiming at the identification and systematic exploitation of the observables has been under way for many years, but its execution is hampered by profound technical difficulties, which have not yet been overcome completely.

For a number of reasons the notion of the world point has become suspect even in Lorentz-covariant field theories. Chiefly, the strictly local relationships implied by partial differential equation systems appear to be the principal cause of the infinities from which current quantum field theories suffer. Non-local theories, which were in vogue a few years ago were intended to ameliorate the situation by the introduction of form factors and similar devices, whose effect is to make the rate of change of the field at one point dependent on the situation in a surrounding four-dimensional domain of non-zero extension. From the point of view of general relativity a different development might be envisaged, one in which the notions of point and neighborhood are modified. From elementary-particle physics we know that experimentally we cannot penetrate into the details of structures below some finite spatial domain, of the order of nuclear distances. In general theory of relativity, we have the greatest difficulty when we attempt to impart to the world point absolute individuality. Logically, we can construct "geometries" whose manifolds are built up of neighborhoods that will not permit decomposition into distinct points. I am not so presumptuous as to venture a definite prediction that the physical theory of the future will get along without the world point. But I submit that such a development is possible, and that physicists, geometers, and philosophers might find this possibility attractive.

RELATIVITY AND CAUSALITY*

PETER HAVAS
*Lehigh University, Bethlehem, Pennsylvania, U.S.A.***

Introduction

One of the most important aspects of the replacement of the conceptual structure of classical physics by that of quantum theory is the change in our outlook on causality and determinism. This was preceded by a similar change in outlook within classical physics brought about by Einstein's theory of relativity. The changes within the classical theory are not superseded by those of quantum theory; rather a thorough quantum mechanical analysis has to start from relativistic concepts instead of Newtonian ones. However, only a beginning of such an analysis exists. Most of the studies of the changes brought about by quantum theory emphasize the relationship with Newtonian theory. Much less work has been done on the changes within classical theory; worse still, much of what has been concluded about their implications is based on misunderstandings. These misunderstandings have grown over the years and, by seemingly restricting the scope of relativistic theories, have delayed the exploration of some alternatives to the classical models underlying the present quantum theory of elementary particles, now widely admitted to be inconsistent.

Before we can discuss the changes brought about by the theory of relativity, we have to consider briefly the prerelativistic situation and the usual connotation of "causality" in physics[1]. We are all familiar with the everyday usage of the words "cause" and "effect"; it frequently implies the interference by an outside agent (whether human or not), the "cause", with a system, which then experiences the "effect" of this interference. When we talk of the principle of causality in physics, however, we usually do not think of specific cause-effect relations or of deliberate intervention in a

---

* Work supported by the National Science Foundation and the Aerospace Research Laboratories of the Office of Aerospace Research, USAF.

** Now at Temple University, Philadelphia, Pennsylvania, U.S.A.

[1] The literature on causality is enormous, and the number of definitions of the term almost equals the number of authors. For the purposes of this talk neither an extensive bibliography nor an attempt to arrive at a generally acceptable definition is necessary. For an elementary discussion of causality from the point of view of a physicist see [1]; a concise discussion of some of the implications of the theory of relativity for causality is given in [2] ; [3] and [4] are extensive discussions by physicists turned philosophers, with numerous references.

system, but in terms of theories which allow (at least in principle) the calculation of the future state of the system under consideration from data specified at a time t_0. No specific reference to "causes" or "effects" is needed, customary, or useful, but it is understood that all the phenomena (or variables) which can influence the system have been taken into account in the initial specification, i.e., that the system is *closed*. Conversely, if the system is *open*, i.e. if interference by an outside agent is allowed, no prediction of the future state of the system from its present one is possible. If the interference is arbitrary, no scientific statement at all can be made; if it is specified as a definite function of time instead, the state of the system at times $t > t_0$ may still be calculated, but it is not a function of the state at t_0 alone.

A clear distinction between open and closed systems is essential if confusion is to be avoided in discussing the problem of causality in physics, within a prerelativistic as well as in a relativistic framework, and this distinction will be fundamental in the considerations presented here.

Causality in Newtonian mechanics

The prototype of a successful causal theory is Newton's mechanics of mass points. In this theory the acceleration of each particle at a given instant times its mass is equated to the total force acting on this particle; the force between any two particles is a function of the positions (and possibly the velocities) of the particles at the instant considered. If these functions are known (e.g. as expressed by Newton's law of gravitation), and the positions and velocities of all particles at time t_0 are specified, the positions (and velocities) of the particles are determined for any later time or at least as long as the system can be considered as closed.

Furthermore, the knowledge of the required initial conditions is in principle available to a single observer at t_0, as no limit is imposed on the velocity of transmission of information. If he also is a speedy calculator, he could use this knowledge to determine the future behavior of the system fast enough actually to predict this behavior before it happens.

This determinism and predictability prevail even if the system was *not* closed at $t < t_0$; for the behavior of a Newtonian system it is irrelevant *how* it reached the state of t_0, whether by natural development (i.e. having been closed at all $t < t_0$), or by the influence of phenomena no longer active at t_0. This allows experimental verification of the predictions of Newtonian mechanics not simply by observation of an initial state as found in nature and later comparison of the results of calculation and observation, but by deliberate setting up of a particular initial state.

But if the system was indeed closed at all $t < t_0$, its past behavior can also be calculated for all earlier times; thus, if we chose to call our initial configuration the "cause" of all the other configurations determined by it mathematically, we would have "effects" at earlier as well as at later times than the "cause", which is not a particularly fortunate choice of language. In a causal, deterministic theory everything is interconnected; it is purely arbitrary to single out a particular component as "the cause" of another component.

The forces of Newtonian point mechanics are instantaneous action-at-a-distance forces; no reference to any medium between the particles is made in the theory, nor to any mechanism of transmission, any gradual spreading of "effects" due to the particles. For such a system the number of initial data required to specify the future development is finite; for n mass points (a system "with $3n$ degrees of freedom") we need $6n$ numbers (the $3n$ components of position and the $3n$ components of the velocities). The main initial success of Newtonian theory was in the realm of celestial mechanics; the stars can be represented by mass points in spite of their enormous size because of their tremendous separation, but can be observed without noticeably disturbing them because of this size. In many terrestrial applications such convenient approximations are not possible; rather than to describe extended bodies as an assemblage of an enormous number of separate mass points, it was found advantageous to treat them as a continuum, and to take forces as acting only between contiguous elements, a development mainly due to Cauchy. The equations of motion of such bodies are partial differential equations rather than the ordinary differential equations of point mechanics; the bodies possess an infinite number of degrees of freedom, and infinitely many numbers are needed to specify the initial state. Thus the detailed verification of the predictions of the theory poses experimental problems essentially different from those of point mechanics[2].

Causality in electrodynamics

In the 19th century the theory of such continuous mechanical systems became the model for the mathematical description of nonmechanical phenomena, in particular for Maxwell's electrodynamics. Maxwell himself thought of electromagnetic phenomena in terms of a mechanical model and a material medium, the ether, which was supposed to fill all of "empty" space; the electromagnetic field corresponded to a state of stress of the ether, in close analogy to the state of stress of an elastic body. This field was

[2] The development of the concepts of action at a distance and of continuous action is described in [5]. The technical questions of Newtonian mechanics are discussed in numerous books; for the most complete recent survey see [6] and [7].

described by four vectors with sources ρ (electric charge density) and **j** (electric current density), a set of sixteen functions linked by eight partial differential equations of the first order, the famous Maxwell equations. As the number of unknown functions exceeds the number of differential equations, further relations between the functions have to be stipulated. If Maxwell's theory is considered to be a macroscopic, phenomenological one (just as the Newtonian mechanics of continua), they are to be taken from experiment; if it is considered to be valid on a microscopic level, they are to be derived from a specific microscopic model of matter[3].

The simplest case from both points of view is a system permanently without any charges and currents. Then Maxwell's equations possess the same causal properties as the equations of continuum mechanics: the specification of an infinity of data (e.g. the values of the electric and magnetic fields everywhere, subject to certain restrictions implicit in Maxwell's equations) at a given instant determines the past and future state of the system uniquely.

However, the detailed verification of the predictions of the theory even in this simplest case poses difficulties beyond those of continuum mechanics, because rather than the determination of the positions of the elements of a continuous, but at least finite and material, mass distribution it would require the determination of fields., i.e of forces on test charges and currents everywhere in space.

In the presence of charges and currents we have to distinguish between the mathematical properties of the macroscopic and the microscopic descriptions. From the macroscopic point of view the equations do not necssarily describe a closed system; the variation of the charges and curreents with time may depend on physical processes not included in the equations (e.g. the presence of an electric generator driven by another machine, the motion of charges due to non-electromagnetic forces, etc.), and must be stipulated by additional equations or as explicit functions of time[4]. From

---

[3] Historically there has been a constant interplay between the macroscopic and the microscopic points of view and this has led to a regrettable lack of clear distinction between them in most textbooks of electrodynamics, unlike those of mechanics. Continuum mechanics is usually presented as a theory in its own right, clearly distinguished from an atomic theory of matter, whether classical or quantum. Macroscopic electrodynamics, with rare exceptions (e.g. [7] and [8]), is not presented as an independent theory, but as if it could not be separated from specific microscopic models or even from a specific microscopic theory, that of Lorentz [9]. Apart from the methodological confusion, such a presentation obscures the fact that the macroscopic Maxwell theory has a clear, important, and probably permanent domain of application, while this is not the case for the microscopic one, as quantum theory is indispensable on the atomic level.

[4] The most commonly made assumption is to put **j** proportional to the electric field strength **E**; any dependence on non-electromagnetic quantities is thereby excluded, and thus the equations describe a closed system. Another common assumption is to specify **j** explicitly as a function of time, thereby implying an open system.

the microscopic point of view, the charges and currents present are those due to the elementary particles which make up all matter. These elementary charges move according to some equations of motion; the field equations and equations of motion taken together again describe a closed system.

The theory of relativity

While the role of Maxwell's theory in the historical development of relativity is well known and need not be discussed here, it should be kept in mind that relativity removed any physical basis for interpreting this theory in terms of a mechanical continuum and only left the mathematical similarities. What remained was a theory which (with only partial justification, as will be discussed later) became a model for a relativistic formalism. It should be noted, however, that Maxwell's theory of light is *not* an essential ingredient of the theory of relativity, which is based on the existence of a maximum signal velocity, the velocity of light in empty space, but is independent of any particular theory of light.

Whereas Newtonian physics had taken the existeuce of an absolute time and of an absolute meaning of simultaneity of distant events for granted, Einstein showed that the existence of a maximum signal velocity implied that the concept of simultaneity of distant events was not absolute, but involved an element of definition; the most convenient way of establishing what is meant by "the same time" at widely separated points is by means of the fastest signals available, namely light in empty space. The same definition is adopted for all inertial systems, and the theory is developed on the basis of Einstein's two postulates: I. When properly formulated, the laws of physics are of the same form in all inertial systems; II. In all such systems, the velocity of light in empty space has the same value c.

The existence of a maximum signal velocity implies a very unexpected result, which is of fundamental importance for the problem of causality. In Newtonian physics all velocities are thought to be possible, Thus a physicist using suitable equipment could in principle communicate with (send a signal to) any point in space, no matter how distant, in an arbitrarily short (even zero) time interval, i.e. if he operates at time t_0, he could in principle influence any event anywhere at all times $t \geqq t_0$ and similarly all events occurring anywhere at all times $t \leqq t_0$ could influence him. According to the theory of relativity, however, the physicist could not send any signal faster than c and thus in a time $t - t_0$ he could only reach points at a distance $\leqq c(t - t_0)$; conversely, he could only be reached by those signals emitted at an earlier time t' which originated at distances $\leqq c(t_0 - t')$. This is represented graphically in Fig. 1 (where the third spatial dimension is omitted). Only the region within the forward part of the cone shown there

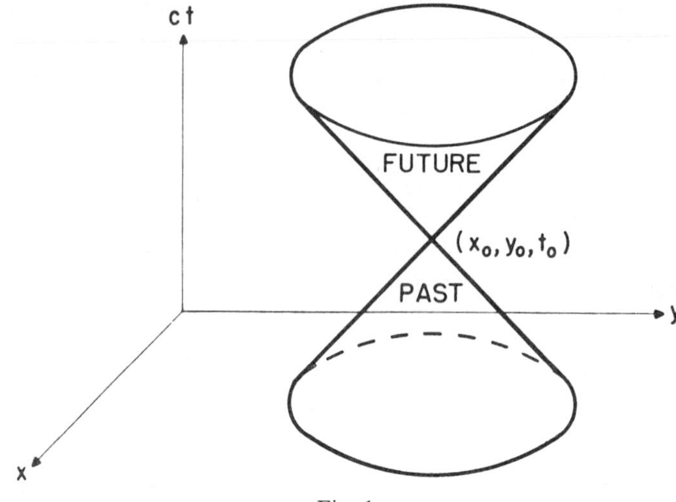

Fig. 1

(the "future light cone") can be reached by signals from (causally connected with) the point (x_0, y_0, t_0) and only signals emitted from within the backward part of that cone (the "past light cone") can reach this point. The region outside the cone can not be causally connected with this point; however, by a suitable coordinate transformation any point within this region can be made simultaneous with the apex of the cone, and thus in a sense the entire region outside the light cone constitutes the "present" for the observer at the apex. In Newtonian physics this region collapses into a plane $t = t_0$ separating the past and the future.

"Causality" and "causal connection"

The appearance of a region which can not be causally connected with a point, i.e. which can not be influenced by any happenings at that point and cannot influence them, is a fundamental new feature of the theory of relativity.[5] But what is its significance for our discussion of causality?

We have just used the words "causally connected" and "influence" rather loosely. But we *had* to use them loosely, in the naive, anthropomorphic sense of "cause" and "effect", in the sense in which the "effects" of a "cause" can be actually traced, because it is only in this sense that signals can be recognized, and it is the necessity of the use of signals to establish the meaning of simultaneity of distant events which is at the

[5] The space-time structure of the special theory of relativity sketched here is discussed in all textbooks of this theory; a particularly detailed account is given in [10]. Detailed further developments are presented in [11] and [12]; these are summarized, discussed, and extended in [13]. For a wider, less detailed, survey of the various ideas on time see [14]; some special questions are discussed in [15].

very basis of the theory of relativity. We were not talking about causality in the sense of predictability in a closed system because, as we discussed earlier, in such a system we can *not* identify one part of a phenomenon as the cause of another, trace the spreading of effects, or recognize signals.

The very concept of signals is based on interference with a system, as it is this interference which allows recognition of physical phenomena as signals, i.e. transmission of information. A signal must therefore involve an element of irreversibility, as has been stressed particularly by Mehlberg [13] and recently by Terletskii [16][6,7].

As the concept of such signals is essential for allowing the definition of the global space-time coordinates fundamental for the special theory of relativity, so is the possibility of a theoretical description of open systems at some level. Only after a framework of space-time coordinates directly interpretable in terms of measurements has been established, can specific physically meaningful special relativistic theories be formulated. In particular, any special relativistic theory which aims to provide a description of the entire universe as a closed system can not do so to the exclusion of theories which are not all-encompassing, but which allow the description of open systems. This point was overlooked in an argument presented by Bondi [20] that "relativity in fact becomes logically tenable only when the classical picture has been abandoned" and replaced by the indeterminism of quantum theory, a conclusion which was based on the assumption that the concept of propagation of information has a physical significance within a "fully deterministic theory" (i.e. a theory treating the universe as a closed system).

A similar neglect to observe the distinction between the concept of causality as applied to closed systems and that of "causal connections", which is inapplicable in such systems, has led to a number of other misunderstandings. Thus it is widely assumed that the interaction of charged particles can be described only in terms of retarded, but not of advanced, fields if

[6] It has been noted repeatedly that in spite of the basic role of the maximum signal velocity c the formalism of the theory of relativity permits the description of velocities exceeding c, nonphysical ones such as phase velocity (discussed in most textbooks) as well as actual particle velocities [16, 17]. However, the theory excludes the possibility of using such velocities as signals. This is obvious (and well known) for nonphysical velocities; for the case of particle velocities it has been pointed out by Terletskii [16]: "Forbidden only is the process in which the emission of such particles is systematically repeated and associated with an increase in entropy of the radiator".

[7] It should be noted that there is no connection between the necessity of entropy increase in the emission of a signal and the frequently made suggestion that the direction of flow of time should be defined by the direction of entropy increase (see e.g. [18] and the criticism offered in [14] and [19]).

acausal behavior is to be avoided[8], and field theories with nonlocal interaction are frequently considered to lack causality simply because of "lack of propagation character of the field equations" [23]. Similarly an almost universal belief has developed that the "causality requirements" of the theory of relativity imply the *necessity* of a theory of near action, for reasons to be discussed in more detail below. Thus the impression was created that this theory is less versatile than the prerelativistic one, and some alternative approaches to a more satisfactory description of nature were neglected, such as the study of the mechanics of interacting mass points. We shall now turn to the examination of some recent advances in this area[9].

The problem of relativistic point mechanics

We consider n mass points performing a certain motion. This can be represented by drawing their "world lines", i.e. their positions as functions of the time as shown in Fig. 2, which also shows the intersection of these

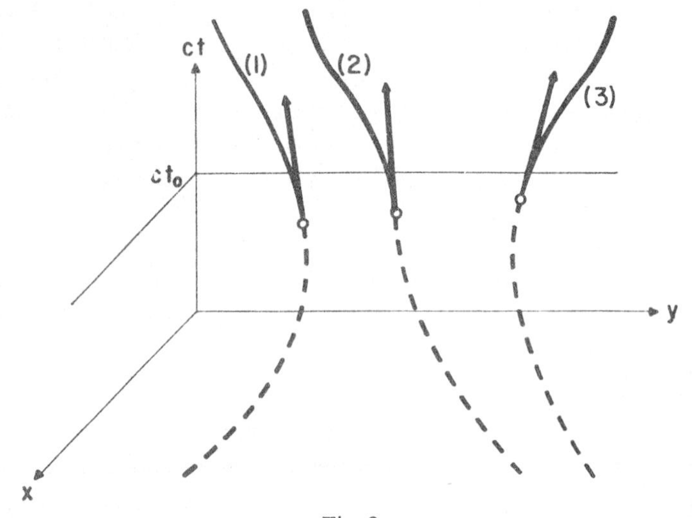

Fig. 2

[8] In most textbooks of electrodynamics this is taken for granted and advanced fields are rejected out of hand without any discussion. Nevertheless, advanced fields have been used by many authors, but unfortunately most of them did not explain the legitimacy of such a use, thus contributing to a breakdown in communications on this point. For a discussion in an earlier lecture see [21] (a detailed paper is in preparation). Part of the discussion given by Wheeler and Feynman in their "absorber theory of radiation" [22] does not seem quite to the point, as it investigates the effect of outside intervention (by a mechanism not included in the basic equations) in a theory designed to describe a closed system.

[9] For a detailed review of the various approaches to this problem see [24]. It should be noted that in the axiomatizations of relativistic mechanics by H. Hermes [25] and by H. Rubin and P. Suppes [26] this problem is ignored entirely, and that the textbooks of relativity restrict themselves to a discussion of the dynamics of a single particle under the influence of an external force.

world lines with the plane $t = t_0$ and the velocities at these points. In Newtonian mechanics, as discussed before, the motion is specified by giving the positions and velocities at this time, since we are dealing with instantaneous action-at-a-distance forces. However, the system of Newton's equations of motion and force laws does not conform to the relativity postulates, and thus such a description is unacceptable. Nevertheless, we can ask ourselves whether it is possible to find a new set of equations of motion and of force laws, which *does* conform to the relativity postulates, and which still allows us to predict the motion of the particles from a set of data (not necessarily the same as in Newtonian mechanics) given at $t = t_0$ in any inertial system.

This problem is the natural generalization of the Newtonian problem of n bodies. Nevertheless the question was apparently not posed in this form when the theory of relativity was created, and not investigated until 1949[10]. This was due to the widespread erroneous belief that a force law which depends on the simultaneous positions of two particles corresponds to an instantaneous spreading of "effects"; by this same misunderstanding it was also frequently assumed that the force exerted by one particle on another could only manifest itself as a retarded action, i.e. only within its future light cone. However, forces as such have no physical significance; they are only auxiliary quantities entering the laws of motion, the physically observable quantities being at best the positions of the particles as functions of time. Thus the problem we are concerned with is that of particles left to move freely under their mutual actions, and the possibility of specifying this motion by a set of data at time t_0. This is *not* the same problem as *starting* the motion at t_0 with the specified initial data, which is indeed incompatible with the requirements of relativity; the determination of the world lines by arbitrary starting data given at spatially separated, but simultaneous, points would indeed imply that the arbitrarily set data at the position of one of the particles influence the motion of the other particles at the same instant, and thus correspond to transmission of information with unlimited speed. But if the motion has been proceeding freely at earlier times, the setting of "arbitrary" initial data at t_0 is only a mathematical arbitrariness; we do not imply that we physically set these data at t_0, but only ask how the motion would proceed in a case in which these data are *observed* at t_0.

Such a distinction between the case of free motion of a closed system

---

[10] There were some early attempts to formulate relativistic generalizations of Newtonian force laws by Poincaré, Sommerfeld, and others, which mostly applied only to particles moving with constant velocity, and which were rapidly abandoned in favor of field theoretical approaches.

with a set of data *existing* at t_0, and a motion *started* at t_0 with the same initial data (which requires that the system was *open* up to that time) does not have to be made in Newtonian mechanics. It is essential for relativity, however.

An alternative to the necessity of introducing this distinction would be to require that the effects of the particular values of the initial data should make themselves felt only after times excluding signals exceeding c. The customary procedure is to specify data on the full surface $t = t_0$, rather than at the points of intersection of the world lines alone, thus introducing infinitely many additional ("field") degrees of freedom, and to require that the laws determining the time development of the variables (which now are functions given at all points of space-time rather than on isolated world lines) should only involve near action, i.e. action between neighboring points, rather than action at a distance, in analogy to Newtonian continuum mechanics rather than point mechanics. Maxwell's electrodynamics was a ready-made example of such a theory, and thus it came to be almost universally considered as *the* model of a relativistic theory.

However, for closed systems there is no need to pattern a theory after this model, and there is no logical reason preventing the development of a relativistic point mechanics with Newtonian causality, i.e. a theory in which the state of the particles can be specified by $6n$ data.

The construction of such a theory was first attempted on the basis of a Hamiltonian formalism, mainly by Dirac [27] and Thomas [28]; this approach corresponds to force laws depending on simultaneous positions. Another approach [29] used force laws depending on positions in space-time of pairs of particles related by a relativistically invariant expression. Both approaches have had partial successes, but have also run into difficulties, mainly of physical interpretation in the first approach, and of extension beyond two-body problems in the second [24]. However, a resolution of these difficulties is to be expected, and our further analysis does not depend on any mathematical details of a particular formalism.

The specification of open or closed systems and subsystems

We shall now discuss in more detail the different cases which may arise in the physics of mass points and of continua, comparing the situation in Newtonian theory and in the special theory of relativity, with some remarks on the general theory. Time does not permit a full justification and discussion of all the results stated; in particular, we can not go into any details concerning general relativity.

We first note that an underlying assumption of one form of Newtonian mechanics is that the universe can be represented as a closed system of mass points. Thus the future behavior of this all-encompassing system is determined if at t_0 the initial data are specified *in an infinite region.* But what if we only have available (or only choose to use) initial data in a *finite* region, i.e. in a *subsystem* of the universe? Without further assumption we have no right to treat this subsystem as closed, and since we have no knowledge of the influence of the rest of the universe, we can not say anything about the future behavior of this subsystem. However, the usual (explicit or implicit) further assumption is that the forces between particles are such that the total system is *separable*, which means that if it is divided into subsystems which are sufficiently far removed from each other, each subsystem can be described in terms of variables referring to it alone. Thus a subsystem can be considered as closed to the extent that we can assume that it is far removed from the other subsystems; clearly this last assumption can at best be valid for a finite time interval (as otherwise the subsystem would *be* our universe). It is a mathematical property of Newtonian point mechanics that separability can be incorporated into its formalism without difficulty.

In the Newtonian mechanics of continua, the behavior of a limited portion of a medium can no longer be calculated just from the initial data, as such a subsystem is in *actual contact* with the rest of the system and thus can never be considered as closed. Thus we must stipulate the conditions at such a boundary explicitly for all time; *without* such data *nothing* can be calculated.

In both point and continuum mechanics the knowledge of all the initial data at t_0 is in principle available everywhere at this time; however, this is not the case for boundary conditions of an open subsystem in continuum mechanics, as they are functions of all $t \geqq t_0$. Thus for such a subsystem actual prediction is *not* possible; instead, boundary conditions can (in principle) either be observed or set up arbitrarily, and the results of a Newtonian calculation can then be compared with observation.

We now turn to a consideration of the corresponding cases in the special theory of relativity. If we again wish to describe the universe in terms of a closed system of particles with direct interactions alone, a Hamiltonian formalism can be set up which allows the calculation of the past and future behavior of the system from the same initial data as in the Newtonian case, as shown by Bakamjian and Thomas [28]. If we consider a finite subsystem, we again need the assumption of separability; however, unlike the Newtonian case, the mathematical description of systems with this property poses a serious problem, as pointed out by Foldy [30][11]. To the extent that a subsystem can not be considered separated from the rest of the system, it is not closed, and initial data alone do not permit us to calculate anything, just as for Newtonian open systems; moreover, as noted before, we can *not* consider a relativistic particle system as open to the extent of permitting setting up of an initial state.

On the other hand, if we start from a continuum description (whether of a material medium or a field) with near action only (thus excluding nonlocal field theories [23]), we *can* calculate something about the past and future behavior of a subsystem in a limited region from initial data alone, provided it is closed (apart from boundary effects). Since we have near action, the effects of changed initial data can propagate only with velocities $\leqq c$. Thus a knowledge of the initial data on the surface I allows the calculation of past and future behavior in the regions P and F indicated in Fig. 3 without any need for boundary conditions, since no effects can cross the light

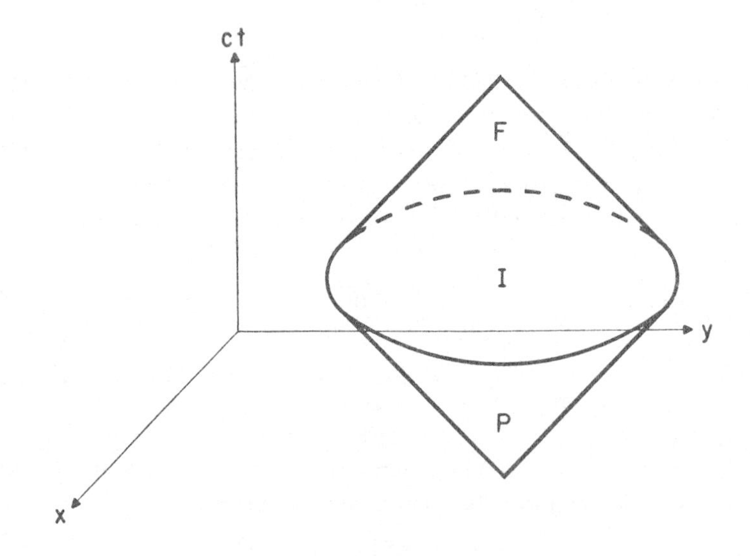

Fig. 3

[11] Although Foldy succeeded in devising a procedure for constructing separable systems, it requires an infinite number of steps; no explicit examples are known. This difficulty is due to the fact that the Bakamjian-Thomas formalism requires abandonment of the concept of invariant world lines, i.e. the variables appearing in the Hamiltonian formalism can not be interpreted as the usual coordinates and momenta transforming under the Lorentz group.

cones[12]. For a calculation in a more extended region, we *do* need boundary conditions, however, just as in the Newtonian case, whether we are dealing with material media or fields. But unlike material media, fields necessarily extend over all space; thus they can never describe a limited closed system, and the prescription of boundary conditions can not be dispensed with.

Due to the finite velocity of propagation of signals, an observer located at a point can *not* obtain information at time t on the initial data at the same time except on those at his own location, in contrast to the Newtonian case. He can *gather* such information from a limited region at an earlier time (within his past light cone), but because other effects may intrude at any later time, the data obtained are not sufficient to permit prediction without some assumptions about these effects[13].

Our previous considerations on special relativity can readily be extended to general relativity. However, our specific conclusions depend on what we *mean* by these theories.[14] In our previous discussion we meant by *special* relativity the theory of space-time structure following from Einstein's two postulates, within which we could have many distinct physical theories, such as point mechanics, or a field theory. Similarly, we might understand by *general* relativity the qualitative statements on space-time structure which follow from Einstein's principles of equivalence and of general covariance coupled with his requirement that in an inertial frame of reference locally the special theory of relativity should hold. Then qualitatively we get the same answers as before. If, on the other hand, we mean the specific theory incorporated in Einstein's field equations, our conclusions are restricted in several ways. We now *only* have a field theory available, and thus nothing can be said about a particle system except if we choose to eliminate the field variables from special solutions of the field equations with sources. Furthermore, unlike the various possibilities available in special relativity, and regardless whether the general theory of relativity is only considered to be applicable to macroscopic systems or to microscopic ones as well, the theory describes a closed system *only*, which in principle should include the entire universe. This is due to the universal nature of gravitation,

[12] The shape of the surface I is arbitrary; a circular patch is chosen here to have simple conical regions which can be pictured easily. The first proof of the uniqueness of the solution of Maxwell's equations with initial data given only on a limited surface is due to Rubinowicz [31]; the proof was extended to other equations by Plebański [32]. While in these papers no explicit distinction is made between open and closed systems, the conditions required for the uniqueness proof are satisfied only for closed ones.

[13] Actually, the region between him and the surface on which the initial data he can gather are specified is precisely that within which the behavior of the system is determined from these initial data if these are specified on a circular patch as in Fig. 3 (otherwise the relation is more complicated).

[14] For a more detailed discussion of the distinctions made in this paragraph see [33].

unlike that of electricity; in contrast to Maxwell's theory, *all* matter and *all* fields act as sources in Einstein's field equations. Although this feature simplifies some of our considerations, it also introduces the problem of describing transmission of information, which has not yet been satisfactorily resolved, and poses difficulties of separability, which have not been investigated in detail.

Time symmetry; quantum theory

Insofar as we were dealing with permanently closed systems in our Newtonian or special relativistic considerations, the laws governing the behavior of our system permit calculation of the past as well as the future time development, regardless of the specific form of these laws. If desired on physical or formal grounds, their form could be chosen to be time-symmetric, so that time reversal of all initial data would result in a time-reversed motion of the system.

Many of our previous considerations could be applied immediately to quantum mechanical systems, if only it is understood that the quantities referred to are not the dynamical variables of classical theory, but only the expectation values of these variables. However, recently arguments were advanced by Davidon and Ekstein [34] that for a system localized in the region I of Fig. 3 the combined requirements of relativity and quantum theory would require the specification of the expectation values in the entire region F rather than on I alone to permit the calculations of future expectation values. If the system could not be localized, no such calculation would be possible.

A possibly more basic limitation may be due to the fact that in quantum theory the interference with the system of an observer and his measuring apparatus plays a fundamental role. Therefore no quantum mechanical system can be considered as completely closed, and it appears that the quantum theory of a system which classically would be required to be rigorously closed (whether based on a special relativistic model or on the field equations of general relativity) would face serious difficulties of interpretation. Furthermore, the investigations of the measuring process have until now all been concerned with systems described either in terms of Newtonian action-at-a-distance forces or of near action. Although a quantum mechanical formalism for special relativistic direct particle interactions has been developed [30], the new problems of interpretation of this formalism have not been investigated.

On the other hand, the time-asymmetry apparently implicit in the quantum theory of measurement has been recently reinvestigated [35], and it appears that it is possible to formulate the basic laws of quantum physics, *including* those referring to measurements, in a completely time-symmetric manner. Similarly the time-asymmetry of the specification suggested by Davidon and Ekstein [34] is only apparent; while specification in region F of Fig. 3 is needed to allow calculation of future expectation values, a similar specification in the time-reversed region P is required instead if past expectation values are to be calculated[15].

Conclusions

In our brief survey of various aspects of causality in relativity we had to make a fundamental distinction between open and closed systems. Causality in the sense in which it is incorporated in Newtonian point mechanics can also be achieved in relativistic point mechanics, but while in Newtonian mechanics this required only that the system be closed from a time t_0 for which its state is specified up to a time t_1 until which prediction is desired, the postulates of relativity require that such a system be closed for *all* time. This is a serious restriction, but it is not a unique one; Einstein's theory of general relativity, which is a field rather than a particle theory, must impose a similar requirement. However, other aspects of both these theories pose difficulties, the consequences of which have not been fully explored. These difficulties are connected with the physical requirements any theory has to satisfy, such as that any *detailed* verification of the predictions of a theory can only involve a study of a *part* of the universe, and must permit the possibility of experimentation. But if a system has to be rigorously considered as permanently closed, no experimentation is possible, and it cannot be rigorously closed if it does not encompass the entire universe.

One possibility of escaping from these difficulties might be to renounce detailed verification. Such a verification is impossible in practice anyhow, even in Newtonian mechanics, if we are dealing with a microscopic theory. It might be that we shall not be able to use the full richness of results apparently contained in the theory, but only some macroscopic averages (as in statistical mechanics) or some statements of asymptotic behavior (as is sometimes assumed in elementary particle physics).

These questions warrant further study; but they are questions which are not restricted to a particle formalism in relativity, but occur in field formalisms as well. Thus they do not provide any reason to pursue one approach to the exclusion of the other. The development of physics has suffered from such a one-sided approach, which was less based on considerations of experimental findings than on a misunderstanding of the implications of the

---

[15] Although it is not stated explicitly in their paper, this conclusion is implicit in the arguments presented there and has indeed also been arrived at by these authors.

postulates of relativity. It is to be hoped that future research will explore the full richness of formalisms which can be fitted into the framework provided by these postulates, and thereby arrive at a more perfect agreement with experiment than present theories.

REFERENCES

[1] V. F. Lenzen, *Causality in Natural Science* (C. C. Thomas Publisher, Springfield, 1954).

[2] H. Weyl, *The Open World* (Yale University Press, New Haven, 1932).

[3] P. Frank, *Das Kausalgesetz und seine Grenzen* (Springer-Verlag, Wien, 1932).

[4] M. Bunge, *Causality* (Harvard University Press, Cambridge, 1959).

[5] M. B. Hesse, *Forces and Fields* (Nelson and Sons Ltd., London, 1961).

[6] J. L. Synge, in *Encyclopedia of Physics* (Springer-Verlag, Berlin, 1960), Vol III/1.

[7] C. Truesdell and R. A. Toupin, in *Encyclopedia of Physics* (Springer-Verlag, Berlin, 1960), Vol. III/1.

[8] J. A. Stratton, *Electromagnetic Theory* (McGraw Hill Book Company, Inc., New York and London, 1941).

[9] H. A. Lorentz, *The Theory of Electrons* (B. G. Teubner, Leipzig, 1909); L. Rosenfeld, *Theory of Electrons* (North-Holland Publishing Company, Amsterdam, 1951).

[10] H. Arzeliès, *La cinématique relativiste* (Gauthier-Villars, Paris, 1955).

[11] A. A. Robb, *A Theory of Time and Space* (Cambridge University Press, Cambridge, 1914); *The Absolute Relations of Time and Space* (ibid. 1921); *Geometry of Time and Space* (ibid. 1936).

[12] H. Reichenbach, *Axiomatik der relativistischen Raum-Zeit-Lehre* (F. Vieweg und Sohn, Braunschweig, 1924); *Die Philosophie der Raum-Zeit-Lehre* (W. de Gruyter und Co., Berlin und Leipzig, 1928).

[13] H. Mehlberg, Studia Philosophica **1**, 119 (1935) and **2**, 1 (1937).

[14] G. J. Whitrow, *The Natural Philosophy of Time* (Nelson and Sons, Ltd., London, 1961).

[15] A. Grünbaum, *Philosophical Problems of Space and Time* (A. A. Knopf, New York, 1963).

[16] Ya. P. Terletskii, Dokl. Akad. Nauk SSSR **133**, 329 (1960) [English translation: Soviet Physics-Doklady **5**, 782 (1960)].

[17] O. M. P. Bilaniak, V. K. Deshpande and E. C. G. Sudarshan, Am. J. Phys. **30**, 718 (1962).

[18] H. Reichenbach, *The Direction of Time* (University of California Press, Berkeley and Los Angeles, 1956).

[19] H. Mehlberg, in *Current Issues in the Philosophy of Science* (Holt, Rinehart and Winston, New York, 1961), p. 105.

[20] H. Bondi, Nature **169**, 660 (1952).

[21] P. Havas, "Relativity and Causelity" (Lehigh University Senior Lecture, 1960).

[22] J. A. Wheeler and R. P. Feynman, Rev. Mod. Phys. **21**, 425 (1949).

[23] C. Bloch, Kgl. Danske Videnskab. Selskab, Mat.-fys. Medd. **27**, No. 8 (1952).

[24] P. Havas, in *Proceedings of the International Conference on Statistical Mechanics and Thermodynamics, Aachen 1964* (North-Holland Publishing Co., Amsterdam, in press).

[25] H. Hermes, *Eine Axiomatisierung der allgemeinen Mechanik* (Forschungen zur Logik und zur Grundlegung der exakten Wissenschaften, Neue Folge, Heft 3, Leipzig 1938).

[26] H. Rubin and P. Suppes, Pac. J. Math. **4**, 563 (1954).

[27] P. A. M. Dirac, Rev. Mod. Phys. **21**, 392 (1949).

[28] L. H. Thomas, Phys. Rev. **85**, 868 (1952); B. Bakamjian and L. H. Thomas, Phys. Rev. **92**, 1300 (1953).

[29] P. Havas and J. Plebański, Bull. Am. Phys. Soc. **5**, 453 (1960).

[30] L. L. Foldy, Phys. Rev. **122**, 275 (1961).

[31] A. Rubinowicz, Phys. Z. **27**, 707 (1926).

[32] J. Plebanski, Acta Phys. Polon. **12**, 230(1953).

[33] P. Havas, Rev. Mod. Phys. **36**, 938 (1964).

[34] W. C. Davidon and H. Ekstein, paper presented at this congress; also J. Math Phys. **5**, 1588 (1964).

[35] Y. Aharonov, P. G. Bergmann and J. L. Lebowitz, Phys. Rev. **134**, B1410 (1964).

SPACE, TIME, RELATIVITY*

HENRYK MEHLBERG
The University of Chicago, Chicago, Illinois, U.S.A.

I. The Present Situation in Relativistic Philosophy.

The aim of this paper is to explore the present philosophical status of the relativistic theory of space and time. More specifically, I would like to formulate the principles of a methodologically and philosophically sound reinterpretation of Einstein's relativistic space-time theory, with a view to adjusting this theory to the new situation in physical science, primarily, in atomic physics and its philosophy. The need for re-interpreting relativistic space-time both within and without the realm of atomic and subatomic phenomena is made apparent by the fact that, while relativistic space-time theory is as essential in science as it ever has been, the only available interpretation of this relativistic theory, provided by Einstein in 1905, is no longer literally acceptable at present.

The explanation of the inadequacy, in 1964, of the interpretation devised by Albert Einstein in 1905 is simple: Einstein has construed his relativistic space-time in essentially macrophysical terms, such as "practically rigid frame of reference," "conventionally defined rigid yardstick" or "natural clock." The trouble is, however, that in the quantal world of elementary particles and their associated fields, there simply are no rigid frames of reference, no rigid yardsticks, no natural clocks. Nevertheless, the Lorentz transformation-formulae for the spatio-temporal coordinates of elementary particles and the postulate of the Lorentz covariance of all laws of physics, related either to the macrocosmos or to the microcosmos, are still the foundation of all our quantal theories, in spite of the literal uninterpretability of both the Lorentz transformation formulae and of the postulate of Lorentz covariance in these theories.

The argument I have just outlined is unfavorable to the literal applicability of relativistic space-time to the quantal world. This reasoning can be reinforced by noticing that, according to the assumptions of relativistic space-time, it is always possible to adjoin to three real numbers specifying the three spatial coordinates of a point-event, one more real number which

* The research involved in this paper was conducted under a grant of the U.S. National Science Foundation.

specifies the time-coordinate. Without this assumption, the Lorentz-transformability of spatio-temporal coordinates would be impossible. Suppose however that a material point is located in a region of temporally constant potential and that, with regard to our frame of reference, this material point is at rest. The energy of this point will then be a function of its spatial location. Hence, since energy and time are canonically conjugated, the quantity consisting of the three spatial and of the one temporal coordinates of the point is non-existent, according to the Uncertainty Principle of quantum mechanics, in contrast to the relativistic space-time theory which implies the existence of a space-time point with any specifiable values of its spatial and temporal coordinates.

In the philosophy of physical sciences, as conceived today, it is out of the question that a philosopher should argue responsibly, e.g., that Lorentz covariance should be dropped now in all quantal theories because the conventional interpretation of this requirement of covariance breaks down in these theories. Lorentz covariance belongs in physics proper; it is a physicist's problem. In the philosophy of physical science, we are trying to build up a science of science, more specifically, a science of physical science. The situation is not without analogy with how the science of mathematical sciences, now usually referred to as metamathematics, has evolved from the philosophy of mathematics, as a result of investigations into the consistency of mathematical theories, their independence, their completeness, completability or incompletability, their categoricity and other model-theoretical properties, their decision-procedures etc. Metamathematics is now a going concern, with headquarters in Berkeley, California.

In the case of relativistic space-time, the situation is now less favorable, mainly because no satisfactory interpretation has so far been made available in the crucial field of atomic and sub-atomic theories. It seems to me that the safest procedure for establishing such an interpretation consists in re-axiomatizing relativistic space-time in terms of concepts applicable both in the microcosmos and the macrocosmos. In the sequel of this paper, I shall outline such an axiomatic system which aims primarily at an interpretation of the relativistic space-time theory with a view to establishing its meaningful applicability to pre-quantumtheoretical and quantumtheoretical universes of discourse. Incidentally, the set of axioms I shall propose has also an additional advantage relevant to its evaluation. The point I am driving at is that pre-quantumtheoretical axiomatizations of relativistic space-time, although carried out several times, e.g., by Carathéodory [1], Milne [2] and Reichenbach [3], are open to serious mathematical objections. Thus, Reichenbach's axiomatic system was reviewed by Herman Weyl [4] and was shown to raise insuperable grouptheoretical difficulties. Recently, L. L.

Whyte [5] pointed out that almost all pre-quantumtheoretical axiomatizations of relativistic space-time, including those of Carathéodory and Milne, are open to Weyl's criticism. The set of axioms to be presented in this paper is *not* open to Weyl's criticism, even if the axioms are applied to macrophysical theories only. I do not have to tell you how I dislike to disagree, particularly with Carathéodory, on matters mathematical. My only consolation is hindsight which he could not possibly have had.

To save time, I shall take advantage of the fact that this paper is being read at a Symposium held in 1964. Accordingly, I shall disregard the vicissitudes of relativistic space-time between 1905 and the present moment. Only one exception must be made in refraining from historical considerations, for reasons which will soon become apparent. This exception is related to the fact that in 1925–26 an array of fundamental findings by Planck, Einstein, Rutherford, Bohr and de Broglie, enabled Heisenberg and Schrödinger to discover their non-Newtonian, non-relativistic quantum mechanics. The spectacular success of this theory soon attracted a galaxy of men like Dirac, Pauli, von Neumann and Wigner and ostensibly eclipsed relativistic space-time. However, the apparent set-back of Einstein's ideas of 1905 was only of short duration, in spite of the fact that philosophers of science, physicists and mathematicians interested in the foundations of atomic theory have kept discussing almost exclusively non-relativistic quantum mechanics, and are still concentrating almost exclusively on this non-relativistic Heisenberg-Schrödinger theory, which is now forty years of age. All these investigators of the foundations of atomic theory seem to have disregarded the fact that during the last two or three decades the non-relativistic theory of quantum mechanics of 1925 has been followed by an ever increasing array of unexceptionably relativistic quantal theories which dealt successfully with momentous problems transcending the scope of non-relativistic quantum mechanics. I have in mind theories like quantumelectrodynamics [6], the quantum theory of fields [7], the quantum theory of elementary particles [8], the quantum theory of nuclear processes [9], associated mainly with the names of Schwinger, Feynman, Dyson, Gell-Man, Tomanaga.

To realize e.g., the rapid advance of relativistic field theory which seems to have made a new start, owing to H. Bethe [10], in 1947, let us point out that as early as 1949, F. J. Dyson [11] succeeded in deriving Feynman's theory from that of Schwinger and in generalizing both theories. It is obvious that any two theories T and T' must have achieved a considerable degree of rigor and precision in order that it should be possible to deduce one of them from the other. A more striking symptom of progress made in relativistic quantum theory consists in the fact that successful and rigorous axiomatizations of quantum field theory have been achieved independently and in different ways by several investigators, e.g., Wightman [12] of Princeton University, Lehman, Szymanczyk and Zimmermann [13] of the Max Planck Institute, Haag and Schroer [14] of the University of Illinois. The axiomatization of a theory is certainly a reliable sign of its advanced stage. Moreover, the interest taken at present by the working physicists in this axiomatic or, as they put it, "abstract" approach to quantum field theory contrasts pleasantly with the reluctant attitude struck by the majority of working physicists three decades ago, towards the greatest achievement in axiomatic quantum physics, due to von Neumann.

It goes without saying that the proliferation of relativistic quantal theories is not the outcome of the pursuit of a scientific ideal of unification, viz. the ideal of unifying the relativistic and the quantal approach to physical reality. All these quantal theories owe their origin primarily to the necessity of coping with a deluge of striking experimental discoveries, which raised the number of known elementary particles from three to over thirty, which disclosed the duality of matter and anti-matter, no less significant than the duality of particles and waves, and which established a tight functional relationship between the individual spin of a particle and its social behavior.

The growth of relativistic quantum physics goes obviously to show that a great theory, like that of relativistic space-time, neither dies nor fades away. However, as already mentioned, the new meaning, function, and validity of relativistic space-time, when extended to the quantal realm, raises serious difficulties since the only available interpretation of this theory due to Albert Einstein breaks down in quantum physics.

II. An Axiomatic System for Relativistic Space-Time.

1. Collision-Connectibility as the only Undefined Term.

Our main issue today is therefore to establish a philosophically and methodologically sound re-interpretation of relativistic space-time theory in order to rid this theory of its crippling, macrophysical restrictions. I have already suggested that the most reliable procedure for establishing such an interpretation seems to be the construction of an axiomatic system all the physically undefinable concepts of which are applicable both to the macro- and the microcosmos. As a matter of fact, the axiomatic system to be outlined in the sequel contains only *one* undefined physical concept related to the *collision* of two physical particles.

Let us therefore proceed with a sketch of the proposed axiomatic system and first characterize its vocabulary. This vocabulary includes, apart from the logical and the mathematical terminology, one undefined physical term derived from a dyadic relation between any two events E and E' which obtains whenever a particle α'', distinct from the particles α and α' in which the events E and E' occur, respectively, collides at different times with the particles α and α'. I shall refer to this dyadic relation between the events E and E' as their connectibility by one pair of collisions.

For the sake of simplicity, it is preferable to replace the concept of connectibility of two events by a single pair of collisions with the more general concept of the connectibility of the events E and E' by any finite number of collisions. The concept of "collision-connectibility" of two events construed in this more general way will actually serve as the only undefined physical term of the proposed set of axioms and will be designated by the letter C. Geometrically speaking, the formula "event E is collision-connectible with event E'" is equivalent to stating that event E is either in the upper or in the lower lightcone of the event E'. Should the proposed axiomatic system be applied to macrophysical phenomena only, then stating that E bears the relation C to E' is tantamount to stating that one of these events coincides with the dispatch and the other with the arrival of a light signal. Instead of stating that E and E' are collision-connectible, we shall often simply say that E and E' are *connected.*

We now come to the principal *defined* terms in the proposed set of axioms. They can be explained as follows:

2. Definitions.

DEFINITION 1. *Space-like betweenness.* The event E occurs spatially between E' and E'' if every event connected with both E' and E'' is also connected with E.

DEFINITION 2. *Collinearity.* The events E, E', E'' are collinear if one of them occurs spatially between the other two events.

DEFINITION 3. *Inertial frames of reference.* By definition, an inertial frame of reference F is any exhaustive partition of all events into mutually exclusive classes which satisfies the following three conditions:

(a) If the events E and E' belong to the same class of the frame F then E and E' are unconnected.

(b) If the event E is not a member of a class t of the frame F then E is connected with an event belonging to t.

(c) If the events E and E' both belong to the class t of the frame F then every event E'' collinear with the afore-mentioned two events also belongs to t.

It should be noticed that the above definition of an inertial frame of reference does not involve any term applicable only to macrophysical events. The reason for identifying the partitions of the set of events which meet the three afore-mentioned conditions with inertial frames of reference will be made obvious by the remaining definitions.

DEFINITION 4. The time-like *instants* associated with an inertial frame of reference F, are, by definition, the classes determining the partition which constitutes F. Any instant associated with F will be denoted by the symbol t_F.

DEFINITION 5. *Temporal coincidence* (simultaneity) relative to a frame F. The events E and E' are simultaneous relative to a frame F if both belong to the same instant t_F.

DEFINITION 6. *Spatial coincidence* relative to a frame F. The events E and E' occur at the same place relative to F if both are collision-connectible with the same events at some instant t_F.

DEFINITION 7. *Spatio-temporal coincidence.* The events E and E' coincide in time and space if both are connected with the same events.

DEFINITION 8. *Asymptotic collision-connectibility.* Events E and E' are asymptotically connected if they are connected but do not coincide spatially relative to any frame of reference.

Obviously, asymptotic connectibility is coextensive with connectibility by a light-beam which travels *in vacuo*.

DEFINITION 9. *Spatial congruence* of two pairs of events relative to a frame F. The pair of events $(E_1, E_2,)$ is spatially congruent with the pair (E_3, E_4) in F if these four events are simultaneous in F, and, moreover, if every pair of events simultanenous in F, e.g. (E, E'), satisfies the condition that the spatial coincidence in F of E with E_1 in conjunction with the asymptotic connectibility of E with E_2 is equivalent to the spatial coincidence in F of E' with E_3 in conjunction with the asymptotic connectibility of E' with E_4.

DEFINITION 10.* *Temporal congruence* of two pairs of events relative to a frame of reference F. The pairs of events (E_1, E_2) and (E_3, E_4) are temporally congruent in a frame of reference F if the events E_k $(k = 1, 2, 3, 4)$ coincide spatially in F with the events E'_k $(k = 1, 2, 3, 4)$, the four events E'_k being all simultaneous in F and the pairs (E'_1, E'_2) and (E'_3, E'_4) being spatially congruent in F.

DEFINITION 11. *An orthogonal system of coordinates* $[X_k]$ $(k = 1, 2, 3, 4)$ is said to be *associated* with an inertial frame of reference F by a bi-unique relation R if X is the class of all ordered quadruples of real numbers and R satisfies the following conditions:

(a) The domain of R is the class of all events and its converse domain is X.

(b) The spatial coincidence of any two events E, E' relative to F is equi-

* Def. 10 applies only if the events E_j, E_{j+1} (where either $j = 1$ or $j = 3$) are asymptotically connected. The extension of Def. 10 to cases involving no twofold asymptotic connection is obvious.

valent to the existence of two quadruples X_k and X'_k such that X_k bears the relation R to E if, and only if, this relation also obtains between X'_k and E', provided that $X_j = X'_j$. The variable j varies here over the first three values of k.

(c) The temporal coincidence in F of any two events E, E' is equivalent to the existence of two quadruples X_k, X'_k, such that R obtains both between E, X_k and E', X'_k, provided that the last number in these quadruples be the same.

(d) If the four events E_1, E_2, E_3, E_4, are simultaneous in F then the spatial congruence of the pairs (E_1, E_2) and (E_3, E_4) relative to F is equivalent to the following two conditions:

(a′) There are four quadruples of real numbers x_k^m (k, m varying from 1 to 4) such that the relation R obtains between E_k and the quadruple x_k^m.

(b′) $$\sum_{j=1}^{3} (x_j^1 - x_j^2)^2 = \sum_{j=1}^{3} (x_j^3 - x_j^4)^2 .$$

DEFINITION 12. *Euclidicity.* Geometry is said to be Euclidean at the instant t_F of the inertial frame F if there is a bi-unique relation R whose domain includes all ordered triples of real numbers, whose converse domain is t_F and which satisfies the requirement that the spatial congruence, relative to F, of the pairs (E_1, E_2) and (E_3, E_4) is equivalent to the following two conditions;

(a′) $x_j^m \,\varepsilon\, \mathrm{D(R)}$ and $j = 1, 2, 3$; $m = 1, 2, 3, 4$.

(b′) $x_j^m \mathrm{RE}_m$ entails $\sum_{j=1}^{3} (x_j^1 - x_j^2)^2 = \sum_{j=1}^{3} (x_j^3 - x_j^4)^2$.

3. Axioms.

A1. There is at least one inertial frame of reference.

A2. If two unconnected events are not simultaneous relative to F then, at every instant t_F of F which does not include either of these events, a third event is collinear with both.

A3. Given any instant t_F of a frame F and any event E, there is an event at t_F which coincides spatially, relative to F, with E.

A4. If the four events E_k spatially coincide, relative to F, with four events E'_k then the spatial congruence in F of the pairs (E_1, E_2) and (E_3, E_4) is equivalent to the spatial congruence in F of the pairs (E'_1, E'_2) and (E'_3, E'_4).

A5. Relative to any inertial frame F there is at least one instant t_F when geometry is Euclidean.

A6. If event E connected with E' is not asymptotically connected with

E' then E' occurs between two events which are both asymptotically connected with E.

A7. If event E is simultaneous with E' relative to the frame F there are at least two events asymptotically connected with E which also spatially coincide with E', relative to F.

III. Philosophycally significant features of the proposed set of axioms.

In this third part of my address, I shall comment on the following features of the proposed axiomatic system which seem to be philosophically significant:

1) The scope of the proposed set of axioms.

2) The role of causal concepts in the proposed re-axiomatization of admittedly non-causal quantal theories.

3) Time's arrow in atomic theory, as axiomatized in the above system.

4) The role of relativistic space-time in atomic theory, in the face of the Copenhagen Interpretation of atomic theory and of von Neumann's exploration of observation and measurement in atomic theory.

1. Scope of the Proposed Axiomatic System.

To characterize the scope of Axioms 1–7, I am going to point out a few significant consequences of these axioms without quoting the available, rigorous proofs:

1) Two events are connected with each other if they are not simultaneous with each other in any inertial frame of reference.

2) If geometry is Euclidean in any instant t_F of F, it remains Euclidean in every t_F.

3) An event which spatially coincides in F with event E is connected with E.

4) The frames of reference F, F' are numerically identical if there are at least two events which spatially coincide in either frame.

5) If the coordinate systems $[X_k]$ and $[X'_k]$ are respectively associated with F and F', then the transformation taking $[X_k]$ into $[X'_k]$ is a Lorentz transformation.

6) If the coordinate system $[X_k]$ associated with a frame F undergoes a Lorentz transformation then the resulting system $[X'_k]$ is associated with another inertial frame.

Hence, the proposed axiomatic system provides a proof of validity and monopoly of the Lorentz transformation group, despite the close analogy of this system with those of Carathéodory, Milne, and Reichenbach, where such a proof is not available. The reason why the proposed axiomatic system, similar to the three aforementioned systems in point of undefined physical terms and of assumed axioms, yet provides a proof of the validity of the Lorentz transformation group (of space-time coordinates) is explained in a companion paper.* I am taking this opportunity to thank Dr. F. J. Dyson of the Princeton Institute for Advanced Study for a recent discussion of the fact that the proposed axioms are not open to H. Weyl's criticism.

However, the monopoly of the Lorentz transformation group for two systems of spatio-temporal coordinates associated with two inertial frames of reference, respectively, does not suffice to prove that equations involving such systems of spatio-temporal coordinates remain unchanged during the transition from one inertial frame of reference to another one, i.e., the general Postulate of Lorentz Covariance. In pre-quantummechanical, relativistic physical theories, the monopoly of Lorentz transformations of coordinates did suffice to establish the Postulate of Lorentz Covariance because all the physical equations involved in these theories dealt only with scalars and tensors of finite rank.

But, in quantal theories, the relevant physical quantities are, in general, (bounded or unbounded) self-adjoint operators on the Hilbert-space associated with the physical system under consideration. These operators depend, e.g., in relativistic quantum field theories, on spatio-temporal coordinates construed as numerical parameters, not as operators (in contrast to the non-relativistic quantum-mechanics of Heisenberg and Schrödinger where spatial coordinates are multiplicative operators). It stands to reason that the Lorentz covariance of a quantum theoretical field equation may require more than the validity of the Lorentz transformation group.

Actually, as I. M. Gel'fand [15] has recently shown, two more conditions are required in quantum field theory to guarantee the Lorentz covariance of the relevant equations: (1) The aforementioned operators must transform according to an infinite-dimensional, unitary representation of the inhomogeneous Lorentz group; (2) the matrix coefficients of the relevant equations must, in addition, satisfy an independent, specifiable condition.

The proposed axiomatic system therefore takes care of the validity of the Lorentz transformations of spatio-temporal coordinates. It is impossible to discuss, within the space allotted to this paper, the additional conditions required to establish Lorentz covariance.

2. The Use of a Central Causal Concept in the Proposed Set of Axioms.

It is usually assumed that the collision-connectibility of two events implies that they are causally related. After all, this relation holds between two events only if one of them is either in the upper or in the lower light cone of the other. This spatio-temporal orientation of two events has always been considered by the followers of relativistic space-time theories as tantamount

* included in the forthcoming volume "Mind, Matter, Method" (University of Minnesota Press). This paper of mine also contains a definition of undirectionality of time as opposed to time isotropy; cf. footnote on p. 376.

to a causal relation between the two events. However, the Law of Causality was generally abandoned in all quantal theories, particularly since von Neumann [16] succeeded in proving the incompatibility of non-relativistic quantum mechanics with the existence of so called "hidden parameters." On closer inspection, his proof of the incompatibility of quantum mechanics with the existence of hidden parameters amounts simply to the incompatibility of quantum mechanics with the Principle of Causality.

I do not deny that the cogency of von Neumann's proof has often been criticized. To the best of my knowledge, however, any close and unbiased examination of his proof is bound to admit its cogency. Accordingly, the use of a causal concept in reaxiomatizing admittedly non-causal atomic theories, has to be justified in view of the quantum theoretical surrender of the Principle of Causality. The most natural way of overcoming this difficulty is simply to point out that while the Principle of Causality has to be surrendered at the quantum theoretical level, this surrender implies only that *some* quantal events have no cause. However, the surrender does not preclude at all the existence of causal relations among a class of quantal events whose cardinality proves sufficient for using a causal concept throughout the quantal world. To support this view, which is by no means a mere conjecture, let us first refer to Schrödinger's time-dependent partial differential equation used by von Neumann as one of his two quantum theoretical axioms. Schrödinger's equation shows simply that, in the case of a closed quantum theoretical system, any two of its non-simultaneous quantum states are causally related. Nor is the quantum state of a closed system the only entity eligible for partnership in a causal relation. All observables of a closed system which are governed by a conservation law are equally eligible. In authoritative works on quantum field theory, the causal relations of elementary particles and their associated fields are considered as the main problem of atomic theory. In this context, it may be worthwhile to refer to the "principle of microcausality" which requires that no two events separated by a space-like interval can be causally related. This principle is both a special case of the relativistic law which precludes the propagation of any causal influence with a speed superior to that of light and contributes much to our present understanding of Heisenberg's Uncertainty Principle. My ultimate conclusion regarding the question whether or not the "causal theory of time" [24], which contributes significantly to clarify the meaning of time in science, can be applied with equal success to explain the meaning of time in quantized theories, is, accordingly: "yes, in principle." This does not rule out the necessity of substantial adjustments, when a philosophical approach to pre-quantumtheoretical time is being applied to time as understood in quantal theories.

3. Time's Arrow and Relativistic Space-Time.

The proposed set of axioms for a universally applicable, relativistic space-time theory is formulated entirely in terms of a single physical concept, viz. collision-connectibility. Now, collision-connectibility is clearly a symmetrical relation: If *A* is collision-connectible with *B*, then so is *B* with *A*. On closer analysis, one finds that the symmetry of this relation is equivalent to the undirectionality of time and supported by the invariance of all laws of nature under time reversal. The scientific status of this invariance has changed remarkably in the last few decades. Sir Arthur Eddington has tried throughout his remarkable career, started by his first book in 1920, to convince the scientific community of the world that the laws of nature are not invariant under time-reversal, or, as he preferred to put it, that "time has an arrow." Strangely enough, Eddington's view was already refuted at the moment of its publication, in a famous paper of Paul and Tatiana Ehrenfest [18] which appeared a few years before Sir Arthur's book. To the best of my knowledge, the analysis of the Ehrenfests was never challenged by any serious investigator. The status of the invariance of laws of nature under time reversal (which Sir Arthur made conditional upon the law of increasing entropy in closed mechanical systems) kept deteriorating under the impact of consecutive scientific discoveries. A major phase in this progressing deterioration was the view held by a supreme authority on the foundations of quantum mechanics, von Neumann, who believed that entropy keeps increasing in one fundamental type of microprocesses, viz. those involving measurement. This view of von Neumann was refuted by G. Ludwig [20] and D. Bohm [21] who had the advantage of hindsight and managed to establish the "principle of microreversibility."

To realize the strength of experimental support in favor of the abandonment of the principle of entropy increase in closed systems, it suffices to notice that the Lorentz group is usually interpreted today as including the sub-group of temporal inversions. According to this terminology, covariance under the Lorentz group automatically entails covariance under time-reversal. Since Lorentz covariance has now become the leading principle of theory formation, this must also apply to covariance under time reversal. Let us also recall Khinchine's theorem [22] that, if precisely defined, the entropy of a closed system is as constant as its energy. This is a far cry from Clausius' famous statement that the energy of the universe remains constant but its entropy keeps increasing.

At this juncture, an important point must be made in order to differentiate between the covariance of universal *laws* under time-reversal and covariance, or lack of covariance, of particular *facts* under such reversal. Monsieur O. Costa de Beauregard [23] states explicitly that the reasons for the universal irreversibility, if formulated in a mathematical idiom, would come to a difference between the relevant boundary conditions. This clarification of his attitude shows simply that there is no intrinsic difference between the views on the isotropy of time now held by Monsieur de Beauregard and by myself. For the theory of temporal isotropy I have been advocating for two decades, virtually alone at the start, claims only that the *laws* of nature are invariant under time-reversal [24]. It never occurred to me that such an invariance could also be asserted of the particular facts which occur in nature. Similarly, the planetary motions around the sun are governed by laws invariant under time reversal, but nobody is likely to claim that he, or even a super-natural being, actually can reverse at will the velocities of planetary motions. Thus, the ostensible difference between the views on the isotropy of time which Monsieur de Beauregard and I are holding, vanishes once the invariance of physical laws claimed by me is distinguished from the irreversibility of particular physical facts, which he advocates. Other developments that took place during the last two decades of scientific evolution and contributed to support the invariance of laws of nature under time-reversal have been analyzed in a few papers of mine [25]. I am taking this opportunity to thank E. P. Wigner and V. Bargmann of Princeton University, for recent discussions of the isotropy* of time and of other quantumtheoretical issues.

4. Copenhagen Interpretation, von Neumann's Measurement Theory, Relativistic Space-Time.

We shall now comment in some detail on the crucial philosophical issue of space-time and relativity in connection with quantal theories. This issue arises because the so called "Copenhagen Interpretation of Quantum Theory," due mainly to N. Bohr [26] and W. Heisenberg [27] and presently dominating almost entirely the world's scientific community, has associated with von Neumann's quantum theory of measurement a new revolutionary epistemology which is clearly incompatible with Einstein's relativistic theory of space-time. A clarification of this issue is therefore vital for a universal reaxiomatization of pre-quantumtheoretical and quantumtheoretical relativistic theories since a cluster of mutually incompatible theories could not possibly be reaxiomatized within a single axiomatic system.

Strangely enough, von Neumann does not mention in his treatise the epistemology derived in Copenhagen therefrom, while his theory of quantal

* Temporal isotropy is a "stronger" time property than the Wigner time invariance [26].

measurement is often used to support the central role of the concept of collision in quantum theory. Thus, Prof. Schweber [28] writes in his leading treatise *Introduction to Relativistic Field Theory*:

> "The act of measurement is fundamental to the formulation and interpretation of quantum-mechanical formalism. An analysis of various kinds of measurements at the microscopic level reveals that almost every such physical measurement can be described as a collision process.... The realization of the central role of collision processes in quantum mechanics was of the utmost importance in the recent development of field theory."

This emphasis on the central role of collisions lends ostensibly a valuable support to our attempt at reaxiomatizing the relativistic theory of space-time within and without the quantal realm in terms of C. However, Prof. Schweber's main reason for attributing this decisive role to collisions is their alleged role in quantal measurements. In turn, the role of quantal measurements is described according to the epistemology which the Copenhagen Interpretation has associated with von Neumann's quantum theory of measurement. This epistemology comes to the following two tenets:

(a) The quantal theory of measurement is an integral part of quantum theory proper.

(b) A quantal magnitude M (an "observable," in Dirac's [29] parlance) takes on a particular value V for a particular quantal system S if an actual measurement of M has been performed on S with the result V. Otherwise, the quantity M takes on no value for the system S.

Tenet (a) is essential for any epistemological claim regarding S if this claim is derivable from measurements of quantal quantities. If tenet (a) were false then no epistemological position regarding quantum mechanics could be derived from quantum mechanical measurement theory. Yet, as a matter of fact, the quantum theory of measurement accounts for a considerable part of von Neumann's treatise. But does this warrant this tenet? Von Neumann succeeded in axiomatizing quantum mechanics and used for this purpose only the following two axioms:

(1) Schrödinger's time-dependent equation;

(2) Born's Expectation Rule to the effect that any magnitude M, corresponding to a hypermaximal self-adjoint operator M^{op} on the associated Hilbert space of vectors Ψ has, for an ensemble E of non-interacting systems S, all of which share the same quantum state Ψ, an expectation value equal to the scalar product of the vectors Ψ, $M^{op}\,\Psi$. The entire theory of quantum

mechanics can therefore be replaced with Schrödinger's time-dependent equation and Born's Expectation Rule. Since none of these two axioms involve the concept of quantum mechanical measurement, no premises dealing with quantum measurement can affect quantum mechanics proper.

This argument against tenet (a) can be reinforced by the consideration that, apart from the controversial case of quantum mechanics, no other major physical theory is known to contain its own measurement theory. Thus, Newton's Three Laws of Motion do not refer at all to the measurement of force or of mass. Maxwell's electromagnetic equations make no mention of mensural procedures for electric and magnetic field strength, for electric and magnetic charge. Why should the non-relativistic quantum mechanics of Heisenberg and Schrödinger contain its own measurement theory? I think that the main reason that induced von Neumann to attribute a central role to quantum mechanical measurement theory was primarily of historical nature. He published his extraordinary treatise a few years only after Heisenberg's discovery of the Uncertainty Principle which focussed the attention of the scientific community on measurement theory in quantum mechanics. More importantly, von Neumann's treatise was preceded almost immediately (in 1930) by Dirac's Principles of Quantum Mechanics which von Neumann opposed strongly (and probably admired to the same extent). In the preface of his treatise, von Neumann states emphatically that Dirac's treatment does not live up to the standard requirements of rigor in mathematical physics, even if allowance be made for some easing up of these requirements in comparison with pure mathematics. No wonder that von Neumann took advantage of his familiarity with the theory of bounded operators on Hilbert spaces and after having succeeded in creating the then non-existent theory of unbounded self-adjoint hypermaximal operators on Hilbert spaces, which operators replaced Dirac's observables in quantum mechanics, von Neumann was able to axiomatize rigorously the theory of quantum mechanics without surrendering a single observational consequence which could be derived from Dirac's mathematically objectionable treatment. Parenthetically, let us add that, by including the measurement theory of quantum theory within quantum theory proper, we would confuse the metalanguage* and the object language of quantum theory. Such confusions are known to generate antinomies, and this is one more reason for opposing tenet (a).

We are now embarking upon the discussion of tenet (b) which is of decisive importance to the philosophy underlying quantum mechanics and forms the gist of the Copenhagen Interpretation of Quantum Theory. This tenet

* The meta-linguistic status of von Neumann's measurement theory is shown by the formulation of his "(*m*) principle." Cf. [16], p. 177.

is also of great importance with regard to the extension of relativistic space-time to the microcosmos. I have already mentioned that the tenet expresses an epistemological view associated by Bohr and Heisenberg with von Neumann's quantum theory of measurement, although it is hardly attributable to von Neumann himself. This second tenet is often referred to, in accordance with a terminology occasionally used by W. Pauli and built up systematically by von Weizsäcker, as the "unobjectifiability" of quantum theoretical concepts. Actually, this tenet has several versions, differing in scope. The most radical version forms, if I am not mistaken, the orthodox version of the Copenhagen Interpretation of Quantum Mechanics. In Bohr's first book, one runs again and again into the view that no value V of any magnitude M is possessed by a physical system S unless a measurement of M with outcome V has actually been performed on S. Thus, a hydrogen atom has no position, i.e., its positional coordinates have no value, unless a measurement of these positional coordinates has actually been carried out.

The Copenhagen version of the unobjectifiability of positional coordinates raises, obviously, the grave objection that, so far, science has found only one satisfactory way of accounting for the solar radiation, viz. in terms of H. Bethe's assumption that the hydrogen atoms inside the sun are being partly transformed into helium atoms releasing thereby the amount of energy responsible for the sun's radiation. Bethe's account of solar radiation is incompatible with the Copenhagen version of the unobjectifiability of positional magnitudes and the underlying quantal measurement theory of von Neumann. For, according to this theory, not a single positional coordinate of a single hydrogen atom inside the sun has ever been measured, or could possibly have ever been measured, since von Neumann's measurement theory tells us that the value of any magnitude m for any physical system S can only be measured if, temporarily, S is coupled with an appropriate macroobject called a measuring instrument for M. A closer analysis shows that the macroinstruments for M could not possibly exist under the conditions of radiation-pressure and temperature prevailing inside the sun, without violating relevant, well established physical laws. Accordingly, in keeping with the Copenhagen interpretation, no positional coordinate of any hydrogen atom inside the sun has any value whatsoever, since the necessary measuring instruments could not exist there. In other words, there are no hydrogen atoms inside the sun according to tenet (b).

Needless to say, this paradoxical view inherent in the Copenhagen school of thought is by no means gratuitous. The classical item of evidence supporting the Copenhagen position is best illustrated by the Stern-Gerlach experiment where a container full of silver atoms and permeated by a non-homogeneous magnetic field is governed by the following quantum theoretical laws:

(1) If the percentage of horizontally oriented silver atoms in the container is measured (by applying presently available techniques) then 100% silver atoms will be found to have a horizontal orientation in space.

(2) However, if, instead of horizontal orientation, the percentage of vertically oriented silver atoms in the container is determined, their percentage will drop to 60%. Now, within and without quantum theory, no atom can be at the same time both horizontally and vertically oriented. Hence, if no measurement of the percentage of atoms endowed with any specifiable spatial orientation is performed, the only consistent answer seems to be: there is no such percentage.

Several leading physicists feel uneasy about the epistemology which Copenhagen associates with von Neumann's measurement theory, and quite a few milder versions of the theory have been suggested. The version due to Dirac is less vulnerable. It states that any magnitude M takes on a definite value V for the physical system S if S is in an eigenstate of M and V is the eigenvalue of M corresponding to this eigenstate, regardless of whether or not a measurement of M on S has been performed. Obviously, Dirac's milder version is still radical enough to cripple the physical universe about us and to deprive every physical entity of any location in space and time.*

For lack of space, I cannot afford to discuss alternative, more liberal versions of a theory which associates von Neumann's theory of measurement with an epistemology that wipes out a larger or a lesser fraction of the physical universe. It seems much more promising to me to explore the problem involved in tenet (b) from the contemporary point of view of relativistic field theory and quantumelectrodynamics. Since these two theories are relativistic, they admit a multiplicity of invariants, i.e., of quantities the value of which remains unchanged if measured several times at reasonable time distances. For instance, the Minkowski interval of two events, the rest-mass of a physical object, the action involved in any physical process (i.e., the time-integral of energy) are invariants in this sense. There is no shortage of such invariant magnitudes in quantal theories. It would be plain nonsense to doubt the existence of an invariant or to make its existence conditional upon the performance of an appropriate measurement since all measurements of the same invariant are bound to yield the same value. Clearly, there is no danger here of the relativistic approach reducing the universe about us to the shadowy creature proposed by Copenhagen, which would provide no analogy with the size and richness of the universe.

On closer analysis, the above considerations do not suffice to refute the Copenhagen Interpretation of non-relativistic quantum mechanics. It is a

---

* Cf. [20] p. 61.

fact of the history of science that, sofar, nobody succeeded in establishing an interpretation of non-relativistic quantum mechanics which would be demonstrably superior to that of Bohr, in a clearly defined sense. I have no doubt that another fact pertaining to the history of science, viz. the fact that an overwhelming majority of working physicists follow the Copenhagen Interpretation, carries also a considerable weight. And I cannot help wondering whether any competent physicist working under exactly the same conditions as Bohr did would wind up with an interpretation of quantum mechanics different from the Copenhagen interpretation.

However, the last point in the above argument in favor of the Copenhagen interpretation is actually self-defeating and proves the opposite of the intended conclusion. For the decisive aspect of the present situation of science is that we are working under conditions which differ appreciably from those under which Bohr worked. Bohr was working on the interpretation of a physical theory obtained by a single quantization of classical physics, and in particular, involving the replacement of spatial quantities with appropriate operators. One could not expect that a spatial quantity represented by such an operator would always take on a specifiable value for any physical system.

However, in quantumelectrodynamics, in relativistic field theory, we presently work under conditions which are not comparable to those of Bohr. These theories are not obtained by a single quantization from a classical theory, but imply two consecutive quantizations. Consequently, in quantumelectrodynamics, the field operators on the appropriate Hilbert space depend, in the functional sense, upon the spatio-temporal coordinates which are not, in turn, operators, but real numbers in the classical, pre-quantum-theoretical sense. In physical parlance they are q numbers, not c numbers. This means that the Lorentz transformations of spatio-temporal coordinates in theories involving second quantization, are literally interpretable and there is no more incompatibility between the validity of the Lorentz transformation group and of the Lorentz covariance of the equations of a theory involving second quantizations on the one hand, and the basic physical implications of relativistic quantal theories on the other.

We may sum up this memorable development in two statements:

(1) The non-relativistic quantum mechanics of 1925/26 deprived science temporarily of relativistic ideas and insights, but the relativistic quantal theories involving a double quantization have again bestowed upon science the use of relativistic achievements.

(2) Any problem, posed in a wider framework than used to be applied in the past, becomes thereby more likely to be solved. The issues of quantal measurement, of the validity of the Lorentz covariance and of the Lorentz

transformation group are posed today in an immensely widened framework of relativistic quantal physics, appreciably more comprehensive than the 1925 quantum mechanics, obtained by a single quantization of the corresponding classical theories. Accordingly, the likelihood that a rational solution of these problems will be achieved has increased in the same proportion.

REFERENCES

[1] G. CARATHÉODORY: Zur Axiomatik der speziellen Relativitätstheorie, *Sitzungsberichte der Preussischen Akademie der Wissenschaften*, *Phys. Math. Klasse*, 1924.
[2] E. A. MILNE, *Kinematic Relativity*, Oxford University Press, 1948.
[3] H. REICHENBACH, *Axiomatik der relativistischen Raum-Zeitlehre*, F. Vieweg & Sohn, 1924.
[4] H. WEYL, Review of [3], *Deutsche Literaturzeitung*, 1925.
[5] L. L. WHYTE, Light Signal Kinematics, *British Journal for the Philosophy of Science*, Vol. IV, 1953.
[6] G. KÄLLÉN, *Quantumelektrodynamik*, Handbuch der Physik, V, 1, 1958.
[7] G. WENTZEL, *Quantum Theory of Fields*, Interscience, 1949.
[8] P. ROMAN, *Theory of Elementary Particles*, 1960.
[9] BLATT and WEISSKOPF, *Theoretical Nuclear Physics*, 1952.
[10] H. BETHE, *Phys. Rev.*, Vol. 72, 1947.
[11] F. J. DYSON, *Phys. Rev.*, Vol. 75, 1949.
[12] A. S. WIGHTMAN, Quantum Field Theory in Terms of Vacuum Expectation Values, *Phys. Rev.*, Vol. 101, 1955.
[13] H. LEHMANN, K. SZYMANZIK and W. ZIMMERMANN, The Formulation of Quantized Field Theories, *Il Nuovo Cimento*, Vol. 1, 1955 and Vol. 6, 1957.
[14] R. HAAG and SCHROER, Postulates for Quantum Field Theory, *American Journal for Mathematical Physics*, Vol. 3, 1963.
[15] I. M. GEL'FAND, R. A. MINLOS and Z. Y. SHAPIRO, *Representation of the Rotation and Lorentz Groups*, 1963, especially p. 271.
[16] JOHANN VON NEUMANN, *Mathematische Grundlagen der Quantenmechanik*, Dover Publications, 1943, p. 157.
[17] A. S. EDDINGTON, *Space, Time, and Gravitation*, 1920.
[18] P. and T. EHRENFEST, *Encyclopädie der mathematischen Wissenschaften*, IV, 2, II.
[19] *loc. cit.*, p. 222.
[20] G. LUDWIG, *Die Grundlagen der Quantenmechanik*, 1954, pp. 122.
[21] D. BOHM, *Quantum Mechanics*, 1951.
[22] A. I. KHINCHINE, *Mathematical Foundations of Statistical Mechanics*, Dover, 1949, p. 139 ff.
[23] O. COSTA DE BEAUREGARD, *Le Second Principe de la Science du Temps*, 1963, p. 105.
[24] H. MEHLBERG, *Essai sur la théorie causale du temps*, 1935, pp. 171 ff.
[25] H. MEHLBERG, Time's Arrow, *Current Issues in the Philosophy of Science*, Holt, Rinehart and Winston, 1960, pp. 105 ff.
[26] E. P. WIGNER, Gray theory and its Application to the Quantum Mechanics of Atomic Spectra, 1959, pp. 325 ff.
[27] N. BOHR, *Atomic Theory and the Description of Nature*, Macmillan, 1934.
[28] W. HEISENBERG, *The Physical Principles of Quantum Mechanics*, Chicago, 1930.
[29] S. S. SCHWEBER, *An Introduction to Relativistic Field Theory*, 1961.
[30] P. A. M. DIRAC, *The Principles of Quantum Mechanics*, §8, 1935.

Section VII

Methodology and Philosophy of Life Sciences

THEORIES OF MEANING AND LEARNABLE LANGUAGES*

DONALD DAVIDSON
Stanford University, Stanford, California, U.S.A.

Philosophers are fond of making claims concerning the properties a language must have if it is to be, even in principle, learnable. The point of these claims has generally been to bolster or to undermine some philosophical doctrine, epistemological, metaphysical, ontological or ethical. But if the arguments are good they must have implications for the empirical science of concept formation, if only by way of saying what the limits of the empirical are.

Often it is asserted or implied that purely *a priori* considerations suffice to determine features of the mechanism, or the stages, of language learning; such claims are suspect. In the first part of this paper I examine a typical example of such a position, and try my hand at sorting out what may be acceptable from what is not. In contrast to shaky hunches about how we learn language, I propose what seems to me clearly to be a necessary feature of a learnable language: it must be possible to give a constructive account of the meaning of the sentences in the language. Such an account I call a theory of meaning for the language, and I suggest that a theory of meaning, whether put forward by philosopher, linguist or psychologist, that conflicts with this condition cannot be a theory of a natural language; and if it ignores this condition, it fails to deal with something central to the concept of a language. Nevertheless, as I try to show in the second part of this paper, a number of current theories of meaning do either conflict with or ignore this condition for being learnable.

I

First we learn a few names and predicates that apply to medium-sized lovable or edible physical objects in the foreground of sense and interest; the learning takes place through some conditioning process describable as ostensive. Next come complex predicates and singular terms for objects not necessarily yet observed, or forever out of sight due to size, date, attenuation or inexistence. Then come theoretical terms, learned perhaps by way of "mean-

* I am much indebted to W. V. Quine, I. Scheffler and John Wallace, with whom I have discussed some of the topics of this paper. Professors Quine and Scheffler amiably tried to correct my understanding of their views, with what I know they will consider incomplete success.

ing postulates" or by dint of being embedded in suitably scientific discourse. Somewhere early in the game the great jump was made from term to sentence, though just how may be obscure, the transition being blurred by heavy dependence on one-word sentences: 'Mama', 'Fire', 'Slab', 'Block', 'Gavagai', and so on.

Thus, in brief caricature, goes the building-block theory of language learning, echoing, dusty chapter by dusty chapter, empiricist epistemology.

The theory is now discredited in most details. For one thing, there is no obvious reason to think the order of learning is related to epistemological priority. For another, some of the claims seem contradicted by experience: for example, a child learns the general terms 'cat', 'camel', 'mastodon', and 'unicorn' in what is, to all telling, a uniform way, paging through a picture book, though the child's relation to the extensions of these terms is altogether different. In some cases, the order of language learning is arguably the reverse of the epistemological order: sense-data may be the basis for our knowledge of physical objects, but talk of sense-data is learned, if learned at all, long after talk of physical objects is achieved. Finally, the underlying epistemology, with its assumption of associationist psychology and its simple reductionist theory of meaning is no longer appealing to most philosophers. In the light of all this it is astonishing that something very like the doctrine of language learning which began as a feeble outgrowth of early empiricism should now flourish while the parent plant wilts. What follows is a single example of confused dependence on this outmoded doctrine, but the example could easily be multiplied from current literature.

In [12] P. F. Strawson attacked Quine's well known view that "the whole category of singular terms is theoretically superfluous" (Quine [8], p. 211). Strawson grants, at least for argument's sake, that within a language already containing singular terms, we can paraphrase "all that we at present say with the use of singular terms into forms of words which do not contain singular terms" ([12], p. 434). What Strawson denies is that from this assumption there follows the theoretical possibility that we could speak a language without singular terms, ". . . in which we never had used them, in which the category of singular terms simply did not exist, but in which we were nevertheless able to say, in effect, all that we are at present able to say with the use of singular terms" ([12], pp. 433, 434). Strawson then sets out to establish independently the theoretical impossibility of such a language.

In order to focus on the point at issue, let me explain that I have no interest in contesting two of Strawson's theses, namely, and roughly, that *eliminability* of singular terms does not follow from *paraphrasability*, and that eliminability, as described, is impossible. Both points are moot, so far as I am concerned, pending clarification of the notion of being able to say the same thing.

Not to try conclusions, then, my concern is entirely with the main argument Strawson uses in the attempt to discredit Quine's eliminability thesis. Two claims essential to this argument are these:

(1) for any predicate to be understood, some predicates must be learned ostensively or by "direct confrontation";

(2) for such learning to take place, the ostensive learning situation must be "articulated in the language" by a demonstrative element which picks out or identifies entities of the sort to which the predicate applies ([12], pp. 445, 446).

Quine has countered (in [9], p. 185) that (1) and (2) do not suffice to establish the necessity for singular terms, because demonstratives may be construed as general terms. This may well be true; my interest however is in the implication of (1) and (2) that substantive questions about language learning can be settled on purely *a priori* grounds.[1]

Summarizing his own argument, Strawson says: "Some universal terms must be connected with our experience if any are to be understood. And these universal terms must be connected with particular bits or slices of our experience. Hence, if they are to be learnt *as predicates of particulars*, they must be learnt as predicates of demonstratively *identified* particulars." ([12], p. 446. Italics in original.) Here it is perhaps obvious that the notion of learning appears vacuously in the conclusion; so let us turn briefly to (1) and (2).

Surely it is an empirical question whether, as a result of certain experiences, a person turns out to have acquired some ability he did not have before; yet (1) and (2) claim it is a purely "logical" matter that everyone who has acquired a linguistic ability of a specified kind has travelled a prescribed route. Strawson apparently equates the ostensive learning of a predicate with learning by "direct confrontation." One can imagine two ways in which such a process is intended to be more special than learning the meaning of a predicate through hearing sentences which couple it with demonstrative singular terms. One is, that ostensive learning may require an intention on the part of the teacher to bring an object to the attention of the learner. It seems however that no such intention is necessary, and in fact most language learning is probably due more to observation and imitation on the part of the learner than to any didactic purpose on the part of those observed and

---

1 Strawson quotes Quine [6], p. 218 in support of (1), and Quine [9] might now be quoted in support of something like both (1) and (2). But aside from the fact that Quine declines to draw Strawson's conclusion, there remains the important difference that Quine at most thinks (1) and (2) are true, while Strawson wants to show they are necessary. I am concerned here only with the claim of necessity, though I doubt very much that either (1) or (2), on any natural empirical interpretation, is an important truth about language acquisition.

imitated. A second difference is that direct confrontation (and probably ostension, at least as Strawson interprets it) requires the presence of an appropriate object, while the correct use of a demonstrative singular term does not. But it is not an *a priori* truth, nor probably even a truth, that a person could not learn his first language in a skillfully faked environment. To think otherwise would, as Strawson says in a later book, but perhaps in much the same connection, "be to limit too much the power of human imagination" ([13], p. 200).

In the defense of (2) Strawson reveals what I think is the confusion that underlies the argument I have been criticizing. On the face of it, there is no reason why the ostensive learning of predicates must be "articulated in the language" in one way rather than another; no reason why the reference to particulars (assuming this necessary) must be by way of demonstratives. Nor does Strawson provide any reason; instead he gives grounds for the claim that "no symbolism can be interpreted as a language in which reference is made to particulars, unless it contains devices for making demonstrative ... references to particulars, *i. e.* unless it contains singular terms for referring to particulars ..." ([12], p. 447). Toward the end of [12] Strawson attacks "... the uncritical assumption that a part only of the structure of ordinary language could exist and function in isolation from the whole of which it is a part, just as it functions when incorporated within that whole ... To this extent, at least, language is organic" ([12], pp. 451, 452). Here the thought emerges, clear of any important connections with language learning, that Quine's eliminability thesis is false because we would make a major conceptual change—alter the meanings of all retained sentences—if we cut them off from their present relations with sentences containing singular terms (or demonstratives, or proper names, etc.). I suggest that from arguments for the conceptual interdependence of various basic idioms, Strawson has illegitimately drawn conclusions concerning the mechanism and sequence of language acquisition.

On the question of conceptual interdependence, Quine seems, up to a point at least, in agreement, writing that "the general term and the demonstrative singular are, along with identity, interdependent devices that the child of our culture must master all in one mad scramble" ([9], p. 102). It remains true, if such claims of interdependence are tenable, that we could not learn a language in which the predicates meant what predicates in our language do mean and in which there were no demonstrative singular terms. We could not learn such a language because, granting the assumption, there could not be such a language.[2] The lesson for theories of language learning is wholly

[2] In [13] Strawson again attacks what he takes to be Quine's eliminability thesis, but does not use the argument from learning which I am criticizing.

negative, but not perhaps without importance: in so far as we take the "organic" character of language seriously, we cannot accurately describe the first steps towards its conquest as learning part of the language; rather it is a matter of partly learning.

II

It is not appropriate to expect logical considerations to dictate the route or mechanism of language acquisition, but we are entitled to consider in advance of empirical study what we shall count as knowing a language, how we shall describe the skill or ability of a person who has learned to speak a language. One natural condition to impose is that we must be able to define a predicate of expressions, based solely on their formal properties, that picks out the class of meaningful expressions (sentences), on the assumption that various psychological variables are held constant. This predicate gives the grammar of the language. Another, and more interesting, condition is that we must be able to specify, in a way that depends effectively and solely on formal considerations, what every sentence means. With the right psychological trappings, our theory should equip us to say, for an arbitrary sentence, what a speaker of the language means by that sentence (or takes it to mean). Guided by an adequate theory, we see how the actions and dispositions of speakers induce on the sentences of the language a semantic structure. Though no doubt relativized to times, places and circumstances, the kind of structure required seems either identical with or closely related to the kind given by a definition of truth along the lines first expounded by Tarski in [14], for such a definition provides an effective method for determining what every sentence means (*i. e.* gives the conditions under which it is true). I do not mean to argue here that it is necessary that we be able to extract a truth definition from an adequate theory (though something much like this is needed), but a theory certainly meets the condition I have in mind if we can extract a truth definition; in particular, no stronger notion of meaning is called for.[3]

These matters appear to be connected in the following informal way with the possibility of learning a language. When we can regard the meaning of each sentence as a function of a finite number of features of the sentence, we have an insight not only into what there is to be learned; we also understand how an infinite aptitude can be encompassed by finite accomplishments. Suppose on the other hand the language lacks this feature; then no matter how many

[3] I am aware that many people, including Tarski ([14], p. 164), have felt that it is hopeless to expect to give a truth definition for even the indicative sentences of a natural language, much less imperatives, interrogatives and so on. Then we must find alternative ways of showing how the meanings of sentences depend on their structure, or give up the attempt at theory.

sentences a would-be speaker learns to produce and understand, there will remain others whose meanings are not given by the rules already mastered. It is natural to say such a language is *unlearnable*. This argument depends, of course, on a number of empirical assumptions: for example, that we do not at some point suddenly acquire an ability to intuit the meanings of sentences on no rule at all; that each new item of vocabulary, or new grammatical rule, takes some finite time to be learned; that man is mortal.

Let us call an expression a *semantical primitive* provided the rules which give the meaning for the sentences in which it does not appear do not suffice to determine the meaning of the sentences in which it does appear. Then we may express the condition under discussion by saying: a learnable language has a finite number of semantical primitives. Rough as this statement of the condition is, I think it is clear enough to support the claim that a number of recent theories of meaning are not, even in principle, applicable to natural languages, for the languages to which they apply are not learnable in the sense described. I turn now to examples.

First Example. Quotation Marks. We ought to be far more puzzled than we are by quotation marks. We understand quotation marks very well, at least in this, that we always know the reference of a quotation. Since there are infinitely many quotations, our knowledge apparently enshrines a rule. The puzzle comes when we try to express this rule as a fragment of a theory of meaning.

Informal stabs at what it is that we understand are easy to come by. Quine says, "The name of a name or other expression is commonly formed by putting the named expression in single quotation marks . . . the whole, called a *quotation*, denotes its interior" ([6], §4); Tarski says essentially the same thing in [14], p. 156. Such formulas obviously do not provide even the kernel of a theory in the required sense, as both authors are at pains to point out. Quine remarks that quotations have a "certain anomalous feature" that "calls for special caution"; Church calls the device "misleading." What is misleading is, perhaps, that we are tempted to regard each matched pair of quotation marks as a functional expression because when it is clamped around an expression the result denotes that expression. But to carry this idea out, we must treat the expressions inside the quotation marks as singular terms or variables. Tarski shows that in favorable cases, paradoxes result; in unfavorable cases, the expression within has no meaning ([14], pp. 159–162). We must therefore give up the idea that quotations are "syntactically composite expressions, of which both the quotation marks and the expressions within them are parts." The only alternative Tarski offers us is this:

> Quotation-mark names may be treated like single words of a language and thus like syntactically simple expressions. The single constituents of these names—the quotation marks and the expressions standing between them—fulfil the same function as the letters and complexes of successive letters in single words. Hence they can possess no independent meaning. Every quotation-mark name is then a constant individual name of a definite expression... and in fact a name of the same nature as the proper name of a man. ([14], p. 159.)

In apparently the same vein, Quine writes in [7], p. 140, that an expression in quotation marks "occurs there merely as a fragment of a longer name which contains, besides this fragment, the two quotation marks," and he compares the occurrence of an expression inside quotation marks with the occurrence of 'cat' in 'cattle' ([7], p. 140) and of 'can' in 'canary' ([9], p. 144).

The function of letters in words, like the function of 'cat' in 'cattle', is purely adventitious in this sense: we could substitute a novel piece of typography everywhere in the language for 'cattle' and nothing in the semantical structure of the language would be changed. Not only does 'cat' in 'cattle' not have a "separate meaning"; the fact that the same letters occur together in the same order elsewhere is irrelevant to questions of meaning. If an analogous remark is true of quotations, then there is no justification in theory for the classification (it is only an accident quotations share a common feature in their spelling), and there is no significance in the fact that a quotation names "its interior." Finally, every quotation is a semantical primitive, and, since there are infinitely many different quotations, a language containing quotations is unlearnable.

This conclusion goes against our intuitions. There is no problem in framing a general rule for identifying quotations on the basis of form (any expression framed by quotation marks), and no problem in giving an informal rule for producing a wanted quotation (enclose the expression you want to mention in quotation marks). Since these rules imply that quotations have significant structure, it is hard to deny that there must be a semantical theory that exploits it. Nor is it entirely plain that either Tarski or Quine wants to deny the possibility. Tarski considers only two analyses of quotations, but he does not explicitly rule out others, nor does he openly endorse the alternative he does not reject. And indeed he seems to hint that quotations do have significant structure when he says, "It is clear that we can correlate a structural-descriptive name with every quotation-mark name, one which is free from quotation marks and possesses the same extension (*i.e.* denotes the same expression) and vice versa" ([14], p. 157). It is difficult to see how the correlation could be established if we replaced each quotation by some other arbitrary symbol, as we could do if quotation-mark names were like the proper names of men.

Quine takes matters a bit further by asserting that although quotations are "logically unstructured" ([9], p. 190), and of course expressions have non-referential occurrences inside quotation marks, still the latter feature is "dispelled by an easy change in notation" (p. 144) which leaves us with the logically structured devices of spelling and concatenation. The formula for the "easy change" can apparently be given by a definition (pp. 189, 190). If this suggestion can be carried out, then the most recalcitrant aspect of quotation yields to theory, for the truth conditions for sentences containing quotations can be equated with the truth conditions for the sentences got from them by substituting for the quotations their definitional equivalents in the idiom of spelling. On such a theory,there is no longer an infinite number of semantical primitives, in spite of the fact that quotations cannot be shown to contain parts with independent semantical roles. If we accept a theory of this kind, we are forced to allow a species of structure that may not deserve to be called "logical," but certainly is directly and indissolubly linked with the logical, a kind of structure missing in ordinary proper names.[4]

Second Example. Scheffler on Indirect Discourse. In [10] and [11] Israel Scheffler proposes what he calls an inscriptional approach to indirect discourse. In [1] Carnap analyzed sentences in indirect discourse as involving a relation (elaborated in terms of a notion of intensional isomorphism) between a speaker and a sentence. Church objected in [2] that Carnap's analysis carried an implicit reference to a language which was lacking in the sentences to be analyzed, and added that he believed any correct analysis would interpret the that-clause as referring to a proposition. Scheffler set out to show this claim overstated by demonstrating what could be done with an ontology of inscriptions (and utterances).

Scheffler suggests we analyze 'Tonkin said that snow is white' as 'Tonkin spoke a that-snow-is-white utterance'. Since an utterance or inscription belongs eternally to the language of its ephemeral producer, Church's reproach to Carnap has no force here. The essential part of the story, from the point of view of present concerns, is that the expression 'that-snow-is-white' is to be treated as a unitary predicate (of utterances or inscriptions). Scheffler (and Quine, reporting the doctrine in [9], pp. 214, 215) calls the 'that' an operator which applies to a sentence to form a composite general term. "Composite" cannot, in this use, mean "logically complex"; not, at least, until more theory is forthcoming. As in the case of quotations, the syntax is clear enough (as it is in indirect discourse generally, give or take

[4] Geach has insisted, in [4] and elsewhere, that a quotation "must be taken as *describing* the expression in terms of its parts" (p. 83), but he fails to face the problem of showing how a quotation can be assigned the structure of a description, perhaps due to the confusion revealed in the claim that "man is mortal" contains "man" as a syntactical part (p. 82).

some rewriting of verbs and pronouns on principles easier to master than to describe). But there is no hint as to how the meaning of these predicates depends on their structure. Failing a theory, we must view each new predicate as a semantical primitive. Given their syntax, though (put any sentence after 'that' and spice with hyphens), it is obvious there are infinitely many such predicates, so languages with no more structure than Scheffler allows are, on my account, unlearnable. Even the claim that 'that' is a predicate-forming operator must be recognized as a purely syntactical comment that has no echo in the theory of meaning.

The possibility remains open, it may be countered, that a theory will yet be produced to reveal more structure. True. But if Scheffler wants the benefit of this possibility, he must pay for it by renouncing the claim to have shown "the *logical form and ontological character*" of sentences in indirect discourse ([11], p. 101, Scheffler's italics); more theory may mean more ontology. If logical form tells us all about ontology, then Quine's trick with quotations won't work, for the logical form of a quotation is that of a singular term without parts, and its manifest ontology is just the expression it names. But the definition that takes advantage of its structure brings in more entities (those named in the spelling). It is only an accident that in this application the method draws on no entities not already named by some quotation.

Third Example. Quine on Belief Sentences. We can always trade problems of ontology raised by putatively referential expressions and positions for problems of logical articulation. If an expression offend by its supposed reference, there is no need to pluck it out. It is enough to declare the expression a meaningless part of a meaningful expression. This treatment will not quite kill before it cures; after all singular terms and positions open to quantification have been welded into their contexts, there will remain the logical structure created by treating sentences as unanalyzable units and by the pure sentential connectives. But semantics without ontology is not very interesting, and a language like our own for which no better could be done would be a paradigm of unlearnability.

Quine has not, of course, gone this far, but in an isolated moment he seems to come close. In [9] near the close of a long, brilliant, and discouraging search for a satisfactory theory of belief sentences (and their relatives—sentences in indirect discourse, sentences about doubts, wonderings, fears, desires, and so forth) remarks that once we give up trying to quantify over things believed

> ...there is no need to recognize 'believes' and similar verbs as relative terms at all; no need to countenance their predicative use as in '*w* believes *x*' (as against '*w* believes that *p*'); no need, therefore, to see 'that *p*' as a term. Hence a final alternative that

> I find as appealing as any is simply to dispense with the objects of the propositional attitudes... This means viewing 'Tom believes [Cicero denounced Catiline]'... as of the form 'Fa' with a = Tom and complex 'F'. The verb 'believes' here ceases to be a term and becomes part of an operator 'believes that'... which, applied to a sentence, produces a composite absolute general term whereof the sentence is counted an immediate constituent. ([9], pp. 215, 216.)

In one respect, this goes a step beyond Scheffler, for even the main verb ('believes' in this case) is made inaccessible to logical analysis. Talk of constituents and operators must of course be taken as purely syntactical, without basis in semantical theory. If there is any element beyond syntax common to belief sentences as a class, Quine's account does not say what it is. And, of course, a language for which no more theory can be given is, by my reckoning, unlearnable. Quine does not, however, rule out hope of further theory; perhaps something like what he offers for quotations. But in the case of belief sentences it is not easy to imagine the theory that could yield the required structure without adding to the ontology.

Fourth Example. Church on the Logic of Sense and Denotation. In the last two examples, loss of a desirable minimum of articulation of meaning seemed to be the result of overzealous attention to problems of ontology. But it would be a mistake to infer that by being prodigal with intensional entities we can solve all problems.

On Frege's informal theory in [5] and elsewhere, we are asked to suppose that certain verbs, like 'believes' (or 'believes that'), do double duty. First, they create a context in which the words that follow come to refer to their usual sense or meaning. Second (assuming the verb is the main verb of the sentence), they perform a normal kind of duty by mapping persons and propositions onto truth values. This is a dark doctrine, particulary with respect to the first point, and Frege seems to have thought so himself. His view was, perhaps, that this is the best we can do for natural languages, but that in a logically more transparent language, different words would be used to refer to sense and to denotation, thus relieving the burdened verbs of their first, and more obscure, duty. But now it is only a few steps to an infinite primitive vocabulary. After the first appearances of verbs like 'believes' we introduce new expressions for senses. Next, we notice that there is no theory which interprets these new expressions as logically structured. A new vocabulary is again needed, along the same lines, each time we iterate 'believes'; and there is no limit to the number of possible iterations.

It should not be thought that there would be less trouble with Frege's original suggestion. Even supposing we made good sense of the idea that certain words create a context in which other words take on new meanings (an idea that only makes them sound like functors), there would remain the task of reducing to theory the determination of those meanings—an infinite number each for at least some words. The problem is not how the individual expressions that make up a sentence governed by 'believes', given the meanings they have in such a context, combine to denote a proposition; the problem is rather to state the rule that gives each the meaning it does have.

To return to our speculations concerning the form Frege might have given a language notationally superior to ordinary language, but like it in its capacity to deal with belief sentences and their kin: the features I attributed above to this language are to be found in the language proposed by Church in [3]. In Church's notation, the fact that the new expressions brought into play as we scale the semantic ladder are not logically complex is superficially obscured by their being syntactically composed from the expressions for the next level below, plus change of subscript. Things are so devised that "if, in a well-formed formula without free variables, all the subscripts in all the type symbols appearing are increased by 1, the resulting well-formed formula denotes the sense of the first one" ([3], p. 17). But this rule cannot, of course, be exploited as part of a theory of meaning for the language; expression and subscript cannot be viewed as having independent meanings. Relative to the expression on any given level, the expressions for the level above are semantical primitives, as Church clearly indicates (p. 8).[5] I don't suggest any error on Church's part; he never hints that the case is other than I say. But I do submit that Church's language of Sense and Denotation is, even in principle, unlearnable.

REFERENCES

[1] R. Carnap, *Meaning and Necessity*, University of Chicago Press, Chicago, 1947. 2nd ed.

[2] A. Church, "On Carnap's Analysis of Statements of Assertion and Belief," *Analysis* **10** (1950), pp. 97–99.

[3] A. Church, "A Formulation of the Logic of Sense and Denotation," pp. 3–24 in Henle, Kallen and Langer (eds.), *Structure, Method, and Meaning*: *Essays in Honor of H. M. Sheffer*, Liberal Arts Press, New York, 1951.

[4] P. Geach, *Mental Acts*, Routledge and Kegan Paul, London, 1957.

[5] G. Frege, "Über Sinn und Bedeutung," *Zeitschrift für Philosophie und philosophische Kritik* **100** (1892), pp. 25–50.

[6] W. V. Quine, *Mathematical Logic*, Harvard University Press, Cambridge, Mass., 1951 (revised ed.).

[7] W. V. Quine, *From a Logical Point of View*, Harvard University Press, Cambridge, Mass., 1953.

[8] W. V. Quine, *Methods of Logic*, Henry Holt, New York, 1960.

[5] If the number of semantical levels is fixed at some known finite number, the problem raised here can be solved, since denotata are functions of senses. In [1] Carnap fixes the number of levels at two, but it is not clear that his language can be given a satisfactory interpretation. John Wallace has given this idea a detailed and interesting application in [15].

[9] W. V. QUINE, *Word and Object*, The Technology Press and John Wiley, New York 1960.
[10] I. SCHEFFLER, "An Inscriptional Approach to Indirect Discourse," *Analysis* **14** (1954), pp. 83–90.
[11] I. SCHEFFLER, *The Anatomy of Inquiry*, Alfred Knopf, New York, 1963.
[12] P. F. STRAWSON, "Singular Terms, Ontology and Identity," *Mind* **65** (1956), pp. 433–454.
[13] P. F. STRAWSON, *Individuals*, Methuen, London, 1959.
[14] A. TARSKI, "The Concept of Truth in Formalized Languages," reprinted in *Logic, Semantics, Metamathematics*, Clarendon Press, Oxford, 1956, pp. 152–278.
[15] J. WALLACE, *Philosophical Grammar*, Ph. D. Thesis, Stanford University, 1964.

"WHAT'S IN THE BRAIN THAT INK MAY CHARACTER?"

WARREN S. McCULLOCH*

Research Laboratory of Electronics, Massachusetts Institute of Technology, Cambridge, Massachusetts, U.S.A.

Since we have come together as scientists who would become a bit wiser as to the process of our art, it is proper for us to ask what are the enduring qualities of our activities and what are our present problems. Whether he would create poetry, fiction, or science, the American is apt to think first of Mark Twain's law: "You have to have the facts before you can pervert them." Which are *the* facts? They are those that puzzle us; and not even all of them, but those that arouse in us one and the same sort of uneasiness in various contexts of experience. From a vague sense of there being something similar in these facts, we become curious as to exactly what it is that is similar in them, and we define them with increasing clarity, doing all of this before we are able to phrase a single question to put to nature. At that stage we are uncertain whether we really have one question or several questions.

You will find this difficulty explicit in the writings of Galileo, who, in founding physics, speaks of two sciences where we now find only one. Kepler, in the act of putting physics into the sky to produce elliptical orbits, was actually up against two questions, one in geometrical optics, and the other in mechanics, where he originally thought them one question. At the end of the last century, it looked as though physics was only a matter of pushing one decimal point to have a tidy theory of the universe. Only three awkward items had to be explained. These were the precession of the perihelion of Mercury, the drag of a moving medium on refracted light, and the absence of an aether drift. They raised three apparently separate questions, and no one expected that he had a single answer in the theory of relativity before that answer was forthcoming. Today there is a similar uneasiness in physics, perhaps foreshadowed by the want of a general field theory. It arises from the multiplication of the strange particles of subatomic physics, from the behavior of ballistic missiles, from transitions from

* This work was supported in part by the Joint Services Electronics Program under Contract DA36-039-AMC-03200(E); the National Science Foundation (Grant GP-2495), the National Institutes of Health (Grants NB-04985-01 and MH-04737-04), the National Aeronautics and Space Administration (Grant NsG-496); and in part by the U. S. Air Force (Aeronautical Systems Division) under Contract AF33(615)-1747.

streamline to turbulent flow, and from reports on an enormous object, a-fifth-of-the-age-of-the-universe away, which pulsates so fast that it requires a physical transmission immensely faster than light to keep it going.

The role of the projectile, and of its impact, in the development of physics may be of more than historical importance. In Galileo's hands, it proved fatal to the Greek doctrine of natural places. It disproved Descartes' attempted solution in terms of a plenum with a conservation of motion, and Leibniz' plenum with a conservation of force. It now threatens Newton's conservation of momentum. For macroscopic projectiles and their impacts, there seems to be an intrinsic time, or τ, during which they absorb or deliver energy but during which they are incapable of a conservation of momentum in the macroscopic sense, and thus require a third temporal derivative. Its introduction has also served to explain both varieties of turbulence, the quasi-periodic, and the hyperbolic or explosive, in our rockets. Davis has pointed out that, without this assumption, these can only be explained away by distinct hypotheses *ad hoc*. Several of my friends have been asking whether or not atoms and particles may have a τ that might account for some of their strange properties and, at the other extreme of size, whether or not the gravitational field, like the electromagnetic field, may propagate, thereby giving a τ to gigantic structures. In short, it looks as though physics is again about to enjoy a new resolution, or at least a new revolution, and whether there be one question or many remains to be seen.

Since this is so in the most advanced of sciences, there is no need to apologize for the state of our own, for we are Johnnies-come-lately into the hypothetical and postulational stage of knowledge. Just as chemistry got off to a bad start in the rigid doctrine of alchemy and was saved only by the "puffers," so psychology was hindered by doctrinaire epistemology and saved only by biologists. To make psychology into experimental epistemology is to attempt to understand the embodiment of mind. Here we are confronted by what seems to be three questions, although they may ultimately be only one. It is these which we would like you to consider.

The three exist as categorically disparate *desiderata*. The first is at the logical level: We lack an adequate, appropriate calculus for triadic relations. The second is at the psychological level: We do not know how we generate hypotheses that are natural and simple. The third is at the physiological level: We have no circuit theory for the reticular formation that marshals our abductions.

Logically, the problem is far from simple. To be exact, no proposed theory of relations yields a calculus to handle our problem. When I was growing up, only the Aristotelian logic of classes was ever taught, and that badly.

The *Organon* itself contains only a clumsy description of the apagoge—perhaps from the notes of some student who had not understood his master. Peirce says that, when he was making the *Century Encyclopedia*, he understood the passage so badly that he wrote nonsense. "The apagoge," ordinarily translated "the abduction," is explained by Peirce as one of three modes of reasoning. The first is *deduction*, which starts from a rule and proceeds through a case under the rule to arrive at a fact. Thus: All people with tuberculosis have bumps; Mr. Jones has tuberculosis; *sequitur*—Mr. Jones has bumps. The second, or *induction*, starts from cases of tuberculosis and patients with bumps and guesses that the rule is that all people with tuberculosis have bumps. Peirce calls this "taking habits"; and properly it leads only to probabilities, coefficients of correlation, and perhaps to factor analysis. The guess at the rule requires something more—a creative leap, even in the most trivial cases. The third, or *abduction*, starts from the rule and guesses that the fact is a case under that rule: All people with tuberculosis have bumps; Mr. Jones has bumps; perhaps Mr. Jones has tuberculosis. This, sometimes mistakenly called an "inverse probability," is never certain but is, in medicine, called a diagnosis or, when many rules are considered, a differential diagnosis, but it is usually fixed, not by a statistic, but by finding some other observable sign to clinch the answer. Clear examples of abduction abound in the Hippocratic corpus but are curiously absent in Aristotle's own writings, where one finds only genus, species and differentia.

What seems even stranger in the Greek writings is a total absence of our notions of *a priori* or *a posteriori* probability. The ancients had only a possibility and a guess. Probability as we know it was still nearly two thousand years in the future. Possibility appears in Aristotle's problematic mode but was even more sharply handled by the Stoics and by the physicians. Both groups questioned whether a possible proposition can be said to be true if it never happens to be fulfilled. One thing is clear, then—the mind makes a leap from the cases and facts to the rule, and Mill's attempt to bridge this gap, and the attempts of all of his followers, slur over it too easily. We do not know how we even make the jump and come up with a simple and natural hypothesis—certainly not from probabilities.

When I was young, it was fashionable to sneer at stoic logic as mere pettifoggery; at that very time it was being slowly and laboriously re-created under the alias of the logic of propositions. Thanks largely to Northrop and to Sambursky, I have recently become familiar with its tenets. Had I known it forty years ago, it would have saved me much wasted labor. In the first place it is, as Peirce points out, both pansomatic and triadic in its propositions. There are always three real related bodies: One is the utterance, the *flatus vocis* of Abelard; one is that which it proposes; one is something in the head like a fist in the hand called the *Lekton.* Shakespeare, at about the age of twenty-five, had it clear and wrote foı a lawyers' club:

> What's in the brain that ink may character,
> Which hath not figur'd to thee my true spirit?
> What's new to speak, what new to register,
> That may express my love or thy dear merit?

The lawyers for whom he wrote it were concerned with writing lawyer's law, which grows out of stoic logic, giving us our contracts, corporations, and constitutions, created as postulated entities and hypothetical relations, much as we inherited this structure from the Greeks to start the renaissance of science. What's in the brain is the stoic *Lekton.* Stoic law contemplates possible alternatives but never probabilities, and time enters, allowing no contact without date of termination, no bond without date of redemption, and no elected office but for a limited term.

Time appears in stoic logic in the relation of the necessary to the possible, and I have heard lawyers discuss this as a probable source of this aspect of contractual law. There are three statements attributed to Diodorus, called the Master, of which any two may be true and the third false: Every possible truth about the past is necessary; an impossible proposition may not follow from a possible one; there is a proposition possible that neither is true nor will be true.

Diodorus rejected the third and defined the "possible" as that which is or will be true. This is in keeping with his notion of implication, which is concerned with time. He held that *A* implied *B* only if, for *all* time, *A*, as a function of time, materially implied *B* as a function of that time. For the last of the great Stoic logicians, Philo, implication was our material implication. There were at least two other forms of implication used by the Stoics, one resembling strict implication, and the other perhaps requiring analyticity. Unfortunately, none of these is the implication that we really want for our purposes, and, as you will see, we have had to turn to biology for the notion of a bound cause. A signal should be said to imply its natural cause, which is bound, and not its casual cause; for when it arises ectopically, it is false for the receiver. The communication engineer calls such a false signal "noise." Again, the trouble is that we are dealing with a triad of Sender, Signal, and Receiver, and with the stoic triad: *A* means *B* to *C*. The signal means to the receiver what the sender intended.

In order to avoid paradoxes and ambiguities, the Stoics not only would not allow any self-reference, as in the famous Cretan's "This statement is a lie," but would not allow a proposition to imply itself and, as an added precaution, would not allow a negation within a proposition. This left them with *implication*, and an exclusive *or*, and with a *not both*, the last of which is one of Peirce's *amfexes*, or a version of Sheffer's stroke. Hence, they needed exactly five figures of argument to form a complete logic of atomic propositions.

About 1920, I attempted to construct a logic to handle the problems of knowledge and action in terms of a logical analysis of propositions involving verbs other than the copulative, and found it worse than modal logic. One has to distinguish those verbs in which the physical activity described by the present tense begins in the object and ends in the subject, such as verbs of sensation, perception, etc.; those in which it begins in the subject and ends in the object, such as the verbs of action; the group of so-called intransitive and reflexive verbs in which the events begin in the subject and end in the subject, called the verbs of behavior; and finally, a group of verbs that in the present tense refer to no action but define some kind of action that will be taken if thus-and-so happens—verbs of sentiment, which are like propositional functions rather than propositions. In perception, time's arrow points to the past; in action, to the future; in behavior, it becomes circular; and in sentiment, it simply does not exist. Literally, one deals with a state. I gave up the attempt because I realized that I had been trapped by the subject-predicate structure of language into supposing I was dealing with diadic relations, whereas they were irreducibly triadic. My hypothesis was simple and natural, but I had mistaken the *flatus vocis* for the *Lekton*.

I next attempted to construct for myself a simplest psychic act that would preserve its essential character; you may call it a "psychon" if you will. It was to be to psychology what an atom was to chemistry, or a gene to genetics. This time I was more fortunate, probably thanks to studying under Morgan of fruit fly fame. But my psychon differed from an atom and from a gene in that it was to be not an enduring, unsplittable object, but a least event. My postulated psychons were to be related much as offspring are to their parents, and their occurrence was in some sense to imply a previous generation that begat them. There is perhaps no better understood triadic relation than family structure. Even the colligative terms are clearly specified. There is scarcely a primitive tribe but has a kinship structure. So I was fortunate in this hypothesis in the sense that it gave a theory of activity progressing from sensation to action through the brain, and even more so in this, that the structure of that passage was anastomotic, whereby afferents of any sort could find their way by intersecting paths to any set of efferents, so relating perception to action. The implication of psychons pointed to the past, and their intention foreshadowed the proposed response. In those days the neuronal hypothesis of Ramon y Cajàl and the all-or-none law of axonal impulses were relatively novel, but I was overjoyed to find in them some embodiments of psychons. There was a *Lekton* in the head like a fist in the hand, but it took me out of psychology through medicine and neurology to ensure my pansomatism. Thereafter, in teaching physiological psychology at Seth Low Junior College, I used symbols for particular neurons, subscripted for the time of their impulse, and joined by implicative characters to express the dependence of that impulse upon receipt of impulses received a moment, or synaptic delay, sooner.

But even then I could not handle circularities in the net of neurons, for which I lacked a genetic model. They were postulated by Kubie, in 1930, to explain memory and thinking without overt activity in the supposititious linked reflexes of the behaviorists. Circles were well known as regulatory devices, as reflexes, in which the action instead of being regenerative was an inverse, or negative, feedback. My major difficulty was having insufficient knowledge of modular mathematics. This, Walter Pitts could handle, and we published our paper on a logical calculus for ideas immanent in nervous activity. Chicago in those days was under the spell of Rudolf Carnap, and we employed his terminology, although it was not most appropriate to our postulates and hypotheses. Quite apart from misprints, this has made it unduly difficult for all but a few like Bar-Hillel, who worked with Carnap, and we shall always be grateful to Kleene for putting it into a more intelligible form. I still feel, however, that he treated closed loops too cavalierly and so left open questions that we had raised, and neglected certain distinctions which, in Papert's hands, may prove a source of new theorems relating nets to the structure of the functions that they compute. The history of the ensuing developments in automata theory is certainly familiar to you.

As geometry ceased to be the measurement of the earth, so automata theory is ceasing to be a theory of automata. Recently, in Ravello, I was told that an automaton or a nerve net, like me, was a mapping of a free monoid onto a semigroup with the possible addition of identity. This is the same sort of nonsense one finds in the writings of those who never understood the *Lekton* as an embodiment. It is like mistaking a Chomsky language for a real language. You will find no such categorical confusion in the original Pitts and McCulloch of 1943. There the temporal propositional expressions are events occurring in time and space in a physically real net. The postulated neurons, for all their oversimplifications, are still physical neurons as truly as the chemist's atoms are physical atoms.

For our purpose of proving that a real nervous system could compute any number that a Turing machine could compute with a fixed length of tape, it was possible to treat the neuron as a simple threshold element. Unfortunately, this misled many into the trap of supposing that threshold logic was all one could obtain in hardware or software. This is false. A real neuron, or Crane's neuristor, can certainly compute any Boolean function of its inputs—to say the least! Also, in 1943, the nets that we proposed were completely orderly and specified for their tasks, which is certainly not true of real brains. So, in 1947, when we were postulating a *Lekton* for the knowing of a universal, we began with a paragraph of precautions, that the function of the net be little perturbed by perturbations of signals, thresholds, and even by details of synapsis. All of this underlies the beginnings of a probabilistic logic to understand the construction of reliable automata from less reliable components, as is apparent in the work of Manuel Blum and Leo Verbeek. Finally, in the work of Winograd and Cowan, it is clear that, for an information-theoretic capacity in computation in the presence of noise, the logic has to be multiple-truth-valued, and the constructions require, for coding without fatal multiplication of unreliable components, not threshold elements, but those capable of computing any Boolean function of large numbers of inputs—that is, they must be somewhat like real neurons. The facts that worried us over the years from 1947 to 1963 were simply that real brains do know universals, are composed of unreliable components, and can compute in the presence of noise. The theory of automata has thus proved more provocative than the automata theory divorced from the automata.

Please note that, to this point, we have considered only deductive processes. The automata were not "taking habits." Our group has not been concerned with induction, either experimentally or theoretically. Soon after World War II, Albert Uttley produced the first so-called probabilistic perceptive artifact. It enjoyed what is now called a "layered computation" and could be trained to classify its inputs. He has stayed with the problem, and I happen to know that he has written, but not yet published, an excellent theoretical paper based on a specific hypothesis as to the events determining the coupling of neurons in succession; and, moreover, that physiological experiments performed by one of his friends indicate that his assumption is probably correct. I take it you are familiar with the writings of Donald MacKay, Oliver Selfridge, Marvin Minsky, Gordon Pask, Frank Rosenblatt, and a host of others on perceptrons, learning and teaching machines, etc., and that you know of the numerous studies on the chemical nature of the engram, which certainly involves ribose nucleic acid and protein synthesis.

The next step would obviously be to postulate a process of concept formation. This is the very leap from weighing probabilities to propounding hypotheses. Marcus Goodall, Ray Solomonoff, Marvin Minsky, and Seymour Papert, among my immediate friends, are all after it, and I think they all feel it requires a succession of subordinate insights organized at successive superordinate levels or types. This is what Hughlings Jackson called "propositionalizing." This certainly cannot be left to variation and selection as an evolutionary process starting from chaos or a random net. That would be too slow, for it can be followed in ontogenesis, as Piaget has shown. The child does form "simple" and "natural" hypotheses, as Galileo called them. "Simple" and "natural" are evaluative terms and are based upon the evolution of the organism and its development in the real world, the natural world in which it finds itself. There again we come up against our logical limitation, for there simply does not exist any proper way to handle the triadic, or n-adic, relations of such relata. We cannot state our problem in a finite and unambiguous manner.

That man, like the beasts, lives in a world of relations, rather than in a world of classes or of propositions, seems certain. He does not know the relative size of two cubes from a measurement of the lengths of their edges, or even from the area of their faces. If he can just detect a difference of one part in twenty of a length, he can do the same for areas and also for volumes. I happen to have spent two years in measuring man's ability to set an adjustable oblong to a preferred shape, because I did not believe that he did prefer the golden section or that he could recognize it. He does and he can! On repeated settings for the most pleasing form he comes to prefer it and can set for it. The same man who can only detect a difference of a twentieth in length, area, or volume, sets it at 1 to 1.618, not at 1 to 1.617 or 1 to 1.619. So the aesthetic judgment bespeaks a precise knowledge of certain —shall I say privileged?—relations directly, not compounded of the simpler perceptibles. A sculptor or painter sometimes told me he had added enough to a square so that the part he had added had the same shape as the whole. This example is pertinent here, for in this case we do have an adequate theory of the relations, namely ratio and proportion. But these apply only to the perceived object, not to its relation to the statement or the *Lekton* in the brain of the aesthete. Clearly, the concept of a ratio must be embodied before the concept of a proportion can be conceived as the identity of the ratio. Once formed, the concept endures in us as the embodiment of an eternal verity, a sentiment, like love. To quote from the same Shakespearian sonnet:

> What's new . . .
> . . .
> Nothing sweet boy, but yet like prayers divine,

I must each day say o'er the very same,
Counting no old thing old, thou mine, I thine,
Even as when first I hallowed thy fair name.
So that eternal love in love's fresh case,
Weights not the dust and injury of age,
Nor gives to necessary wrinkles place,
But makes antiquity for aye his page,
Finding the first conceit of love there bred
Where time and outward form would show it dead.

Sonnet CVIII

Such is the beauty we still find, the pure form, the golden section, in the ruins of a Greek temple.

The golden section is a ratio which cannot be computed by any Turing machine without an infinite tape or in less than an infinite time. It is strictly incomprehensible. Yet it can be apprehended by finite automata, including us. Nor does it arise from any set of probabilities, or from a factor analysis of any data or correlation of observations, but as an insight—a guess, like every other hypothesis that is natural and simple enough to serve in science. It is nearer to the proper notions of classical physics than to the descriptive laws, the curve-fittings, that bedevil psychology.

This brings us to the problem of abduction, the apagoge. Evolution has provided us with reflexive arcs organized for the most part by what are called "half-centres," whose activities may alternate, as in breathing or walking, or synchronize, as in jumping. These are then programmed for more complicated sequences, and all of these are marshaled into a few general modes of behavior of the whole man. Psychologists and ethologists count them on their fingers or at most on their fingers and toes. These modes of behavior are instinctive, and only the manner of their expression and their manner of evocation are modified by our experience. The structures that mediate them have evolved in all linear organisms, like us, from an original central net, or reticulum, and while they may be very dissimilar from phylum to phylum, the central core of that reticulum has remained curiously the same in all of us. It is distributed throughout the length of the neuraxis and in each segment determines the activity of that segment locally, and relates it to the activity of other segments by fibers, or axons, running the long way of the neuraxis. The details of its neurons and their specific connections need not concern us here. In general, you may think of it as a computer to any part of which come signals from many parts of the body and from other parts of the brain and spinal cord. It is only one cell deep, on the path from input to output, but it can set the filters on all of its inputs and can control the behavior of the programmed activity, the half-centers and the reflexes. It gets a substitute for depth by its intrinsic fore and aft connections. Its business, given its knowledge of the state of the whole organism and of the world impingent upon it, is to decide whether the given fact is a case under one or another rule. It must decide for the whole organism whether the rule is one requiring fighting, fleeing, eating, sleeping, etc. It must do it with millisecond component action and conduction velocities of usually less than 100 meters pei second, and do it in real time, say, in a third of a second. That it has worked so well throughout evolution, without itself evolving, points to its structure as the natural solution of the organization of appropriate behavior. We know much experimentally of the behavior of the components, but still have no theory worthy of the name to explain its circuit action. William Kilmer, who works on this problem with me, is more sanguine than I am about our approach to the question. Again, the details of our attempts are irrelevant here. The problem remains the central one in all command and control systems. Of necessity, the system must enjoy a redundancy of potential command in which the possession of the necessary urgent information constitutes authority in that part possessing the information.

The problem is clearly one of triadic or n-adic relations, and is almost, or perhaps entirely, unspecifiable in finite and unambiguous terms without the proper calculus.

We see, then, the same theme running throughout.. We lack a triadic logic. We do not know how to create natural and simple hypotheses: we have, at present, no theory to account for those abductions which have permitted our evolution, ensured our ontogenesis, and preserved our lives. The question remains:

What's in the brain that ink may character?

THE KINEMATICS AND DYNAMICS OF CONCEPT FORMATION*

PATRICK SUPPES
Stanford University, Stanford, California, U.S.A.

1. Introduction

Analyses of concept formation can be found in disciplines that seem superficially unrelated. The two oldest traditions are in philosophy and mathematics, the one reaching back to Plato and Aristotle and having a continuous history in the theory of knowledge, and the other going back to Eudoxus, Euclid and their several mathematical contemporaries and successors. The logical status of the analyses of concept formation given by philosophers ranging from Aristotle through Hume to Kant and on to Russell has a complex and ambiguous history. It is common in contemporary discussions, for example, to say that Hume was badly confused about the distinction between the logic and psychology of concept formation, but it is also characteristic of the people who say these things that they do not offer a very precise or literal definition of the logic of concepts, and the word "logic" is used by them in a way that is itself tantalizingly vague.

In the case of the logical analyses of concepts in a mathematical context, particularly as questions have come to be put in terms of precisely characterized notions of definability, a quite finished logical theory of concept formation has developed. Given any theory and given a concept it is possible to ask in a quite definite and precise way whether or not this concept is definable in the theory or, given the concepts of a theory, it is possible to ask if one of the concepts is definable in terms of the others. It is true that problems of definability have often not been discussed as problems of concept formation, and yet it is obvious that there is a close logical relation between the two subjects. If a concept is not definable in terms of a set of other concepts, then in one sense that concept cannot be formed from them. For example, by application of Padoa's classical method for establishing the independence of concepts it is easy to show that for most standard axiomatizations of classical particle mechanics, the concept of mass cannot be defined in terms of the concepts of particle and position, Mach's famous proposal to the contrary. The ordinary-language philosophers who talk about the logic of concepts

* This paper has grown out of research supported by the U.S. Office of Education, Department of Health, Education and Welfare.

are not talking about the application of methods like those of Padoa to the solution of well-defined problems, and it is my own suspicion that there exists no well-defined subject matter corresponding to much of their discussion of the logic of concepts, unless it is indeed the psychology of concept formation.

In spite of the temptation, I do not here attempt to make a case for the dissolution of the logic of concepts into the psychology of concept formation, but rather concentrate on a critique of the current status of concept formation in psychology, with particular reference to questions that seem to have philosophical interest.

I would like to end this paper with at least a sketch of a detailed scientific solution to the problems of concept formation along the lines laid down by Hume in Book I of his *Treatise*. Unfortunately one does not have to dig very far into the psychological literature to find that nothing like an adequate solution has been found. On the other hand, there is one aspect of the problem that I think is now quite well understood, namely, what I have termed the kinematics of concept formation, and in the next section I give a survey, albeit brief, of the main results now available.

2. Kinematics of Concept Formation

Within mechanics, kinematics refers to the descriptive theory of motions. Because of the remoteness of most teaching of mechanics to any complex problems of data analysis, it is often not realized that from an empirical standpoint kinematics can be a quite complicated subject. It was, for example, and still is no simple task to decide from astronomical observations what closed figure represents to a high degree of accuracy the orbit of any one of the planets. A fully detailed statistical discussion of the question is extremely sophisticated and certainly is never mentioned in any of the standard textbooks on mechanics. The corresponding descriptive theory of concept formation has received a great deal of analysis in the recent psychological literature on learning, and a quite reasonable account in descriptive terms of the learning of many concepts can be given. The intended meaning of "reasonable" and "descriptive" needs remarking upon before these terms can mean much to those unfamiliar with the recent psychological literature. In the first place, it is a characteristic of the recent learning literature to abandon the hope of giving a deterministic description of the process of forming a concept on the part of an organism and to settle for a probabilistic description, but it is a mistake to think that it becomes a simple matter to find an adequate probabilistic description as opposed to a deterministic one. In actual fact, for large bodies of data it is a demanding task to satisfy with any strictness tests of goodness of fit. Moreover, for large bodies of data for which a probabilistic descriptive theory is postulated, many probabilistic relations assume a deterministic character at one remove from the data via application of the law of large numbers.

To make matters more concrete, it will perhaps be wise to sketch one simple experiment and the kind of descriptive theory applied to it. The experiment is one in which a young child is learning the concept of identity of sets. The children were of ages running from five to seven years. The sets depicted by the stimulus displays consisted of one, two or three elements. On each trial two of these sets were displayed. Minimal instructions were given the children to press one of two buttons when the stimulus pairs presented were "the same" and the alternative button when they were "not the same". In order to prohibit explaining the learning of the concept by a simple principle of stimulus association, a different stimulus display was shown on each trial. Because of this change of the stimulus display on each trial, no models at the level of simple stimulus associations can be applied to the response data of the children in any straightforward fashion. However, if we move from a stimulus-response association to a concept-association, the simple models used in quite elementary and primitive stimulus-response experiments work extremely well. Perhaps the simplest model is a so-called one-element model which postulates that the subject enters the experiment in the unconditioned state, i.e., the appropriate association or connection between the concept and the correct response is not established. On each trial, there is a constant probability c that the correct association will be established between the concept and the response, and thus that the subject, in this case the child, will enter the conditioned state. When the child is in the unconditioned state there is simply a guessing probability p of making a correct response, but when the conditioned state is entered, the probability of a correct response is one. A simple matrix may be used to describe transitions from the unconditioned (U) to the conditioned state (C).

	C	U
C	1	0
U	c	$1-c$

Other assumptions of a simple and natural sort are added to what has been stated in order to make the postulated sequence of conditioning states a first-order Markov chain. (It is worth noting, however, that we do not have such a chain in the observable responses themselves.) Once we are given the guessing probability p and the conditioning probability c, then all probabilistic questions about the response data are uniquely and completely determined. This means that after estimating these two parameters from the data a wide variety of predictions maybe made.

The strongest prediction of the one-element model I have just described is that prior to the last response error, there is no evidence of learning. It is a characteristic of the model that the guessing probability p is constant prior to the last error. In contrast to this assumption, the central assumption of the simple linear incremental model is that there is an increase in the probability of a correct response on each trial. The simplest way to formulate this incremental model is the following. Let p_n be the probability of a correct response on trial n. Then the probability of an error, q_n, is simply $1—p_n$. It is postulated that $q_{n+1} = \alpha q_n$, where α, the learning parameter, is a real number between 0 and 1. A number of experiments on which these two models have been compared are described in Suppes and Ginsberg (1963). (For some related applications to concept identification see Bower and Trabasso (1964).)

Although the bulk of the simple experiments on concept formation favor very strongly the one-element model, there are several situations in which a compromise between the two most satisfactorily explains the observed response data. This compromise consists in postulating that instead of having simply a single element that is conditioned or unconditioned, the concept-response association is best represented by a two-element model. The two elements may be interpreted as aspects or characteristics of the concept itself. A number of different formulations of two-element models have been published in the literature; a typical and simple extension of the one-element model is that described by the following matrix:

	2	1	0
2	1	0	0
1	b	$1-b$	0
0	0	a	$1-a$

Here, the conditioning parameters a and b play the role of c in the one-element model. It is assumed that the subject starts in the unconditioned state with 0 elements' conditioned as represented in the matrix by the 0 state. The probability of moving from the state of 0 elements' being conditioned to the state of 1 element's being conditioned is a, and correspondingly the probability of moving from state 1 to state 2 is b. Moreover, the probability of a correct response when in state 0 is p_0, and the probability of giving the correct response when in state one is p_1. As before, the probability is one of giving a correct response when all elements are conditioned, i.e., when the state is 2. It should be apparent that part of the greater success of this two-element model is simply the fact that it has four parameters, namely, a, b, p_0 and p_1, to be estimated from the data rather than two, as in the case of the one-element or linear incremental model. All the same, independent of parameter estimations there are some qualitative features of the data in many simple concept formation experiments that support the two-element model. For example, considerable evidence is presented in Suppes and Ginsberg (1963) to show that for many experiments the mean learning curve for response data is concave from above and quite apart from estimation of any parameters, such a curve is consistent with the two-element model, but not with the one-element or linear model. The essential point for the present discussion is that the one-element or two-element sort of model does predict with considerable accuracy the probabilistic characteristics of response data in simple concept formation experiments.

A typical prediction of the one-element model is shown in Table 1. These data are drawn from an experiment in which six- and seven-year old children were

TABLE 1.

Empirical and Theoretical Frequency Distribution of Response Errors in Blocks of Four Trials for Children's Learning of A System with Four Production Rules

Number of Errors	*Empirical Frequency*	*Theoretical Frequency*
0	9	8.15
1	59	58.30
2	161	156.46
3	172	186.63
4	92	83.48

being taught the simplest sort of mathematical proofs (Suppes (1961),(1964)). The experiment was performed in collaboration with John M. Vickers. The mathematical systems used in the experiment deal with production of finite strings of 1's and 0's. The single axiom is the single symbol 1. The rules of production are of the simplest sort. For example, given a string then one rule permits the addition of two 1's on the right. Another permits the deletion of a 1 on the right. The language used with the children was not, as you might expect, that used here. A child was shown a horizontal panel of illuminated red and green squares. Below this panel was a second panel with matching squares. The first square on the left in the lower panel was always illuminated red, corresponding to the single axiom, 1. Corresponding to each rule of production the child was given a button that he could use to light up additional squares or remove squares from the lower panel. His problem was to match the lower panel to the top panel. The theorem being proved was shown in the top panel. Each child was presented with 17 theorems per session for a total of 72 trials. The one-element model predicts that prior to learning how to use the rules of production the child simply guessed the correct response. Moreover, these guesses are drawn from the binomial distribution with parameter p. Table 1 compares the theoretical and empirical distributions for the number of errors in blocks of four trials, for one of the two groups in this experiment. The predictions are quantitatively quite good, and on a standard chi-square test the differences are, as you might expect, not significant.

I have chosen just this one sample of data. Many other similar instances can be found in the recent literature. The kind of predictions exemplified by Table 1 are the sort of descriptive predictions I have labeled *kinematical* in analogy with mechanics. What Hume attempted in Book I of his *Treatise* and what we all desire, namely, an adequate causal explanation, is certainly not given by the kinematical theory of the one-element and two-element models I have described thus far. I now turn to this more complicated problem.

3. Dynamics of Concept Formation

From a philosophical standpoint, the solution to the "kinematical" problems of concept formation are of only limited interest, just as in the case of mechanics it is dynamics and not kinematics that has stimulated so much philosophical discussion of the nature of scientific theories in physics. In many respects, a paradigm example of a dynamical theory of concept formation is provided by Hume in Section VII of Book I of his *Treatise*, the section treating abstract ideas. Following Berkeley, Hume attempts to reduce the formation of abstract concepts or ideas to the process of collecting around a term a number of particular ideas. As one sort of modern discussion of these topics would put it, Hume was concerned to characterize the process by which abstract ideas are coded. To give a complete account of the coding process is certainly in one sense to provide an adequate dynamical or causal theory of concept formation. I said that Hume's theory is a paradigm example, but this is true only in broad outline. It is far from being a paradigm example in its lack of detail and the difficulty of developing a substantial systematic theory from the general notions thrown out by Hume.

The modern theory closest to Hume is that which suggests that the process central to concept formation is the process of verbal mediation. There is a very extensive literature in psychology on verbal mediation, but if one scrutinizes this literature for the hard-core theoretical assumptions, it is difficult to find anything substantial that goes much beyond what Hume had to say. A good way of pin-pointing the problem of verbal mediation theories is to move on to the approaches that have arisen in attempts to solve various practical and theoretical problems involved in constructing intelligent machines. This approach to concept formation can probably most aptly be labeled the theory of artificial intelligence. The superficiality of our understanding of how concepts are formed immediately becomes evident when we examine what help any particular theory in concept formation can give us in programming a computer to play a reasonably adequate game of chess, or to solve simple perceptual problems of pattern recognition. The fact is that verbal mediation theory and its kin are too fuzzy and indefinite to provide any serious scientific help in solving these problems. We all can agree in general terms that there must be a coding process which the brain uses to represent concepts and to store information, but the details of how this coding process works have not been successfully elucidated in current theories of verbal mediation. It could of course be that this elucidation has taken place and the difficulty facing the scientist who wants to apply the theory to problems of artificial intelligence is that the computer he has at hand is not of adequate capacity, but this is not at all the situation. It is simply that the psychological theories of verbal mediation are lacking in systematic scientific content.

From a mathematical standpoint undoubtedly the simplest and neatest dynamical theory of concept formation would be one formulated in terms of an algebra of concepts. The intuitive idea is that an organism is able to apply certain operations to his repertoire of concepts at a given instant in order to produce a new concept. From a formal standpoint such a set-up would be characterized in terms of an algebra in which the elements were the initial concepts and the operations corresponded to operations the organism could perform. A natural first start is to think in terms of Boolean operations on concepts, but it does not take much additional reflection to make clear that this is certainly not an adequately rich apparatus for forming concepts of any complexity. The difficulty, of course, is evident at once when we consider what range of concepts can be defined by use of Boolean operations. Certainly we cannot build the imposing structure of concepts possessed by all higher organisms. The proof, if one is desired, follows by direct application of Padoa's method, and indicates the kind of link that may be forged, once a systematic theory of concept formation is considered, between mathematical and psychological theories of concept formation.

One can continue to push the algebra of concepts by introducing a richer set of operations. There is unfortunately very little, if any, constructive literature to be cited on this line of development, but there is one line of attack that seems to be so intuitively promising that I want to describe it even if it is not clear at the present time how the details are to be worked out. I have in mind the single primitive binary relation of set theory, namely membership, and the operation on concepts corresponding to the membership relation. We know that from a mathematical standpoint it would be a very powerful method of attack. It is also clear that this approach has close connections with verbal mediation theory. The forming of a set, or the assertion that an object is a member of a set, corresponds closely in a psychological sense to the notion of establishing usage for a general term. Admittedly talk of sets of sets of sets does not have any clear psychological meaning or reference, but if we talk about a chaining of verbal mediators, as would arise from the successive notation for sets of sets of sets, we then have immediately at hand in the notation a device that can be linked to the theory of verbal mediation, and also to general ideas of coding. It is my own hunch that this is probably one of the most promising directions in which to work in developing an adequate dynamical theory of concept formation. On the other hand, many treacherous and difficult problems have got to be solved and it is certainly not clear at the present time how to solve them. One interesting aspect of this approach is that if it could be worked out adequately, I am sure it would have repercussions on the foundations of mathematics itself. From a psychological standpoint, talk about sets and algebraic operations sounds rather like medieval talk about mechanics. Not that the talk is wrong. It is just that it seems hopeless in this vein ever to achieve an adequate solution to the problem being investigated. In every case, psychologically we want to turn at once from "abstract" talk about a set to immediate and direct questions about how notation for these sets is coded, but the implications of this line of thought for the foundations of mathematics cannot be explored in the present paper.

As still another inadequately worked-out theory of concept formation, I would like to mention some recent work I have been pursuing with some younger colleagues (particulary Madeleine Schlag-Rey). The central ideal is to extend the kinematical models discussed earlier by imposing several levels of conditioning, the most obvious way of describing two levels being that of rule and instance conditioning. Let me illustrate this distinction by a simple example. Suppose a subject is asked to classify objects that exemplify a number of complex properties, for example, shape, size, color and orientation. A simple example of a rule at one level would be the rule that the correct classification depends on exactly one of these properties. The instances in this case would be the various hook-ups between the positive and negative instances of each property and the classification. A second simple example of a rule, or as we sometimes say, second-order hypothesis, would be the hypothesis that exactly two properties of the list given above are required for correct classification of the objects. According to the theory we have attempted to apply to experimental data, it is postulated that conditioning of rules changes very slowly in comparison to the conditioning of instances and generally there is a high probability that most, if not all, of the instances of a given rule will be run through before the rule is rejected. An *a priori* probability distribution, to be used in the selection of a rule, is also postulated, and in fact the present evidence strongly points toward the desirability of assuming, and then attempting to work out the details of, a hierarchy of rules that is imposed by the organism on the basis probably of both past experience and innate abilities, in order to avoid combinatorial chaos—for example, the number of rules for a two-way classification of 100 stimulus items is just the number of subsets, i.e., 2^{100}, and no unstructured or brute-force attack on this number of rules is the least bit feasible.

From many standpoints the current central problem of concept formation is to find the principles that lead organisms out of the combinatorial jungle that is uncovered in any purely logical or mathematical analysis of complex problem solving. An understanding of these principles would lead to an enormous gain in our understanding of human thinking. And the present problems are not dependent for their solution on tomorrow's news from the neurophysiological front. It would, for example, be a big step forward to be able to lay down general principles for getting about with a computer in this combinatorial jungle of logical possibilities, even if the principles used were not at all those used by any living organisms. What we seem to lack are the right conceptual ways of looking at either concept formation or complex problem solving, and the finding of new and more powerful approaches is bound to have repercussions in philosophy because of the closeness of the subject matter to much of the classical tradition in the theory of knowledge, and the fundamental importance of the processes of concept formation for all human thinking and action.

REFERENCES

[1] BOWER, G. H., and T. R. TRABASSO, Concept identification. In R. C. Atkinson (Ed.), *Studies in Mathematical Psychology*. Stanford: Stanford University Press, 1964, pp. 32–94.

[2] SUPPES, P., Towards a Behavioral Foundation of Mathematical Proofs. Technical Report No. 44, Psycholcgy Series, Institute of Mathematical Studies in the Social Sciences, Stanford University, 1961.

[3] SUPPES, P., Mathematical Concept Formation in Children. Technical Report No. 64, Psychology Series, Institute for Mathematical Studies in the Social Sciences, Stanford University, 1964.

[4] SUPPES, P. and R. GINSBERG, A fundamental property of all-or-none models, binomial distribution of responses prior to conditioning, with application to concept formation in children. *Psychological Review*, **70** (1963), 139–161.

Section VIII

History of Logic, Methodology and Philosophy of Science

THE HISTORY OF THE QUESTION OF EXISTENTIAL IMPORT OF CATEGORICAL PROPOSITIONS

ALONZO CHURCH
Princeton University, Princeton, N. J., U.S.A.

This paper is not an original historical investigation in the sense that I have consulted all the original sources at first hand, but is mainly a report, utilizing the work of others in order to select and organize the history of a particular topic. I must acknowledge my indebtedness especially to the publications of I. M. Bocheński, Philotheus Boehner, and Ernest A. Moody.

Since this is a historical report, concerned in large part with traditional logic, I shall use the word "proposition" in its traditional sense. According to explicit definitions that were usual, a proposition is, in effect, a (declarative) sentence taken in conjunction with its meaning. In actual use the word often seems to vacillate between this (explicitly avowed) sense and the more abstract sense according to which different sentences may sometimes express the same proposition. But for the immediate purpose it is not necessary to fix the meaning of the word more precisely than this.

At the beginning of a certain elementary branch of logic (as it is now considered), i.e., the traditional theory of categorical propositions, a decision has to be made as to the meaning to be attached to sentences of categorical form in certain extreme cases in which one of the terms (subject or predicate) is either empty or universal. Especially the case of an empty subject term has proved troublesome or has led to controversy. As the usage of everyday language, out of which the traditional theory arose, is partly vacillating and partly unclear on this point, one simply has to make some convenient decision, subject to obvious conditions of adequacy and internal consistency, and then get on to more important matters. Technically there should be no great difficulty. But the history of the matter has a great deal of human interest, as to how the point could at first be overlooked, then later engender so much heat of controversy, so much plain confusion and stubborn clinging to preconceptions.

Aristotle never considers the question of existential import in connection with categorical inference, and the same is true of the early Scholastics, e.g. Petrus Hispanus.

Perhaps the best rendering of the Aristotelian logic in a way to conform to modern standards of rigor in the statement of it is that by Łukasiewicz.[1] Łukasiewicz's system is internally consistent, as he shows. But in the application it is required that empty terms must not be substituted for the variables. And for this reason the Łukasiewicz version of Aristotelian logic is inadequate for purposes for which that logic was originally intended.[2]

The fact is that ordinary purposes do require reasoning with empty and universal terms, and with terms about which we do not know in advance but what they may be empty or universal. For example, consider the astronomer who infers conditions which life on Mars must satisfy—he does not know whether such life exists but his conclusions may influence our estimate of the probability—or the mathematician who infers one after the other a series of properties that all exponents not obeying Fermat's Last Theorem possess. Or consider the premisses that all men are mortal and that I am a man; it would seem that the usual inference from these premisses might properly and desirably be drawn, even by one who is genuinely in doubt whether immortal beings exist.

However, Łukasiewicz's system is of course not intended as a logic for use, but for a historical purpose, to render the logic of Aristotle into logistic form in such a way as to remain as close as possible to the original. From this point of view, Łukasiewicz may have the best possible formulation (or one such). But there do inevitably remain some questions and uncertainties. In particular, Łukasiewicz takes the position that contraposition and obversion are not Aristotelian; and it is because these are not included in his system that Łukasiewicz needs only the presupposition that terms are not empty, and not the additional presupposition that terms are not universal. Actually this is a little debatable, just because Aristotle is not unmistakably clear and consistent. In one passage Aristotle characterizes a negative name such as "non-man" as being not a name but an ὄνομα ἀόριστον (an infinite name, as it is usually translated—but this is quite a different meaning of "infinite" from that familiar in modern mathe-

---

1 *Elementy Logiki Matematycznej*, Warsaw 1929, and *Aristotle's Syllogistic*, Oxford 1951, second edition 1957.

2 Referring to a suggestion of von Freytag-Löringhoff, and in connection with Łukasiewicz's system, Paul Lorenzen proposes that the particular affirmative and particular negative propositions be defined by $PiQ = (\exists R)(RaP \wedge RaQ)$, $PoQ = (\exists R)(RaP \wedge ReQ)$ (*Archiv für Mathematische Logik und Grundlagenforschung*, vol. 2 (1956), pp. 100–103). Evidently these definitions require the same presupposition of non-emptiness that the Łukasiewicz system does, or else, as indicated in Lorenzen's *Formale Logik* (Berlin 1958, second edition 1962), they require "dass nur Prädikate aus einer vorgegebenen Klasse (etwa einer sog. Begriffspyramide) betrachtet werden." In either case they lead to difficulties in the actual use of the logic as distinguished from its internal consistency. Moreover in the latter work Lorenzen indicates that, under these definitions, the particular negative PoQ is no longer the contradictory of the universal affirmative PaQ, so that the square of opposition is not preserved.

matics). In spite of this restriction or qualification, examples of contraposition do occur: "If man is an animal, what is not animal is not man," "If the pleasant is good, the non-good is not pleasant." Bocheński points out that these propositions are not quantified — at least not explicitly. Yet the only reasonable understanding of them is as universal affirmative; and the contraposition then seems to be of the sort which changes a universal affirmative into a universal negative. Probably from this source, contraposition appears in Petrus Hispanus as being presumably Aristotelian. But for Petrus Hispanus, contraposition changes a universal affirmative into a universal affirmative, and a particular negative into a particular negative. And I shall hereafter use the term "contraposition" in this sense.

There are also examples of obversion in Aristotle, e.g. the inference of "No man is just" from "Every man is non-just" and the inference of "Not every man is non-just" from "Some man is just." But it is conspicuous that these inferences are sanctioned in only one direction. Obversion of a negative proposition never occurs, and in this respect the Scholastic logic, wittingly or not, definitely goes beyond Aristotle.

P. F. Strawson's proposal[3] for validating the traditional and Aristotelian logic is so close to Łukasiewicz's in spirit that it is convenient to consider it here, out of its chronological order. In this proposal, the "presupposition" of non-emptiness affects only the subject term instead of all terms. It is partly on this account and partly because Strawson, unlike Łukasiewicz, wishes to preserve all the traditional immediate inferences, including contraposition and obversion, that Strawson's system proves to be internally inadequate. For example, it is held that *No A is B* presupposes rather than asserts that there are *A*'s. There is no presupposition that there are *B*'s, and *No A is B* is allowed to be true even when nothing is *B*. In consequence if we wish to make the traditional inference from *No A is B* to *No B is A*, we must introduce the additional presupposition that there are *B*'s. Strawson is aware of this, but holds that the traditional immediate inferences are nevertheless satisfactorily preserved. Of course this cannot be accepted, and Strawson's position betrays a lamentable lack of concern for the formality of logic. Adequate formality requires that any matter of extralinguistic fact which must be known before an inference can be made shall be stated as a premiss of that inference. And failure to observe this raises difficulties which become more and more serious as one goes on to more advanced parts of logic and more complicated applications of logic.

However, let us return to the task of the historical treatment of the question of existential import in more nearly chronological order.

[3] *Introduction to Logical Theory*, London and New York 1952.

As far as presently known the first logician to consider the question of existential import or to propose a tenable theory of it was William of Ockham, who holds that the affirmative categorical propositions are false and the negative true when the subject term is empty. The same idea occurs also in Buridan, as Moody points out, but I do not know that either Ockham or Buridan ever explicitly considers the particular negative in this regard. Details of the process by which this doctrine became generally accepted among the later Scholastics are a chapter in the history of logic that has still to be written. But it will be convenient to look at the culmination of it in the *Ars Logica* of John of St. Thomas, precisely because John is not an innovator in this area but follows predecessors, as he himself indicates.

John makes it explicit that a particular negative proposition is true if the subject term is empty, and even gives one example of this in which the subject term is self-contradictory.

Prima facie, Ockham's doctrine would invalidate contraposition, and also obversion of negative propositions, but John maintains these inferences by the device of "constantia." The "constantia" is namely the same as Strawson's "presupposition," except that it must be separately stated each time it is required. For example, in order to infer "Everything non-white is a non-man" from "All men are white" we must add the constantia: "And the non-white exists." In modern eyes, the constantia looks much like a second premiss so that contraposition and also obversion of a negative proposition become mediate inferences.

To avoid misleading in regard to the logic of John of St. Thomas it is necessary to add that John has, under the head of ampliation and restriction, a doctrine that rather strongly suggests the "universe of discourse" of the nineteenth century algebraists of logic, although John does not of course use this term or take altogether the point of view that is suggested by it. In particular, in the case of "All white men are men," if the verb is understood as referring to present time, the universe of discourse is that of presently existing things, and if there are no presently existing white men, the proposition is false. If the verb is understood as abstracting from time in order to make a timeless statement, then the universe of discourse is that of things "in intellectu," of essences rather than actualities, and the truth of the proposition requires only the existence of white men "in intellectu." This latter is the case of what John calls natural supposition. But in both these cases, and in the case of other universes of discourse (e.g. the universe of things existing in the past if the verb is in the past tense), the often disputed inference from "All white men are (were) men" to "Some men are (were) white men" is valid, on condition that the universe of discourse remains unchanged.

Of course natural suppositon faces all the familiar difficulties in the logic of unactualized possibles, and it would seem that this particular universe of discourse would better be deleted in favor of a variety of others. But the idea of combining the Ockham doctrine of existential import with the admission of various different universes of discourse is entirely sound in itself.

There exists another, much more recent, attempt to save the traditional syllogisms and immediate inferences entire — which, though not all its proponents do this, is most conveniently described in direct reference to John of St. Thomas. Namely the proposal is that, for purposes of logic, all propositions shall be taken in natural supposition. Proponents of this usually add (what is undoubtedly the fact) that in everyday colloquial discourse the existential import of statements that are made varies from case to case and must be judged from the context. Some hold that logic has no business with these extra assertions of existence and should simply ignore them, taking all propositions in natural supposition. Others hold, more reasonably, that these assertions of existence should be taken as separate assertions, which are to be separately stated and treated. But even in this latter form, the restriction which the doctrine imposes upon the application of traditional logic is really quite intolerable — e.g., "Some men (presently existing) are wise" is not adequately represented by "Some men (*in intellectu*) are wise," even if we supply as additional propositions "Men presently exist" and "Wise beings presently exist."

The doctrine of existential import which was just described is referred to by Lewis Carroll in 1897, but he does not quote his source. The earliest proposal of the doctrine that is known to me is in the Cambridge dissertation of Abraham Wolf in 1900. Jacques Maritain attributes a somewhat similar idea to Lachelier (1907).

Maritain's own account[4] of existential import agrees in its main outlines with that of John of St. Thomas, but by comparison with it is seriously deficient in several respects. Namely (1) Maritain's account of the constantia is defective, and in particular he fails to remark that contraposition requires the constantia. (2) Maritain's examples in the discussion of existential import are entirely of affirmative propositions, with one exception; the matter of negative propositions with an empty or non-existent subject is illustrated only by a singular proposition, and it is said only that such a proposition *may* be true, rather than that it *must* be true (as consistency

[4] *Petite Logique* (Éléments de Philosophie, IIe fascicule), Paris 1923. In preparing this paper I have used the English translation by Imelda Choquette, which in various printings or editions has sometimes the title *Formal Logic* and sometimes the title *An Introduction to Logic.*

would require, and as John in fact holds). (3) Maritain fails to mention at all the crucial matter of propositions with self-contradictory subjects; it should have been said that, even in natural supposition, these propositions are false if affirmative, true if negative.

Some neo-Scholastics, being perhaps misled by Maritain's discussion of natural supposition, have misunderstood John of St. Thomas more radically, and in such a way that they obtain doctrines of existential import which to a greater or less extent resemble that of Wolf.

In particular F. C. Wade, in the introduction to his translation of the first part of the *Ars Logica*, proposes a doctrine which is quite close to Wolf's. It is to Wade's credit that he observes that there is a difficulty about propositions with self-contradictory subjects. But he seeks to meet the difficulty by excluding such propositions as not being propositions at all, failing to see that the self-contradictory is not always immediately recognizable in a way to make this feasible. Indeed it is often important, especially in mathematics, to be able to reason about a subject which for all one knows may be self-contradictory. And it seems that both Wolf and Wade would do better to have recourse to the Ockham doctrine of existential import in at least the extreme case of a self-contradictory subject.

What is often considered the modern doctrine of existential import, that the universal propositions are true and the particular false when the subject term is empty, appears implicitly in a short paper which was published by the mathematician Arthur Cayley in 1871.[5] In this paper Cayley uses the algebra of logic to work out a scheme of syllogisms in what is actually the obvious way. But he does not point out that the resulting scheme is at variance with the traditional one, and it seems at least possible that he was just not familiar with the traditional scheme. The explicit introduction of the modern doctrine was made independently by Franz Brentano[6] and C. S. Peirce[7] some years later. Brentano maintains some sort of absolute correctness for this account, because he maintains that the categorical propositions are, in an absolute sense, *really* either existential propositions or contradictories of existential propositions. But Peirce, after stating the senses of the categorical propositions "which are traditional" (i.e., Ockham's doctrine) explains the different senses of them which he will

---

[5] *The Quarterly Journal of Pure and Applied Mathematics*, vol. 11 (1871), pp. 282–283. In the first display in the paper, "No *X*'s are *X*'s" is evidently a misprint for "No *X*'s are *Y*'s." The same correction applies also to the reprint of the paper in Volume 8 of Cayley's *Collected Mathematical Papers*.

[6] Book II Chapter VII of *Psychologie vom Empirischen Standpunkte*, Leipzig 1874. Compare also J. P. N. Land, in *Mind*, vol. 1 (1876), pp. 289–292, and Franz Hillebrand, *Die Neuen Theorien der Kategorischen Schlüsse*, Vienna 1891.

[7] *American Journal of Mathematics*, vol. 3 (1880), pp. 15–57.

adopt — with no justification offered, but presumably just on the ground that he finds them more convenient for his purpose.

It is tempting to conclude that no satisfactory doctrine of existential import can preserve all the traditional immediate inferences and syllogisms, and the square of opposition, since in fact neither that of Ockham nor the modern doctrine does so.[8] The closest approach is the doctrine of Ockham and the late Scholastics, which does preserve all the properly Aristotelian inferences, including even those kinds of obversion and the inference akin to contraposition which Aristotle seems to allow. In this latter respect, and quite apart from the matter of application of the logic (which was discussed above), the late Scholastic version is in fact superior to Łukasiewicz's.[9]

If one attaches importance to the purely historical faithfulness of preserving the traditional scheme, either unchanged, or with as little change as possible, then one must decide whether it is the Aristotelian logic proper that is to be preserved, or the Scholastic logic (e.g. of Petrus Hispanus), or one of the post-Scholastic versions of traditional logic. And one must decide how much importance to attach to the linguistic oddity that the late Scholastic doctrine makes it true that "Some chimaeras are not animals" — on the ground alone that there are no chimaeras.[10]

---

[8] Such a conclusion is debatable only because of the vagueness of the adjective "satisfactory." Strawson points out that all the traditional immediate inferences and syllogisms, and the traditional square of opposition, can be preserved in the context of a standard (modern) logic of quantifiers if we define PaQ as $\sim(\exists x)(Px \sim Qx) \,.\, (\exists x)Px \,.\, (\exists x)\sim Qx$, and PeQ as $\sim(\exists x)(PxQx) \,.\, (\exists x)Px \,.\, (\exists x)Qx$, and PiQ as $\sim .\, PeQ$, and PoQ as $\sim .\, PaQ$. This loses the traditional law of identity PaP. But even the law of identity can be maintained too, if we follow a slightly more complicated scheme proposed by H. B. Smith (*The Journal of Philosophy*, vol. 21 (1924), pp. 631–633). In effect, we obtain Smith's scheme as a modification of Strawson's if we allow PaQ to be true also when P and Q are *both* empty, and when P and Q are *both* universal — and modify Strawson's meanings of PeQ, PiQ, and PoQ correspondingly. The same idea was later suggested independently by Stanisław Jaśkowski (*Studia Societatis Scientiarum Torunensis*, section A, vol. 2 no. 3 (1950), pp. 77–90).

But it seems that all those who have proposed or considered schemes of this sort put them forward as mere curiosities and immediately reject them as being too artificial for actual adoption.

[9] It would indeed be possible to modify the Łukasiewicz version by introducing the additional presupposition that terms are not universal, and then to allow all the usual cases of obversion and contraposition. Cf. Ivo Thomas, *Dominican Studies*, vol. 2 (1949), pp. 145–160, and Anders Wedberg, *Ajatus*, vol. 15 (1949), pp. 299–314. But the resulting modified Łukasiewicz system is even more deficient than the original, from the point of view of its application as a formalized or systematized version of the logic of ordinary discourse.

[10] One might mitigate the oddity by reading the particular negative with the words "not all" — as e.g. "Not all chimaeras are animals" must evidently be counted true if "All chimaeras are animals" is counted false. But the interchangeability of "not all" and "some not" (or more exactly, the analogous interchangeability in Latin) was already laid down by Petrus Hispanus and, as far as I know, was not questioned by later Scholastics. On the other hand, John of St. Thomas finds it necessary to defend the linguistic oddity by remarking that it's as if one had said "The chimaera is not an animal."

Finally it should be said that if one allows the constantia at appropriate places, in effect a second premiss for certain immediate inferences, then both the late Scholastic version of existential import and the modern version preserve the entire traditional scheme in (what is otherwise) the broadest interpretation.

A MEDIEVAL DISCUSSION OF INTENTIONALITY

PETER GEACH
The University, Birmingham, England

In this paper I shall critically examine the way a fourteenth-century logician, Jean Buridan, dealt with certain puzzles about intentional verbs. The class of verbs I shall be considering will all of them be expressions that can be completed into propositions by adding two proper names; the class will include, not only ordinary transitive verbs, but also phrases of the verb-preposition type like 'look for' or 'shoot at', and furthermore constructions like 'hopes—will be a better man than his father' or 'believes—to be a scoundrel', which turn into propositions as soon as we add mention of who hopes or believes this and about whom he does so. In modern grammar, the term 'a verbal' rather than 'a verb' is used for this wider class; following a suggestion of Professor Bar-Hillel, I adopt this term.

In either or both of the proper-name places that go with such a verbal, it is possible, without destroying the propositional structure (*salva congruitate*, as medieval logicians say), to substitute a phrase of some such form as 'some *A*' or 'every *A*' or 'the (one and only) *A*'; the letter '*A*' here represents a simple or complex general term which is grammatically a noun or noun-phrase. The peculiarity of certain verbals that presently concern us comes out when such a phrase formed from a general term stands in object position, in a construction 'b F'd an *A*' or the like. Consider for example the sentence 'Geach looked for a detective story'. This sentence is ambiguous: in ordinary conversation we might sucessfully resolve the ambiguity by asking the question 'Was what Geach was looking for a particular detective story, or was it just *a* detective story?' It is an odd psychological fact that this question would convey the intended distinction of meanings; for logically the words of the question leave it wholly obscure what is intended. After all, nothing in this world or in any possible world could be "just *a* detective story" without being "a particular detective story"; and even if such an *individuum vagum* could somehow have being, Geach could not read it, so it certainly is not what he looked for.

A similar example in Buridan himself is 'I owe you a horse'. Here again one would quite naturally distinguish between owing a particular horse —the handing over of that very horse would be part of the bargain—and "just" owing *a* horse. Whatever we are to make of this distinction, it is clear that what the customer wants is a real live horse; not a possible horse, or an indefinite horse that is literally not all there. Buridan does have some very dubious passages in which he quantifies over *possibilia*, such as possible horses and possible men; he has, for example, a doctrine that when a general term '*A*' stands as subject to a modalized predicate, it is "ampliated" or stretched so that it relates to possible *A*s as well as actual *A*s. On this view, to use an example formally similar to Buridan's own, 'Some man is necessarily damned' would be exponible as 'Some actual or possible man is necessarily damned', which appears to draw the sting of the doctrine of reprobation. Anyhow, when it's a question of owing a horse Buridan will not bring in possible horses.

It may well appear equally out of place to bring into the picture such intentional entities as the senses of expressions; the sense of the term 'horse' or 'detective story' is assuredly not something whose possession would content a man when he is owed a horse or is looking for a detective story. All the same the sense, or as Buridan calls it the *ratio*, given us by an object expression really is somehow involved when the verbal is an intentional one; Buridan says the object expression *appellat suam rationem*, where we may perhaps render the verb '*appellat*' (perhaps the most obscurely and multifariously employed of all medieval semantic technicalities) by saying the expression "calls up" or "evokes" its own *ratio*. Buridan's main point is in any case clear: that the truth-value of a sentence whose verbal is an intentional one may be changed if we change the *ratio*, the sense, given in the object expression, even if this expression still relates to the same thing in the world. This point is clearly correct, as many of Buridan's examples show; thus, if a body is both white and sweet and I see it, I may truly say 'I have discerned something white by my sense of sight', but not 'I have discerned something sweet by my sense of sight'.

Can't I, though, say I have discerned something sweet by my sense of sight, if the body I have discerned is in fact sweet? Well, it's better to say 'There is something sweet that I have discerned by sight' if this is the case I have in mind. This means, according to Buridan, that the sweet thing is something which my sense of sight discerns under some *ratio* or other, not at all necessarily under the *ratio*: sweet—this *ratio* comes in only *sub disiunctione ad alias rationes*, as one possible alternative *ratio*.

Buridan here observes a tendency of Latin idiom, which comes out in much the same way in English: when a verbal is intentional, an object phrase is differently construed when it preceeds the main verb that is a part or the whole of this verbal, and when it follows that verb. In general, taking 'to F' as representative of an intentional verbal, we shall distinguish between the forms 'b F's an *A*' and 'There is an *A* that b F's'. The latter form if construed in the spirit of Buridan will mean 'There is some actual *A* of which it holds that b F's it under some *ratio* or other' — possibly, though, under some other *ratio* than that of being an *A*. (In Buridan's Latin examples there are no words corresponding to 'there is' and 'that' in the form 'There is an *A* that b F's'; this difference between English and Latin idiom is obviously trivial). Of course this distinction of word order is not *strictly* observed in ordinary English or Latin; we have to consider what is actually meant in the case under consideration. An instructive example, to which I shall presently return, is one I borrow with slight alteration from John Austin: 'I saw a man born in Jerusalem' *versus* 'I saw a man run over in Oxford'. But we may very well decide for present purposes to stick to Buridan's convention of word order, as a way of marking the distinction he quite properly wishes to draw.

Buridan rightly regards the inference from 'There is an *A* that b F'd' to 'b F'd an *A*' as an invalid form; for if 'There is an *A* that b F'd' is true, the general term represented by '*A*' need not give us the aspect of *ratio* under which such an *A* was the object of b's F'ing. Moreover, though Buridan himself does not consider (the Latin equivalent of) the verbal 'was looking for', his machinery helps to clear up our puzzles about this verbal. We should use 'There is a detective story Geach was looking for' for the case where one would say I was looking for "a particular" detective story, and reserve 'Geach was looking for a detective story' rather for the case where I was "just" looking for *a* detective story. In the former case, if the book I was looking for was a book printed in Baskerville type, one might equally well say 'There is a book printed in Baskerville type that Geach was looking for'; but one could not infer 'Geach was looking for a book printed in Baskerville type', since the *ratio* under which I was looking for the book will have been the one expressed by 'detective story', not the one expressed by 'book printed in Baskerville type'.

Buridan assumes that from 'b F'd an *A*' we may infer 'There is something that b F'd'. In fact, as we shall presently see, he assumes more than this; and even this much is doubtful. But we may use this assumption to show how Buridan's theory looks in terms of modern quantification theory. The intentional verbal represented by 'to F' would answer, not to a two-termed relation between a person and an object of his F'ing, but to a three-termed relation between a person, an object of F'ing, a *ratio*; the forms 'b F'd an *A*' and 'There is an *A* that b F'd' would respectively come out as:

For some *z*, b F'd *z* under the *ratio: A*

For some *z*, and for some *w*, *z* is an *A*, and b F'd *z* under the *ratio w*.

It may be helpful to work out in this way the concrete example from John Austin that I mentioned just now. 'I saw a man run over in Oxford' would come out as:

For some *z*, *z* is a man and I saw *z* in Oxford under the *ratio*: run over, and 'I saw a man born in Jerusalem' would come out as:

For some *z*, *z* is a man and *z* was born in Jerusalem and, for some *w*, I saw *z* under the *ratio w*.

Buridan's analysis, seen in this light, already raises one severe difficulty. The form with the term preceding the intentional verbal essentially involves quantifying over *rationes*. Now I am not at all impressed by rhetoric about mysterious entities, by questions in the peculiar (anti-) metaphysical tone of voice 'But what *are* they?'; it is all right to quantify over entities if you can supply a sharp criterion of identity for them. The paradigm of this is Frege's sharp criterion of identity for numbers. But for *rationes* we can get out of Buridan no hint of such a criterion. He gives us many examples of *rationes* he takes to be patently different, like the *ratio*: white and the *ratio*: sweet; but we have not even one example of the same *ratio* differently expressed, from which we might divine a criterion of identity. This lacuna makes Buridan's account schematic at best.

A more serious defect is Buridan's acceptance of the inference from 'b F'd an *A*' to 'There is an *A* that b F'd'. Notice that this would still be unacceptable even if we let pass the inference to 'There is something that b F'd'; for this does follow from:

For some *z*, b F'd *z* under the *ratio*: *A*

but there is still no obvious necessity that what is F'd under the *ratio*: *A* should in fact be an *A*. And in concrete examples Buridan's pattern of inference, from the form "with the term after" to the form "with the term before", often seems patently invalid. He himself chooses examples that look all right, e.g. from 'I see something white' to 'There is something white I see', or from 'I recognize somebody coming' to 'There is somebody coming whom I recognize'. But of course we may not pass from 'James is looking for a universal solvent' to 'There is a universal solvent James is looking for;' we may not even pass from 'Geach is looking for a detective story' to 'There is a detective story Geach is looking for', because, as we saw, the premise is compatible with my not having been looking for "a particular" detective story.

Even among the examples Buridan himself supplies, there are some that hardly favour his thesis. Suppose poor old Socrates is a competent astronomer immured in a dungeon where he cannot tell night from day. Then

Buridan is willing to pass from the premise 'Socrates knows some stars are above the horizon' to the conclusion 'There are some stars Socrates knows are above the horizon'. You naturally ask which stars Socrates knows are above the horizon; Buridan has a ready answer—the ones that are. To be sure, it is in a rather attenuated sense that the constellation Aries (say) is then known by Socrates to be above the horizon: namely, not in respect of the complex *ratio* expressed by 'Aries is above the horizon', but only in respect of the one expressed by 'Some stars are above the horizon'. But at least Buridan can hardly be forced to pass from premises that would in this case be true to a conclusion that would in this case be false. Although in his view *one* true answer to the question 'Which stars does Socrates know are above the horizon?' might be 'There is the constellation Aries that Socrates knows is above the horizon', he can on his own principles disallow the inference from this to 'Socrates knows that the constellation Aries is above the horizon'; this shift of the term, *from* before the main-verb part of the verbal 'knows that—is above the horizon' *to* after that verb, is just what Buridan disallows, as we saw. And Buridan's motive for wishing to pass to the conclusion 'There are some stars that Socrates knows are above the horizon' is that even the abstract knowledge expressed in 'Some stars are above the horizon' ought in his eyes to latch on to some actual, and therefore definitely specifiable, stars.

This example is rather a cheat, though, because Buridan is trading on a peculiarity of the verb 'to know'. What you know is so; if Socrates knows some stars are above the horizon, then some stars are above the horizon, and we may ask which ones are. But suppose Socrates believes some lizards breathe fire—not being so good a zoologist as an astronomer. May we pass to the conclusion 'There are some lizards that Socrates believes breathe fire'? and if so, which lizards are these? Certainly not the ones that do breathe fire, for Socrates' belief is false. Moreover, it may be impossible to find any principle for picking out, from among actual lizards, the ones Socrates believes to breathe fire. Buridan often resorts, when faced by such an embarrassing choice, to a parity-of-reasoning argument; what is true of some so-and-so, if there is no reason to think it is true of one so-and-so rather than another, must be true of each and every so-and-so. We should then have to pass on from 'There are some lizards that Socrates believes breathe fire', to the further conclusion 'As regards each lizard, Socrates believes that it breathes fire'. From this, indeed, we could on Buridan's principles not infer 'Socrates believes that each lizard breathes fire'; Socrates' belief would latch on to each and every lizard, but only *via* the complex *ratio* expressed by 'Some lizards breathe fire'. Still, in Buridan's own words, *videtur durum*, it seems tough to swallow.

I now come on to a similar puzzle of Buridan's own, in which his own principles raise severe difficulties. A horse dealer, he supposes, currently owns just three thoroughbreds, Brownie, Blackie, and Fallow. The customer accepts the dealer's promise 'I will give you one of my thoroughbreds', but the dealer fails to deliver, and denies that he owes the purchaser anything. 'If I owe you any horse', he argues, 'I must owe you either Brownie or Blackie or Fallow; they're the only horses I have, and you can't say I owe you some other horse, let us say H. M. the King's charger. Now what I said didn't relate specially to Blackie rather than Fallow, or the other way round; what goes for one goes for the other; and similarly for Brownie. So either I owe you both Brownie and Blackie and Fallow, or I owe you not one of them. You'll surely not have the nerve to say I owe you all three, when I only said I'ld let you have one; so I don't owe you even one. *Good* morning.'

Part of the trouble Buridan has in saying what is wrong with the horse-coper's argument comes from his accepting the inference from 'I owe you a horse' to 'There is a horse I owe you'; and we have seen reason to doubt the validity of the rule by which Buridan has to allow this step. But even without this inferential move, Buridan can give the horse-coper a longer run for his (or rather his victim's) money. Even if in general we cannot validly pass from 'b F's an *A*' to 'There is an *A* that b F's', it seems plausible to accept this particular instance of a term-shift ng inference: 'I owe you something: *ergo*, there is something I owe you'. And we can consistently accept it without accepting Buridan's invalid rule; many philosophers seem to think that any instance of an invalid form is an invalid argument, but this is a gross logical error. Let us then have our horse-coper arguing again. 'If I owe you a horse, then I owe you something. And if I owe you something, then there is something I owe you. And this can only be a thoroughbred of mine: you aren't going to say that in virtue of what I said there's something else I owe you. Very well then: by your claim, there's one of my thoroughbreds I owe you. Please tell me which one it is'.

Buridan has an ingenious way out of the difficulty. it can be said that *x* is owing by me to *y* if and only if by handing over *x* to *y* I should be quits with him (*essem quittus*). Now whichever thoroughbred (*bonus equus*) *x* may be, if the horse-coper hands over *x* to the purchaser then they are quits. Buridan concludes that, whichever thoroughbred *x* may be, the dealer owes *x* to the customer. This goes even for H. M. the King's charger; for if the horse-coper bought or were given this horse, he could be quits with his customer by handing it over, as he could not be if he had not handed over exactly what was owing.

We can now clear up the "parity or reasoning" argument about the three horses the dealer already has. It is true of Brownie, and it is true of Blackie, and it is also true of Fallow, that this is a horse the dealer owes his customer. So, considering just Brownie and Blackie, we may in a sense say these are two horses the dealer owes. But Buridan rightly warns us to avoid confusing collective and distributive predication; it is not that there are two horses together which the dealer owes, but that of each one of the two it holds good that he is a horse the dealer owes. Moreover, by Buridan's general principle we may not pass from 'There are two horses the dealer owes' to 'The dealer owes two horses'; that would call up the *ratio*: two horses, which did not come into the dealer's original promise. Similarly, from 'Brownie is a horse the dealer owes' we may not infer 'The dealer owes Brownie'; this would be true only if the dealer's original promise had called up the *ratio* expressed by the name 'Brownie', as *ex hypothesi* it did not.

It is with great reluctance that one faults Buridan's brilliant argument, securing obvious justice for the customer without bending the laws of logic. All the same, the inference from 'I owe you something' to 'There is something I owe you' cannot be let pass unchallenged. No doubt many an investor in the Post Office Savings Bank has inferred from 'The Post Office owes me something' to 'There is something the Post Office owes me'; his thought is 'The Post Office has it all stacked up somewhere—*aes alienum*, other people's brass—and some of it's mine'. The inference is invalid, and the conclusion is false; and the mistake is far from trivial—as those unscrupulous politicians were well aware who made use of this false impression in their highly successful propaganda to the effect 'The Socialists were going to take *your* money out of the Savings Bank and pay it over to the unemployed'!

This difficulty is not confined to the present example; there are, I think, very many instances in which from 'b F's an *A*' or 'b F's something or other' we may by no means infer that there is an identifiable something-or-other which b is F'ing. On this point Buridan's theory stands in need of radical reconstruction. I think I can see in part how such a reconstruction would be possible; but there are difficulties that still remain.

Let us take the example 'Geach was looking for a detective story'. On our former analysis, which closely followed Buridan, this would come out as:

For some *x*, Geach was looking for *x* under the *ratio*: detective story.

But this would have the unwelcome consequence that, even when I was "just" looking for *a* detective story, there was some identifiable *x*—not necessarily a detective story, to be sure—that I was looking for. We need rather a dyadic relation between Geach and a *ratio*: 'Geach was-looking-for-something-under the *ratio*: detective story', where I have hyphenated the words of the verbal to show that logically it is just one indivisible relative term.

We may achieve considerable clarification if we expand this analysis to:

> Geach was looking for something under the *ratio* evoked (*appellata*) by the expression 'detective story'

and then treat 'was looking. . . by' as a single relative term, which we may abbreviate to 'was L'ing.' We now have a verbal flanked by a name of a person and a quotation, rather than by a name of a person and a designation of a *ratio*. This dodge, suggested to me by Professor R. Montague, avoids our difficulties about quantifying over *rationes*; we may now analyse 'There is a detective story Geach was looking for' as:

> For some x, x is a detective story, and, for some w, w is a description true just of x, and Geach was L'ing w (Geach was-looking-for-something-under-the-*ratio*-evoked-by the definite description w).

Here we quantify over forms of words, whose criterion of identity, if not completely clear, is a lot clearer than that of *rationes*.

A similar method suggests itself for dealing with problems of what may be called *intentional identity*. In a way parallel to the Buridan convention we may distinguish between 'There is a poet whom both Smith and Brown admire' and 'Smith and Brown both admire the same poet'; the latter would cover the case where both Smith and Brown are victims of the same literary fraud as to the existence of a poet, as well as the more normal case where they both admire (say) Wordsworth's poetry. Let us use the expression 'AP' as short for 'admire as a poet someone conceived under the *ratio* evoked by'; then 'There is a poet whom both Smith and Brown admire' would come out as:

> For some x, x is a poet and, for some w, w is a description true just of x, and both Smith and Brown AP w

whereas 'Smith and Brown both admire the same poet', taken as conveying only intentional identity, would come out in the simpler form:

> For some w, w is a definite description, and Smith and Brown both AP w.

Unfortunately, the line of solution we have been following leads us into difficulties. Suppose we use 'D'd' as short for the verbal 'dreamed of someone under the *ratio* expressed by'. Then in our present view we should have to paraphrase 'There is a red-head Harris dreamed of' as:

> For some *x*, *x* is a red-head and, for some *w*, *w* is a description true just of *x*, and Harris D'd *w*.

Now suppose we take *w* to be the description 'the fattest woman in the world'. The paraphrase would be true if Harris dreamed of the fattest woman in the world and the fattest woman in the world is in fact a red-head; but the proposition paraphrased might then quite well be false, because in Harris's dream there may have been no red-head, and the fattest woman he saw in his dream may have been as bald as an egg. (I owe this counter-example to my pupil Mr. David Bird.) Similar difficulties arise for our account of intentional identity: for if c and d each worshipped something under the *ratio* expressed by 'the deity of the Sun', it does not follow that c and d both worshipped the same deity—c might be an ancient Egyptian worshipping the ancestor of Pharaoh, and d a Japanese worshipping the ancestress of the Mikado.

I hope this paper shows why modern logicians still need to take medieval logicians seriously. In great measure their problems are ours; while for some of them, like the problems of *suppositio*, modern logic provides adequate solutions, there are other problems, about modal and intentional contexts for example, that are still wide open; and the talent that was shown by medieval logicians in wrestling with their problems demands our deepest admiration.

PROGRAM

1964 INTERNATIONAL CONGRESS FOR LOGIC, METHODOLOGY AND PHILOSOPHY OF SCIENCE

Hebrew University, August 26–*September* 2, 1964

WEDNESDAY MORNING, AUGUST 26

Opening Ceremonies

YEHOSHUA BAR-HILLEL, Secretary of the Organizing Committee, Presiding
Addresses by
ABBA EBAN, Deputy Prime Minister, State of Israel
ELIAHU ELATH, President, Hebrew University
MORDECAI ISH-SHALOM, Mayor of Jerusalem
AHARON KATZIR, President, Israel Academy of Sciences and Humanities
GEORG H. VON WRIGHT, President, Division of Logic, Methodology and Philosophy of Science, International Union of History and Philosophy of Science
Ceremony honoring Professor S. H. Bergmann on the occasion of his eightieth birthday.

WEDNESDAY AFTERNOON, AUGUST 26

Invited Hour Address — Section IV.

A. KATZIR (Israel), Chairman

M. POLANYI (England), "Science, Tacit and Explicit"

Invited Symposium on Probability and Rational Belief — Section V

G. H. VON WRIGHT (Finland), Chairman

R. B. BRAITHWAITE (England)
J. HINTIKKA (Finland)
H. E. KYBURG, Jr. (U.S.A.)
R. C. JEFFREY (U.S.A.)

Contributed Papers — Section I. *Mathematical Logic*

H. B. CURRY (U.S.A.), Chairman

K. DE BOUVÈRE (Netherlands), "Definability and Synonymity"
R. FRAÏSSÉ (France), "Une Généralisation de l'Ultraproduit" (read by J.-P. Bénéjam)
D. KUREPA (Yugoslavia), "On the Rectangle Tree Hypothesis"

M. Machover (Israel), "Contextual Determinacy in Lésnieswki's Grammar"

J. Reichbach (Israel), "Verification of Theses by Finite and Infinite 0-1 Sequences"

B. van Rootselaar (Netherlands), "Homomorphisms of Recursive Word-Arithmetics"

THURSDAY MORNING, AUGUST 27

Invited Symposium on Applications of the Theory of Mathematical Machines to Mathematical Logic — Section I

L. Kalmár (Hungary), Chairman

J. R. Büchi (U.S.A.)
A. Ehrenfeucht (Poland)
J. C. Shepherdson (England)

Contributed Papers — Section VII. *Methodology and Philosophy of Life Sciences*

M. Masterman (England), Chairman

S. Amarel (U.S.A.), "On the Formation of the Concept of a Transformation by Computer"

M. R. A. Chance (England), "The Status of Ethological Method in Biology"

G. Gunther (U.S.A.), "Self Reflection Systems and Value Designation" (read by A. Stern)

W. Haas (England), "Why Linguistics Is Not a Physical Science"

T. H. Mott, Jr. and B. M. Ross (U.S.A.), "A Logical Analysis of Concept Attainment" (read by B. M. Ross)

N. Prywes (U.S.A.), "An Algorithmic Approach to Classification" (read by Ḣ. Hiz)

F. S. Rothschild (Israel), "Experimental Confirmation of Biosemiotic Hypotheses"

D. K. Székély (Israel), "General Purpose Unifying Automata"

THURSDAY AFTERNOON, AUGUST 27

Invited Hour Address — Section V.

A. W. Burks (U.S.A.), Chairman

M. Black (U.S.A.), "Outstanding Problems in the Philosophy of Induction"

Invited Symposium on Logical and Pragmatical Aspects of Explanation and Prediction — Section IV

I. Scheffler (U.S.A.), Chairman

D. Føllesdal (Norway)
B. Mandelbrot (U.S.A.)
M. B. Hesse (England)
W. Sellars (U.S.A.)

Contributed Papers — Section II. *Foundations of Mathematical Theories*

D. Kurepa (Yugoslavia), Chairman

K. Bing (U.S.A.), "A Special Case of the Axiom of Choice"

M. Davis (U.S.A.), "Diophantine Real Numbers and Functions"

J. DERRICK (England), "Diagonal Completeness in Axiomatic Set Theory"

H. GAIFMAN (Israel), "Further Consequences of the Existence of Measurable Cardinals"

F. W. LAWVERE (U.S.A.), "A First-Order Theory of the Category of Sets and Functions"

L. LÖFGREN (Sweden), "A Constructivistic View on the Accessibility of Sets"

R. M. SMULLYAN (U.S.A.), "Superinductive Classes — I"

E.-J. THIELE (West-Germany), "Finite Axiomatisability of Restricted Separations"

FRIDAY MORNING, AUGUST 28

Invited Symposium on Space, Time and Relativity — Section VI

S. SAMBURSKY (Israel), Chairman

P. BERGMANN (U.S.A.)
P. HAVAS (U.S.A.)
H. MEHLBERG (U.S.A.)

Contributed Papers — Section III. *Philosophy of Logic and Mathematics*

A. CHURCH (U.S.A.), Chairman

P. BERNAYS (Switzerland), "Revision of the Programme of Proof Theory" (read by E. SPECKER)

CH. BOASSON (Israel), "The Use of Logic in Legal Reasoning"

L. CAUMAN (U.S.A.), "On Indirect Proof"

A. ESHKENASI (Bulgaria), "On the Problem of the Specific Character of the Formal Logical Systems"

R. HARRÉ (England), "The Principles of Formalisation"

C. POLITIS (Greece), "Limitations of Formalisation"

A. L. STERN (U.S.A.), "Economy and Reduction: Multi-Valued Logics"

FRIDAY AFTERNOON, AUGUST 28

Invited Hour Address — Section VII.

S. ADLER (Israel), Chairman

W. S. MCCULLOCH (U.S.A.), "What's in the Brain that Ink May Character?"

Tour of Jerusalem

Banquet

SATURDAY, AUGUST 29

Day on the Beach, Guided Tour to Caesarea and Natanya

SUNDAY MORNING, AUGUST 30

Invited Hour Address — Section I.

A. A. FRAENKEL (Israel), Chairman

R. L. VAUGHT (U.S.A.), "The Löwenheim-Skolem Theorem"

SUNDAY AFTERNOON, AUGUST 30

Invited Symposium on Selected Problems in Greek and Medieval Logic — Section VIII

W. KNEALE (England), Chairman

P. GEACH (England) N. RESCHER (U.S.A.)

Contributed Papers — Section VI. *Methodology and Philosophy of Physical Sciences*

E. I. J. POZNANSKI (Israel), Chairman

A. BRESSAN (Italy), "A General Many-Sorted Modal Language ML"

L. N. BRILLOUIN (U.S.A.), "The Arrow of Time"

J. L. DESTOUCHES (France), "General Mathematical Physics and Schemes; Application to Theory of Particles"

W. D. DAVIDON and H. EKSTEIN (U.S.A.), "Observables in Relativistic Quantum Mechanics" (read by H. Ekstein)

R. ENGLMAN (Israel), "Measurements in Quantum Mechanics and the Mind of the Experimenter"

A. P. POLIKAROV (Bulgaria), "Structures of Solutions of Scientific Problems"

Reception by the Mayor of Jerusalem
Meeting of the Executive Committee of the IUHPS/DLMPS

MONDAY MORNING, AUGUST 31

Invited Symposium on the Justification of Formalization — Section III

P. LORENZEN (West-Germany), Chairman

A. APOSTEL (Belgium) F. KAMBARTEL (West-Germany)
S. KÖRNER (England)

Contributed Papers — Section V. *Foundations of Probability and Induction*

H. FREUDENTHAL (Netherlands), Chairman

J. E. FENSTAD (Norway), "Algebraic Logic and the Foundation of Probability"

H. A. FINCH (U.S.A.), "Bayesian Rules for the Rational Reconstruction of a Hypothesis Weakened by Adverse Observations"

S. MCCOLL (U.S.A), "Connexive Implication and the Syllogism"

J. J. MEHLBERG (U.S.A.), "An Extension of the Frequency Interpretation of Probability Theory"

K. WALK (Austria), "Extension of Inductive Logic to Ordered Sequences"

MONDAY AFTERNOON, AUGUST 31

Invited Hour Address — Section VI.

L. N. BRILLOUIN (U.S.A.), Chairman

O. COSTA DE BEAUREGARD (France), "Irreversibility Problems"

Invited Symposium on Concept Formation — Section VII

P. SUPPES (U.S.A.), Chairman

D. DAVIDSON (U.S.A.) S. PAPERT (England)

Contributed Papers — Section VIII. *History of Logic and Methodology of Science*

A. SZÁBO (Hungary), Chairman

A. JOJA (Roumania), "Les Propositions sur les Futurs Contingents et le Déterminisme"
I. LAKATOS (England), "Proof-Generated Concepts in Infinitesimal Calculus"
M. L. PERL (Israel), "Newton's Justification of the Laws"
S. SAMBURSKY (Israel), "Proclus' Explanation of the Planetary Motions"
J. F. STAAL (Netherlands), "Indian Logic and Sanskrit Syntax"

Invited Hour Address — Section II

S. K. KLEENE (U.S.A.), Chairman

A. TARSKI (U.S.A.), "Metamathematical Properties of Some Affine Geometries"

General Assembly of IUMPS/DLMPS

TUESDAY MORNING, SEPTEMBER 1

Invited Symposium on Foundations of Set Theory — Section II

A. TARSKI (U.S.A.), Chairman

C. C. CHANG (U.S.A.)
P. J. COHEN (U.S.A.)
A. LÉVY (Israel)
R. MONTAGUE (U.S.A.)
S. TENNENBAUM (U.S.A.)

Contributed Papers — Section IV. *General Problems of Methodology and Philosophy of Science*

M. HESSE (England), Chairman

M. BUNGE (Argentine), "A Measure of Predictive Performance"
F. DENK (West-Germany), "Can There Be an 'Ars Inveniendi'?" (read by Mrs. Székély)
T. GRÜNBERG (Turkey), "Phenomenalism and Observation"
J. HOROVITZ (Israel), "Three Related Aspects of Scientific Arguments"
K. D. IRANI (U.S.A.), "On Criteria of Acceptance"
C. JOJA (Rumania), "A Propos de la Philosophie des Sciences"
J. RUYTINX (Belgium), "Kinds of Empirical Languages and the Unity of Science"

TUESDAY AFTERNOON, SEPTEMBER 1

Invited Hour Address — Section III

A. HEYTING (Netherlands), Chairman

A. ROBINSON (U.S.A.), "Formalism 64"

Tour of Jerusalem
Reception by the Israel Academy of Sciences and Humanities

WEDNESDAY MORNING, SEPTEMBER 2

Invited Half-Hour Addresses — Section I

A. COBHAM (U.S.A.), "The Intrinsic Computational Difficulty of Functions"
S. C. KLEENE (U.S.A.), "Classical Extensions of Intuitionistic Mathematics"
M. O. RABIN (Israel), "A Direct Method for Undecidability Proofs"
E. P. SPECKER (Switzerland), "The Calculus of Partial Propositional Functions"

Contributed Papers — Section IV. *General Problems of Methodology and Philosophy of Science*

Y. LEIBOWICZ (Israel), Chairman

U. FOA (Israel), "The Semantic Principal Components"
R. SUSZKO (Poland), "Formalization Without Bound Variables and Natural Language"
E. TALMOR (Israel), "Meaning and Exchangeability"
L. TONDL (Czechoslovakia), "Semantic and Pragmatic Aspects of the Explanation Model"
H. A. WEISSMAN (Israel), "On the Concept of 'Universals'"

WEDNESDAY AFTERNOON, SEPTEMBER 2

Invited Half-Hour Addresses — Section II. *Foundations of Mathematical Theories*

M. O. RABIN (Israel), Chairman

H. J. KEISLER (U.S.A.), "Ultraproducts"
W. SCHWABHÄUSER (East-Germany), "Metamathematical Methods in Foundations of Geometry"
G. TAKEUTI (Japan), "Recursive Functions and Arithmetical Functions of Ordinal Numbers"

Contributed Papers

P. GEACH (England), Chairman

A. S. DELLAL (Israel), "Negation, Exact and Blurred"
J. MORGENSTERN (France)
I. TAMMELO and L. PROTT (Australia), "Legal and Extralegal Justification" (read by J. Horowitz)

Discussion on Meaningful Linguistics

S. KÖRNER (England), Chairman

F. KIEFER (Hungary)
M. MASTERMAN (England)
S. GORN (U.S.A.)

Invited Hour Address — Section VIII

S. H. BERGMANN (Israel), Chairman

A. CHURCH (U.S.A.), "The History of the Question of Existential Import of Categorical Propositions"

Gala Performance of Inbal Dance Theatre